Structure/Reactivity and Thermochemistry of Ions

NATO ASI Series

Advanced Science Institutes Series

A series presenting the results of activities sponsored by the NATO Science Committee, which aims at the dissemination of advanced scientific and technological knowledge, with a view to strengthening links between scientific communities.

The series is published by an international board of publishers in conjunction with the NATO Scientific Affairs Division

A Life Sciences B Physics	Plenum Publishing Corporation London and New York
C Mathematical and Physical Sciences	D. Reidel Publishing Company Dordrecht, Boston, Lancaster and Tokyo
D Behavioural and Social Sciences E Engineering and Materials Sciences	Martinus Nijhoff Publishers Dordrecht, Boston and Lancaster
F Computer and Systems Sciences G Ecological Sciences	Springer-Verlag Berlin, Heidelberg, New York, London, Paris, and Tokyo

Series C: Mathematical and Physical Sciences Vol. 193

Structure/Reactivity and Thermochemistry of Ions

edited by

Pierre Ausloos

and

Sharon G. Lias

Center for Chemical Physics,
National Bureau of Standards,
Gaithersburg, Maryland, U.S.A.

D. Reidel Publishing Company

Dordrecht / Boston / Lancaster / Tokyo

Published in cooperation with NATO Scientific Affairs Division

Proceedings of the NATO Advanced Study Institute on
Structure/Reactivity and Thermochemistry of Ions
Les Arcs, France
June 30-July 11, 1986

Library of Congress Cataloging in Publication Data

NATO Advanced Study Institute on Structure/Reactivity and Thermochemistry of Ions (1986 : Les Arcs, France)
Structure/reactivity and thermochemistry of ions.

(NATO ASI series. Series C, Mathematical and physical sciences; vol. 193)
"Published in cooperation with NATO Scientific Affairs Division."
Includes index.
1. Ions—Congresses. 2. Thermochemistry—Congresses. 3. Chemistry, Physical and theoretical—Congresses. I. Ausloos, Pierre J. II. Lias, Sharon G., 1935– . II. North Atlantic Treaty Organization. Scientific Affairs Division. IV. Title. V. Series: NATO ASI series. Series C, Mathematical and physical sciences ; no. 193)
QC701.7.N39 1986 541.3'9 86–31416
ISBN-13:978-94-010-8284-6 e-ISBN-13:978-94-009-4021-5
DOI: 10.1007/978-94-009-4021-5

Published by D. Reidel Publishing Company
P.O. Box 17, 3300 AA Dordrecht, Holland

Sold and distributed in the U.S.A. and Canada
by Kluwer Academic Publishers,
101 Philip Drive, Assinippi Park, Norwell, MA 02061, U.S.A.

In all other countries, sold and distributed
by Kluwer Academic Publishers Group,
P.O. Box 322, 3300 AH Dordrecht, Holland

D. Reidel Publishing Company is a member of the Kluwer Academic Publishers Group

---

Softcover reprint of the hardcover 1st edition 1987

CONTENTS

PREFACE

This volume presents the proceedings of a 1986 Advanced Study Institute entitled "Structure/Reactivity and Thermochemistry of Ions", held at Les Arcs, France, June 30 to July 11, 1986. The format of a NATO Institute is ideally suited to in-depth communications between scientists of diverse backgrounds. Particularly in the field of ion physics and chemistry, where on-going research involves physicists, physical chemists, and organic chemists - who use a variety of experimental and theoretical techniques - it is found that in the relaxed but stimulating atmosphere of a NATO ASI, each professional group provides unique insights, leading to a better definition and solution of problems relating to the properties of gas phase ions. This book presents chapters based on the lectures presented at the Les Arcs ASI. The participants took the initiative to organize a number of specialized workshops - informal discussion groups which considered questions or problem areas of particular interest. The accounts of these sessions, which are also included in this book, make stimulating reading, and include considerable useful information.

This Advanced Study Institute is the fourth in a series of NATO-sponsored institutes devoted to the chemistry and physics of ions in the gas phase. The first, in 1974, in Biarritz, France, focussed on "Interactions between Ions and Molecules". This was followed by the 1978 Institute, "Kinetics of Ion/Molecule Reactions" in La Baule, France, and the 1982 ASI on "Ionic Processes in the Gas Phase" in Vimeiro, Portugal, in which emphasis was given to unimolecular processes of ions.

A comparison of the Proceedings of the four NATO Advanced Study Institutes presents a clear record of the dramatic experimental and theoretical advances which have been made over the past 12 years in the field of ion physics and chemistry. It has now become possible to elucidate structural, thermochemical, and spectral properties of ions as well as the dynamics of ionic processes, with a degree of detail which was unthought of only a decade ago.

Organization of the Les Arcs ASI was initiated by David Dixon, who unfortunately, was unable to attend. We would like to acknowledge his active participation in planning the meeting, as well as the contributions of Keith Jennings, T. B. McMahon, and Rose Marx in planning and running the day-to-day activities at Les Arcs.

The Organizing Committee of the 1986 Advanced Study Institute at Les Arcs would like to acknowledge the continuing support and guidance of Dr. Craig Sinclair, Director of the ASI Programme of the Scientific Affairs Division of the North Atlantic Treaty Organization.

Pierre Ausloos
Sharon G. Lias

Gaithersburg, Maryland, U. S. A.
October, 1986

CAPTURE COLLISION THEORY

Douglas P. Ridge
Department of Chemistry
University of Delaware
Newark, DE 19716

INTRODUCTION

A proper treatment of the conductivity of ionized gases lead to Langevin's discovery of the phenomenon of capture in ion molecule collisions (Langevin, 1905). The background of Langevin's work was the work of Maxwell (1860a, 1860b) and Boltzmann (1902) on the transport properties of gases. Maxwell first showed that it was possible to proceed from intermolecular force to collision trajectories and cross sections to collision rates to transport coefficients. He considered in particular the case of a repulsive potential which varied as the fifth power of the intermolecular distance. He found that in this case the diffusion collision frequency is independent of the velocity which simplifies the computation of the transport coefficients immensely. Commenting on this work Boltzmann observed:

> Even as a musician can recognize his Mozart, Beethoven, or Schubert after hearing the first few bars, so can a mathematician recognize his Cauchy, Gauss, Jacobi, Helmholtz, or Kirchhoff after the first few pages. The French writers reveal themselves by the extreme formal elegance, while the English, especially Maxwell, by their dramatic sense. Who, for example, is not familiar with Maxwell's memoirs on his dynamical theory of gases?...The variations of the velocities are, at first, developed majestically; then from one side enter the equations of state; and from the other side, the equations of motion in a central field. Ever higher soars the chaos of formulae. Suddenly, we hear, as from kettle drums, the four beats "put n = 5." The evil spirit V (the relative velocity of the two molecules) vanishes; and, even as in music, a hitherto dominating figure in the bass is suddenly silenced, that which had seemed insuperable has been overcome as if by a stroke of magic...This is not the time to ask why this or that substitution. If you are not swept along with the development, lay aside the paper. Maxwell does not write programme music with explanatory

P. Ausloos and S. G. Lias (eds.), Structure/Reactivity and Thermochemistry of Ions, 1–13.

> notes...One result after another follows in quick succession till at last, as the unexpected climax, we arrive at the conditions for thermal equilibrium together with the expressions for the transport coefficients. The curtain then falls (Chandrasekhar, 1979)!

Maxwell's work captured the imagination of many besides Boltzmann. One case where an intermolecular force could be deduced from physical principles was the ion-molecule case. If the ion is taken as a point charge and the molecule as a hard sphere of isotropic polarizability, then the intermolecular force is attractive and proportional to the polarizability and the inverse fifth power of the intermolecular distance. Langevin was interested in using Maxwell's method to compute the conductivity of ionized gases. This seemed a particularly promising test of the work of Maxwell and Boltzmann.

In the course of this work, Langevin calculated trajectories for ion molecule collisions. The singularity in the scattering angle at the critical impact parameter clearly signaled the capture phenomenon. He provided diagrams of the collision trajectories and clearly describes the capture collision phenomenon. For a given relative velocity there is a critical impact parameter. At larger impact parameters, the scattering angle decreases with impact parameter. At smaller impact parameters, the trajectories spiral inwards until they are reflected off the repulsive hard sphere. In the absence of a short range repulsive potential, these trajectories would have unbounded scattering angles. Langevin determined diffusion cross sections as a function of the radius of the hard sphere. The calculations involved multiple levels of integration and represented a substantial computational achievement for the time.

THE ION INDUCED DIPOLE INTERACTION

The first determination of a capture collision rate was done by Eyring, Hirschfelder and Taylor (1936). They were interested in the rate of radiation induced equilibration of ortho hydrogen. They concluded that ion molecule reactions were involved and used Eyring's Absolute Rate Theory to calculate a capture rate. They realized that capture might be the key to the kinetics of these ion molecule reactions. Their calculations will be described in more detail below.

Gioumousis and Stevenson in 1957 took a different approach to calculating a capture rate (Gioumousis and Stevenson, 1957). They used the formalism of classical mechanics. At the critical impact parameter, b_o, the trajectory results in a metastable orbit with radius r_o. Two conditions must be met by the metastable orbit. First, the sum of the fictitious "centrifugal" force and the ion neutral attractive force must be zero:

$$F_r = -\left(\frac{\partial V}{\partial r}\right)_{r_o} + \frac{m v_o^2 b_o^2}{r_o^3} = 0 \qquad (1)$$

Second, the sum of the ion neutral potential energy and the rotational energy of the orbiting ion neutral pair must equal the initial relative kinetic energy:

$$V(r_o) + \frac{mv_o^2 b_o^2}{r_o^2} = 1/2mv_o^2 \tag{2}$$

In Eqs. (1) and (2), V(r) is the interaction potential, v_o, is the relative velocity and m is the reduced mass of the colliding pair. The first condition provides that there is no net force in the radial direction. The second condition provides that there is no radial velocity and that energy is conserved. The mathematical statements of these conditions constitute two equations which can be solved simultaneously to find b_o and r_o as a function of the relative initial velocity v_o. Johnston shows that these equations can be solved for any attractive intermolecular potential that varies as the nth power of intermolecular separation if n is greater than 2 (Johnstone, 1966). Gioumousis and Stevenson solved the equations for the ion induced dipole potential:

$$V(r) = \frac{-\alpha e^2}{2r^4} \tag{3}$$

In Eq. (3), α is the angle averaged polarizability of the neutral and e is the magnitude of the electron charge. With b in hand, the capture cross section can be determined by taking b_o as its radius:

$$b_o = \left(\frac{4\alpha e^2}{mv_o^2}\right)^{1/4} \tag{4}$$

$$\sigma = \pi b_o^2 = \frac{2\pi e}{v_o}\left(\frac{\alpha}{m}\right)^{1/2} \tag{5}$$

Integration of the product of the cross section with v_o over a Maxwell-Boltzmann distribution of velocities gives the capture collision rate:

$$k_L = 2\langle v_o \sigma \rangle = 2\pi e\left(\frac{\alpha}{m}\right)^{1/2} \tag{6}$$

The Gioumousis and Stevenson model can be represented graphically as shown in Fig 1. The intermolecular potential corresponding to the attractive intermolecular force is V(r). Adding the orbital rotational energy, $E_J(r)$, gives the effective potential $U_J(r)$:

$$U_J(r) = \frac{J(J+1)\hbar^2}{2mr^2} + V(r) \tag{7}$$

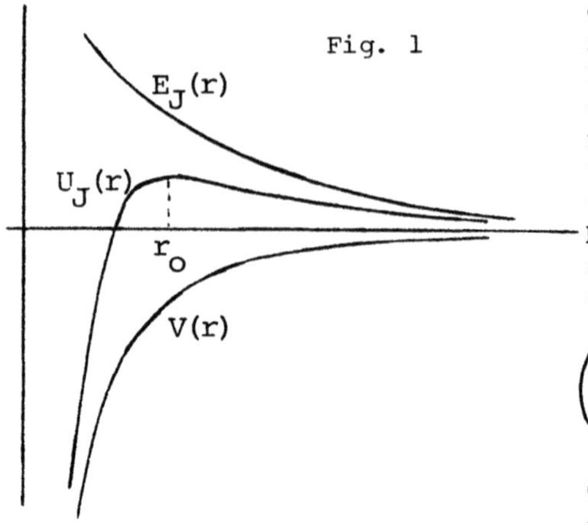

Fig. 1

There is a maximum in the effective potential. This is the key to the Absolute Rate Theory determination of the capture rate. This maximum is taken as the transition state. The transition state can then be located by setting the derivative of the effective potential to zero:

$$\left(\frac{\partial U_J(r)}{\partial r}\right)_{r_o} = 0 \tag{8}$$

The value of r for which this is true is r_o, the value of r for the transition state. The partition function for the transition state is then taken as that for a rigid rotor with the reduced mass of the ion neutral pair and length r_o. The internal modes are assumed to be unchanged in the transition state. The expression for the rate constant from absolute rate theory (or activated complex theory), k_{ACT}, is then given by:

$$k_{ACT} = \frac{kT}{h} \frac{q^{\neq}}{q_A q_B} = \frac{kT}{h}\left(\frac{\hbar^2}{2\pi mkT}\right)^{3/2} \sum_{J=0}^{\infty} (2J+1) e^{-U_J(r_o)/kT} \tag{9}$$

The sum is over the rotational quantum number of the rotational states of the ion neutral pair. The rotational quantum number is determined by the initial relative velocity and the impact parameter. The relationship between the two is given by:

$$E_J = \frac{mv_o^2 b^2}{2r^2} \tag{10}$$

Using this relationship, the Gioumousis and Stevenson and the Eyring, Hirschfelder and Taylor methods can be shown to be identical in general.

THE ION DIPOLE INTERACTION

These methods apply only to central potentials. The interaction between an ion and a polar molecule must lead to something analogous to capture, however. Reaction rate constants for ion polar molecule reactions exceed the capture rate calculated using only the polarizability of the molecule and neglecting the ion dipole interaction. A number of approximate methods have been devised to calculate capture collision rates for ion polar molecule systems. These include the "locked dipole" theory of Moran and Hammill (1963), the "average dipole orientation" theory of Su and Bowers (1973a, 1973b), the "average energy" theory of Barker and Ridge (1976), the "average free energy" or "activated complex" theory of Celii, Weddle and Ridge (Ridge, 1979; Celii, Weddle and Ridge, 1980), the thermodynamic theory of Turluski and Forys (1979), the "microcanonical variational transition state" theory of Chesnavich, Su and Bowers (1980), the parametrized trajectory calculations of Su and Chesnavich (1982), the "adiabatic invariance" theory of Bates (1981, 1982, 1984), the "perturbed rotational state" theory of Sakimoto (Sakimoto and Takayanagi, 1980; Sakimoto, 1981, 1984), the "total energy and angular momentum" theory of Hsieh and Castleman (1981), the "centrifugal sudden approximation" theory of Clary (1985) and the "statistical adiabatic channel" theory of Troe (1985). In discussing these theories we will emphasize those which lead to simple expressions for capture collision rates.

The expression for the interaction potential for a point charge with a point polarizable dipole is given by:

$$V(r,\theta) = \frac{-\alpha e^2}{2r^4} - \frac{\mu e \cos\theta}{r^2} \qquad (11)$$

where μ is dipole moment of the molecule. The length of the vector between the charge and the dipole is r and the angle between that vector and the dipole is θ. The general approach in the various theories is to convert the anisotropic ion dipole term into a central potential using some approximation. The simplest approximation is to set θ to zero. That is we assume that the dipole is "locked" in the direction of the approaching ion. This leads to the locked dipole result of Moran and Hammill (1963). In this approximation, the capture collision rate is given by:

$$k/k_L = 1 + .7979\ P \qquad (12)$$

$$P = \mu/(\alpha kT)^{1/2} \qquad (13)$$

where kT is Boltzmann's constant times the temperature.

Typically, however, a polar molecule is rotating as the ion approaches and does not lock onto the direction of the approaching ion. Most of the remaining theories assume that θ assumes some distribution of values. That distribution is assumed to vary with r. Once the distribution is specified then the ion dipole potential becomes effectively a central potential. A capture rate can then be determined as for other central potentials.

The average dipole orientation theory of Su and Bowers assumes that the sum of the ion dipole energy of interaction and the rotational energy of the dipole is a constant (Su and Bowers, 1973a, 1973b). This might be expressed explicitly as:

$$E = 1/2I\dot{\theta}^2 - \frac{\mu e \cos\theta}{r^2} \tag{14}$$

$$E = \text{constant}$$

where I is the moment of inertia of the dipole and $\dot{\theta}$ is the angular velocity of the rotating dipole. Holding E fixed requires the dipole to rotate rapidly through favorable alignments with the ion and slowly through unfavorable alignments. It also requires that for smaller values of r the dipole must "librate" through an angle less than 2π. Both of these features had been observed in exploratory classical trajectory calculations (Dugan and Magee, 1967; Dugan, Rice and Magee, 1969; Dugan and Canright, 1972). With E fixed, an explicite relationship between $\dot{\theta}$ and θ could be determined and an average angle of alignment calculated. The theory thus had the appeal of physical authenticity and computational accessibility. The results could also be stated in a simple form. A tabulated "locking parameter" was introduced into the locked dipole expression. There were problems with the theory, however. Bohme (1974) and others (Barker and Ridge, 1976) pointed out that comparison with experimental rate constants suggested that the theory underestimated capture rates. Castleman and Hsieh (Hsieh and Castleman, 1981) pointed out that E is not rigorously conserved. Soon, other approaches were tried.

Barker and Ridge (1976) suggested using a Boltzmann distribution to calculate an average potential energy as a function of r:

$$\langle V(r)\rangle = \frac{\int_0^{\pi} e^{-V(r,\theta)/kT}\, V(r)\sin\theta d\theta}{\int_0^{\pi} \sin\theta d\theta} = E_{ave} \tag{15}$$

This traded a dependence on $\dot{\theta}$ for a dependence on T. Taking T as the gas temperature gives a closed form expression for a central potential from which a capture rate can be calculated. The result agreed quite well with momentum transfer cross sections, but appeared to overestimate capture rates.

Turluski and Forys used statistical mechanics in a somewhat different way (1979). Instead of calculating the average energy, they calculated the average free energy:

$$\langle V(r)\rangle = -kT \ln q = F_{ave} \tag{16}$$

$$q = \frac{\int_0^{\pi} e^{-V(r,\theta)/kT} \sin\theta d\theta}{\int_0^{\pi} \sin\theta d\theta} \tag{17}$$

The free energy is the appropriate potential for thermodynamic forces under isothermal conditions ($dF = -SdT - w_{rev}$). The energy is the potential of a force under isoentropic conditions ($dE = TdS - w_{rev}$).

This also gives a closed form T dependent central force from which capture rates can be readily calculated. These capture rates are only slightly larger than the largest experimental rate constants indicating that they are quite good approximations to the actual capture rate.

Celii, Weddle and Ridge independently did the same calculation done by Turluski and Forys (Ridge, 1979; Celii, Weddle and Ridge, 1980). They also pointed out that the same result can be obtained from an approach analogous to that used by Eyring, Hirschfelder and Taylor in determining capture rates for non-polar molecules. It is only necessary to include the average free energy, F_{ave}, in V(r) in Eq. (7). This approach to activated complex theory is referred to as variational transition state theory since the transition state is found using variational calculus.

Chesnavich, Su and Bowers did a related treatment of the problem using microcanonical variational transition state theory (1980). This is similar to the activated complex theory except that microcanonical partition functions rather than canonical partition functions are used. The results of these variational methods are in quite good agreement with one another as shown in Fig. (2).

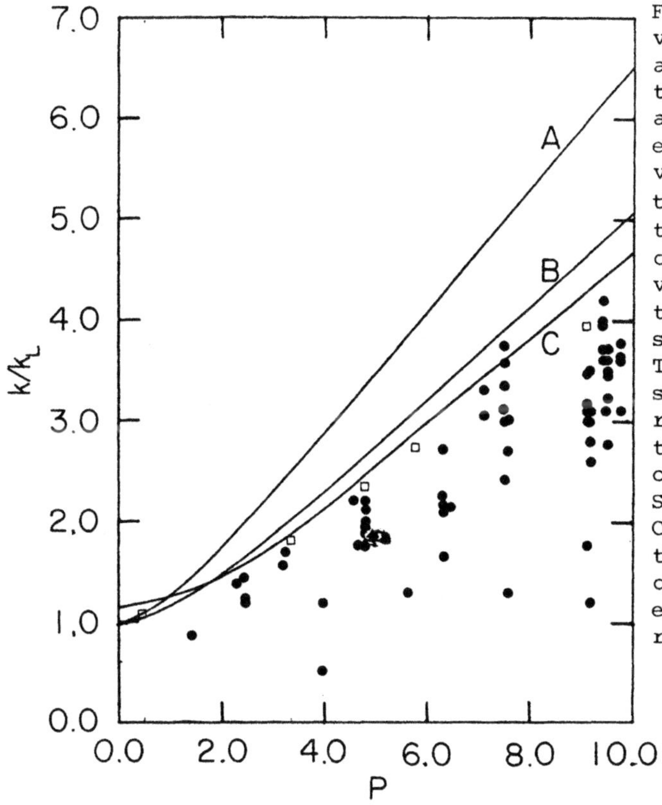

Fig. 2. k/k_L vs P. A is the average energy theory, B the average free energy or activated complex theory and C the microcanonical variational transition state theory. The open squares are the results of trajectory calculations by Su and Chesnavich and the closed circles are experimental rate constants.

In search of a definitive solution to the problem, Su and Chesnavich have carried out extensive trajectory calculations (1982). These are rigorous numerical solutions to the classical equations of motion. The results have been parameterized in the following simple equations:

$$k/k_L = \frac{(P/\sqrt{2} + 0.509)^2}{10.526} + 0.9754 \qquad 0 < P < 2\sqrt{2} \tag{18}$$

$$k/k_L = 0.3371P + 0.62 \qquad P > 2\sqrt{2} \tag{19}$$

These are essentially "experimental" results and should be viewed as the definitive solution to the problem as posed. That is, these are the correct classical capture rates for a point dipole interacting with a point charge where the dipole is allowed to have rotational motion and a moment of inertia. It is still profitable to pursue other theories further, however. Other theories may be computationally much simpler and more easily extended to other intermolecular potentials. The magnitude of quantum mechanical effects may be indicated by other theories. The success or failure of other theories to give results in agreement with the trajectory calculations and experimental results on real systems may lead to useful physical insights.

Bates pointed out that there is a constant of the motion (an "adiabatic invariant") that had been overlooked in the various theories (1981, 1982, 1984). In the limit of slow approach and rapid rotation of the polar molecule, the action of the angular momentum conjugate to θ is a constant. That is

$$\oint P_\theta d\theta = \text{constant} \tag{20}$$

The action of p_θ is an average of p_θ over the range of accessible values of θ. Given this relation, Bates was able to calculate capture rates for the case that the orbital motion of the colliding pair is in the plan of rotation of the dipole. These results agree well with the variational transition state theory results. In fact, Bates' results are in between the two variational results for most dipole moment values. All three results, however, are significantly larger than the trajectory calculations as shown in Fig. (2).

Sakimoto incorporated the adiabatic invariant pointed out by Bates in a three-dimensional semiclassical calculation (Sakimoto and Takayanagi, 1980; Sakimoto, 1981, 1984). Bates had done his calculations on a system where the motion was confined to a plane. Sakimoto removed that restriction. In addition to the adiabatic invariant used by Bates, Sakimoto noted the rigorous invariance of the component of the rotational angular momentum of the rotor that falls along the vector between the ion and the neutral. This is best illustrated by considering the approach of an ion along a direction perpendicular to the plane of rotation of the dipole. The ion dipole interaction energy is zero and does not change as the dipole rotates. Hence the angular momentum of the rotor is conserved. Thus Sakimoto quantized the

the component of the angular momentum parallel to the direction of approach, and he quantized the adiabatic invariant described by Bates. The quantum numbers of both motions were fixed throughout the collision. The ion dipole interaction energy can then be determined as a function of separation. Capture rate constants can then be determined as a function of the two quantum numbers. A thermal rate constant can then be determined by integrating over a thermal distribution of rotational states and assuming that distribution initial directions of approach is isotropic. Sakimoto's results are compared with the trajectory calculations in Table I. The agreement is excellent. This verifies the

TABLE I. Theoretical Values of k/k_L[a]

P $(=\mu/(\alpha k_B T)^{1/2})$	T(HCl)[b] K	Transition State Theory (present)	Classical trajectory (Su and Chesnavich)	Semi-classical (Sakimoto)
2.83	409	1.55	1.57	1.58
5.66	102	2.48	2.53	2.52
8.48	45	3.49	3.48	3.48
11.31	26	4.48	4.43	4.49

a) k_L is defined in Eq. (6).

b) The temperature in K of HCl gas corresponding to the indicated value of P.

validity of Sakimoto's assumptions and further supports the validity of the trajectory calculations.

It is useful and instructive to attempt to apply the invariants noted by Bates and Sakimoto to the activated complex theory approach. The relevant coordinates and momenta are shown in Fig. (3). The ion dipole interaction potential in terms of the coordinates of Fig. (3) is

$$V(r,\theta,\Psi) = \frac{-\mu e \cos\theta \sin\psi}{r^2} \tag{21}$$

The component of the angular momentum designated p_ρ is rigorously invariant as noted by Sakimoto. The average magnitude of the angular momentum component designated p_ϕ is invariant in an average sense as noted by Bates. If those two quantities are fixed then ψ, the angle between the interparticle vector and the angular momentum of the rotor, is fixed as shown in Fig. (3b). That means that the sin term in the expression for V(r) in Eq. (21) is fixed by the initial direction of approach. It is then a constant throughout a collision. A partition function for the transition state analogous to the one in Eq. (17) can be calculated using the $V(r,\theta,\psi)$ in Eq. (21). A capture rate constant

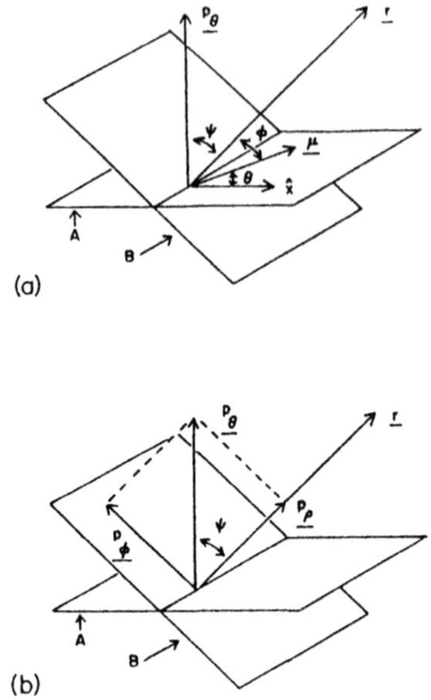

Fig. 3. Geometric relationships for an ion slowly approaching a rapidly rotating polar rotor.

(a) $\underline{p}_\Theta$ is the angular momentum vector of the rotating dipole $\underline{\mu}$. $\underline{p}_\Theta$ is normal to plane A. Therefore $\underline{\mu}$ lies in A. $\underline{r}$ is the vector originating at the center of mass of the rotor and terminating at the approaching ion. Plane B is perpendicular to $\underline{r}$. $\hat{x}$ is a unit vector in the direction of the projection of $\underline{r}$ on A. Ψ is the angle between $\underline{p}_\Theta$ and $\underline{r}$. Θ is the angle between $\hat{x}$ and $\underline{\mu}$.

(b) $\underline{p}_\rho$ is the component of $\underline{p}_\phi$ along $\underline{r}$. $\underline{p}_\phi$ is the projection of $\underline{p}_\Theta$ in plane B. Other symbols as in (a).

can then be calculated which will be a function of sin ψ. The resulting rate constants can then be integrated over an isotropic distribution of angles:

$$k = \frac{\int_0^{\pi/2} k(\psi)\,\sin\psi\, d\psi}{\int_0^{\pi/2} \sin\psi\, d\psi} \qquad (22)$$

The resulting values of the capture rate are compared to the trajectory calculations in Table I. The agreement is excellent. This substantiates the validity of the activated complex approach.

The calculations required by the activated complex approach are particularly simple. This together with the agreement obtained with the trajectory calculations suggests the usefulness of further calculations using the approach. For example, capture rate constants for the situation that the translational and rotational temperatures differ are readily done. The results of such calculations are illustrated in Fig. (4) and tabulated in Table II (Starry and Ridge). The activated complex approach has been used to calculate charge quadrupole capture rates (Celii, Weddle and Ridge, 1980). Incorporation into this calculation of the dynamical restrictions noted by Bates and Sakimoto is straightforward. The resulting capture rates are in good agreement with those determined by Bates and Mendas (1985).

Finally, there have been two recent quantum mechanical calculations of ion dipole capture rates. Troe has used a "statistical adiabatic

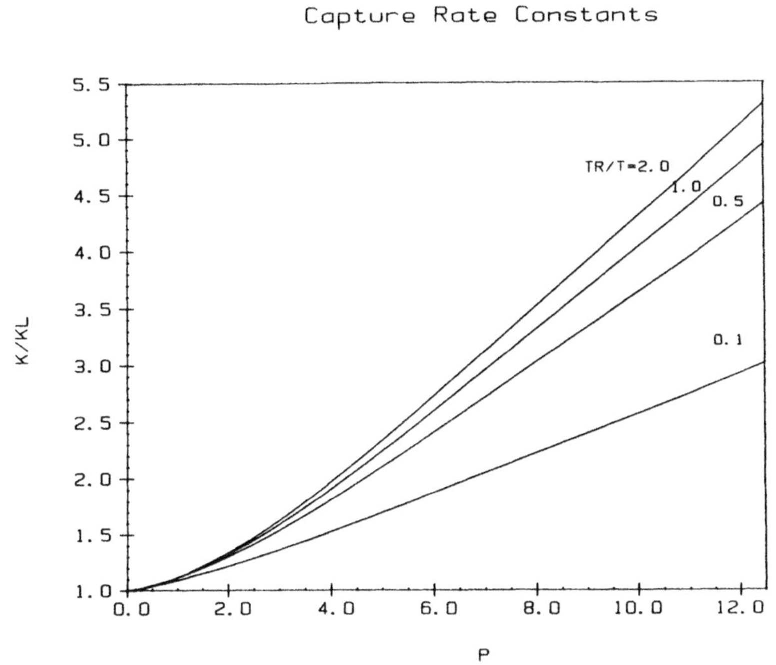

Fig. 4. k/k_L vs P for several values of the ratio of the rotational temperature T_R to the translational temperature T.

TABLE II. k/k_L at Various Temperatures[a]

P[b]	T_R/T[c]						
	0.2	0.4	0.6	0.8	1.0	1.5	2.0
1.0	1.083	1.089	1.091	1.093	1.094	1.095	1.096
2.0	1.250	1.283	1.300	1.310	1.316	1.326	1.331
3.0	1.446	1.522	1.560	1.584	1.601	1.626	1.640
4.0	1.656	1.781	1.847	1.889	1.918	1.961	1.986
5.0	1.875	2.053	2.149	2.210	2.252	2.317	2.354
6.0	2.098	2.333	2.461	2.542	2.599	2.686	2.735
7.0	2.325	2.618	2.779	2.881	2.953	3.063	3.124
8.0	2.555	2.907	3.101	3.226	3.312	3.445	3.520
9.0	2.786	3.199	3.427	3.574	3.676	3.832	3.919
10.0	3.020	3.494	3.756	3.924	4.042	4.222	4.322

a) Calculated using activated complex theory (Celii, Weddle and Ridge, 1980; Starry and Ridge)

b) $P = \mu/(\alpha k T_R)^{1/2}$

c) T_R is the rotational temperature and T the translational temperature channel model" to calculate capture rates for the H_3^+ + HCN reaction that agree quite well with the trajectory calculations for temperatures above 10K (Troe, 1985). Clary has done a partial wave analysis using the centrifugal sudden approximation to obtain capture rates (Clary, 1985). For the H_3^+ + HCN case, Clary's results are in good agreement with the trajectory calculations above ca. 100K. At lower temperatures, Clary's numbers exceed the trajectory numbers substantially. At very low temperatures, Clary's numbers approach the locked dipole numbers. It is evident from a comparison of Eq. (19) with Eq. (12) that the numbers from the trajectory calculations (and the numbers from Sakimoto's calculations) do not approach the locked dipole limit at low temperatures. The numbers from the activated complex theory do not approach the locked dipole limit at low temperatures either. The reason for this is that in these models the capture cross section is so large at low temperatures that at separation r_o the rotational energy of the rotor, though small, is still significant compared to the charge dipole interaction. There are a few experimental results, however, which suggest the possibility that the trajectory calculations underestimate the capture rate at very low temperatures (Viggiano, Paulson and Dale, see report of workshop on capture collsion theory in this volume).

This overview suggests that the parameterized equations obtained from the classical trajectory calculations of Su and Chesnavich are the best means to obtain an estimated capture rate for comparison with experimental reaction rate constants. There may remain some doubt about the best approach below 100K where quantum mechanical treatments may have an advantage, but at this writing the parameterized trajectory calculations seem the result of choice. When reporting the theoretical capture rate in the literature, it might bes be referred to as simply k_c with an appropriate reference to Su and Chesnavich (1982) (see Eqs. (18) and (19)). The ADO designation properly refers to Su and Bowers original result (1973a, 1973b). Confusion is inevitable if that designation is used for other theoretical capture rates.

REFERENCES

Barker, R.A. and Ridge, D.P. (1976) J. Chem. Phys. **64**, 4411.

Bates, D.R. (1981) Chem. Phys. Lett. **82**, 396.

Bates, D.R. (1982) Proc. R. Soc. Lond. A 384, 289.

Bates, D.R. (1984) Chem. Phys. Lett. **111**, 428.

Bates, D.R. and Mendas, I. (1985) Proc. R. Soc. Lond. A 402, 245.

Bohme, D.K. (1974) "The Kinetics and Energetics of Proton Transfer" in Interactions Between Ions and Molecules, ed. P. Ausloos, Plenum Press, p. 489.

Boltzmann, L. (1902) Theorie des gaz.

Celii, F., Weddle, G. and Ridge, D. P. (1980) J. Chem. Phys. **73**, 801.

Chandrasekhar, S. (July 1979) Physics Today, p. 25.

Chesnavich, W.J., Su, T. and Bowers, M.T. (1980) J. Chem. Phys. **72**, 2641.

Clary, D.C. (1985) Molec. Phys. **54**, 605.

Eyring, H., Hirschfelder, J.O. and Taylor, H.S. (1936) J. Chem. Phys. 4, 479.
Gioumousis, G. and Stevenson, D.P. (1957) J. Chem. Phys. 29 294.
Hsieh, E.T. and Castleman, Jr., A. W. (1981) Int. J. Mass Spectrom. and Ion Phys. 40, 295.
Johnstone, H.S. (1966) Gas Phase Reaction Rate Theory, Ronald, New York, p. 142ff.
Langevin, P. (1905) Ann. Chim. Phys. Ser. 8, 5, 245.
Maxwell, J.C. (1860a) Phil. Mag. Series 4, 19, 19; Ibid. (1860b) Series 4, 20, 21.
Moran, T.F. and Hamill, W.H. (1963) J. Chem. Phys. 39, 1413.
Ridge, D.P. (1979) "Comments on Intermolecular Potentials for Polyatomic Ions and Molecules" in Kinetics of Ion Molecule Reactions, ed. P. Ausloos, Plenum Press, New York, P. 55.
Sakimoto, K. (1981) Chem. Phys. 63, 419.
Sakimoto, K. (1984) Chem. Phys. 85, 273.
Sakimoto, K. and Takayanagi, K. (1980) J. Phys. Soc. Japan 43, 2076.
Starry, S. and Ridge, D.P. unpublished results.
Su, T. and Bowers, M.T. (1973a) J. Chem. Phys. 58, 3027.
Su, T. and Bowers, M.T. (1973b) Int. J. Mass Spectrosc. Ion Phys. 12, 347.
Su, T. and Chesnavich, W.J. (1982) J. Chem. Phys. 76, 5183.
Troe, J. (1985) Chem. Phys. Lett. 122, 425.
Turluski, J. and Forys, M. (1979) J. Phys. Chem. 83, 2815.
Viggiano, A.A., Paulson, J.F. and Dale, F., see report of workshop on capture collision theory in this volume.

COLLISION THEORY: SUMMARY OF THE PANEL DISCUSSION

Douglas P. Ridge
Department of Chemistry
University of Delaware
Newark, DE 19716

A positive conclusion from the discussion at the workshop was that the ion dipole capture rates from the parameterized classical trajectory calculations are generally the best choice for comparison to experimental rate constants. Quantum effects may be important at temperatures below ca. 100K, however. Questions were raised about the role in determining collision rates of terms in ion-molecule potentials other than the ion dipole and ion induced dipole terms. The applicability of the various capture theories at translational energies above ca. 0.1 eV was questioned. A number of these conclusions and questions are concisely summarized in Henchman's contribution below. Viggiano, Paulson and Dale report below intriguing evidence of low temperature rate constants larger than expected from classical collision theory.

Chava Lifschitz raised the question of the relationship between capture and unimolecular decomposition. Both involve passage over a centrifugal barrier. Studies of unimolecular decomposition may also provide insight into the dynamics associated with passage over a centrifugal barrier. Dissociation of chlorobenzene and carbon dioxide cations were cited as possible examples. The need for more work on the problem was pointed out.

COMPARING THEORY AND EXPERIMENT

Michael Henchman
Department of Chemistry
Brandeis University
Waltham, MA 02254

There is an increasing need to calculate accurate collision rate constants. One obvious application is in modelling the synthesis of interstellar molecules at temperatures below 100 K.[1] Another application is in using RRKM theory to model nucleophilic displacement reactions. There, an experimental rate constant, normalized to the calculated collision rate constant, can be expressed as a reaction

P. Ausloos and S. G. Lias (eds.), Structure/Reactivity and Thermochemistry of Ions, 15–21.

efficiency; and the bimolecular reaction can be treated in terms of the unimolecular reaction of the reaction intermediate.[2] Such investigations, studied as a function of temperature, are currently providing a more stringent test of simple models that have proved so successful at 300 K.[3,4]

Several different theories have appeared in the past five years. Many experimenters, rather than choose between these new theories, have persisted with the theories which have served them in the past, e.g. ADO, AADO etc. Is this incorrect? If so, by how much? Which is the best theory and why? How accurate is it? These are some of the questions that a handler of data rather than theory, must continue to face; and incomplete answers to these questions, are presented here, for those who may be similarly perplexed. The principal focus is on the collision of ions with polar molecules.

1. WHICH THEORY AND WHY?

At temperatures above 100 K, the Su/Chesnavich theory[5] (SC)--or the Ridge theory[6] which gives very similar results--should be used. Neither the ADO nor the AADO theory is as accurate. Both the SC and the ADO/AADO theories are derived from the same classical electrostatic potentials. The ADO/AADO theories derive solutions making plausible but empirical approximations. The SC theory produces "exact", "experimental" solutions--in the form of parameterized solutions to extensive trajectory calculations using the potentials.

Clary's ACCSA theory (adiabatic capture centrifugal sudden approximation),[7] which is based on quantum theory, is more accurate than the SC theory, which is purely classical. Both give results which are indistinguishable above 100K. The SC solution is given in analytic form whereas the ACCSA result has to be derived for each particular case--as too does the approach of Takayanagi and Sakimoto (both pioneers in the field).[8] Troe's preliminary results[9] have been modified[10] and agree with the SC and the ACCSA predictions above 100 K.

To summarize, the SC theory is the treatment of choice above 100 K. It represents accurately and conveniently the results of three different approaches.

Below 100 K, the SC approach underestimates the rate constant because it treats the rotational energy distribution classically. In the ACCSA treatment, as the temperature is reduced, the polar molecule is increasingly relegated to the J=0 state, being there "locked" into the electric field of the approaching ion. While the SC result certainly provides a lower bound, there is no present alternative to a complete ACCSA treatment for each case, although Troe's new solution[10] may modify that. The rate constant may possibly show a maximum value at very low temperatures ($\sim$1 K).[10]

The ACCSA theory assumes that the rotational state of the polar molecule cannot change during the course of the collision. In more extended calculations, Clary has removed that constraint and found no appreciable effect on the results of the original calculations.

2. HOW GOOD IS THE BEST?

The accuracy of the rate constant can be no better than the validity of the potential used to derive it. How valid is the potential for separations beyond the location of the centrifugal barrier? From the theoretical viewpoint, the potential must include terms, as needed, to treat electron repulsion, electron exchange, dipole-dipole, dipole-induced dipole, ion-quadrupole, anisotropic polarizability, and so on. The increasing number of components for the potential and the increasing complexity of the calculation do not recommend this as a useful or feasible exercise.

A more tractable approach is to compare experiment with theory under conditions where the simplest theory should apply. The simplest candidate is the transfer of a proton from one atom to another: $XH^+ + Y \longrightarrow X + YH^+$ or $X^- + HY \longrightarrow XH + Y^-$. As part of a search to

TABLE 1
Rate Constants (10^{-9} cm^3/molec s) at 300 K

Reaction	Theory[a]	Experiment	
		JILA[b]	AFGL[c]
$F^- + HCl \longrightarrow HF + Cl^-$	1.87	1.55±0.31	1.56±0.31
$F^- + HBr \longrightarrow HF + Br^-$	1.53	1.19±0.25	1.24±0.25

[a]Reference 5. [b]Reference 11. [c]Reference 12.

establish calibration standards for reaction rate measurements, we have carefully repeated the measurements of Weisshaar et al. shown in Table 1, obtaining excellent agreement in both cases. The agreement between experiment and theory is less than satisfactory, because strongly exothermic proton transfer reactions generally proceed with unit efficiency, in the absence of competing channels.[13] The agreement may actually be worse than indicated. The reactant ion is not, as assumed, a point charge but polarizable; and this should add a dipole/induced dipole term to the potential, increasing the theoretical value and accentuating the disagreement.

In a test at the next level of complexity, we have examined the reaction $OH^- + HF \longrightarrow H_2O + F^-$ over a temperature range 200-450 K.[13] There the agreement is excellent between experiment and theory, for the simple ion/dipole potential. Even for this four-atom system, there will however be a dipole/dipole term in the potential but its effect on the rate constant is hard to predict. Apparently it may be easier to estimate the magnitude of this using the statistical theory of Ridge.[6]

There is a continuing need to compare experiment and theory for the simplest systems.

3. LIMITS OF ENERGY AND SIZE

Any theoretical description that relies solely on an attractive potential, must begin to fail as the centrifugal barrier approaches the potential minimum. This can occur in two different ways--for small ions at elevated energies and for large ions at normal temperatures.

Many years ago an upper energy limit was established at $\sim$0.1 eV.[1] It is still possible to see, in the literature today, comparisons between experiment and theory (using only attractive potentials) that extend to 1 eV. These comparisons have no physical basis and they do not serve their intended purpose. They are actually counterproductive. We are encouraged to think that we "understand" where in fact we don't.

Theory treats the ion as a point charge. One can imagine making such large ions, for example by solvation, that physically extend to the region of the centrifugal barrier. Here again the repulsive part of the potential must play some role, even at thermal energies. This is one motive for studying proton transfer reactions as a function of cluster size.[13,14] In the future this will provide further tests of our understanding of collision theory.

4. UNANSWERED QUESTIONS

In conclusion, there are some intriguing questions:
(1). Why does theory often work so well for complicated reactions but poorly for the simple ones? One would predict the reverse--that the theory would give the most accurate representation of the simplest systems. This question is raised again by Armentrout's extensive data on simple systems that are among the most accurate ever obtained. In only one system, $O^+ + H_2$, was the Langevin prediction observed.[15]
(2). Why are the reaction efficiencies of several simple reactions constant, independent of translational energy? The excitation functions of these reactions, measured either using the guided-ion beam technique[15] or the merged-beam technique,[16] show a $\sigma(E) \propto E^{-1/2}$ energy dependence; and the cross sections are a constant fraction of the Langevin value. What are the dynamical implications of this unexpected result?
(3). Why, for several reactions, does the energy dependence expected for the Langevin theory, extend to an energy range where the Langevin theory cannot possibly apply?[15,16] What causes, for several systems, a $\sigma(E) \propto E^{-1/2}$ energy dependence in an energy range where the centrifugal barrier cannot be sampling a r^{-4} potential?

I thank D. C. Clary, P. M. Hierl, D. P. Ridge and A. A. Viggiano for instructive discussion.

REFERENCES

1. Adams, N. G., Smith, D. and Clary, D. C.: 1985, Ap. J. **296**, L31.
2. Dodd, J. A. and Brauman, J. I.: 1986, J. Phys. Chem. **90**, 3559.
3. Caldwell, G., Magnera, T. F. and Kebarle, P.: 1984, J. Am. Chem. Soc. **106**, 959.

4. Hierl, P. M., Ahrens, A. F., Henchman, M., Viggiano, A. A., Paulson, J. F. and Clary, D. C.: 1986, J. Am. Chem. Soc. 108, 3142.
5. Su, T. and Chesnavich, W. J.: 1982, J. Chem. Phys. 76, 5183.
6. Ridge, D. P.: unpublished results.
7. Clary, D. C.: 1984, Mol. Phys. 53, 3; 1985, Mol. Phys. 54, 605. Clary, D. C., Smith, D. and Adams, N. G.: 1985, Chem. Phys. Lett. 119, 320.
8. Sakimoto, K.: 1985, Chem. Phys. Lett. 116, 86.
9. Troe, J.: 1985, Chem. Phys. Lett. 122, 425.
10. Troe, J.: presented at the 9th International Symposium on Gas Kinetics, Bordeaux, France, 20-25 July, 1986.
11. Weisshaar, J. J.; Zwier, T. S. and Leone, S. R.: 1981, J. Chem. Phys. 75, 4873.
12. Hierl, P. M., Ahrens, A. F., Henchman, M., Viggiano, A. A. and Paulson, J. F.: unpublished results.
13. Hierl, P. M., Ahrens, A. F., Henchman, M., Viggiano, A. A., Paulson, J. F. and Clary, D. C.: 1986, J. Am. Chem. Soc. 108, 3140.
14. Viggiano, A. A., Dale, F. and Paulson, J. F.: see this chapter.
15. P. Armentrout, see the chapter in this volume.
16. Gentry, W. R.: 1979, in Gas Phase Ion Chemistry (Bowers, M. T., Ed.), Academic Press, New York, Vol. 2, p. 221.

REACTIONS OF $H^+(H_2O)_n$ WITH POLAR MOLECULES

A. A. Viggiano, J. F. Paulson and F. Dale
AFGL/LID
Hanscom AFB, MA 01731-5000

For the purpose of calculating ion-molecule collision rate constants, it is generally assumed that the ion can be treated as a point charge. This assumption can break down for at least two reasons. First, the charge may no longer be located at the center of the ion. Second, the ion may become sufficiently large that the "hard sphere" radius is larger than the capture radius. Some recent results taken in our laboratory may be applicable to this problem.

We have measured the proton transfer rates of $H^+(H_2O)_n$ (n=1-10) with several polar neutrals, including acetonitrile, ammonia, methanol, pyridine and acetone. The measurements were made in a variable temperature flowing afterglow/selected ion flow tube apparatus. The ions were made in a sidearm of the apparatus operated as a flowing afterglow at high pressure. A diaphragm with a hole 0.5 cm in diameter separated the ion source region from the flow tube. Typical operating pressures were 20 torr in the ion source region and 0.5 torr in the flow tube. The neutrals were injected into the flow tube through a heated injector. This injector consisted of a 1/16" stainless steel tube welded concentrically to a 1/4" stainless steel tube. When a current was passed through one tube to the other, the smaller inner tube was heated significantly while the outer tube remained approximately at the flow tube temperature. This enabled measurements of rate constants in cases where the neutral had a vapor pressure of less than 10^{-6} torr at the temperature of interest.

The measured rate coefficients for $H^+(H_2O)_4$ reacting with CH_3CN versus temperature are shown in Figure 1. The present experimental results are shown as plusses, arrows represent lower limits (due to

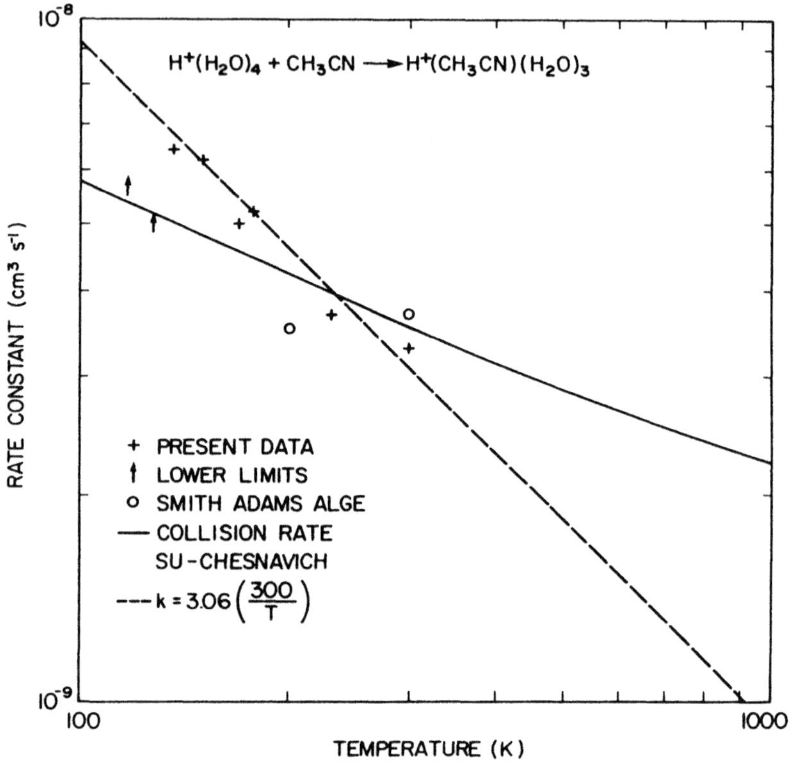

Fig. 1. Plot of rate constant vs T for the reaction of $H^+(H_2O)_4$ reacting with CH_3CN. Plusses are the present data, arrows represent lower limits, circles are the data of Smith et al., the solid line is the capture rate calculated according to the theory of Su and Chesnavich and the dashed line is a least squares fit to the data.

difficulty in injecting CH_3CN at low temperatures), circles represent the work of Smith et al., the solid line is the collision rate constant as calculated from the theory of Su and Chesnavich, and the dashed line represents the least squares fit to the data. It is quite evident that the present results show a much steeper temperature dependence than predicted by theory. This can be caused by the repulsive part of the potential or by the $H^+(H_2O)_4$ charge being redistributed during the course of the collision. Methanol and pyridine have a less dramatic effect than CH_3CN but still appreciable ($T^{-0.75}$). Acetone follows the calculated curve, although the absolute magnitude of the rate coefficient is somewhat lower. Ammonia shows little temperature dependence.

In conclusion, proton transfer reactions of $H^+(H_2O)_n$ with several neutral molecules proceed on essentially every collision. The temperature dependence of some of these reactions is considerably steeper than predicted by theory. This may be due to the notion of a point charge no longer being valid.

REFERENCES

1. D. Smith, N. G. Adams, and E. Alge, Planet. Space Sci., **29**, 449 (1981).
2. T. Su and W. J. Chesnavich, J. Chem. Phys., **76**, 5182 (1982).

OPTICAL STUDIES OF PRODUCT STATE DISTRIBUTIONS IN THERMAL ENERGY ION-MOLECULE REACTIONS

Veronica M. Bierbaum and Stephen R. Leone*
Joint Institute for Laboratory Astrophysics
National Bureau of Standards and University of Colorado
and Department of Chemistry and Biochemistry
University of Colorado, Boulder, Colorado 80309-0440

ABSTRACT. Product state distributions of thermal energy ion-molecule reactions are determined by the sensitive optical methods of infrared chemiluminescence and laser-induced fluorescence detection. Experiments to obtain detailed vibrational state populations are carried out in a flowing afterglow reaction vessel, and measurements to extract rotational state distributions are performed in a single-collision, crossed-beam apparatus that uses a flowing afterglow ion source. Product state information is obtained for a series of proton transfer reactions and charge transfer reactions, which reveals many aspects of the dynamical behaviors of these processes. Measurements are also presented for polyatomic ion-molecule reactions, for optically-determined rates of ion collisional excitation and deactivation, and on visible chemiluminescence yields and branching fractions for reactions important in the aurora.

1. INTRODUCTION

The ability to make accurate product state determinations plays an important role in the elucidation of the details of reaction dynamics. Through the combined use of theoretical trajectory calculations, experimental measurements, and chemical insight, it is now possible to describe the detailed dynamical motions of many chemical events (1,2). Although highly accurate methods have been available for a long time to study most neutral chemical systems, the field of ion-molecule chemistry has only recently enjoyed the influx of techniques that provide the same kind of detail. One reason for the delay in applying optical detection techniques to the study of ion reactions is that ion

*Staff Member, Quantum Physics Division, National Bureau of Standards.

P. Ausloos and S. G. Lias (eds.), Structure/Reactivity and Thermochemistry of Ions, 23–55.

densities are typically smaller than densities for comparable neutral reactants, making it more difficult to achieve adequate signals for detection. Although there are many highly detailed crossed beam studies of ion reactions in the literature (3,4), the effects of space charge and stray fields usually limit the range of kinetic energies to several eV and higher.

Of special interest to us is the study of ion chemistry via optical state detection (5); we are also interested in reactions carried out at thermal or near thermal energies. At these lower kinetic energies, the dynamics of the collision events will be more sensitive to the long range attractive forces and to the underlying subtleties of the potential energy surfaces that govern the reactions. Some ion reactions involve transformations that are nearly identical to corresponding neutral systems. Examples are the proton transfer reactions, $F^- + HX \rightarrow HF(v) + X^-$, which have similar exothermicities and mass combinations and are exactly analogous to the neutral $F + HX \rightarrow HF(v) + X$ systems. However, the ion reactions proceed over potential surfaces that have deep attractive wells and no barriers, whereas the neutral reactions have activation barriers and smaller long range attractive forces. Thus, new measurements on the product states of these ion reactions can be combined with the tremendous amount of knowledge available on the neutrals to provide a powerful study of how changes in the potential surface affect the reaction dynamics. It is important, though, that the kinetic energy be in a range where the molecular motions can be affected by the potential surface features. Other ion reactions have no neutral counterparts, and thus it is possible to study a number of phenomena via product state determinations for the first time.

Optical methods of product state determination are an excellent means of studying the dynamics of a large variety of collision processes. Through vibrational state distributions, the mechanisms of energy transfer are explored, the locations of nonadiabatic surface crossings can sometimes be determined, and evidence for long-lived dynamical behavior can be inferred from the degree of statistical behavior. Rotational state information is an exquisite indicator for reactions that have several microscopic branches or mechanisms (1).

Molecular beam techniques are important complementary methods that can provide information such as the lifetime of the reactive complex, the mechanistic behavior as a function of collision energy, and information about specific product channels from the forward/backward asymmetry and translational energy release. Reactive studies of state-selected ions (e.g. the TESICO technique (1)) provide another complementary method in the array of techniques used to study ion dynamics. Through the principle of detailed balancing, there are already reassuring correspondences from the results of reactions carried out on specific vibrational states in both the forward and backward directions.

In the work described in this chapter, variations of the flowing afterglow technology (6) are used together with the optical techniques of infrared chemiluminescence (7) and laser-induced fluorescence (8) for product state detection of ion-molecule reactions (5). In one apparatus, the afterglow flowtube is modified to incorporate the laser excitation beam for laser-induced fluorescence or the infrared or visible detector for wavelength-resolved chemiluminescence studies. In this device, accurate vibrational and electronic state information is obtained for proton transfer reactions, charge transfer systems, and heavy atom transfer reactions at thermal energy (0.03 eV). Several rates of vibrational deactivation are obtained using optical state detection, and new experiments are probing the dynamics of state excitations for ions in electric drift fields. Rotational states are unfortunately completely thermalized in the determinations in this apparatus; however, this factor also becomes an important benefit in the accurate and efficient determination of the vibrational populations.

In another device, a small portion of a flowing afterglow source is extracted to form a free jet expansion into a high vacuum chamber to create a high flux, low density "beam" of ions. The ions are crossed with an effusive spray of neutrals for reaction. Laser-induced fluorescence detection in this system provides product state information with single rotational state detection. The apparatus is used for determinations of a number of charge transfer reactions carried out at kinetic energies of 0.1-0.2 eV; the energy is elevated slightly from thermal because of the supersonic expansion. Because of attendant uncertainties in assigning the final laboratory velocities of the individual product states and the necessity to integrate over rotational states to obtain vibrational populations, this apparatus gives important information about vibrational distributions, but with greater uncertainty than the flow tube device. In several cases, elegant dynamical details are obtained from rotational state information, and significant changes in vibrational distributions are observed at these increased collision energies.

2. EXPERIMENTAL

2.1. Characteristics of the Flowing Afterglow for Optical State Detection

The flowing afterglow technique was developed in 1963 by Ferguson, Fehsenfeld and Schmeltekopf to study the ion chemistry and physics of the earth's atmosphere (6). In the conventional instrument, ions are generated in a fast flow of buffer gas, they are allowed to react with added neutral reagents and they are analyzed and detected with a quadrupole mass filter. Increasingly sophisticated modifications of the technique have dramatically expanded its capabilities and applications over the last two decades. We have coupled the flowing afterglow method with the optical detection techniques of infrared and visible chemiluminescence and laser-induced fluorescence to permit detailed studies of the product state distributions of thermal energy ion-molecule reactions. The flowing afterglow represents an ideal technique for these studies for several reasons (5). The high ion densities permit detection and resolution of relatively weak emissions. The high collision frequency of the ions with the buffer gas and other neutrals insures that the reactant ions possess well-defined thermal energy distributions. The availability of flow-drift techniques allows the study of reactions and energy transfer phenomena as a function of kinetic energy. The intrinsic chemical versatility of the technique permits optical study of a wide range of ionic and neutral reactants. Finally, the ease of kinetic analysis allows quantitative determination of rate constants for reactive and energy transfer processes.

2.2. Infrared and Visible Chemiluminescence Detection in the Flowing Afterglow

Detection of infrared chemiluminescence in the flowing afterglow (5) requires modification of the flow tube to allow collection of light, use of infrared detectors, filters and associated electronics, and incorporation of a shutter to provide modulation of ions. The instrument is shown schematically in Figure 1. Helium buffer gas is pumped at high velocity and moderate pressure (80 m s^{-1} and 100 Pa, respectively) through a 7.3 cm i.d., 1 m long flow tube. Ions are formed at the upstream end by electron impact ionization of small flows of added gases. The ions flow about 50 cm before introduction of neutral reagents to ensure that diffusion and flow properties are well-defined and that the ions are collisionally relaxed. The ion-molecule reaction of interest is initiated by addition of the neutral reactant through a movable, fixed or radial inlet, and the ion chemistry is

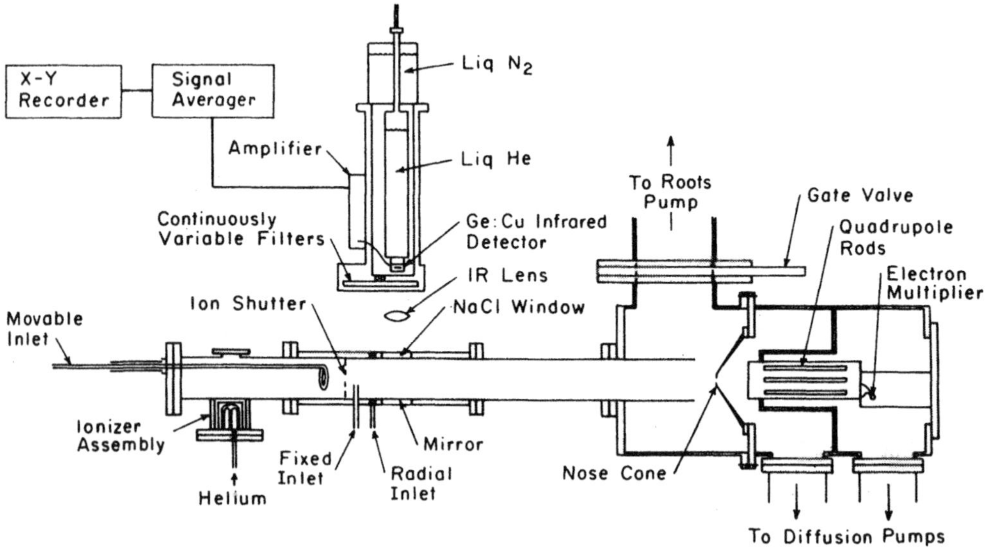

Fig. 1. Flowing afterglow apparatus modified for infrared chemiluminescence detection.

monitored downstream by a quadrupole mass filter. Infrared emission from the products of the ion-molecule reaction is emitted from the flow tube through a NaCℓ window just downstream of the reaction region. The light is focused by a CaF_2 lens, dispersed by filters and detected by an infrared detector. Two infrared detector assemblies have been employed: 1) an indium antimonide detector for the reactions of $Cℓ^-$ and CN^- and 2) a copper-doped germanium detector for the reactions of F^- and N^+. The photovoltaic InSb detector is a liquid-nitrogen cooled element with a wavelength response of 1.0-5.5 μm. The large area, high detectivity and ease of cooling make this detector best suited for initial studies. Wavelength resolution of the transmitted light is provided by a series of room temperature fixed frequency interference filters mounted below the detector. The photoconductive Ge:Cu detector is a liquid helium cooled element with a wavelength response of 2-30 μm. Sapphire and/or magnesium fluoride windows at 4 K provide a long-wavelength blackbody cutoff. The detector views emission through a wavelength-selective rotatable circular variable interference filter assembly (CVF) and a 1.5 mm wide slit, both cooled to 77 K. The wavelength calibration and resolution function of the filter/slit combination are determined using a broadband infrared source and a monochromator; most spectra have been recorded with a resolution of 0.06-0.11 μm. The sensitivity of the detector and filter as a function of wavelength and resolution is measured by chopping and imaging the output of a calibrated blackbody source onto the detector; the output versus CVF setting is normalized to the calculated blackbody emission curve (9).

In addition to the reactant ion of interest, the ion source generates a variety of reactive neutral species which can contribute to the observed infrared emission. To isolate the chemiluminescence arising from the ion-molecule process, the ions are modulated by repetitively applying a repulsive potential to a high transmission mesh stretched across the flow tube just upstream of neutral reactant addition (10). The reactive neutral species are unaffected by this potential. However, the infrared emission arising from the products of the ion-molecule reaction is modulated, reflecting the periodic absence of the reactant ions. Typically 10^4-10^5 modulation periods are accumulated in a signal averager at each of a series of wavelengths to construct a complete infrared emission spectrum. Since several hours are required to obtain a spectrum, the data are taken nonsequentially in wavelength to prevent systematic errors due to variations in the ion currents or flow rates. The emission at a selected reference point is also checked periodically. For fixed interference filter experiments, populations are deduced from the transmission through several filters and known values of the Einstein coefficients, A_v. For the CVF data, a stick spectrum consisting of the wavelength and relative emission intensity of the various P and R lines is calculated for each $v \rightarrow v-1$ band from the known Dunham coefficients and the expression

$$I_{em}(vJ \rightarrow v'J') \propto \left|R_{vJ \rightarrow v'J'}\right|^2 \nu^3_{vJ \rightarrow v'J'} (J+J'+1)\left(Q_v^{rot}\right)^{-1} \exp\left(-E_{vJ}^{rot}/kT_{rot}\right) .$$

This equation gives the emission intensity (photons/s) for a Boltzmann distribution of upper state rotational levels at temperature T_{rot}. $|R_{vJ\rightarrow v'J'}|^2$ is the square of the transition moment; ν^3 is the frequency factor for a given line; (J+J'+1) is the rotational line strength factor; Q_v^{rot} is the rotational partition function of vibrational level v at temperature T_{rot}; and E_{vJ}^{rot} is the rotational energy of the vJ level. In most cases, a thermal distribution of rotational states is observed and used in the computer fitting routine (9). At each CVF wavelength λ_i, the stick spectrum for each band is convolved with the experimentally determined filter transmission function to yield the calculated emission intensity per unit population, $I_v(\lambda_i)$. The relative vibrational state populations N_v are derived by a linear least-squares fit of the calculated spectrum

$$I_{calc}(\lambda_i) = \sum_v N_v I_v(\lambda_i)$$

to the experimental spectrum with N_v as the adjustable parameters in the fit. Figure 2 shows the experimental data and calculated best fits for the reactions $F^- + HBr,DBr \rightarrow Br^- + HF(v),DF(v)$ (11). Small corrections to the resulting vibrational populations are then applied to account for radiative cascading, collisional quenching and reactive loss between the time of formation and detection of the vibrationally excited product. These corrections are generally small, typically <10% of the raw population. When possible, vibrational surprisal plots are extrapolated to obtain estimates of the v=0 population and of the average fraction of available energy deposited into product vibration.

The experimental setup for detection of visible chemiluminescence (12) in the reaction of $N^+ + O_2$ is very similar to that described above. The visible emission is collected through a quartz window, focused with a quartz lens and detected with a cooled S-20 photomultiplier. Fixed frequency interference filters are used to isolate the emission features and photon counting techniques are used to achieve the maximum detection sensitivity. To reduce visible background emissions, increased reagent gas (N_2) pressures, reduced emission currents and increased ion source-to-window distances are employed.

2.3. Laser-Induced Fluorescence Detection in the Flowing Afterglow

Detection of laser-induced fluorescence in the flowing afterglow involves modification of the flow tube (Fig. 3) to include high quality light baffles for the introduction of the laser beam, replacement of the infrared detector and its viewing port with a suitable photomultiplier and lens collection optics, and alteration of the ion source region to incorporate blackened elbows and a 1 cm diameter aperture to create a high pressure region in the source (13,14). The latter modification increases the number of collisions of the excited metastable species within the source region, so that excited states that would cause light emission either directly or by reaction in the flow tube are completely quenched before entering the photomultiplier detection region. These several changes effect dramatic reductions in the background scattered light from the laser and the flowing afterglow gases,

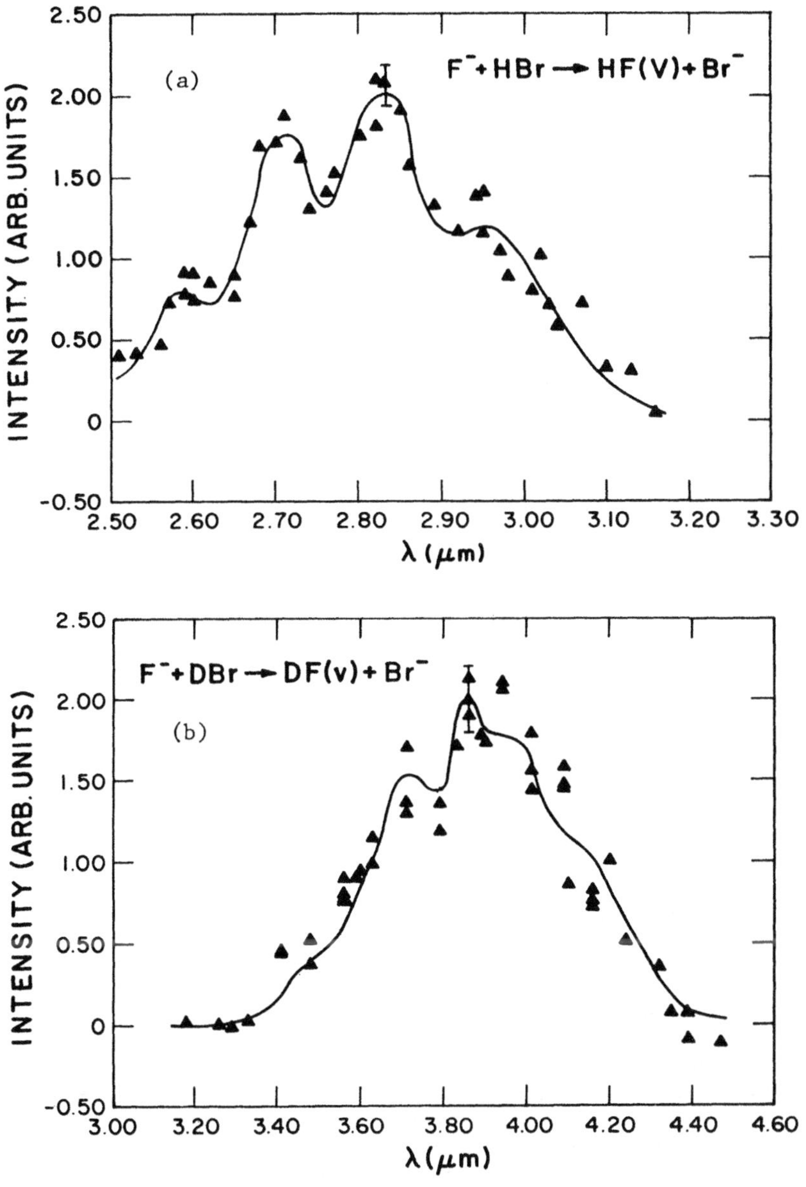

Fig. 2. Infrared chemiluminescence spectra for HF (Fig. 2a) and DF (Fig. 2b) formed in the reactions $F^- + HBr,DBr \rightarrow Br^- + HF(v),DF(v)$. The experimental data have been corrected for relative detection sensitivity. The solid lines are the least-squares fits to the spectra for the best vibrational population distributions.

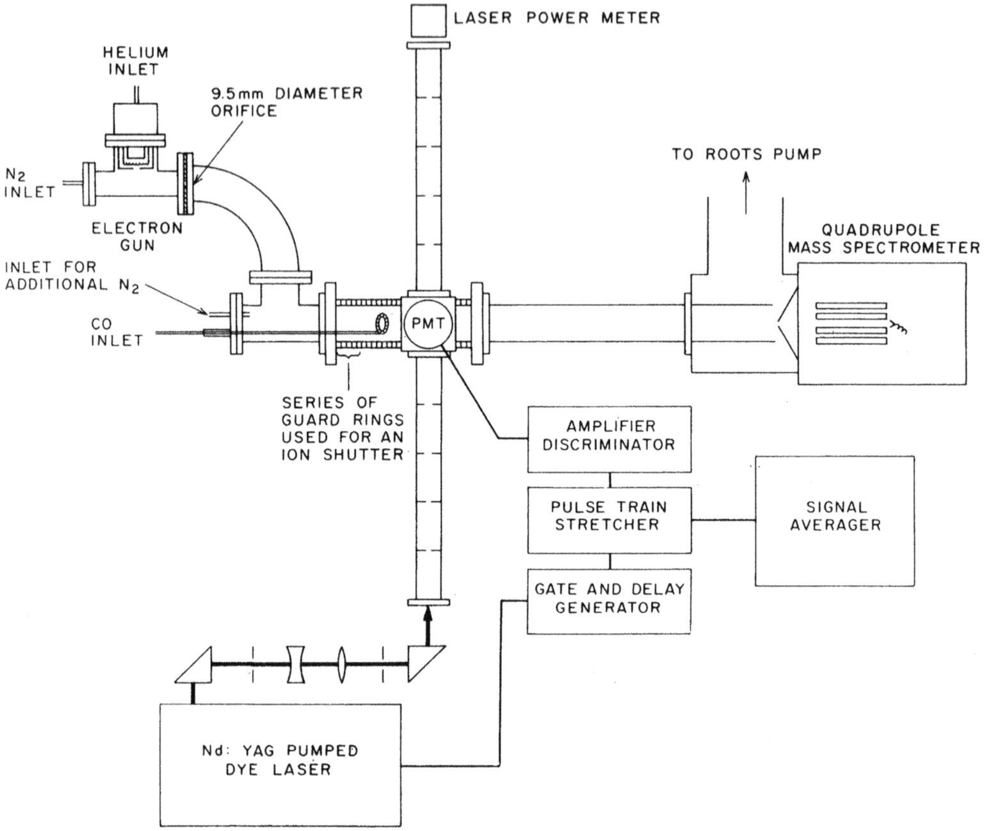

Fig. 3. Flowing afterglow apparatus modified for laser-induced fluorescence detection.

allowing detection of ion densities as low as 10^4 cm^{-3} per product state.

Ions are produced in 10^8–10^9 cm^{-3} densities in the high pressure source and flow down the tube through the laser interaction region. A few cm before the laser beam crosses the flow tube, the neutral reactant is introduced by an effuser in the center of the flow tube. In order to obtain unrelaxed product state information, the distance between reagent introduction and probing and the reagent density are crucial. This is especially true for ion products, which undergo rapid charge transfer reactions with their parent neutral gas. Typically, the reagent density must be maintained low enough so that only 5% of the ions are reacted in the short distance before probing. Under these conditions, relaxation of the product states by collisions with the neutral reagent gas is negligible.

Experiments with the pulsed Nd:YAG-pumped dye laser use time-gated detection, so that photons from the laser-induced fluorescence are counted only during the short time window of the excited state radiative lifetime. There is no need for ion modulation, as was the case in the infrared experiments, because of the "molecular" specificity of the laser probing. However, an ion shutter, consisting of a 30 V potential applied to a short section of flow tube, is used to test whether the molecular signal is due to a reaction of an ion or neutral metastable state, or to subtract any background. Experiments with an argon ion pumped cw ring dye laser use continuous photon counting electronics; the ions or the frequency of the laser can be modulated for additional signal discrimination. The photon counting equipment or an analog boxcar averager is used as the pre-electronics to the input of the multichannel analyzer for spectral scanning or counting of the signal photons on a line.

Laser-induced fluorescence detection is accomplished with a high quantum efficiency photomultiplier tube. The fluorescence light is collected with a 5 cm diameter f/1 quartz lens and passes through an interference filter, which transmits the vibrational band of interest from the molecular emission. A slit is used between the lens and the photomultiplier to allow only light from the zone of the laser excitation beam to be imaged on the photocathode. This discriminates significantly against the background light from the afterglow gases. The experiments are carried out by exciting two different vibrational states to the same vibrational state in the electronically excited manifold. In this way, emission is observed from the same upper state for the determination of a pair of relative vibrational populations. Thus any differences in quenching or radiative rates of the emitting states do not affect the population determinations. For example, for the determination of the v=1/v=0 ratio in CO^+, the laser is used to excite the (1,1) and (1,0) bands and the (1,2) band is observed in emission.

All measurements in the flowing afterglow apparatus use the linear (unsaturated) regime of excitation in laser-induced fluorescence. Laser fluences of less than 1 mJ cm^{-2} are typical. The analysis of the signals to obtain the populations in the linear and saturated limits has been considered in detail (8). In our studies, the high density of helium buffer gas insures that the rotational populations will be fully equilibrated, while the vibrational populations, even for ion products, are almost unaffected by these collisions (15). Thus Boltzmann distributions in rotation are assumed in the reduction of the data, and these are confirmed by rotational analysis of scans of entire spectra taken under reaction conditions. Because the rotational levels are equilibrated, the vibrational populations can be determined from a small portion of a bandhead, or only a few rotational lines, as long as all the states that are being excited by the laser can be identified. This greatly reduces the amount of time needed to determine the vibrational populations.

The buffer gas collisions serve another important function. Product molecules are born with varying amounts of translational kinetic energy release. This requires a flux-to-density conversion in a crossed beam apparatus, with attendant uncertainties about the forward/backward scattering details that are needed to determine the laboratory velocities (16). In the flow tube device, the buffer gas immediately thermalizes the translational velocities of all product states to the same value. Thus no correction is necessary to account for the fact that the laser measures densities in this case. The accuracy of the determinations of vibrational populations is considerably improved by this advantage.

Figure 4 shows a scan of the R_{21} branch bandhead region of CO^+ for three different vibrational bands used to probe the populations in the v=0, 1, and 2 states that are produced by the N^+ + CO reaction. As can be seen, the signal-to-noise in these laser-induced fluorescence experiments is excellent. The rotational lines that are probed in each bandhead are indicated. From the qualitative intensities of the three spectral features, it is readily apparent that v=0 has the largest population, and that the population in v=2 is very minor. The integral of these bandhead features is used together with the Boltzmann populations in the rotational states that contribute to the bandhead region to extract accurate population distributions. Equation (1) is used to carry out the conversion from signal intensity to populations:

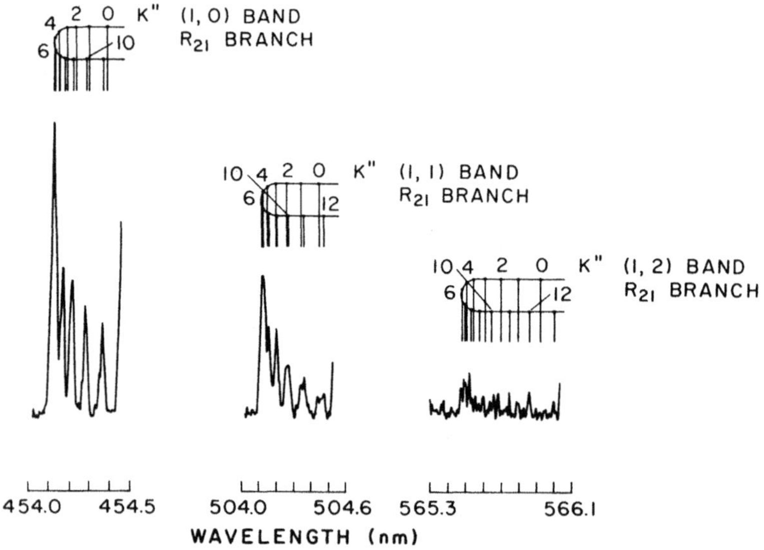

Fig. 4. Laser-induced fluorescence spectra of the R_{21} branch bandhead regions of CO^+ (v=0, 1, and 2) produced from the N^+ + CO charge transfer reaction.

$$\frac{N_1}{N_2} = \frac{I_1}{I_2}\,\frac{q_2}{q_1}\,\frac{\left[\sum_{K''=0}^{K''_{max}} S_{K''}\, e^{-BK''(K''+1)/kT}/Q\right]_2}{\left[\sum_{K''=0}^{K''_{max}} S_{K''}\, e^{-BK''(K''+1)/kT}/Q\right]_1}\,\frac{(n\lambda^2)_2}{(n\lambda^2)_1} \quad . \qquad (1)$$

Here, N_i is the population of the ith vibrational state, I_i is the laser-induced fluorescence intensity measured for this state, and q_i is the Franck-Condon factor for the excitation transition. The quantities in the sums account for the specific fractions of rotational population that are sampled by each bandhead region and $S_{K''}$ is the rotational line strength. Q is the rotational partition function for the given state. Finally, the factor $n\lambda^2$ accounts for the fact that the dye laser scan is linear in wavelength, whereas the necessary formulae for line strengths require frequency (14).

During the population determinations, tests are performed to assign the degree of vibrational deactivation that has occurred at the point of laser-induced fluorescence probing. Empirically, this is done by varying the reagent gas pressures and changing the neutral inlet distance from the laser detection zone to measure changes in the ratios of the vibrational populations (14). Under our typical conditions, the extent of deactivation, e.g. the loss of $CO^+(v=4)$ by collisions with CO is about 15% (14). This is the worst case because the vibrational deactivation of $CO^+(v)$ with CO occurs by rapid symmetric charge transfer. With other gases, the effects of relaxation are much smaller. In the course of the population determinations, both the correct zero-pressure population ratios are extracted as well as the rates of vibrational deactivation with different reagent and ion precursor gases.

2.4. Single Collision Apparatus

While the flowing afterglow is an excellent device for many measurements of vibrational populations that will be described here, it does have a shortcoming: all information contained in rotational state details of the reactive process is lost because of rapid rotational relaxation by collisions with the high density of helium carrier gas. In addition, some particularly fragile products may also be lost by collisional isomerization, or collisions with He might redistribute vibrational excitation in polyatomic products. For these reasons, it is desirable to have an apparatus that is capable of producing high fluxes of thermal ions for reaction under low pressure conditions, where no further collisions with either the reagents or buffer gas will perturb the state distributions.

An apparatus designed to satisfy such requirements is pictured in Figure 5. A flowing afterglow source is the method of choice for production of the high fluxes of thermalized ions because its electrically neutral plasma affords the highest densities of ions possible. The

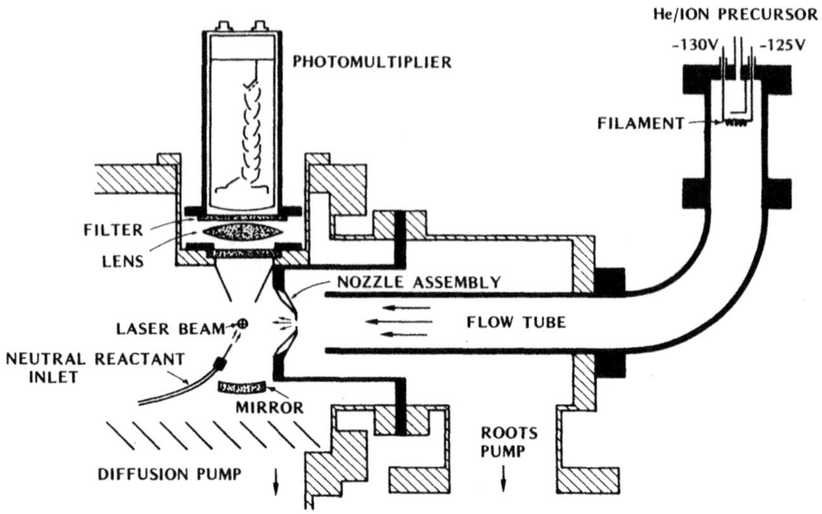

Fig. 5. Schematic of the flowing afterglow-supersonic expansion apparatus for studying thermal energy ion-molecule dynamics under single collision conditions.

large number of collisions of the ions with both the buffer gas and the ion precursor molecules insures complete relaxation of the ion states before reaction. The ions in the central portion of the flow tube are allowed to undergo a mild free jet expansion through a 3 mm diameter orifice into a high vacuum region. This produces a "beam" of ions with a kinetic energy that is elevated only slightly above thermal; the reactions are carried out at 0.1-0.2 eV relative kinetic energy. The ions are crossed with an effusive spray of reagents, and the products are probed by laser-induced fluorescence under single collision conditions. At present, the ion densities are too small to make state measurements by infrared chemiluminescence. In addition, in order to overcome the signal-to-noise limitations of the very low ion densities after the expansion, the technique of saturated laser-induced fluorescence is used to increase the signal strength appreciably (8).

The flow tube in this apparatus is a 5 cm diameter by approximately 40 cm long stainless steel tube, with an ion source at the upstream end and a blackened 90° bend to reduce scattered light from the afterglow. Most of the gas is pumped out by a high capacity blower pump, which establishes a flow velocity of 1×10^4 cm s^{-1} of the helium carrier gas and the entrained ions. The ion source is divided by a stainless steel plate that has a 1 cm diameter orifice in it to create a higher pressure, lower velocity zone to quench the metastable states fully. After the ions undergo this stage of pressure reduction, they encounter the 3 mm diameter orifice, and part of the ions and helium buffer gas is extracted by a supersonic expansion into the main chamber. The main reaction vessel consists of a 50 cm diameter stainless steel chamber with baffle arms to introduce the laser and a fluorescence port for the photomultiplier detection. The neutral reagent is added through a linear array of 9 small stainless steel tubes to create an effusive spray of reactant. The pulsed Nd:YAG pumped dye laser excitation beam excites the products of the ion-molecule reaction and single-photon counting electronics pick up the signal within a short time gate after the laser pulse. A differentially pumped chamber opposite to the ion beam houses a quadrupole mass filter for diagnostics of the ion production and reaction.

Calculations of the supersonic expansion of the helium-entrained ions give a terminal Mach number of about 4, a relative translational temperature of 40 K, and a final velocity of 1.6×10^5 cm s^{-1} (17). Assuming that there is no slip between the ions and the helium expansion, the typical relative kinetic energies, which are dependent on the masses of the ion and neutral reagent, are 0.1-0.2 eV. Measurements of the rotational state distributions of diatomic ions undergoing the supersonic expansion show somewhat incomplete rotational accommodation, with bimodal distributions of 60 and 160 K for the low and high rotational levels, respectively (17). The ion densities before and after the expansion are estimated to be 10^9 and 10^6 cm^{-3}, respectively. Thus, experiments are carried out routinely on as few as 10^4 molecules cm^{-3} per product state.

The subtleties of using saturated laser-induced fluorescence to extract population distributions has been discussed in detail (8). Difficulties can arise when probing two bands with very different transition strengths, e.g. if the weaker band is not saturated and the stronger one is. In addition, the laser beam intensity profile can include substantial intensity in the wings of a Gaussian beam where the molecules may not be saturated. In our experiments, a set of irises is used to clean up the intensity profile of the laser beam to achieve a rectangular intensity versus distance across the beam. In addition, bands are probed that have very similar transition strengths so that equal degrees of saturation can be achieved. Finally, the saturation signal versus laser power is measured to insure that the experiments are carried out in a regime where a high degree of saturation is achieved. Typically energy densities of 10-30 mJ cm^{-2} in a 1 cm^{-1} bandwidth are used.

Under these conditions, the signal intensity depends on the ratio of degeneracies for the upper and lower state that are pumped, but not on the Franck-Condon factor for the pumping transition. The expression for the intensity of laser-induced fluorescence signal from a single level becomes:

$$I(J) = C \frac{g_{2J}}{g_{1J} + g_{2J}} N_{1J} \tag{2}$$

where g_1 and g_2 are the degeneracies of the upper and lower states, respectively, C includes all the detection efficiency factors, and

N_{1J} is the density in the lower state that is being probed. A further check on the proper use of the saturated laser-induced fluorescence is made by plotting the Boltzmann populations of a thermal sample of CO^+ using both the degeneracy factors and the Honl-London factors, which would be applicable if saturation is not achieved. The reduction of the data using the degeneracy factors and assuming saturation gives a temperature of 287 K, in good agreement with the correct 295 K temperature of the sample. In contrast, a treatment that assumes that saturation is not achieved gives a plot with large scatter and an erroneously high temperature of 369 K (17).

In order to achieve single collision conditions, collisions with both the helium buffer gas and the reagent gases are considered. After the supersonic expansion, the mean free path for collision of a product molecule with the main constituent, helium, is estimated to be 10 cm. Thus, rotational state-changing collisions with the helium buffer gas are unlikely. Similarly, since the ion precursor density is at least 1000 times smaller than the helium, there is no possibility for collisions of the product ions with that gas. The most important species that is involved in collisional relaxation of the product states is the neutral reagent used for the reaction partner. Relatively high densities of this reagent are desirable to react a significant fraction of the primary ion beam; however, then the probability of a secondary collision of the product molecule with the reactant becomes large.

Ideally, if the extent of reaction of the primary ion beam is maintained below 5%, the probability of a secondary collision is low. However, the tradeoff of signal-to-noise for reacting such a small fraction of the ions is great. Therefore, data are usually taken at several reagent flows and the populations that result are back extrapolated to the low pressure limit. Figure 6 shows the dramatic relaxation that occurs for the $N_2^+(v=1)$ state with N_2 flow in the $Ar^+ + N_2$ reaction. This is evident immediately from the change in the spectral features in the figure at two different N_2 flows and is quantified in the plot of the v=0/v=1 ratio as a function of N_2 flow. The extrapolation to the low pressure limit yields the final v=0/v=1 ratio. Similar tests are performed on the rotational populations to determine the correct pressure limits under which nearly single collision conditions are achieved and the data are back-extrapolated to the zero pressure limit.

A final consideration in the analysis of the data concerns the fact that the laser measures the densities of product states, whereas the population ratios are related to the fluxes into each state of the products. The conversion from density to flux requires a knowledge of the laboratory velocities of each product state. Unfortunately, this is not often known and is not simply related to the translational energy release, unless some assumptions are made about whether the products are scattered isotropically, forward or backward. In an unusual circumstance, one product state might be forward scattered and another back scattered. This could result in a large ambiguity concerning the lab velocities needed to reduce the data. In a worst-case example, for $Ar^+ + N_2$, this difference in the product detection efficiency can

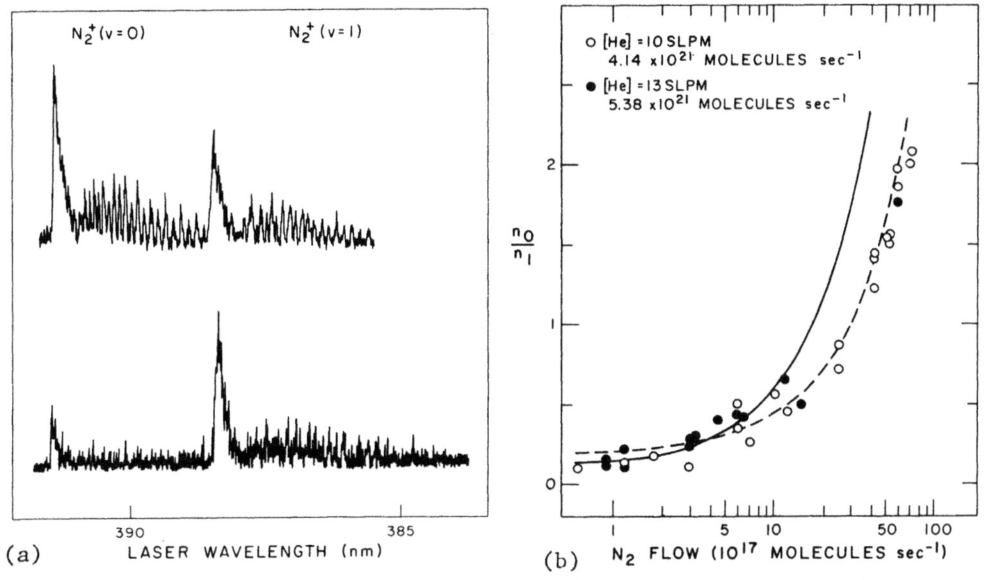

Fig. 6. (a) Laser-induced fluorescence spectra at a high flow of N_2 (upper) and at a very low flow, showing the severe vibrational relaxation that occurs. (b) Plot of the v=0/v=1 ratio of populations as a function of N_2 flow. The solid and dashed lines are two different models to extrapolate the data to zero pressure.

amount to a factor of ten uncertainty in the product ratio (16). However, if more realistic angular distributions of scattering are assumed, the uncertainty in the product state determinations is reduced considerably. The single collision apparatus is an excellent technique for obtaining rotational product state information, and is a valuable but less accurate method for complete vibrational information.

3. SPECIFIC SYSTEMS

3.1. Proton Transfer Reactions

Proton transfer reactions $Y^- + HX \rightarrow X^- + HY$ are ubiquitous throughout chemistry as they occur in a wide variety of natural and man-made environments. While the kinetics and thermodynamics of these processes have been extensively investigated, essentially nothing was known about energy disposal in the thermal energy processes. These systems are especially attractive for study due to their relative simplicity; most occur adiabatically on a single potential energy surface and are thus relatively tractable theoretically. Moreover, many of the analogous hydrogen atom transfer reactions Y + HX → X + HY have been studied in detail; the product vibrational and rotational state distributions have been obtained for many systems and the relative importance of such features as total available energy, kinematic effects and shape of the potential surface are at least qualitatively known.

Table 1 gives the exothermicities and rate constants for the reactions

$$F^- + HBr,DBr \rightarrow Br^- + HF(v),DF(v)$$

and for their neutral analogs (11,18). The overall energetics and the reduced masses for the ion and neutral processes are very similar. In contrast, the rate constant for the ion-molecule reaction is about thirty times larger, reflecting the dramatically different potential energy surfaces. The ionic surface has a long range ion-dipole and ion-induced dipole attractive potential in the entrance channel and a deep well in the exit channel. The neutral surface has a small barrier in the entrance valley followed by a monotonic decrease in energy as the reagents evolve to products. These different potential energy surfaces might give rise to very different product vibrational energy distributions. Energy disposal in the neutral reaction can be described as "mixed energy release" where the H atom recoils from the Br while the HF bond is still extended (19). The presence of the deep potential well in the ion reaction might be expected to increase the probability of long-lived complexes and thus might result in a more statistical population of states in the products. The effect of deuteration is of particular interest since two opposing effects on the vibrational distribution might be expected. Deuteration may decrease the propensity for vibrational excitation since the heavier D atom will recoil less rapidly from the Br. Alternatively, deuteration may decrease the probability of secondary encounters and result in a hotter product vibrational distribution.

The experimental infrared chemiluminescence data for the F^- + HBr,DBr reactions are shown in Figure 2 along with the least-squares computer fit (11). The resulting nascent vibrational distributions are given in Table 1, with v=0 populations deduced from surprisal analysis. The vibrational populations for the neutral reactions are taken from the fast flow reactor data of Setser and coworkers (18).

TABLE 1. Vibrational distributions of proton and deuteron transfer reactions and their neutral analogs measured by infrared chemiluminescence detection.

Reaction	ΔH (eV)	k ($cm^3\ s^{-1}$)	N_0	N_1	N_2	N_3	N_4	N_5	N_6	$\langle f_v \rangle$	Ref.
$F^- + HBr \rightarrow Br^- + HF$	-2.08	1.2(-9)	(0.02)	0.09	0.29	0.34	0.28			0.60	11
$F + HBr \rightarrow Br + HF$	-2.11	4.5(-11)	(0.04)	0.09	0.21	0.33	0.33			0.59	18
$F^- + DBr \rightarrow Br^- + DF$	-2.10		(0.02)	0.05	0.12	0.16	0.25	0.22	0.20	0.63	11
$F + DBr \rightarrow Br + DF$	-2.13		(0.04)	0.06	0.13	0.16	0.21	0.23	0.17	0.58	18

The striking similarity of the ionic and neutral distributions can be understood in terms of qualitative kinematic arguments. For collinear collisions on potential surfaces with small barriers and no potential wells, the exchange of a light particle between two heavier ones favors product vibration no matter whether the energy release occurs before the new bond is formed ("attractive energy release") or after the old bond is broken ("repulsive energy release") (20,21). This follows from the strong inertial coupling between products and reagents; the light hydrogen atom cannot impart much momentum in collisions between two heavier particles. In mass-weighted coordinates which diagonalize the kinetic energy, the angle between the entrance and exit channels for the F + HBr mass combination is only 15 degrees, which implies very strong inertial coupling. Most of the reaction exothermicity must therefore end up as product vibration. These dynamical considerations follow from the reactant mass combination and therefore apply to both the ion and neutral reactions. The large value of $\langle f_V \rangle$, the average fraction of the available energy deposited in product vibration, for both the F^- and F reactions suggests that secondary collisions within the reaction complex either do not occur or are not efficient in removing vibrational energy from the HF or DF products.

If the above arguments correctly describe both the ionic and neutral reactions, a significant isotope effect is not expected for either reaction. Slightly colder vibrational distributions could occur for the DBr reactions since the skew angle opens up to 21 degrees and the inertial coupling may be less efficient. Slightly hotter vibrational distributions would result from fewer secondary collisions due to the more slowly moving D atom. The similar experimental results for the deuterated and undeuterated reactions demonstrate that such effects are very small and well within the experimental uncertainty.

The high probability for disposal of reaction exothermicity into product vibration in proton transfer reactions is further confirmed in the systems (9,10,22)

$$F^- + HC\ell, HI \rightarrow C\ell^-, I^- + HF$$

$$C\ell^- + HBr, HI \rightarrow Br^-, I^- + HC\ell$$

$$CN^- + HC\ell, HBr, HI \rightarrow C\ell^-, Br^-, I^- + HCN \quad .$$

In these reactions population of the maximum vibrational level accessible by the exothermicity is observed; product distributions are similar to those for the analogous neutral reactions where these data are available for comparison. Trajectory calculations (23) for the $C\ell^-$ + HBr reaction predict a strong propensity to deposit energy into product vibration. These studies demonstrate that most collisions are indeed direct in spite of the potential well that arises from the $BrHC\ell^-$ complex stability; this is intuitively reasonable since the well is small relative to the exothermicity.

Precise vibrational state distributions are possible for the proton transfer reaction (13)

$$O^- + HF \rightarrow F^- + OH(v=0,1)$$

without the need for surprisal analysis. Both vibrational states of the OH product are directly detected by laser-induced fluorescence. The resulting distribution ($N_0 = 0.82$ and $N_1 = 0.18$) with its low v=1 population is characteristic of product states whose energies are close to the exothermicity; in this case, the energy in excess of v=1 is only 140 cm^{-1}. Nevertheless, the result indicate strong preference for energy disposal into vibration since the value expected by a statistical "prior" distribution is only 0.07. This reaction poses a more serious challenge to detailed interpretation since both the O^- reactant and the OH product possess two closely lying spin-orbit states which may differ substantially in reactivity and detailed dynamics. No theoretical study of this process is currently available.

3.2. Associative Detachment Reactions

Associative detachment reactions, $A^- + B \rightarrow AB + e^-$, are intriguing ion-molecule processes, unparalleled in the realm of neutral chemistry. These reactions play a critical role in the balance between electrons and ions in the earth's upper atmosphere, as well as in lasers and other plasma phenomena. Associative detachment reactions proceed through the formation of unbound or quasibound AB^- molecular negative ion states which then decay to products. These collision complexes are the key intermediates for a wide variety of physical processes including elastic and inelastic electron-molecule scattering, dissociative attachment and collisional detachment. The extensively studied electron scattering processes probe the AB^- state over a small range of internuclear separations where AB is strongly bound. In contrast, associative detachment explores the long-range side of the internuclear potential, from separations near infinity to those corresponding to the bound neutral. Therefore, a study of the product state distributions of the associative detachment reactions should provide important complementary information for a more detailed description of the transient AB^- species.

In the associative detachment reaction, the production of the electron imposes important kinematic constraints on the final energy distribution in the product molecule. Due to its low mass the departing electron can remove very little of the initial orbital angular momentum of the colliding reactants; this momentum must then be retained as rotational excitation in the product molecule. This results in a direct and predictable relationship between incoming impact parameter and the product rotational state (24,25). Angular momentum conservation when coupled with the constraint of available reaction energy, then restricts formation of the highest product vibrational states to collisions of small impact parameter (low rotational states). Thus, population of the high vibrational states is partially controlled by kinematics.

A second major determinant of energy disposal in the associative detachment reaction is the fundamental dynamical branching of reactants into product vibrational states. The AB^- complex is expected to detach over a range of internuclear distances each of which corresponds to a distinct distribution of product vibrational states, based on the Franck-Condon principle. The observed distribution then is an integral of the detachment probabilities into each state as a function of internuclear separation, convoluted with the fraction of reagents remaining at each separation.

To explore the kinematic and dynamical effects in associative detachment reactions we have studied infrared chemiluminescence from the products of $O^- + CO$ (26), $C\ell^- + H$ (24), $CN^- + H$ (22) and $F^- + H,D$ (27,28). These latter reactions

$$F^- + H \rightarrow HF(v=0-5) + e^-$$

$$F^- + D \rightarrow DF(v=0-7) + e^-$$

are the most informative due to the isotope substitution and the high vibrational levels that are energetically accessible. Figure 7 shows the product vibrational state distribution for these reactions. In both cases there is extensive vibrational excitation extending to the highest allowed levels. This suggests high probability for detachment of the electron at large internuclear separations of the reagents. In the HF results in particular there is a dramatic truncation in the highest vibrational level (v=5) due to the kinematic constraint of angular momentum conservation, with concomitant production of high rotational excitation. Unfortunately, the nascent rotational distribution cannot be detected in helium buffer gas since collisional relaxation of rotations by helium is known to be rapid. We have, however, confirmed that high rotational states are populated by partially resolving infrared emission from DF in argon buffer gas (27); extended P and R branch tails in the red and blue ends of the spectrum are clearly evident, indicating a highly nonthermal distribution in rotation. A similar effect is expected for HF.

The highly inverted vibrational distribution in HF is explained by a virtual state model developed by Gauyacq (29,30). In this picture transitions from HF^- into the $HF + e^-$ continuum become possible over a range of internuclear separations much greater than the value where the HF and HF^- curves cross. This is possible because the binding energy becomes small enough that electronic motion begins to couple effectively with nuclear motion, i.e., the Born-Oppenheimer approximation breaks down. This mixes the electronic wave function with the continuum states allowing for dynamically induced transitions to occur from HF^- to $HF + e^-$ far outside the adiabatic $HF + e^-$ well. Because large internuclear separations are sampled first in the $F^- + H$ collision and the transition probability is large, the virtual state picture predicts population of the highest vibrational states. Moreover, small internuclear distances are not effectively sampled, causing small populations in the low vibrational levels. The vibrational distribution for $F^- + H$ calculated by Gauyacq is also shown in Figure 7; agreement with the experimental data is excellent.

Although both the $F^- + H$ and $F^- + D$ reactions produce high vibrational excitation, the hydrogen atom reaction channels a higher fraction

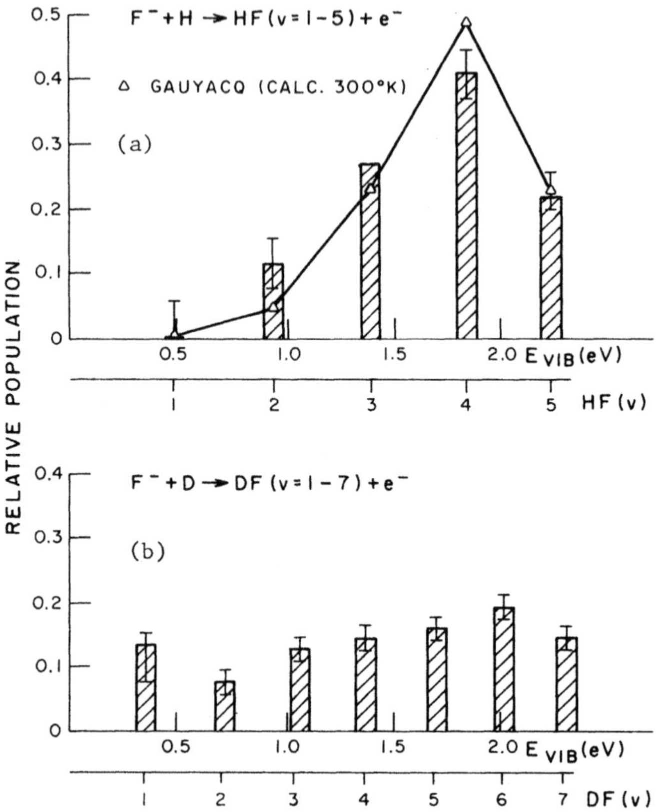

Fig. 7. Vibrational population distributions for HF (Fig. 7a) and DF (Fig. 7b) formed in the reactions $F^- + H,D \rightarrow HF(v),DF(v) + e^-$. The theoretical results of Gauyacq (Ref. 29,30) for the hydrogen atom reaction are shown for comparison.

of the reaction energy into vibration ($\langle f_v \rangle = 0.72$) than does the deuterium atom reaction ($\langle f_v \rangle \lesssim 0.61$). This is not a result of kinematics; it can be shown that the angular momentum constraint diverts nearly identical fractions of the reaction exothermicity away from vibration in both the HF and DF cases (27). Furthermore, the virtual state model of Gauyacq does not explain this difference. Calculations indicate that the HF and DF distributions should be very similar. There is currently no adequate explanation for the observed difference in the HF and DF distributions.

Associative detachment reactions thus represent a superb example of the interplay of experiment and theory in understanding the detailed mechanisms of ion-molecule processes. Nevertheless, it is clear that increasingly precise data as well as refinements of theory are essential for a full description of the kinematics and dynamics of associative detachment.

3.3. Polyatomic Ion-Molecule Reactions

The proton transfer and associative detachment reactions described above have considered very simple systems involving a total of only two, three or four atoms in the reactants. In all cases, efficient channeling of reaction energy into product vibration was observed and vibrational levels up to the limit of the exothermicity were populated. These reactions appear to be direct, and intermediate complexes, if formed, must be short-lived.

Systems involving large polyatomic reactants offer a potentially different dynamical situation since the number of degrees of freedom is increased. This may enhance the probability of "sticky" collisions in which the complex survives for several rotational, and possibly, vibrational periods. If the available energy flows readily throughout the intermediate then the reaction exothermicity may be partitioned in a statistical fashion among the product degrees of freedom (31). A randomized product state distribution is then the signature of a long-lived intermediate complex.

To probe for these statistical effects in polyatomic systems we have studied infrared chemiluminescence from the heavy atom transfer reactions (28)

$$SF_6^- + H \rightarrow SF_5^- + HF(v=0-12)$$

$$SF_6^- + D \rightarrow SF_5^- + DF(v=0-17) \quad .$$

Figure 8a shows the nascent HF vibrational distribution from the H atom reaction; there is relatively efficient disposal of energy into vibration ($\langle f_v \rangle = 0.37$) and levels between v=2 and v=7 are populated. The results for the D atom reaction are similar with $\langle f_v \rangle = 0.38$ and populations extending from v=2 to v=9 in DF. Although there is only moderate vibrational excitation in these reactions the results are far from statistical. Figure 8b shows the "prior" statistical distribution for the SF_6^- + H reaction where the calculated $\langle f_v \rangle$ is only 0.04. Thus the lack of energy randomization in the HF and DF products suggests that a long-lived complex is not formed along the reaction coordinate.

The key to interpretation of the vibrational distributions is provided by analogy with neutral reactions. The mass combination for these reactions is somewhat special since they are **L + HH' → LH + H'** systems (21). Extensive studies of such neutral reactions have demonstrated that the average fraction of available energy appearing in the **LH** vibration $\langle f_v \rangle$ correlates very well with the amount of attractive energy release (i.e., energy that is released early along the reaction pathway when **L-H** bonds are highly extended and **H-H'** bonds are still near equilibrium distances) (32,33). The experimental $\langle f_v \rangle$ values for the SF_6^- + H,D systems show that a moderate portion of the energy release is attractive; however, the larger fraction is repulsive energy

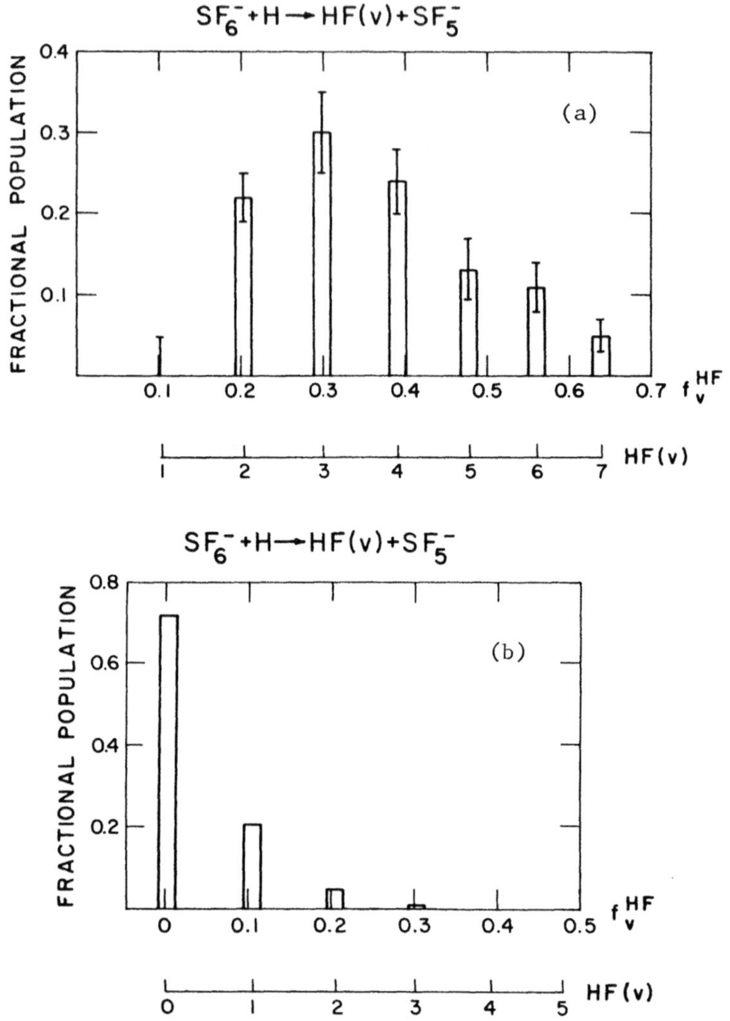

Fig. 8. (a) HF(v) nascent experimental distribution of the SF_6^- + H reaction. (b) HF(v) predicted statistical distribution of the SF_6^- + H reaction.

release as the reactive S-F bond breaks. Due to their light mass and high velocity the H or D atom can closely approach an F atom before a significant amount of repulsive release occurs. This favors high deposition of energy into recoil translation of the products and possibly also vibrational excitation of SF_5^-.

Vibrational state distributions have been determined by Setser and co-workers for a number of neutral reactions involving hydrogen atom and polyatomic molecules (32,33). Our HF and DF results are remarkably similar to those for the neutral systems both in the shape of the distributions and in the values of $\langle f_v \rangle$ for the diatomic product. It appears therefore that neither the entrance nor exit wells in the ion-molecule reaction are very important in determining energy disposal. This is not surprising when it is realized that the entrance valley is very shallow due to the large size of SF_6^- and small polarizability of H or D. The attractive energy in the exit channel, although greater, might not affect the population distribution because of the much larger recoil translational energy release of the products. As a result the SF_6^- + H,D reactions may closely resemble their neutral counterparts.

This analogy with neutral systems provides an intuitively satisfying explanation of the data; however, the existence of long-lived complexes and secondary collisions which deplete the higher vibrational states of HF and DF cannot be rigorously excluded. Secondary encounters may be significant for these higher vibrational states where the recoil translational energy must be low, by energy conservation. The SF_5^- and HF or DF may then be trapped within the attractive well, allowing for energy transfer to the SF_5^- bath of vibrational states. From an alternative perspective, it has been noted (34) that the occurrence of a non-statistical product state distribution does not necessarily preclude the existence of long-lived collision complexes. These species may be constrained to sample only a small fraction of the available phase space despite a substantial lifetime.

3.4. Auroral Reactions

A global manifestation of the importance of gas phase ion chemistry and physics occurs in the earth's ionosphere. This weakly ionized plasma embedded in the atmosphere between 50 km and 400 km influences radio waves and produces magnetic disturbances on the earth's surface. The ions, although only trace constituents, play a major role in the channeling and redistribution of energy in this region of the atmosphere. This occurs directly through ion-molecule reactions and indirectly through the coupling with neutral chemistry. Chemiluminescence from the products of these reactions generates auroral and airglow emissions. However, modeling of these emissions demonstrates that neither their intensities nor wavelength distributions are fully understood; in particular, the electronic and vibrational product state distributions of several important ion-molecule reactions are not known.

An ion-molecule reaction of critical importance in the earth's upper atmosphere is the destruction of N^+ by the highly exoergic reaction with O_2:

$$N^+ + O_2 \xrightarrow{43\%} NO^+ + O + 6.7\ \text{eV}$$

$$\xrightarrow{51\%} O_2^+ + N + 2.5\ \text{eV}$$

$$\xrightarrow{6\%} O^+ + NO + 2.3\ \text{eV} \quad .$$

The first product channel releases sufficient energy to populate NO^+ vibrational levels up to v=28; this would generate infrared emission between 4.3 and 7.0 μm, a region in which there are unidentified atmospheric emissions. Moreover, in addition to formation of the ground state $O(^3P)$, the electronically excited states $O(^1D)$ and $O(^1S)$ are energetically accessible; this would give rise to visible red and green emissions at 630 nm and 557.7 nm, respectively.

To fully characterize energy disposal in the $N^+ + O_2$ reaction we have studied wavelength resolved infrared (35) and visible chemiluminescence (12) from the reaction products. Figure 9 displays the observed infrared emission spectrum, the least-squares computer fit of the data and the resulting $NO^+(v)$ distribution. The distinct bimodal population of vibrational levels suggests that these may correspond to production of the atomic oxygen in different electronic states. If no

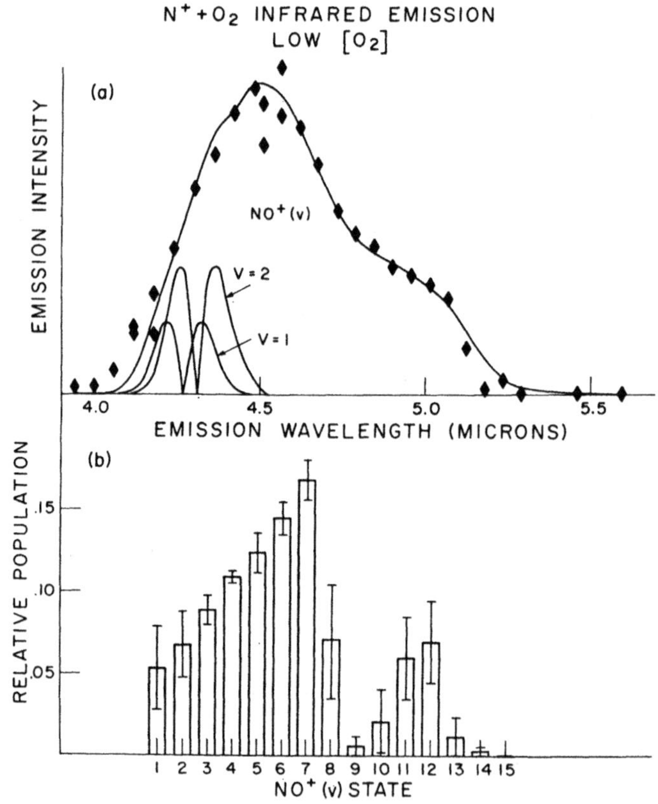

Fig. 9. (a) The observed fluorescence spectrum due to $NO^+(v)$ from the reaction $N^+ + O_2$. The solid line is the least-squares fit to the data. (b) The resulting $NO^+(v)$ distribution.

energy is deposited into translation, production of $NO^+(v \lesssim 8)$ can be accompanied by formation of $O(^1S)$, production of $NO^+(v \lesssim 18)$ can be accompanied by formation of $O(^1D)$, and production of $NO^+(v \lesssim 28)$ can be accompanied by formation of $O(^3P)$. Therefore, one explanation for the shape and vibrational cutoffs in the distribution is that substantial amounts of $O(^1S)$ and moderate amounts of $O(^1D)$ are formed in the reaction, with little translational energy release.

The electronic state product branching can be probed directly, however, with the visible chemiluminescence studies. The $O(^1S)$ product was monitored by the $O(^1S)$-$O(^1D)$ emission at 557.7 nm. The $O(^1D)$ product was monitored via sensitized fluorescence at 760 nm from $O_2(b^1\Sigma_g^+)$ formed by energy transfer from $O(^1D)$ to $O_2(X^3\Sigma_g^-)$. Absolute yields were then inferred by comparison to the known $O(^1S)$ and $O_2(b^1\Sigma_g^+)$ emission intensities from the reaction of $Ar(^3P)$ with O_2 (36). The results indicate production of $\lesssim 0.1\%$ $O(^1S)$, $70 \pm 30\%$ $O(^1D)$, and by conservation, $30 \pm 30\%$ $O(^3P)$. Combination with the infrared results then strongly suggest that formation of $O(^1D)$ is accompanied by $NO^+(v \lesssim 8)$ and formation of $O(^3P)$ is accompanied by $NO^+(v=9-15)$. This assignment implies that ~30% of the available energy is deposited into vibration, ~20% appears as electronic excitation of the oxygen atom while the remaining 50% must be channeled into translational and rotational excitation of the products. These results are fully consistent with those of Bowers and coworkers (37) who studied this reaction using the kinetic energy release ICR technique.

A qualitative correlation diagram for the reaction of N^+ with O_2 explains the experimental observations (12). No adiabatic pathways directly link the initial reactants to the products. This implies that an initial non-adiabatic electron jump to one of the surfaces correlating with the charge transfer products is required. These surfaces correlate with channels forming NO^+ and $O(^3P)$ or $O(^1D)$ but not with the channel forming $O(^1S)$.

The determination of the electronic state branching of the atomic oxygen product has important consequences for the modeling of auroral emissions (38). These results demonstrate that the $N^+ + O_2$ reaction is not a significant source of the $O(^1S)$ green emission at 557.7 nm and thus resolves the controversy surrounding this issue (38). However, the high production of $O(^1D)$ by $N^+ + O_2$ is responsible for ~10% of the auroral red line emission at 630 nm; this contribution was previously not recognized.

3.5. Charge Transfer Reactions

Charge transfer reactions involve nonadiabatic electronic surface crossings. Our experiments consider how the electronic energy of the reaction, which is released in the transfer of the light electron, manifests itself in the motions of the heavy nuclei. We consider only systems involving atomic ions reacting with diatomic neutral molecules at near thermal energies in this first work. We find that determinations of the product state distributions in these charge transfer reactions can elucidate new details about the locations and nature of the surface crossings that govern the transformations.

Several simple models for charge transfer processes have been discussed. One is a Franck-Condon model that considers the Franck-Condon factors for ionization of the isolated diatomic molecule. For the systems studied here involving CO and N_2 molecules, the Franck-Condon factors would predict formation of v=0 with 95% probability; thus this model would predict disposal of energy primarily into product translational energy release. Another model considers the importance of internal energy resonance principles; thus the exothermicity of the reaction would be routed maximally into vibrational and rotational excitation of the products. A third idea considers the formation of a long-lived complex with a statistical partitioning of the energy between translation, vibration and rotation states according to phase space principles. We find that no single, simple mechanism can explain all the results observed and that some charge transfer processes are remarkably selective in the partitioning of their energy. The same high degree of specificity also manifests itself in the back reactions of selected states (1).

The three reactions considered here are the charge transfers between N^+ and CO and Ar^+ with N_2 and CO (14,16,17,39-42). These reactions have some interesting similarities and differences (Table 2). The overall exothermicity is sufficient to populate up to v=2 in the CO^+ product of the N^+ + CO reaction (14,17,41). For Ar^+ + CO the maximum vibrational state is v=6 (39,40), and for Ar^+ + N_2 only the v=0 and v=1 states are accessible (16,42). The diatomic molecules, CO and N_2, have very similar electronic structures and almost identical Franck-Condon factors, which favor production of the v=0 state. The atomic ions are both P states, however, the Ar^+ is a nearly filled P-shell, whereas N^+ has one empty location in each of the three P orbitals. Most important of all is that we observe remarkable differences in the product state distributions in vibration and rotation for each of these three reactions. This is convincing evidence that a simple picture of an atomic ion P state interacting with a diatomic molecule is not detailed enough and that there is an opportunity to elucidate chemical subtleties of the potential surface interactions with these experiments.

Table 2 shows the results of the determinations of the vibrational populations for the three charge transfer reactions, N^+ + CO, Ar^+ + N_2, and Ar^+ + CO. One very apparent difference among the three reactions is that the N^+ + CO reaction favors the formation of v=0 at thermal energies, whereas both the Ar^+ + N_2 and Ar^+ + CO reactions produce high vibrational excitation. The Ar^+ + N_2 reaction deposits energy into the v=1 product channel with very high specificity. This has also been observed over a wider range of collision energies in crossed beam reaction studies (43).

In the N^+ + CO reaction, the vibrational population distribution obtained at thermal energy appears to adhere to a Franck-Condon type of mechanism. The Franck-Condon factors for the CO diatomic are 0.96, 0.04, and 0.0 for v=0, 1, and 2, respectively, which is very close to the observed distribution of 0.71, 0.27, and 0.02. The observed distribution has been fit with some success to a simple model assuming that two mechanisms may contribute: a long range electron transfer

TABLE 2. Vibrational distributions of charge transfer reactions measured by laser-induced fluorescence.

Reaction	Kinetic Energy (eV)	v_{max}	Populations								Approximate Uncertainty	Reference
			v=0	v=1	v=2	v=3	v=4	v=5	v=6	v=7		
$N^+ + {}^{12}CO \rightarrow {}^{12}CO^+(v) + N$	0.03	2	0.71	0.27	0.02						10%	(10)
	0.16	2	0.40	0.57	0.03						20%-30%	(16)
$N^+ + {}^{13}CO \rightarrow {}^{13}CO^+(v) + N$	0.03	2	0.70	0.26	0.04						10%	(10)
$Ar^+ + N_2 \rightarrow N_2{}^+(v) + Ar$	0.28	1	0.11	0.89							10%	(12,17)
$Ar^+ + CO \rightarrow CO^+(v) + Ar$	0.03	6	0.06	0.07	0.09	0.15	0.21	0.27	0.15		15%	(14)
	0.2	7	0.07	0.05	0.09	0.10	0.15	0.31	0.20	0.03	30%-100%	(15)

that results in a Franck-Condon distribution and a collision complex mechanism, in which phase space statistical partitioning of energy may take place (14).

Experiments in an ion trap suggest that the degree of vibrational excitation of the N^+ + CO reaction may increase sensitively with both kinetic energy and by isotopic substitution, which shifts the position of a possible resonance to the v=2 state of the CO^+ product (44). In our experiments in the flowing afterglow apparatus at thermal energy, a change in the isotope of CO to ^{13}CO showed no difference in the population distribution, even though the v=2 level is shifted 95 cm^{-1} lower in energy. A possible explanation for why there might be no change is that the v=2 state is already energetically accessible even without the isotopic shift. Experiments with the single collision apparatus do confirm the fact that the vibrational excitation in the CO^+ product is a sensitive function of the kinetic energy. At 0.16 eV kinetic energy, the v=1 population increases by a factor of two at the expense of v=0, while the v=2 population remains the same. At the same time, the total cross section is not varying significantly. The fact that the population in v=2 does not change proves definitively that energy resonance is not a key determinant in the mechanism of this ion reaction.

While energy resonance is not important in N^+ + CO, it does seem to be a crucial consideration in the $Ar^+ + N_2$ reaction. This reaction produces the v=1 state with extremely high specificity over a wide range of kinetic energies. Since the v=1 level is slightly endothermic, the reaction cross section increases dramatically with kinetic energy (45, 46). At the slightly elevated kinetic energy of the single collision apparatus used here, the v=1/v=0 ratio is 10, and a similar effect is reported in crossed beam angular scattering measurements (43). The vibrational results are clearly identifying a special region of the electronic potential surface crossings, which has to be accessed in order for the charge transfer to take place.

The Ar^+ + CO reaction does not exhibit as high a degree of vibrational specificity as $Ar^+ + N_2$. All vibrational levels are populated to some extent, but the vibrational distribution is extremely skewed to the highest vibrational levels. This reaction is the most exothermic of the three studied, and its large exothermicity forces the greatest contest between the Franck-Condon mechanism and energy resonance criteria. The extremely hot vibrational distribution suggests that energy resonance is a very important factor. The observed vibrational distribution cannot be a result of a statistical or phase space partitioning of the energy. In order to get the high vibrational excitation, the mechanism must involve an elongation of the CO bond before the electronic surface hopping can occur. It is interesting, though, that at least in a small fraction of the reactive collisions which produce low vibrational states, nearly 1.7 eV of reaction energy is partitioned into translation and rotation. In this reaction, there is no dramatic change in the population distribution with kinetic energy, although a slightly higher vibrational excitation is observed at the elevated kinetic energy of the single collision apparatus.

The Ar^+ + CO reaction has been studied in an ion trap by both kinetic energy release and laser-induced fluorescence probing (47-49).

However, the results of those studies are inconsistent with the observations here. In the translational energy release determinations, a strong feature corresponding to the formation of only v=4 is reported. The laser-induced fluorescence measurements see evidence for both v=0 and v=4. However, from our measurements on the population distributions and the rates of relaxation of vibrationally excited CO^+, it is unlikely that the large v=0 component observed in the ion trap experiments is directly produced in the charge transfer reaction, and the exclusive formation of the v=4 state is not observed here.

Some of the most interesting results to come out of our studies with the single collision apparatus involve information about the rotational state distributions. This information provides important details on the mechanisms of the charge transfer process. If a vibrational state is formed by a long range electron transfer such as would occur in a Franck-Condon type of process, the rotational levels will not be appreciably excited above the level of thermal excitation of the reagent diatomic. If, on the other hand, a mechanism to form a vibrational state involves collision complex formation, then in the ensuing break up of the complex, the rotational states may become highly excited. The close interaction distance ensures that the departing atom will be able to impart a torque on the diatomic ion product.

Figure 10 shows the dramatic results that come from the rotational product state data in the N^+ + CO reaction (41). The rotational excitation in the v=0 state is minimal, giving a rotational temperature of only 400 K in the CO^+ product when starting with 300 K CO reagent. In contrast, the rotational excitation in v=1 is 2000 K and is highly non-Boltzmann. This strongly suggests a dual mechanism for the charge exchange process: the v=0 state may be formed by a

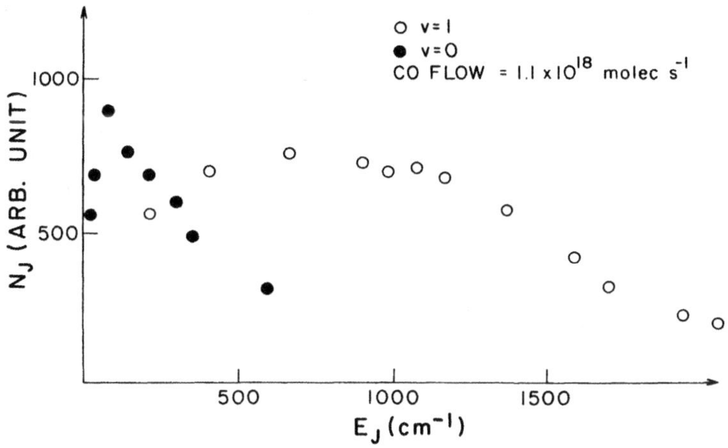

Fig. 10. Rotational population distributions in the v=0 and v=1 states of the N^+ + CO charge transfer reaction.

Franck-Condon like mechanism, whereas the v=1 state must involve a more complex mechanism that results in both the excitation of v=1 and the high rotational excitation in that state.

A similar effect has also been observed for the $Ar^+ + N_2$ reaction (42). Once again the v=0 state is rotationally colder than the v=1 product state, in this case 710 versus 980 K, respectively. This is especially surprising because there is so little energy left for rotation after v=1 is excited. The results of rotational determinations for the $Ar^+ + CO$ reaction find that both the v=5 and v=6 levels have similar rotational temperatures of 780 K. No measurements could be made on lower vibrational states because of signal limitations. In all cases, a significantly elevated rotational temperature is taken as an indication that a "complex" mechanism is operating, and a very low rotational excitation signifies the possibility of a direct, Franck-Condon like electron transfer.

From the totality of data on the charge transfer reactions, a number of generalizations concerning the mechanisms have been made (14,16,39-42). The results have been considered in light of molecular orbital pictures and electronic correlation diagrams. Unfortunately, very little information is available to help construct detailed potential surfaces at several angles of approach. Much more theoretical work needs to be done in this regard. From our results, it seems likely that both the Ar^+ reactions with N_2 and CO involve a close-encounter collision in which the interaction of the ion with the molecule causes a major distortion of the diatomic bond length, most likely through elongation. It is only after this elongation that the potential surface crossing is "found" and the electron transfer occurs. The specific vibrational distribution formed at the crossing is an indicator of the regions of access from one surface to the other.

In $N^+ + CO$ and in $Ar^+ + N_2$, there is definite evidence that two mechanisms occur simultaneously in the same ion reaction, resulting in microscopic branches. One mechanism gives the v=0 state and the other gives the v=1 and higher states. A reason for the two different mechanisms has been described by Gerlich (50). He suggests that the basic curve-crossing electron jump can occur either on the way "in" or on the way "out" after the region of closest approach. For a reaction like $N^+ + CO$, if the curve crossing occurs on the way in, the reactants then sample a deep attractive potential well. This can result in a collision complex that produces the v=1 state and the high rotational excitation. If the curve crossing occurs on the way out, the reagents only sample a repulsive state, and this would result in v=0 and the low rotational excitation, resembling the Franck-Condon mechanism. The two different mechanisms are also observed in the forward/backward scattering details of merged beam experiments (50).

3.6. Vibrational Relaxation of Ions

Recently new techniques have been developed by Ferguson and co-workers to measure the vibrational deactivation rates of ions by using charge transfer reactions whose rates are sensitive to the additional energy of the vibrational excitation as a monitor (15). The optical probing method of laser-induced fluorescence is also able to measure rates of ion vibrational deactivation, and in the course of our studies on reaction dynamics, a number of deactivation rates are required to obtain the nascent vibrational distributions. Perhaps the most interesting new aspect of our method is the capability to measure the rates for a whole series of vibrational states in a single ion and to monitor not only the disappearance of the vibrationally excited state but also to ascertain which state is formed in the deactivation process.

Vibrational deactivation of $CO^+(v)$ with both CO and N_2 is investigated (14,39). For $CO^+(v)$ with CO, detailed measurements are carried out on v=4 and v=1. The rates are nearly identical, 6×10^{-10} and 5×10^{-10} cm^3 $molecule^{-1}$ s^{-1}, respectively (39). Similar rates are detected for v=5 and v=6 as well. In addition, the v=4 state is found to deactivate directly to v=0 in a single relaxation collision. From the values of the absolute rate coefficients and the similarities of the rates, these data strongly suggest that the deactivation of all states of $CO^+(v)$ with CO occur by a mechanism of symmetric charge transfer. In this case, even high vibrational levels would deactivate directly to v=0, and most likely the vibrational quanta are left behind with the neutral CO after the charge transfer.

Another result considers the deactivation of $CO^+(v=1)$ with N_2, for which the rate coefficient is found to be 1.3×10^{-10} cm^3 $molecule^{-1}$ s^{-1} (14). The large size of this rate coefficient prompted Ferguson, Adams and Smith (51) to investigate the three body association rate for CO^+ with N_2 to see if this value would fit their model for vibrational deactivation via vibrational predissociation. They find an anomalously large three body association rate coefficient for $CO^+ + N_2 + He$ as well, indicating that the magnitude of the vibrational deactivation rate fits very well with the models of collision complex formation and vibrational deactivation upon break up. In addition, it indicates that there is a strong chemical affinity of CO^+ for N_2 that had not been observed before, much like the stabilization energy of a charge transfer complex in symmetric systems.

3.7. Drift Studies of Ion Excitation and Alignment

The optical detection capabilities described here allow a number of new kinds of studies to be explored involving ion excitation processes when ions are subjected to electric drift fields. We briefly mention here the results of this work. In a first experiment, the rotational excitation of N_2^+ drifted in He is measured at different field strengths (52). The rotational distributions are essentially Boltzmann with a rotational temperature that is found to be in agreement with the total collision energy, as described by the theory of Viehland, Lin and Mason (53). In new experiments, the alignment of the plane of rotation of the N_2^+ molecules when drifted in He is being explored. In preliminary experiments a small alignment effect is observed, indicating that the anisotropy of the ion-atom potential is sufficient to provide some m_J selectivity in the cross sections for rotational energy transfer (54). Classically, the process can be thought of as the tipping of the plane of rotation to present a minimum cross section to the oncoming He.

Acknowledgment. We gratefully acknowledge support of this work by the Air Force Office of Scientific Research and the National Science Foundation.

REFERENCES

1. S. R. Leone, Ann. Rev. Phys. Chem. **35**, 109 (1984).
2. S. R. Leone, Science **227**, 889 (1985).
3. Ch. Schlier, ed. Molecular Beams and Reaction Kinetics, Proc. of the International School of Physics, Enrico Fermi (Academic, New York, 1970).
4. See other chapters in this book.
5. V. M. Bierbaum, G. B. Ellison and S. R. Leone, in: Gas Phase Ion Chemistry, Vol. 3, Ions and Light, ed. by M. T. Bowers (Academic, New York, 1984).
6. E. E. Ferguson, F. C. Fehsenfeld and A. L. Schmeltekopf, Adv. Atom. Mol. Phys. **5**, 1 (1969).
7. S. R. Leone, Acc. Chem. Res. **16**, 88 (1983).
8. R. Altkorn and R. N. Zare, Ann. Rev. Phys. Chem. **35**, 265 (1984).
9. J. C. Weisshaar, T. S. Zwier and S. R. Leone, J. Chem. Phys. **75**, 4873 (1981).
10. T. S. Zwier, V. M. Bierbaum, G. B. Ellison and S. R. Leone, J. Chem. Phys. **72**, 5426 (1980).
11. A. O. Langford, V. M. Bierbaum and S. R. Leone, J. Chem. Phys. **83**, 3913 (1985).
12. A. O. Langford, V. M. Bierbaum and S. R. Leone, J. Chem. Phys. **84**, 2158 (1986).
13. C. E. Hamilton, M. A. Duncan, T. S. Zwier, J. C. Weisshaar, G. B. Ellison, V. M. Bierbaum and S. R. Leone, Chem. Phys. Lett. **94**, 4 (1983).
14. C. E. Hamilton, V. M. Bierbaum and S. R. Leone, J. Chem. Phys. **83**, 601 (1985).
15. E. E. Ferguson, J. Phys. Chem. **90**, 731 (1986).
16. L. Hüwel, D. R. Guyer, G. H. Lin and S. R. Leone, J. Chem. Phys. **81**, 3520 (1984).
17. D. R. Guyer, L. Hüwel and S. R. Leone, J. Chem. Phys. **79**, 1259 (1983).
18. K. Tamagake, D. W. Setser and J. P. Sung, J. Chem. Phys. **73**, 2203 (1980).
19. C. A. Parr, J. C. Polanyi and W. H. Wong, J. Chem. Phys. **58**, 5 (1973).
20. J. C. Polanyi, Acc. Chem. Res. **5**, 161 (1972).
21. P. J. Kuntz, in Dynamics of Molecular Collisions, Part B, edited by W. H. Miller (Plenum, New York, 1976).
22. M. M. Maricq, M. A. Smith, C. J. S. M. Simpson and G. B. Ellison, J. Chem. Phys. **74**, 6154 (1981).
23. S. Chapman, Chem. Phys. Lett. **80**, 275 (1981).
24. T. S. Zwier, M. M. Maricq, C. J. S. M. Simpson, V. M. Bierbaum, G. B. Ellison and S. R. Leone, Phys. Rev. Lett. **44**, 1050 (1980).
25. T. S. Zwier, J. C. Weisshaar and S. R. Leone, J. Chem. Phys. **75**, 4885 (1981).

26. V. M. Bierbaum, G. B. Ellison, J. H. Futrell and S. R. Leone, J. Chem. Phys. **67**, 2375 (1977).
27. M. A. Smith and S. R. Leone, J. Chem. Phys. **78**, 1325 (1983).
28. C. E. Hamilton, V. M. Bierbaum and S. R. Leone, J. Chem. Phys. **80**, 1831 (1984).
29. J. P. Gauyacq, J. Phys. B **15**, 2721 (1982).
30. J. P. Gauyacq, J. Phys. B **16**, 4049 (1983).
31. R. G. Macdonald and J. J. Sloan, Chem. Phys. **31**, 165 (1978).
32. R. J. Malins and D. W. Setser, J. Chem. Phys. **73**, 5666 (1980).
33. M. A. Wickramaaratchi, D. W. Setser, B. Hildebrandt, B. Korbitzer and H. Heydtmann, Chem. Phys. **84**, 105 (1984).
34. J. C. Rayez, IXth International Symposium on Gas Kinetics, Bordeaux, France, July 20-25, 1986.
35. M. A. Smith, V. M. Bierbaum and S. R. Leone, Chem. Phys. Lett. **94**, 398 (1983).
36. J. Balamuta and M. F. Golde, J. Phys. Chem. **86**, 2765 (1982).
37. A. O'Keefe, G. Mauclaire, D. Parent and M. T. Bowers, J. Chem. Phys. **84**, 215 (1986).
38. A. O. Langford, V. M. Bierbaum and S. R. Leone, Planet. Space Sci. **33**, 1225 (1985).
39. C. E. Hamilton, V. M. Bierbaum and S. R. Leone, J. Chem. Phys. **83**, 2284 (1985).
40. G. H. Lin, J. Maier and S. R. Leone, J. Chem. Phys. **82**, 5526 (1985).
41. G. H. Lin, J. Maier and S. R. Leone, J. Chem. Phys. **84**, 2180 (1986).
42. G. H. Lin, J. Maier and S. R. Leone, Chem. Phys. Lett. **125**, 557 (1986).
43. A. L. Rockwood, S. L. Howard, W. H. Du, P. Tosi, W. Lindinger and J. H. Futrell, Chem. Phys. Lett. **114**, 486 (1985).
44. A. O'Keefe, D. Parent, G. Mauclaire and M. T. Bowers, J. Chem. Phys. **80**, 4901 (1984), and private communication.
45. W. Lindinger, F. Howorka, P. Lukac, S. Kuhn, H. Villinger, E. Alge and H. Ramler, Phys. Rev. A **23**, 2319 (1981).
46. D. Smith and N. Adams, Phys. Rev. A **23**, 2327 (1981).
47. R. Marx, G. Mauclaire and R. Derai, Int. J. Mass Spectrom. Ion Phys. **47**, 155 (1983).
48. J. Danon and R. Marx, Chem. Phys. **68**, 255 (1982).
49. R. Marx, in: Ionic Processes in the Gas Phase, ed. by M. A. Almoster Ferreira (Reidel, Boston, 1984), p. 67.
50. D. Gerlich, in: Symposium on Atomic and Surface Physics, ed. by F. Howorka, W. Lindinger and T. D. Märk (Studia, Innsbruck, 1984).
51. E. E. Ferguson, D. Smith and N. Adams, Chem. Phys. Lett. (in press).
52. M. A. Duncan, V. M. Bierbaum, G. B. Ellison and S. R. Leone, J. Chem. Phys. **79**, 5448 (1983).
53. L. A. Viehland, S. L. Lin and E. A. Mason, Chem. Phys. **54**, 341 (1981).
54. R. A. Dressler, H. Meyer, V. M. Bierbaum and S. R. Leone (unpublished).

CROSSED-MOLECULAR BEAM STUDIES OF CHARGE TRANSFER REACTIONS AT LOW AND INTERMEDIATE ENERGY

Jean H. Futrell
Department of Chemistry
University of Utah
Salt Lake City, Utah 84112

1. INTRODUCTION

Investigations of the state-to-state reaction dynamics of ion molecule reactions have evolved steadily over the past several years, parallelling in many respects similar developments concerning the kinematics of neutral reactions. The main goal of these studies is the elucidation of the microscopic mechanism of reaction which describes the precise way in which reactants are converted into products. Intimately connected with this question is the disposal of energy--internal, translation and reaction exo- or endothermicity--in the reaction products. We shall summarize in this report recent results obtained primarily in this laboratory using the crossed beams technique.

The energetics of two reaction systems which we shall discuss are shown in Figure 1--namely, $N_2^+(N_2,N_2)N_2^+$ and $Ar^+(N_2,Ar)N_2$. The first of the obviously has an energy-resonant channel available. The second, which we have investigated as the state-selected reaction (1) is exothermic by about

$$Ar^+ ({}^2P_{3/2}) + N_2(X\,{}^1\Sigma_g, v=0) \rightarrow N_2^+ (X\,{}^2\Sigma_g, v=0) + Ar ({}^1S) \quad (1)$$

0.18 eV while generation of the product ion $N_2^+(X\,{}^2\Sigma_g, v=1)$ is endothermic by 0.09 eV. The upper state of the argon ion doublet, $Ar^+({}^2P_{1/2})$ and vibrational levels of ground state N_2 are not shown in Fig. 1, which is drawn to illustrate our experimental conditions, in which a state-selected $Ar^+({}^2P_{3/2})$ beam crosses a supersonic jet of N_2 at 7K. Not shown in the diagram is the kinetic energy which may be utilized to drive endothermic reactions; on the scale of Fig. 1 this would be represented as a band extending far off the scale considered. The relative lengths of the lines shown as $(Ar^+ + N_2)$ reactants corresponds to the population of rotational states of N_2 from J=0 to J=3 calculated for a 7K expanded jet. The N_2 energy level scale shown on the right is shifted by 15.58 eV, the adiabatic ionization potential of N_2, with respect to the other energy levels.

P. Ausloos and S. G. Lias (eds.), Structure/Reactivity and Thermochemistry of Ions, 57–80.

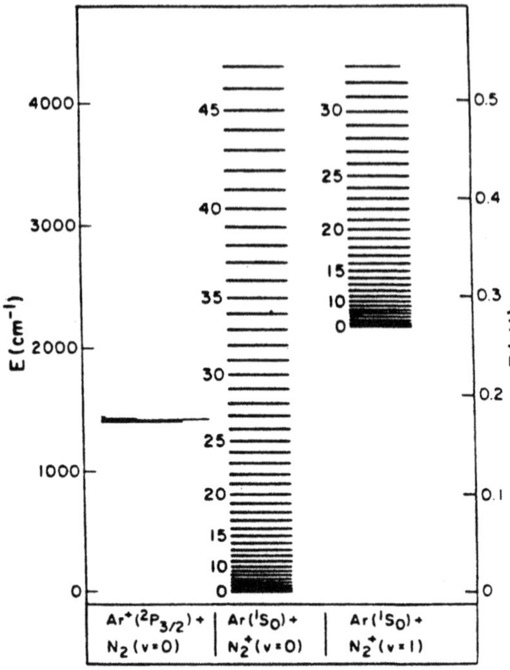

Figure 1. Energy level diagram of reactants and products of the $Ar^+ + N_2$ charge transfer reaction referenced to $N_2^+(^2\Sigma_g^+, v=0, J=0)$ as the zero of energy. The Boltzmann distribution of $N_2(v=0)$ rotational levels of 7K is indicated qualitatively by the lengths of the respective rotational lines.

2. EXPERIMENTAL

The apparatus used in our laboratory and its salient characteristics have been described elsewhere (1,2). For the present study the most relevant features are the high pressure ion source which generates relaxed $Ar^+(^2P_{3/2})$ and $N_2^+(X\,^2\Sigma_g^+, v=0)$ reactant ions and the supersonic neutral beam source which provides a narrow beam of monoenergetic N_2 species at laboratory energies of the order of 0.5 eV; this permits the efficient detection and ready characterization of the products from resonant charge transfer reactions.

The mass-selected ion beam is decelerated to the chosen collision energy and crossed at 90° with the neutral beam generated by supersonic expansion through three differentially-pumped vacuum chambers. The energy spread of the ion beam is approximately 0.25 electron volts with an angular spread of less than 1° (above 2 eV laboratory energies), while the molecular beam has an angular width of about 2°. Laboratory angular resolution is two degrees vertically and three degrees horizontally. Energy resolution of the differential analyzer is about 2.5% the transmitted energy. In the present experiments, the ions are accelerated to about 3.5 eV after being resolved by the detector defining slits, resulting in an energy resolution of about 0.09 eV.

Laboratory angular and energy distributions are taken at a series of angles from 0° to 90° for the product ion. These data are then transformed into Cartesian probabilities, $P_c(u_1,u_2,u_3)$--the probability density of finding the product with a velocity defined by the Cartesian coordinates infinitesimally close to u_1,u_2,u_3--and scattering contour diagrams are constructed. Using the relation

$$P(T') \propto \frac{u'}{m} \int P_c(u_1,u_2,u_3) \sin \chi d\chi \quad (2)$$

where χ is the CM scattering angle, probability densities $P(T')$ of product relative translational energies, T', are also calculated.

The supersonic beam of neutral particles is critically important to the present study. First, the well-defined velocity vector distribution is important for distinguishing different mechanisms for charge transfer. Secondly, the supersonic acceleration of neutrals allows us to detect low-velocity ionic products moving with essentially the velocity of the reactant neutral. A seeded beam is used in the present study, consisting of about 90% He mixed with N_2. This ratio was established empirically to provide the best compromise of beam flux and narrow angular distribution. The beam energy, flux and angular distributions are measured by ionizing a small fraction of the neutral beam using an electron gun mounted orthogonally to the ion and neutral beams at the collision center.

3. REACTION DYNAMICS OF THE CHARGE TRANSFER REACTION OF N_2^+ WITH N_2 AT LOW AND INTERMEDIATE ENERGIES.

3.1. Introduction

Symmetrically resonant charge exchange is fairly well understood for atomic systems. However, very little information exists on the same reaction between molecular ions and their parent molecules. There are several similarities to the atomic case, particularly regarding the large total charge-exchange cross-sections which are observed for these reactions. However, in contrast with atomic collisions, the exact symmetry of the problem is broken by molecular orientation with respect to the collision axis. The possibility of vibrational and rotational energy transfer further complicates the basic dynamics of charge-exchange collisions in molecular systems.

It will be shown that the nitrogen system constitutes a "textbook example" of the Rapp and Francis (3) conjecture that there is a fundamental change in mechanism of charge-transfer in the low-collision-energy regime where the possibility of the formation of transient orbiting complexes is anticipated. At high energy a two-state, parameterized treatment of charge-transfer adequately describes the symmetrical resonance interaction of two identical systems which pass each other in rectilinear trajectories. As collision energies are reduced a significant number of the charge transfer interactions proceed via "orbiting" or "capture" trajectories; at this point charge-transfer becomes a typical ion-molecule reaction channel. With decreasing energy a smooth transition to this low-energy limiting case is anticipated. The high stability of the putative N_4^+ complex (which is estimated theoretically to be bound by 1.4 eV (4) and has been measured in recent experiments to be 0.9 eV (5,6)) is also an important parameter. Provided the system accesses the ground-state configuration of N_4^+, it is expected on statistical grounds that reducing the total energy as much as possible will increase the lifetime of the N_4^+ intermediate.

3.2. Results and Discussion.

Figure 2 is the CM probability contour plot measured for the charge transfer reaction at 0.74 eV collision energy (7). At this relative kinetic energy two distinct mechanisms are observed which are cleanly separated from each other in terms of the disposal of angular momentum in the reaction. These reactions may be described as a direct, or impulsive, mechanism and one proceeding via a reaction complex. For the direct mechanism the locus of product vectors is nearly identical to that of the N_2 neutral-beam velocity vectors. Almost no angular momentum is exchanged and the mechanism is well-described as corresponding to a long-range "electron jump". Since our experiments involved only ground-state N_2^+ ions [excited species having been relaxed by collisions inside the high pressure ion source], the reaction may be written as follows:

$$\underline{N_2^+(X\,^2\Sigma_g,v=0)} + N_2(X\,^1\Sigma_g,v=0) \rightarrow \underline{N_2(X\,^1\Sigma_g,v=0)} + N_2^+(X\,^2\Sigma_g,v=0) \qquad (3)$$

where the underlined species are those moving with higher laboratory velocity before and after reaction. It therefore corresponds exactly to the symmetric resonance case which is well-known for atomic systems.

The second, "statistical complex", mechanism is defined by the complete symmetry, within experimental error, about the location of the center-of-mass velocity vector. This high degree of forward-backward symmetry of the reaction complex products implies a lifetime which exceeds the rotational period of the N_4^+ intermediate. The clean separation of these two mechanisms at 0.74 eV collision energy is indicated in Figure 2.

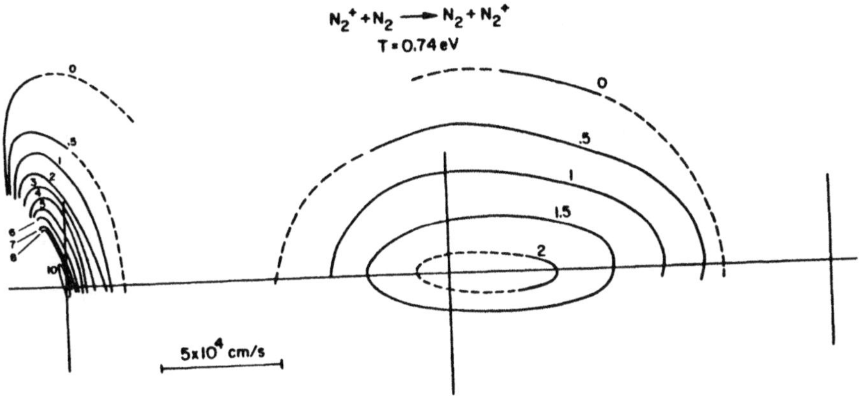

Figure 2. Scattering diagram for the charge transfer reaction $N_2^+(N_2,N_2)N_2^+$ at 0.74 eV collision energy. The crosses mark the velocity of the center-of-mass of the system and the CM velocity of N_2 prior to collision. Contours define the relative probability densities of product velocities (in Cartesian velocity space).

It is also of interest that the product ions from the second mechanism leave the center-of-mass with significantly reduced kinetic energy, implying that much of the collision energy is disposed in internal modes of the products of reaction. This is evidence for very strong coupling of the internal modes in the collision complex. Strong coupling is anticipated from the theoretical paper of deCastro, Shaefer, and Pitzer (4) which describes the relatively strongly bonded N_4^+ intermediate which appears to be formed in the charge-transfer reaction. These workers calculated that the linear configuration of the N_4^+ was the ground-state species, bound by about 1.4 eV with respect to the separated nitrogen molecular ion and nitrogen molecule. The structure gave a bond distance associated with this geometry of equal distances of 1.08 Å for the end nitrogen pairs and a separation distance of 1.931 Å for the central pair of nitrogens. Their calculation predicted 0.6 of the unit charge is located in the central pair of atoms and 0.4 on the end nitrogens. Trapezoidal and T-shaped geometries were also considered and found to be slightly less bound intermediates than the linear configuration. They also noted that the linear configuration gave the proper dissociation asymptote and that there were some problems with the dissociation asymptote for the other geometries.

Although scattering diagrams such as Fig. 2 contain the most informative information regarding reaction mechanisms, it is important to note that they provide a distorted view of the relative importance of different mechanisms. This results from the methods used in constructing these diagrams from the laboratory data obtained by energy analysis of products scattered into a selected laboratory angle. The true relative importance of the direct and complex mechanisms is obtained by applying Equation (2) to the data of Figure 2. Figure 3 shows the results, demonstrating that the two mechanisms are of comparable importance at 0.74 eV collision energy.

In addition to defining angular scattering--the forte of beam experiments--conservation of energy provides details on energy conversion in these reactions. The total energy E_{TOT} which is present in the reactants is the sum of the relative translational energy, T, (i.e., collision energy) and the energy in the internal degrees of freedom, U. Similarly, the total energy present in the system after reaction is given by the sum of the relative translational energy of the products, T', and by the energy in the appropriate internal degrees of freedom of the products, U. Consequently, conservation of energy

$$E_{TOT} = T + U = T' + U' \tag{4}$$

leads to the relationship between translational endoergicity ($\Delta T = T - T'$) and the internal energy change ($\Delta U = U' - U$).

$$\Delta T = \Delta U \tag{5}$$

Hence a careful measurement of ΔT also measures the change in internal energy in the reaction.

We superpose a ΔT scale on the bottom and a ΔU scale on the top of Figure 3 which express the energy conversion in terms of vibrational quanta of $N_2^+(X\ \Sigma_g)$ or of $N_2(X\ \Sigma_g)$ [the vibrational spacings are nearly identical, differing by about 150 cm^-]. Clearly the direct mechanism corresponds to translationally resonant charge transfer [$\Delta T = \Delta U = 0$], while the complex mechanism involves extensive internal excitation of the products. The internal energy can be accounted for if about three quanta of vibration (distributed between the N_2^+ and N_2 products in an unknown manner) are deposited in the products; more plausibly a distribution of vibration and rotational states is formed. If all of the energy were disposed in rotation, about 150 quanta of rotation would be required, comparable to but slightly higher than is allowed by conservation of angular momentum. It is our speculation -- based on the extensive conversion of translational to internal energy observed--that the internal energy disposition is partioned approximately statistically in all vibrational and rotational degrees of freedom of the N_2^+ and N_2 products.

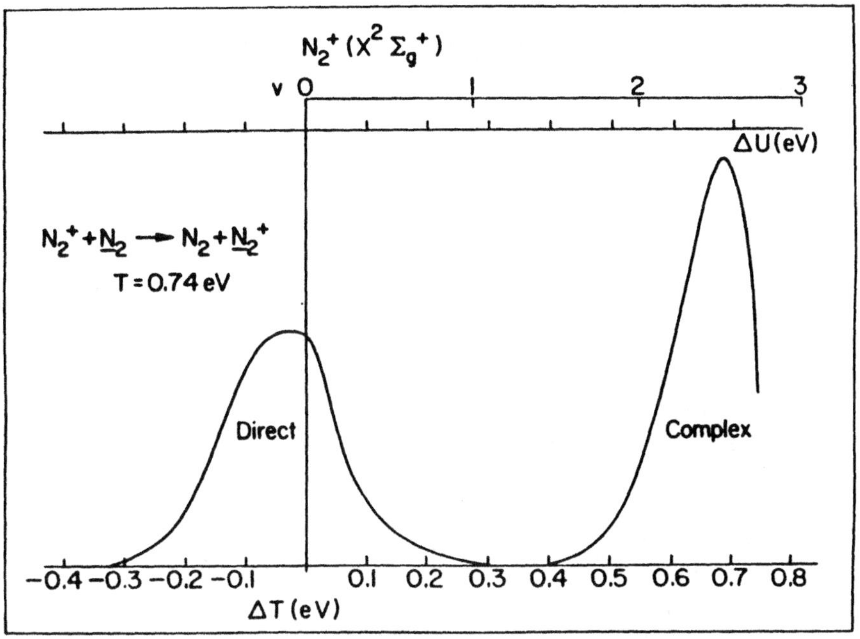

Figure 3. Relative translational energy distributions of products expressed as translational exoergicity (see defining equations in text) at 0.74 eV collision energy for the reaction $N_2^+(N_2,N_2)N_2^+$. The correlation with ΔU is shown with a superimposed scale for v=0, 1, 2, 3, 4 levels of $N_2^+(X\ ^2\Sigma_g^+)$.

Figure 4 is the scattering diagram at the significantly higher energy of 9.96 eV (7). Under these conditions the collision complex mechanism has completely disappeared. While the collision complex mechanism will be the dominant one at collision energies below 0.7 eV CM, the direct mechanism dominates high-energy collisions. This is consistent with a dramatic decrease in the lifetime of the N_4^+ intermediate with increasing total energy of the system. Under these conditions the available energy exceeds the binding energy of the N_4^+ complex by more than an order of magnitude, and the scattering diagram provides no evidence for a contribution from the persistent complex model demonstrated in Figs. 2 and 3.

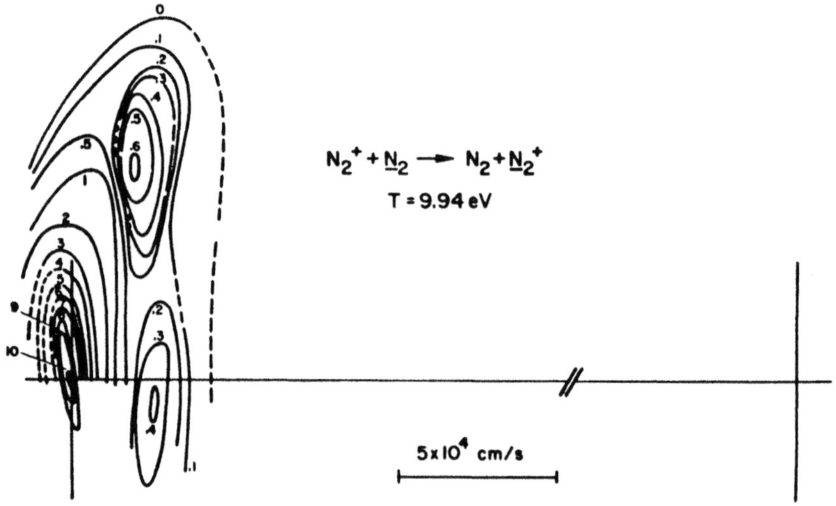

Figure 4. Scattering diagram for the charge transfer reaction $N_2^+(N_2,N_2)N_2^+$ at 9.94 eV collision energy. The crosses mark the velocity of the center-of-mass of the system and the CM velocity of N_2 prior to collision. Contours define the relative probability densities of product velocities (in Cartesian velocity space). The broken line indicates that the vector is not to scale: the CM velocity is equal to the distance between crosses multiplied by 2.38.

The direct, or impulsive, mechanism which is dominant at high energy is the same as that observed at low energy--namely, the translationally-resonant, $\Delta v=0$), reaction (4). However, we also note the formation of satellite peaks representing scattering with significant exchange of angular momentum at center-of-mass scattering angles of approximately five and eight degrees. Both of these ridges of intensity fall inside the TRCT circle, demonstrating that they are translationally endoergic processes. The translational endoergicity noted corresponds to $\Delta T(\equiv\Delta U)$ of about 0.5 and 0.8 eV respectively. These ridges of intensity could therefore be associated with the population of v=2 (ΔU = 0.545 eV) and v=3 (ΔU = 0.815 eV) products, respectively (23). Alternatively, this could represent favored distributions of rotational energy in the products of reaction or some special distribution of rotational and vibrational energy in the two products.

It should be pointed out that the contour diagram of figure 4, while it provides interesting and precise information about the mechanism of reaction, presents a somewhat distorted view of the relative importance of the scattering processes generating ground state and vibrationally excited N_2^+. The integration of the differential cross section of Figure 4 using Eq. (2) indicates that the intensity maximum at v'=2 accounts for about 18% of the total scattering intensity at this collision energy. The beginning development of a propensity for the formation of v'=2 products is of special interest in comparison with the data of McAfee et al. shown in Figure 5 (8). This is a computer enhanced deconvolution of the experimental data for charge transfer at a collision energy of 55.54 eV. Pronounced alternation of relative intensities with the preference for exchange of an even number of vibrational quanta is evident in the figure. [We note parenthetically that the experimental data shown in this figure involve both ground state and vibrationally excited N_2^+. Charge transfer processes which are superelastic because of the transfer of vibrational energy into translation are responsible for the scattering into negative laboratory angles shown in the figure.]

The result of the McAfee et al. (8) is rationalized on the basis of approximate quasi-diatomic potential energy curves and a Landau-Zener-Stueckelberg curve crossing model which utilizes the orientation effect on the size of the coupling constant to rationalize the high probability for even numbers of vibrational quanta to be exchanged. The quadrupole-like character of the wave function accounts for the coupling strength that enters into the estimation of internuclear distances for curve crossings of the order of 4 Å and gives a transition probability of the order of 0.5 for v=2. It is concluded that the coupling terms at large R between N_2^+ and N_2 are symmetric in the internal vibrational coordinates. This model is very different from the one utilized by Flannery, Cosby and Moran (9) which ignores orientation contributions to the interaction potential. The Flannery et al. model results in a set of curve crossings at smaller internuclear separations between states of the same, approximate g-u

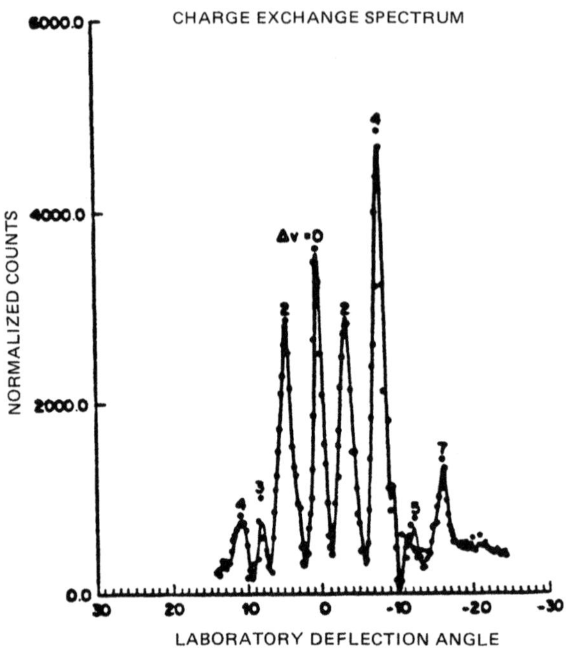

Figure 5. Relative intensity of low energy scattered ions from the reaction of 111 eV N_2^+ with a neutral supersonic beam of N_2 (55.5 eV CM). Points are experimental; the solid lines are Gaussian fits using a fifth order spline function to locate the transition center positions designated. A broad background scattering contribution has been subtracted. Negative angles represent exothermic (superelastic) charge transfer of vibrationally-excited N_2^{+*}, which positive angles are the endothermic charge transfer reactions of $N_2^+(X\ ^2\Sigma_g,$ v=0). The pronounced preference for transfer of even numbers of quanta is clearly evident in the figure.

symmetry. Such a model is dominated by $\Delta\nu=0$ with a sharply declining probability of charge exchange with increasing $\Delta\nu$.

It would appear that the Flannery et al. model (9) provides a better description for the low energy charge exchange results while the McAffee et al. model (11) provides the best explanation for the high energy ones. This difference most likely relates to the significant changing of collisional interaction times as the energy is varied from 0.7 to 55 electron volts. At high energies there is no opportunity for any reorientation of the molecules du. collision and the electron exchange frequency is determined simply by the "resonance" between electron exchange and internal vibrational motion. Impulsive energy transfer is not involved in the vibrational excitation observed and the transitions are induced by the distortion of the molecular cloud in the process of charge exchange.

It is interesting to speculate that these shoulders of intensity on the main peak in Figure 4 may represent the same small-impact-parameter collisions which are associated with the complex mechanism at low collision energies. Although the "complex" mechanism has two signatures at low energy--forward-backward symmetry and energy-exchange--only energy-exchange is observed at higher energy. The development of forward-backward symmetry requires a lifetime significantly longer than a rotational period. It is also indicative that the collision samples an attractive potential during the collision process. The lifetime of N_4^+ is shortened dramatically with increasing energy; at higher collision energy the centrifugal potential also overwhelms the true potential and large angle scattering can only occur from the repulsive part of the potential. This is clearly the case for the 111 eV (LAB) beam study of McAfee, et al. (9) which shows that significant vibrational excitation accompanies high energy charge-transfer in this system.

The Langevin model (point charge-polarizable molecule) is often invoked to rationalize the dynamics of low energy ion-molecule collisions. However, no beam experiment with the objective of constructing scattering diagrams can be carried out using currently-developed techniques in the energy range where the Langevin mechanism is clearly expected to be dominant. (The results would be quite boring anyway.) The theoretical study of N_4^+ geometry and energetics by deCastro, Schaeffer and Pitzer (4) calculates 1.4 eV binding energy while the Langevin model predicts only 0.17 eV at the same separation distance. This strongly-attractive true potential accounts for the orbiting trajectories actually observed at 0.74 eV CM, while the existence of several geometries with similar binding energies helps to rationalize the efficient coupling of energy into internal modes. It therefore follows that the "complex" model demonstrated in Fig. 2 can be explained without invoking the low-energy Rapp-Francis-Langevin model for charge transfer.

The detailed mechanism of charge and energy exchange during the collision therefore remains a matter of conjecture. The coupling modes we have qualitatively invoked for efficiently leaking translational energy into internal modes would trap the collision partners inside the centrifugal barrier during the first passage (at low energy such that a well exists in the effective potential) and several "encounters" are likely prior to separation of the products. This may also be responsible, in part, for the clean separation of mechanisms observed.

A curve-cr' 'ng mechanism between states of approximately the same g-u symmetry ... small internuclear distances (1-2 Å) is invoked for N_2^+/N_2 in the theoretical treatment of charge-transfer by Flannery, Moran and Cosby (9). This model predicts that $\Delta v=0$ will be the dominant channel; this product appears predominantly at zero scattering angle, while significant angular deflections are predicted for vibrationally-excited products. These predictions are largely consistent with our experiments. Multiple encounters occurring for orbiting collisions were not discussed by these researchers; however, angular information would clearly be lost and the small impact parameter collisions responsible for vibrational excitation might be significantly enhanced. Alternatively coupling of modes in the various geometries of N_4^+ can be invoked to explain the essentially statistical mixing of energy in all modes, including translation, suggested by Figs. 2 and 3 for the collision complex model. Our results at 9.9 eV can also be interpreted as suggesting a transition to the Landau-Zener-Stueckelberg mechanism proposed by McAfee et al. (8) for high energy collisions.

3.3. Conclusions

The conceptual picture suggested by the results discussed is as follows: At thermal energies the Langevin cross-section is quite large and separates ion-molecule trajectories into two classes--those having very large impact parameters so that only modest deflection occurs and those with impact parameters less than the critical value for orbiting which causes the two scattering partners to undergo one or more intimate collisions. The Langevin cross-section for charge transfer, according to the hypothesis of Rapp and Francis (3), is $(\Pi/2)b_L^2$, where b_L is the impact parameter for orbiting trajectories and the factor 1/2 is the probability each of the collision partners will retreat as the charged particle. Consideration of the much stronger binding energy of N_4^+ relative to that resulting from ion-induced dipole interactions extends the energy range for orbiting trajectories to ca. 1 eV, encompassing the range currently accessible to the crossed-beam technique.

The mechanism associated with the small impact parameters necessary to intermix the internal energies thoroughly is therefore manifested at low energies by both of the characteristic signatures of the orbiting-complex mechanism--scattering symmetric about the CM vector and energy exchange. At the high energies of Figure 4 the cross-section for orbiting collisions is smaller than the hard-core collision cross-section associated with the N_4^+ intermediate itself--equivalent to impact parameters of ca. 2-3 angstroms. Hence the longer-range electron-jump mechanism is the dominant one at high energy. The small fraction of collisions which occur at internuclear distances comparable to those of the N_4^+ species may be responsible for the satellite intensities observed in addition to the main peak. Under these circumstances extensive mixing of vibrational and rotational energy is expected. A mixture of curve-crossing and coupling of modes in N_4^+ rationizes the detailed mechanisms satisfactorily.

4. THE CHARGE-TRANSFER REACTION OF Ar^+ WITH N_2

4.1. Introduction

The selection-ion drift tube study of this reaction by Lindinger et al. (10) and the 80 and 300 K flow tube study of Smith and Adams (11) show that the overall rate coefficient for argon ions reacting with nitrogen increases about two orders of magnitude with increasing collisional energy over the range from thermal to about 3 eV. They interpret their results in terms of the endothermicity of the reaction channel generating $N_2^+(X\,\Sigma_g^+, v=1)$ from $Ar^+(P_{3/2})$. They assert that $Ar^+(P_{1/2})$ is rapidly converted to $Ar^+(P_{3/2})$ in drift tubes and that the rate coefficient for the exothermal reaction of $Ar^+(P_{3/2})$ with N_2 to generate $N_2^+(X\,\Sigma_g^+, v=0)$ declines from 10^- cm mol$^-$ s$^-$ at 80 K to $7\times10^-$ cm mol$^-$ s$^-$ at 140 K. The rate of the endothermic reaction of $Ar^+(P_{3/2})$ generating $N_2^+(X\,\Sigma_g^+, v=1)$ rises from 10^- cm mol$^-$ s$^-$ at 300 K to the Langevin rate at 3 eV relative translational energy (1). The minimum in the k(E) curve is obtained by extrapolating the $N_2^+(v=0)$ and $N_2^+(v=1)$ curves.

These experiments have resolved the orders of magnitude discrepancies in reported values for the rate coefficient of this reaction. The apparent low probability for generating ground state ions at low collision energy, the differing reactivities for the $Ar^+(P_{3/2})$ and $Ar^+(P_{1/2})$ substates and the large rate coefficient for the reverse reaction exhibited by vibrationally excited N_2^+ make rate constant measurements extremely sensitive to experimental conditions. These intriguing kinetics features also identify this reaction system as an interesting candidate for a detailed dynamics study. It will be shown that a very high degree of quantum-state specificity is exhibited by this reaction; this specificity is responsible for the interesting kinetics features noted. The chapter in this volume by Steve Leone and the move detailed report of the laser-induced fluorescence (LIF) study of this system by Hüwel, Guyer, Lin and Leone (12) at 0.24 eV has demonstrated the state-specificity of this reaction. They reported that $N_2^+(\chi\,\Sigma g,\ \nu = 1)$ is the dominant reaction product, accounting for 87 ± 10% of the total product flux at 0.24 eV collision energy.

It is well established that LIF is the definitive technique for determining product internal state distributions, while the beam method provides overall information on angular and linear momentum exchange. Together they provide tight specifications for the detailed mechanism of reaction and a challenging data set for theoretical investigation of the potential energy hypersurfaces directing the course of charge-exchange reactions. The major uncertainty is estimating the average product laboratory velocities $\bar{v}_0$ and $\bar{v}_1$ for the two vibrational states, v=0, 1. For a given pair of initial and final rovibronic states of the reagent N_2 and product N_2^+, the center-of-mass velocity of this particular final state can be calculated from the known exothermicity and the initial translational energy. However, for each state the transformation from center-of-mass velocity, u, to laboratory velocity, v, depends critically on the CM scattering angle. No experimental information exists for the differential cross section of the Ar^+-N_2 charge transfer reaction at this low collision energy and it is not presently possible to carry out well-defined beam studies at this energy. The most definitive results obtained to date are described below.

4.2. Beam Experiments

A CM scattering diagram generated by this laboratory (13) which presents the Cartesian probability (the probability of locating a product with a velocity defined by the Cartesian coordinates of the product ion) for the reaction of $Ar^+(^2P_{3/2})$ at a collision energy of 1.73 eV is shown in Fig. 6. The concentric circle defines the locus of velocity vectors which would be observed if the experiment were conducted with monochromatic ion and neutral beams in their ground states and if the product ions were formed exclusively in the $(X\ ^2\Sigma_g^+, v=1)$ state with no rotational energy. The inference which we reach from this diagram is that the products are formed predominantly in the v=1 vibrational level with a small number of rotational quanta. Further, the fact that the product is forward scattered (with reference to the initial neutral beam velocity vector) is evidence that very little momentum is transferred in the collision.

The product ion state assignment is further elaborated in Fig. 7, which shows the integration of the Cartesian velocity space diagram of Fig. 6 over angle using Eq. (2) and analyzes the data in terms of the observed translational endoergicity, ΔT, and the corresponding change in internal energy, ΔU. Also shown in Fig. 7 as the cross-hatched area is the combined dispersion of the ion and neutral beam energies; comparison with the observed area of product ion intensity and the superposed vibrational energy scale for $N_2^+(^2\Sigma_g^+, v)$ at the top of the figure demonstrates that the product is generated mainly in the v=1 level with a modest distribution of rotational energy. This result is in excellent agreement with the conclusions reached in the LIF study at 0.24 eV CM.

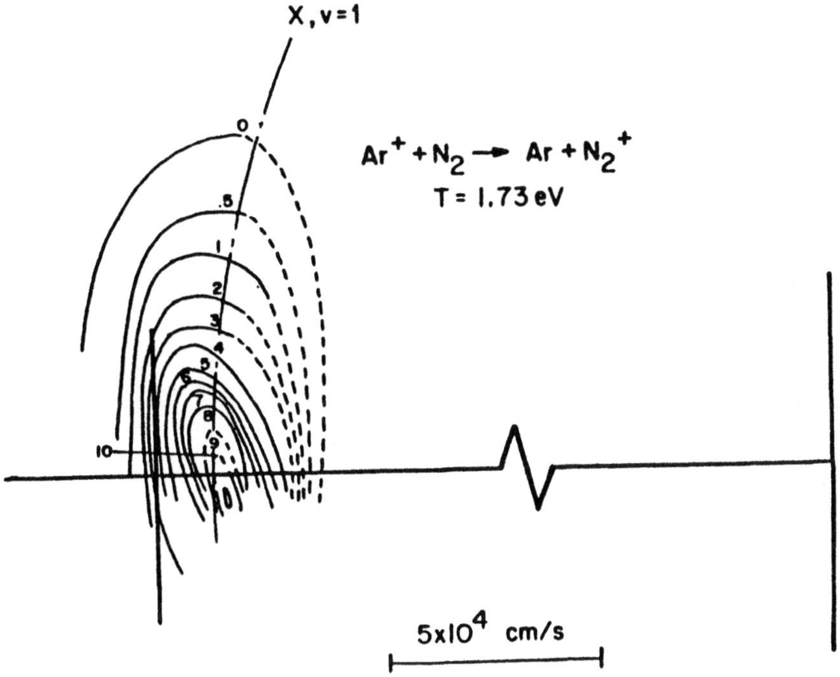

Figure 6. Scattering diagram for the charge-transfer reaction $Ar^+(N_2,Ar)N_2^+$ at 1.73 eV collision energy. The crosses mark the center-of-mass velocity and the CM velocity vector of N_2 prior to collision. Break in line indicates the velocity vector difference between the N_2 neutral and center-of-mass is not to scale. The distance between the two crosses should be multiplied by about 1.7 to correspond to the indicated velocity scale. Contours define the probability densities of product velocities (in Cartesian coordinates) and demonstrate that a direct mechanism populates a slightly endothermic channel. The inscribed radius is the locus of vectors corresponding to the translational to internal energy conversion required to drive the reaction $Ar^+(^2P_{3/2})$ + + $N_2(X,v=0)$ = $Ar(S_0)$ + N_2 + (X,v=1) with no energy deposited as rotation.

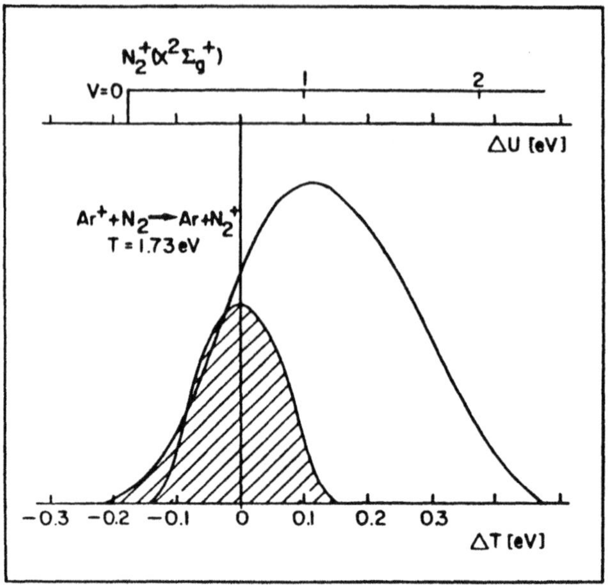

Figure 7. Relative translational energy distribution of products expressed in terms of the translational exoergicity observed for the reaction $Ar^+(N_2,Ar)N_2^+$ at 1.73 eV collision energy. The correlation with ΔU is shown with a superposed scale for v=0, 1, 2, 3 vibrational levels for $N_2^+(X\ \Sigma_g^+)$. The crosshatched peak is the energy distribution of the neutral N_2 beam convoluted with the analyzer apparatus function and integrated over scattering angle.

Leone et al. (Ref. 11 and Chapter) pointed out that the principal uncertainty in estimating the relative cross-sections for forming the v=1 and v=0 states of $N_2^+(\ \Sigma_g)$ resides in the unknown angular scattering reaction dynamics for this system. We have also noted that it is presently impossible to carry out a crossed-beam study at a collision energy of 0.24 eV. Not only would the energy spreads in the two beams exceed the collision energy but also angular spreads, space charge and energy resolution problems preclude the meaningful measurement of differential cross-sections at very low ion energy. For the Ar^+/N_2 system there is the further complication that the rate coefficient drops dramatically with decreasing energy (10).

In an attempt to reach the lowest energy pragmatically attainable we have carried out the scattering experiment at 0.6 eV CM whose results are displayed in Fig. 8. As before, the concentric circles define the locus of vectors corresponding to the generation of v=0 and v=1 products with no rotational energy. Just as in the higher energy experiments already discussed the dominant product is formed in v=1. There is also some evidence in this diagram for several percent of forward-scattered v=0 product.

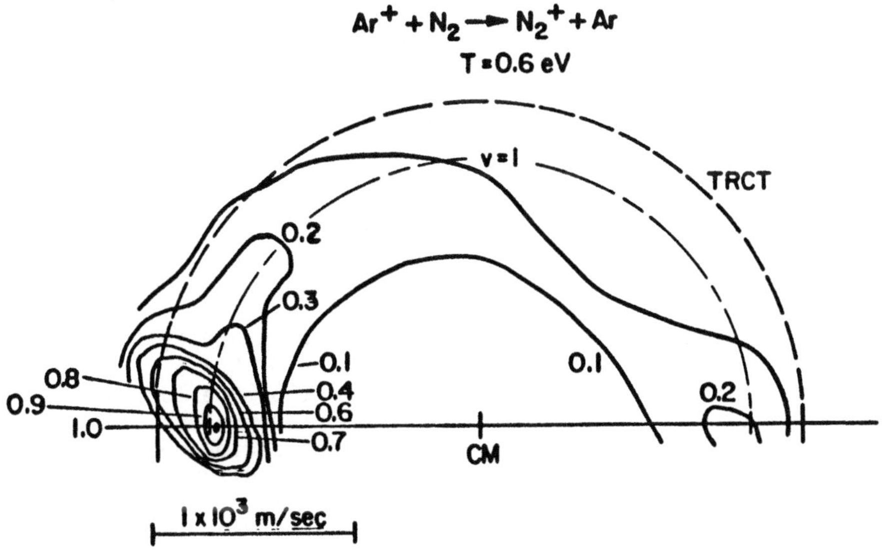

Figure 8. Scattering diagram for the charge-transfer reaction $Ar^+(N_2,Ar)N_2^+$ at 0.06 eV collision energy. The maximum intensity at 0° and the 20% contour at 180° (with respect to the initial N_2 velocity vector) both correspond to $-\Delta U_1=0.1$ eV and match the endothermicity of the reaction $Ar^+(^2P_{3/2}) + N_2(X\ ^1\Sigma_g^+, v=0) \rightarrow Ar(^1S) + N_2^+(X\ ^2\Sigma_g^+, v=1)$. The symmetry beginning to develop in this diagram suggests ca. 40% of all reactive collisions result in orbiting of the center-of-mass.

The most interesting new feature is the back-scattered maximum which connects by a ridge of large-angle scattered intensity to the predominant forward peak. This beginning development of backward-forward symmetry is most plausibly interpreted as arising from a collision intermediate whose lifetime is about one rotational period. the observation of this partial symmetry at 0.6 eV (and its disappearance with increasing collision energy) is evidence for stronger attractive forces for the ArN_2^+ species than can be accounted for by ion-induced dipole and ion-quadrupole forces. Speaking in pseudo-diatomic terms, a potential well-depth of the order of 1 eV is required to rationalize the angular scattering observed, while the quantum-state specificity observed requires a direct or impulsive mechanism for the actual charge-exchange step. The well-depth sampled by 0.6 eV collisions inferred from this scattering measurement is consistent with the 1.2 eV binding energy estimated for ArN_2^+ by Teng and Conway in a high pressure mass spectrometry study (14).

We have also investigated this reaction at several intermediate energies. The results obtained at 1.1 eV collision energy are presented in Fig. 9 (15). At this intermediate collision energy, the results are strikingly different. The efficient utilization of translational energy to populate several endothermic product channels is evident, as is the dominance of large-angle scattering under these experimental conditions. The dominance of large-angle scattering and the observation of much more extensive conversion of translational energy into internal energy in the formation of products are in striking contrast to the higher-energy results.

There are three clearly resolved centers of intensity in the scattering diagram. The probable interpretation of these maxima is indicated by the superimposed circles drawn to show the loci of center-of-mass velocity vectors which correspond to products scattered with exactly the translational-to-internal energy conversion required to generate the v=0,1,2,3, and 4 levels of the ground electronic state (assuming no rotational excitation). Since rotational energy levels are closely spaced and not resolvable in our scattering experiment, and since the centers of intensity in Fig. 9 are both well resolved and an excellent match to the vibrational spacing of N_2^+, we interpret our results as evidence for vibrational rather than rotational excitation of the product ion. The shading of intensity into the region between the superposed v=n circles is evidence for some rotational excitation (particularly for v=3) but we suggest that the overall rotational excitation is modest.

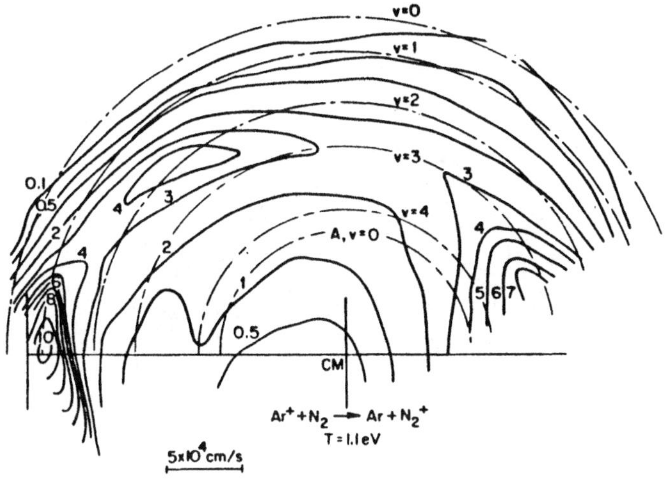

Figure 9. Scattering diagram for the charge-transfer reaction $Ar^+(N_2,Ar)N_2^+$ at 1.1 eV collision energy. The crosses mark the center-of-mass velocity and the CM velocity of N_2 prior to collision. Contours define the probability densities of product velocities (in Cartesian coordinates) while the inscribed circles define the loci of velocity vectors for N_2^+ energy conversion corresponding to the formation of $N_2^+(X\ \Sigma_g^+)$ with 0,1,2,3,4, quanta of vibration and of $N_2^+(A\ \Pi_u, v=0)$ with no energy deposited in rotation.

Recently we have extended our crossed beam studies of charge exchange and inelastic scattering for this system which defines the boundaries of this "window" in which the mechanism changes dramatically more precisely (16). The lower bound is approximately 0.8 eV; Fig. 10 shows the structured diagram obtained at this energy. Improved resolution in both energy and angle permits us to resolve the apparent vibrational structure more clearly here than in Fig. 9. The upper bound is below 1.5 eV, Fig. 11 summarizes our results at this collision energy. Here the "window" has closed and $N_2^+(X\ \Sigma_g^+, v=1)$ once again dominates the charge transfer channel. Our improved resolution separates a small superelastic shoulder along the relative velocity vector which is indicative that a small amount of $N_2^+(X\ \Sigma_g, v=0)$ is also formed.

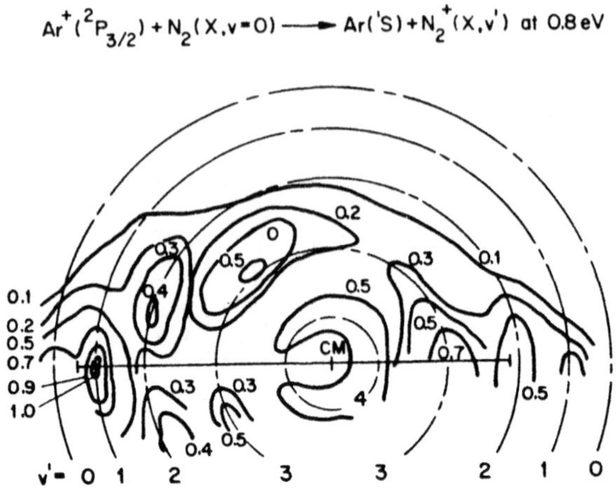

Figure 10. Scattering diagram for the charge-transfer reaction $Ar^{+}(N_2,Ar)N_2^{+}$ at 0.8 eV collision energy. The crosses mark the center-of-mass velocity and the CM velocity of N_2 prior to collision. Contours define the probability densities of product velocities (in Cartesian coordinates) while the inscribed circles define the loci of velocity vectors for N_2^{+} products leaving the collision center with translation-to-internal energy conversion corresponding to the formation of $N_2^{+}(X\ \Sigma_g)$ with 0,1,2,3,4 quanta of vibration with no energy deposited in rotation.

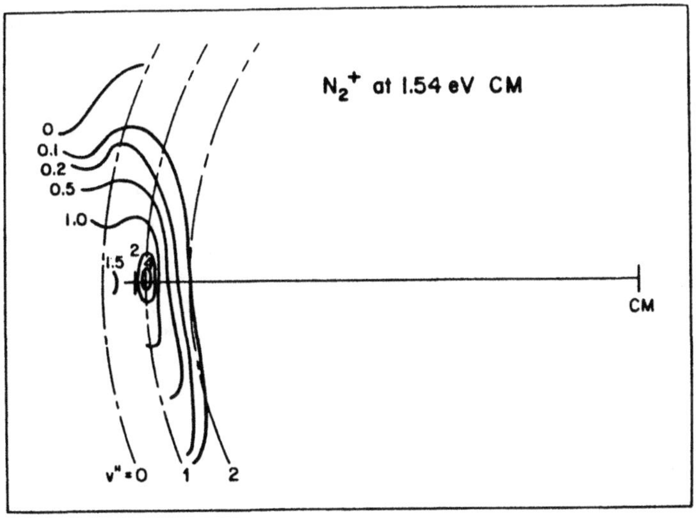

Figure 11. Scattering diagram for the charge transfer reaction $Ar^+(N_2,Ar)N_2^+$ at 1.54 eV collision energy. The crosses mark the center-of-mass velocity and the CM vector of N_2 prior to collision. The contours define the probability densities of product velocities (in Cartesian coordinates) while the circular radii correspond to the translational energy changes required to drive the reactions $Ar^+(^2P_{3/2}) + N_2(X,v=0) = Ar(S_0) + N_2^+(X,v=0,1,2)$ with no energy deposited in rotation.

4.3. Conclusions

The general conclusion reached from the beam experiments at 0.6 eV and at collision energies of 1.5 eV and higher, some of which have been discussed in this chapter, and from the laser-induced fluorescence study of Leone, et al. is that the predominant product from the charge-transfer of $Ar^+(P_{3/2})$ with $N_2(X\ \Sigma_g^+,v=0)$ is the vibrationally excited product $N_2^+(X\ \Sigma_g^+,v=1)$ with a modest degree of rotational excitation. This interesting conclusion is supported by microscopic reversability arguments applied to the study of the reverse reaction using state-selected N_2^+ by Govers, Guyon, Baer, Cole, Frohlich and Lavollee (17).

The latter workers used a synchrotron UV light source and time-of-flight coincidence techniques to measure the reaction cross-sections for ground state N_2^+ excited to various vibrational levels and for the first electronically-excited A-state. Within experimental error the cross-section for reaction of $N_2^+(X\ \Sigma_g, v=0)$ with Ar is 0, while the cross-section rises rapidly for v=1, reaches a broad maximum of about 30 square angstroms at v=2, and then falls off.

The first steps toward achieving a realistic theoretical characterization of the ArN_2^+ system have recently been taken in the papers by Spalburg and co-workers (18,19). Their time-dependent quantum mechanical treatment accurately predicts experimental results for the total reactive cross section as a function of collision energy, the specific generation of $N_2^+(X\ \Sigma_g, v=1)$ as the dominant channel at most energies (their prediction is that this would be true at all energies) and the absence of significant angular scattering at energies above 1.7 eV. However, the predictions in these papers (18,19) of the reactive <u>and</u> inelastic channels at low energy (1.2 eV collision energy) appear to be incorrect (16). Finally, no angular scattering predictions are possible with this theory in its present form, which assumes rectilinear trajectories. Some of the problems with the theory are obvious. It ignores long-range ion-induced dipole and ion-quadrupole forces and any valence forces responsible for the <u>ca</u>. 1.2 eV well-depth for ArN_2^+. Adding these features, plus carrying out appropriate trajectory calculations to deduce angular scattering for all channels as a function of energy, are logical extensions which will precede more serious calculations involving the coupling between potential hypersurfaces. These extensions are currently in progress (20).

In conclusion, the first detailed dynamical investigation of charge transfer for a simple triatomic system which has utilized state-specific measurements to determine the probability of charge transfer as a function of collision energy and angular scattering properties has revealed a wealth of experimental details. Clearly simplistic models invoking energy resonance Franck-Condon factors and simplified treatments of non-adiabatic coupling must be abandoned along with any conjecture that a statistical model is the most appropriate model for charge transfer at low energy - at least for three atom systems. It is suggested that non-abiatic transitions in localized crossing seams of the reactant product potential energy surfaces is responsible for the quantum state specificity which is observed. Understanding the details of these processes which exhibit pronounced quantum state specificity along with angular-scattering specificity will depend upon the interplay between detailed theoretical studies and further detailed experiments.

5. ACKNOWLEDGEMENTS

The author gratefully acknowledges many helpful discussions of these processes with Professors Steve Leone and Eric Gislason and with faculty colleagues at Utah, especially Professors Shirts, Simons and

Wight. Finally I thank graduate students Howard, Anderson and Sohlberg, and Postdoctoral Fellows Friedrich, Shukla and Rockwood for carrying out the work from our laboratory described in this paper. We are grateful to the National Science Foundation for support of most of our work on charge transfer reaction dynamics.

REFERENCES

1. M. L. Vestal, C. R. Blakley and J. H. Futrell, Phys. Rev. A, 23, 2327 (1981).

2. C. R. Blakley, P. W. Ryan, M. L. Vestal and J. H. Futrell, Rev. Sci. Instrum., 47, 15 (1976).

3. D. Rapp and W. E. Francis, J. Chem. Phys., 37, 2631 (1961).

4. S. C. deCastro, H. F. Schaefer, III and R. M. Pitzer, J. Chem. Phys., 74, 550 (1981).

5. S. H. Linn, Y. Ono and C. Y. Ng, J. Chem. Phys., 74, 3342 (1981).

6. K. Stephan, T. D. Mark, J. H. Futrell and H. P. Helm, J. Chem. Phys., 80, (1984).

7. B. Friedrich, S. L. Howard, A. L. Rockwood, W. E. Trafton, Jr., Du Wen Hu and J. H. Futrell, Int. J. Mass Spectrom. Ion Proc., 59, 203 (1984).

8. K. B. McAfee, Jr., C. R. Szmanda, R. S. Hozack and R. E. Johnson, J. Chem. Phys., 77, 2399 (1982).

9. M. R. Flannery, P. C. Cosby and T. F. Moran, J. Chem. Phys., 59, 5494 (1973).

10. W. Lindinger, F. Howorka, P. Lukac, S. Kuken, H. Villinger, E. Alge, and H. Ramler, Phys. Rev. A, 23, 2327 (1981).

11. D. Smith and N. G. Adams, Phys. Rev. A, 23, 2327 (1981).

12. L. Hüwel, D. R. Guyer, G-H Lin and S. R. Leone, J. Chem. Phys., 81, 3520 (1984).

13. B. Friedrich, W. Trafton, A. Rockwood, S. Howard and J. H. Futrell, J. Chem. Phys., 80, 2537 (1984).

14. H. H. Teng and D. C. Conway, J. Chem. Phys., 59, 2316 (1973).

15. A. L. Rockwood, S. L. Howard, Du Wen-Hu, P. Tosi, W. Lindinger and J. H. Futrell, Chem. Phys. Lett, 114, 486 (1985).

16. S. L. Howard, A. R. Shukla, K. Birkinshaw and J. H. Futrell (to be published).

17. T. R. Govers, P. M. Guyon, T. Baer, K. Cole, H. Fröhlich and M. Lavollée, Chem. Phys., **87**, 373 (1984).

18. M.R. Spalburg, J. Los and E.A. Gislason, Chem. Phys., **94**, 327 (1985).

19. M.R. Spalburg and E.A. Gislason, Chem. Phys., **94**, 339 (1985).

20. E. A. Gislason, personal communication.

PRODUCTION, QUENCHING AND REACTION OF VIBRATIONALLY EXCITED IONS IN COLLISIONS WITH NEUTRALS IN DRIFT TUBES

Eldon E. Ferguson
Aeronomy Laboratory
National Oceanic and Atmospheric Administration
Boulder, Colorado U.S.A.

1. INTRODUCTION

The Flowing Afterglow Technique (Ferguson et al., 1969) provided the first chemically versatile method for measuring ion-molecule reactions at room temperature, later extended from 80K to 900K (Lindinger et al., 1974). Positive and negative ions of great variety can be studied, including metal ions, heavily solvated ions, doubly charged positive ions, electronically excited ions and now vibrationally excited ions. A variety of neutrals including H, O, N atoms, OH radicals, O_3, H_2O_2, HNO_3, H_2SO_4 and N_2O_5 have been reacted with ions in the Flowing Afterglow. Reactions of ions with vibrationally excited N_2 (Schmeltekopf et al., 1968) and H_2 (Jones et al., 1986) have been carried out. The chemical versatility is unmatched by any other technique.

The subsequent Flow Drift technique (McFarland et al., 1973) allowed an extension of the energy range from thermal to several electron volts relative kinetic energy, bridging the gap between thermal energy and higher energies accessible to beam experiments. An inherent problem with drift tube measurements, in which kinetic energy is supplied to the ions by means of an applied electric field, is that the energetic ions do not have Maxwellian distributions of velocities. Thus one does not obtain true thermal rate constants. This problem has been solved for atomic ions in atomic buffer gases (Lin and Bardsley, 1977; Viehland and Mason, 1977) so that cross sections have been deduced from ~ 0.04 to 2 eV for several O^+ and O^- reactions, carried out in helium buffer gas. The measured mobilities of the ions in the buffer gas supply sufficient information about the ion-neutral interaction to allow either a deduction of the ion velocity distribution from which reaction cross-sections can be derived from experimental rate constants (Lin and Bardsley, 1977), or they allow measured rate constants to be corrected to thermal ones (Viehland and Mason, 1977). The velocity distribution problem has not yet been solved for either molecular ions or molecular buffer gases because of the complexity added by the internal degrees of freedom, which are

P. Ausloos and S. G. Lias (eds.), Structure/Reactivity and Thermochemistry of Ions, 81–95.

excited in collisions.

Another aspect of drift tube measurements is that the internal degrees of freedom of molecular ions may not be in equilibrium with translation; in general they are not. Usually rotational excitation is equilibrated with translation but vibrations are not, depending on the buffer gas, the ion, the drift tube length, the pressure and the applied electric field strength. Recently theory has shown that molecular ion internal excitation in atomic buffer gases will be in equilibrium with translation at steady state (Viehland et al., 1981). In practice, particularly with He buffer gas, steady state is not usually reached in conventional drift tubes involving ~ 10^3 collisions between the molecular ions and the He buffer gas.

This situation must of course be reckoned with; it can be either an advantage or a disadvantage. If one is simply interested in thermodynamic rate constants (as one often is) then it is clearly a disadvantage. On the other hand, it does offer an experimental possibility of separating the contributions of kinetic energy and vibrational energy to reaction rate constants, i.e. the possibility of a certain limited amount of state-to-state chemistry. It is this aspect of flow drift tubes that is addressed in this discussion, namely the manner in which vibrational excitation and deexcitation rate constants may be obtained, and demonstrations of specific effects on reaction rate constants due to ion vibrational excitation. Such studies offer insight into molecular reaction mechanisms and dynamics and into the nature of molecular potential surfaces.

2. VIBRATIONAL EXCITATION

A clear indication of ion vibrational excitation in drift tubes has been given by the finding of quite different rate constants versus energy for many ion neutral reactions carried out in different buffer gases. The most usual experience has been to find that a reaction rate constant is larger in Ar buffer than in He buffer. This kind of result could also be due to the differing velocity distribution of the ions in the two buffer gases. However, the effect of different velocity distributions is quantitatively very much less and is only significant for a very steeply increasing rate constant with energy, in which case the velocity distribution with the greatest high energy tail will produce the largest rate constant. This situation has been worked out in detail for the reaction $O^+ + N_2 \rightarrow NO^+ + N$, for which the rate constant increases by a factor of ~ 40 between thermal energy and ~ 1eV (Albritton et al., 1977). This is about as extreme a case as can be encountered for velocity distribution effects, and yet is quite small compared to commonly observed vibrational enhancement effects. A convincing proof of the vibrational nature of many enhancements has often been obtained by vibrationally quenching the reactant ion by addition of a trace of its parent neutral. The quenching by charge-transfer normally occurs near the collision rate, while the velocity distribution would be scarcely perturbed by trace gas addition. Large rate constant changes due to added trace quantities of ion reactant neutral establish the occurrence of vibrational enhancement of the rate constant.

When a vibrational enhancement is established for an ion molecule reaction, and such enhancements are very common for positive ion reactions which occur with rate constants less than Langevin, this can be used to establish at least lower limits on vibrational excitation rate constants in the buffer gas. In this way the ions O_2^+, CO_2^+, N_2O^+, NO_2^+, H_2O^+ and SO_2^+ have been found to be vibrationally excited, very near the vibrational energy threshold, in ~ 10^3 collisions with Ar and in some cases with Ne (Durup-Ferguson et al., 1983). This implies a rate constant for vibrational excitation greater than 10^{-3} k_L (collision rate constant) or $\geq 10^{-12}$ cm^3 s^{-1}. By an application of the principle of detailed balance the deexcitation rate constant at thermal energy must exceed the excitation rate constant at threshold, so that one also knows that thermal energy quenching rate constants in these cases exceed 10^{-12} cm^3 s^{-1}.

On the other hand, these ions are found not to be significantly vibrationally excited in He buffer gas at threshold and at energies somewhat above. This is deduced from the fact that vibrationally sensitive reactions do not change rate constant in He buffer when a trace of their parent is added (which would quench any ion vibrational excitation).

This technique could be used to determine excitation rate constants quantitatively by varying the number of collisions in order to find out how many are required. The easiest way to do this would be to vary the drift tube length. This has not as yet been done.

Lindinger and his students at Innsbruck have done a detailed study of vibrational excitation and deexcitation of N_2^+ ions in He buffer at higher energies (higher E/N in the drift tube) using Ar charge-transfer as a monitor for vibrationally excited N_2^+, a technique that will be discussed in detail below. The Innsbruck studies offer great promise for probing the subject of vibrational excitation of ions in flow drift tubes.

The rate constant limits already obtained have been useful and have necessitated a reinterpretation of some earlier laboratory data obtained in stationary afterglows, e.g. data that had been erroneously applied to the interpretation of the ion chemistry of planetary atmospheres.

3. ION-ATOM INTERCHANGE REACTIONS

As an example (Lindinger et al., 1975) of an ion-atom interchange reaction, i.e. a conventional chemical reaction in which chemical bonds are broken and made, the slightly exothermic proton transfer reaction

$$N_2OH^+ + CO \underset{k_{-1}}{\overset{k_1}{\rightleftarrows}} COH^+ + N_2O \qquad (1)$$

has been studied in the forward and reverse directions in He and Ar buffer gases. It is observed that the reverse (-1) endothermic reaction is enhanced in Ar relative to He, i.e., enhanced by vibrational excitation. It is enhanced even more strongly by direct heating in the low energy range of overlap between the drift tube and a heated flowing afterglow system (300-900K). On the other hand, the forward exothermic reaction rate constant is slightly decreased by vibrational excitation. This can be readily understood as a reaction occurring on a potential surface with the barrier toward the N_2OH^+ + CO reactant channel. The late barrier for the reverse endothermic reaction then leads to vibrational enhancement, while the early barrier in the forward exothermic reaction (necessarily early from detailed balance) leads to vibrational hindrance for that reaction. This kind of study has been little pursued. Such studies, of course, are only qualitative, and so perhaps of little interest.

4. CHARGE-TRANSFER REACTIONS

A very striking finding (Lindinger et al., 1975) was that the exothermic charge-transfer reaction

$$CO_2^+ + O_2 \rightarrow O_2^+ + CO_2 \qquad (2)$$

was considerably enhanced by vibrational excitation of the CO_2^+. An enhancement of almost an order of magnitude is observed at 0.3 eV relative kinetic energy in Ar as compared to He. N_2 buffer gas gave a rate constant intermediate between He and Ar. There is probably a simple mass effect on vibrational excitation, such that vibrational excitation generally increases with buffer gas mass.

The vibrational enhancement increases with relative kinetic energy but this reflects the increasing vibrational population in Ar with increasing energy, rather than an increasing vibrational enhancement with kinetic energy, presumably. Derai et al (1985) showed that the vibrational enhancement remains large at near thermal energy in ICR experiments using CO_2^+ ions produced in different vibrational states by different charge-transfer reactions. The low energy onset of the enhancement in the flow drift tube is near the low frequency CO_2^+ bending mode, ~ 500 cm^{-1}, and this clearly implicates the bending mode in the enhancement, although not exclusively at higher energies where stretching modes can be excited and might contribute to the vibrational enhancement.

The fact that the measurements are only qualitative is not a very great liability, since the effect is very large and the mechanism has not even been qualitatively established theoretically. A mechanism involving coupling of vibrational angular momentum in the degenerate bending mode with orbital angular momentum to increase the complex lifetime has been suggested (Durup-Ferguson et al., 1983) but has not been established as correct. The idea is that a transfer of some of the orbital angular momentum (~ 100 ħ) into vibrational angular momentum (~ ħ) would be a transfer of relative kinetic energy into internal energy, thereby increasing the complex lifetime and consequently the probability for reaction.

Vibrational enhancement of slow ($k < k_L$) charge-transfers turns out to be quite general. The charge-transfers of CO_2^+, N_2O^+, NO_2^+, SO_2^+ and H_2O^+ with NO are all found to be significantly enhanced in Ar or Ne buffer gas (due to ion vibrational excitation in these buffers) relative to He (Durup-Ferguson et al., 1983). Two of the ions, H_2O^+ and SO_2^+, are non-linear and so do not have degenerate vibrations and therefore do not have vibrational angular momentum. The angular momentum coupling model cannot be exclusively correct, therefore.

The reason for vibrational enhancement of charge-transfer is quite mysterious and unexpected and should yield an interesting insight into the nature of molecular interactions when it is understood. No cases in which vibrational excitation decreases a charge-transfer rate constant have yet been reported.

The flow drift technique has the capability of providing rapid surveys of qualitative vibrational effects, which can subsequently be probed more quantitatively in other ways. We shall later describe some quantitative studies which can be carried out, however, specifically the vibrational dependence of various $O_2^+(v)$ reactions as a function of v.

5. VIBRATIONAL QUENCHING MEASUREMENTS

Recently a new flow drift tube technique has been developed (Böhringer et al., 1983; Dobler et al., 1983) for vibrational relaxation measurements of ions. So far the diatomic ions O_2^+, NO^+ and N_2^+ have been measured. This involves production of vibrationally excited ions in low pressure electron impact sources, injection of the ions by an aspirator arrangement (SIFT source) into non-relaxing helium buffer gas, and detection of the ions by injection of a monitor gas in front of the final mass spectrometer detection. The monitor must be able to distinguish the ion vibrational state, normally on the basis of energy. For example, CH_3I has an ionization potential greater than NO^+ and so does not react with $NO^+(v=0)$, but charge-transfer with

$NO^+(v\geq1)$ is exothermic and fast so that the signal of CH_3I^+ in the mass spectrometer is a measure of the concentration of $NO^+(v\geq1)$ in the tube. Then various vibrational quenchers can be added and the relative changes of CH_3I^+ signals give directly the quenching rate constants, in the usual manner of rate constant measurements. Different monitors can be used to detect different vibrational levels, e.g. Xe for $O_2^+(v>0)$, SO_2 for $O_2^+(v>1)$ and H_2O for $O_2^+(v>2)$. The monitor ion technique was used earlier at NOAA to measure reactions of the electronically excited ions, $O_2^+(a^4\Pi_u)$, $NO^+(a^3\Sigma)$ and O^{+*}.

Results of vibrational relaxation rate constants determined in this way are listed in Table I. Several generalities appear. (1) The rate constants are larger than those for vibrational relaxation of neutral diatomic molecules of comparable frequency. The usual experimental lower limit rate constant, $k_q \sim 10^{-12}\ cm^3\ s^{-1}$, corresponding to a probability $\sim 10^{-3}$, is exceeded in most cases, indeed in all cases of quenching by neutral non-rare gases except NO^+ quenching by O_2. (2) There is a strong correlation between k_q and the electrostatic interaction between the ion and the quencher. In every case polar neutrals are particularly effective quenchers. He is always the least efficient quencher. (3) In most cases k_q decreases with increased relative KE.

These features clearly indicate that the long range attractive forces are responsible for the vibrational quenching. Vibrational relaxation of high frequency neutral diatomics is generally inefficient and the efficiency increases with collision energy. In this case the quenching is driven by the short range repulsive interaction. The adiabatic principle applies to this situation, leading to a probability of vibrational relaxation $P \sim \exp\left(-\frac{4\pi^2 a\nu}{v}\right)$, which in turn leads to the familiar Landau-Teller (1936) equation. In some cases attractive forces dominate neutral vibrational relaxation, e.g. hydrogen halide self-relaxation, CO_2 quenching by H_2O, NO quenching by NO (Lambert, 1977; Yardley, 1980), cases in which a significant attractive potential exists, either because of chemical bonding or hydrogen bonding. In these cases the quenching probability decreases with increasing temperature to a minimum, then increases with further increase in temperature.

In the case of molecular ions the dominance of the attractive interaction in collision with neutrals is far more common due to the long range electrostatic interactions. One expects the quenching rate constant to decrease to a minimum with relative collision energy and then increase with further energy increase in the region dominated by the repulsive interaction. The measure of energy, and presumably the location of the minimum, should be related to the attractive well depth. The only case so far in which the minimum has been observed is for $O_2^+(v)$ quenching by Kr (Kriegl et al., 1986), where the broad minimum is somewhat less but is in the neighborhood of the potential well depth, 0.33 eV.

The weak quenching of ions by helium is due to the very low polarizability of He, leading to a weak electrostatic interaction. At room temperature $O_2^+(v)$ is not more readily quenched by He than is neutral $O_2(v)$! The quenching of $NO^+(v)$ by all rare gases is below detectability, i.e. $k_q < 10^{-12}$ cm^3 s^{-1}. This is due both to the relatively weak interaction and probably also, particularly in the case of the quite polarizable Xe, due to the isotropy of the interaction (Ferguson, 1986). It is known from neutral trajectory studies that anisotropy facilitates neutral vibrational relaxation by enhancing the V → T, R energy transfer.

The weak quenching of $NO^+(v)$ by O_2 is an interesting case. This appears to be a case in which a repulsive chemical interaction dominates the attractive electrostatic interaction, presumably because $NO^+(^1\Sigma)$ + $O_2(^3\Sigma)$ approach on a repulsive triplet curve. It was speculated that $O_2(^1\Delta_g)$ would approach NO^+ on the attractive ground state curve, corresponding to the stable NO_3^+, and lead to very efficient quenching and this turned out to be the case (Dotan et al., 1985). This is, so far, the only case in which molecular ion vibrational relaxation has been found to be inhibited by repulsive chemical interaction. Similar situations arise where exothermic ion-molecule reactions are inhibited by repulsive chemical interaction curves, e.g. the charge-transfer between O^+ and NO (Ferguson, 1975), but such situations have only rarely been identified. Usually one simply does not have enough information about the intermediate complex to deduce whether potential curves are attractive or repulsive.

6. VIBRATIONAL RELAXATION MODEL

The following model (Ferguson, 1984, 1986; Dobler et al., 1983) has been developed to describe the vibrational relaxation data. In the collision of an ion and a neutral an orbiting complex is formed due to the long range electrostatic interaction. The lifetime depends on the well depth and anisotropy, vibrational frequencies, etc., more or less independently of the vibrational state of the ion, at least for diatomics. During the collisional lifetime of the complex the vibrational state of the ion might change due to coupling of the vibrational motion with the relative ion-neutral motion, causing the ion-neutral pair to fly apart. This is analogous to vibrational predissociation which is now being so intensively studied in weakly bound neutral van der Waals molecules. This vibrational predissociation would have the consequence of vibrational relaxation and the vibrational quenching rate constant k_q would be given by

$$k_q = k_c k_{vp} \tau \tag{3}$$

where k_c is the rate of formation of complexes, k_{vp} is the unimolecular vibrational predissociation rate and τ is the lifetime of the complex. This is valid if $k_{vp} \tau << 1$ so that quenching is rate limited by vibrational predissociation and not complex formation. If $k_{vp} \tau \gtrsim 1$, then $k_q = k_c$; quenching will be rate limited by the collision rate. Such cases do exist for strongly attractive interactions, as seen in Table I.

In order to deduce the dynamically interesting parameter k_{vp}, we determine the lifetime from the three-body association of the same ion and neutral. This is given by $k^3 = k_c k_s \tau$, where k_s is the stabilization rate constant for the transient complex by the third body M. Eliminating the product $k_c \tau$ one obtains a relationship between k_{vp}, k^3 and k_q,

$$k_{vp} = \frac{k_q k_s}{k^3} . \tag{4}$$

Values of k_{vp} can be determined from experimental measurements of k_q and k^3. The stabilization rate constant k_s is essentially the collision rate constant $k_L = 2\Pi e\sqrt{\alpha/\mu}$ for heavy third bodies and is typically about 1/4 this for the common case of He third body.

The values of k_{vp} so far obtained are listed in Table II. The remarkable finding is that the values of k_{vp} are rather confined in range, lying mostly in the range 10^9-10^{10} s^{-1}. This contrasts with the situation of vibrational predissociation of van der Waals molecules where a much wider range of values is found, from less than 10^6 s^{-1} to greater than 10^{12} s^{-1}. The situation is of course different in the case of van der Waals vibrational predissociation; there the weakly bound cluster is in the ground vibrational state of the weak cluster bond when a vibrational level of one of the partners is excited, with an energy exceeding the weak cluster bond. In the present case of vibrational relaxation, the two partners are orbiting in their dissociation continuum with the ion in a vibrationally

TABLE I. Vibrational Quenching Rate Constants ($cm^3 s^{-1}$) at 300K

	Quencher	k_q^1	k_q^2	k_c	Z_{10}	Z_{21}
O_2^+	He	<2(-15)	<6(-15)	5.0(-10)	>3(5)	>9(4)
	Ne	<1(-14)	<2(-14)	4.2(-10)	>3(4)	>2(4)
	Ar	1.0(-12)	2.5(-12)	7.1(-10)	710	280
	Kr	1.1(-11)	1.7(-11)	7.6(-10)	69	45
	H_2	2.5(-12)	5.0(-12)	1.5(-9)	610	305
	D_2	6.5(-13)	2.6(-12)	1.1(-9)	1700	430
	N_2	1.9(-12)	5.4(-12)	8.0(-10)	420	150
	CO	4.4(-11)	6.5(-10)	9.3(-10)	21	14
	CO_2	1.0(-10)	2.0(-10)	8.7(-10)	8.7	4.4
	H_2O	1.2(-9)	--	1.4(-9)	1.2	-
	CH_4	6.0(-10)	--	1.1(-9)	1.9	-
	SO_2	5.7(-10)	--	1.2(-9)	2.1	-
	SF_6	1.1(-10)	2.0(-10)	9.7(-10)	8.8	4.9
	O_2	3.0(-10)	4.0(-10)	7.4(-10)	2.5	1.9
		k_q		k_c	Z	
NO^+	NO	5(-10)		7.5(-10)	1.5	
	N_2	7(-12)		8.0(-10)	114	
	CO_2	4(-11)		8.7(-10)	22	
	CH_4	3(-11)		11.4(-10)	38	
	NO_2	1.5(-10)		11.8(-10)	8	
	He	<1(-13)		5.6(-10)	>5600	
	Ar	<1(-12)		7.1(-10)	>710	
	Kr	<1(-12)		7.6(-10)	>760	
	NH_3	1.5(-9)		1.5(-9)	1	
	C_2H_4	8.7(-10)		1.3(-9)	1.5	
	H_2O	9(-11)		1.4(-9)	16	
	CO	1.0(-11)		9.3(-10)	93	
	$O_2(^3\Sigma)$	<1(-12)		7.4(-10)	>740	
	$O_2(^1\Delta)$	3±2(-10)		~7.4(-10)	~2	
	H_2	<1(-12)		1.5(-9)	>1500	
	D_2	<1(-12)		1.1(-9)	>1100	
	Ne	<1(-12)		4.2(-10)	>420	
	N_2O	~2(-10)		9.5(-10)	5	

TABLE II. Vibrational Predissociation Rate Constants

Ion	Neutral	$k_q cm^3 s^{-1})$	$k^3 cm^6 s^{-1})$	$k_{vp}(10^9 s^{-1})$
$O_2^+(v=1)$	Ar	1.0(-12)	3.5(-31)[a]	2.1
	Kr	1.1(-11)	9.4(-31)[a]	7.9
	H_2	2.5(-12)	5.3(-32)[a]	25
	N_2	1.9(-12)	1.0(-30)	1.4
	CO_2	1.0(-10)	2.6(-29)[b]	2.7
	SO_2	5.7(-10)	5.5(-29)[b]	0.7
	H_2O	1.2(-9)	2.5(-28)	$\geq 3.3, k_q \sim k_c$
$NO^+(v=1)$	Kr	<1(-12)	9.7(-32)[a]	<7
	O_2	<1(-12)	9.0(-32)	<7
	N_2	7.0(-12)	3.0(-31)	17
	CO	1.0(-11)	1.9(-30)	3.7
	CO_2	4.0(-11)	1.4(-29)	2.0
	CH_4	3.0(-11)	<5.8(-29)	>0.4
	NH_3	1.5(-9)	8.8(-28)	$\geq 1.2, k_q \sim k_c$
	H_2O	9(-11)	1.5(-28)	0.4
$O_2^-(v=1)$	CO_2	~2.7(-11)	4.7(-29)	~0.35
$CO^+(v=1)$	N_2	1.3(-10)	8.0(-29)	0.9

[a] extrapolated from M = He, 80K to M = N_2, 300K by $3.8(3.75)^{-(\ell/2 + 1)}$.

[b] extrapolated from M = He, 200K to M = N_2, 300K by $3.8(1.5)^{-(\ell/2 + 1)}$.

where ℓ = number of rotational degrees of freedom of reactants, 2 for neutral atoms, 4 for linear neutrals, 5 for non-linear neutrals.

excited state.

Other cases of vibrational transfer to proximate but non-bonded neighbors also occur at similar rates. For example, the vibrational relaxation of OH groups adsorbed on silica in a variety of solvents have rates in the 10^9-10^{10} s^{-1} range (Heilweil et al., 1985) and the relaxation rates of five benzene vibrations in the crystal (Velsko and Hochstrasser, 1985) span the range of k_{vp}'s in Table II.

Table II includes two values of k_{vp} not obtained from the flow drift tube, a value for $CO^+(v=1) + N_2$ quenching reported by Hamilton et al. (1985) utilizing a three-body rate constant of Ferguson et al. (1986), and the single negative ion value, for $O_2^-(v=1) + CO_2$, from an unpublished measurement of Miller and Lineberger combined with a literature value of the three-body association rate constant. The value of k_q for $CO^+(v=1)$ quenching by N_2 is extremely large, about 70 times larger than $O_2^+(v=1)$ quenching by N_2 and about 20 times larger than $NO^+(v=1)$ quenching by N_2. This implies a deeper attractive well than the purely electrostatic well depth ~ 0.2eV. The three-body association rate constant is extremely large, reflecting the large well depth also. These observations seem to require a "chemical" bonding between CO^+ and N_2, with a well depth as large as ~ 1eV (Ferguson et al., 1986).

7. VIBRATIONAL RELAXATION BY NEAR RESONANT CHARGE-TRANSFER

The vibrational relaxation of ions by their parent neutrals is a special case. Vibrational relaxation usually occurs by near resonant charge transfer, e.g.

$$O_2^+(v=1) + O_2(v=0) \rightarrow O_2^+(v=0) + O_2(v=1) + 325\ cm^{-1} \quad (5)$$

In this case the rate constant and its energy dependence are identical to that of the charge-transfer, which has been determined with isotopic O_2. The rate constant is approximately Langevin; the same is true for $N_2^+(v) + N_2$, $NO^+(v) + NO$, $CO^+(v) + CO$, $CO_2^+ + CO_2$, and $N_2O^+ + N_2O$. The only exceptions expected are when vibrational relaxation by charge-transfer is quite endothermic ($\Delta H > kT$). This occurs for example for $Cl_2^-(v) + Cl_2$, which does not occur near the Langevin rate because it is endothermic by 310 cm^{-1} ($> kT$) (Ferguson, 1986).

8. COMPETITIVE REACTION AND QUENCHING

Studies of the competitive reaction and quenching of $O_2^+(v)$ have been carried out with CH_4, SO_2 and H_2O (Durup-Ferguson et al., 1984). In the case of SO_2 and H_2O, the situation is simple and obvious, vibrational quenching is fast for the $O_2^+(v)$ levels for which reaction, in this case charge-transfer, is endothermic whereas charge-transfer dominates the quenching when it is exothermic, i.e. for v=2 for SO_2 and v=3 for H_2O. The CH_4 case is much more complex. Reaction is exothermic, and occurs, for all vibrational levels of $O_2^+(v)$. The reaction produces CH_2OOH^+ + H (van Doren et al., 1986) for v=0 with a rate constant which is low, ~ 5 x 10^{-12} cm^3 s^{-1} at 300K, increasing at both higher and lower energy. For v=1 the collisions are dominated by a fast vibrational relaxation, k_q ~ 5 x 10^{-10} cm^3 s^{-1}, although the reactive path is slightly enhanced by v=1. For v=2 an endothermic path to produce CH_3^+ + HO_2 becomes accessible and dominates the collisions and for v=3 charge-transfer becomes energetically accessible and dominates. In a detailed, concerted set of investigations, carried out in Boulder, Birmingham and Meudon and involving O_2^+ vibrational state dependence, temperature dependence from 20K to 560K, kinetic energy dependence from 0.01 to 2 eV and involving the five CH_nD_{4-n} isotopes, a detailed mechanism for this reaction has been deduced (Barlow et al., 1986). The reaction proceeds through a reversible, loose electrostatic complex $O_2^+ \cdot CH_4$, followed by irreversible hydride ion transfer to produce CH_3OOH^+, and finally a rapid ejection of a hydrogen atom to produce methylene hydroperoxy cation. The reaction occurs on a double minimum potential surface whose parameters have been rather well established. This is one of the most detailed unravelings of a complex ion-molecule reaction yet carried out.

One point of interest concerns the lifetime of the reversible $O_2^+ \cdot CH_4$ weakly bound (9 kcal mol^{-1}) complex. From the vibrational relaxation rate k_q ~ 1/2 k_L it was deduced that $\tau \sim k_{vp}^{-1}$ ~ 10^{-9}-10^{-10} s, in accord with the regularities of Table II. Subsequently the three-body reaction of O_2^+ and CH_4 was measured by Böhringer and Arnold (1986) and this allows an independent estimate of τ from $k^3 = k_c k_s \tau$. This value, extrapolated from 87K and with an estimate of 1/4 for the He third body efficiency, also leads to $\tau(300K) = 10^{-9}-10^{-10}$ s, consistent with the lifetime deduced from vibrational relaxation, or otherwise stated, establishing $k_{vp} \sim 10^{9}-10^{10}\ s^{-1}$.

From a dynamical viewpoint, the $O_2^+ + CH_4$ study established that when sufficient vibrational energy was available to provide exothermicity, the order of rates (or efficiency of processes) in the $O_2^+ \cdot CH_4$ complex was: electron transfer > hydride ion transfer > vibrational predissociation.

9. VIBRATIONAL TRANSFER FROM NEUTRALS TO IONS

The first measurements of vibrational energy transfer from a neutral to an ion were recently made in Birmingham (Ferguson et al., 1984). The point investigated was whether the anomalously fast (Tables I and II) vibrational transfer from NO^+ to N_2 was due to the near resonant V→V channel,

$$NO^+(v=1) + N_2(v=0) \rightarrow NO^+(v=0) + N_2(v=1) + 14\ cm^{-1} \quad (6)$$

Since the reverse (essentially thermoneutral) process was not observed, it follows that (6) does not occur, $k_6 < 10^{-13}\ cm^3\ s^{-1}$ compared to the measured $k_q = 8 \times 10^{-12}\ cm^3\ s^{-1}$ which must therefore be V→T, R. This supports the model of complex formation followed by predissociation. In the complex formation step, presumably, relative translational energy is converted into internal rotational energy of the N_2, and perhaps also the NO^+. A minimum energy transfer involves $\Delta J = 2$, which yields $\sim 70\ cm^{-1}$, so that exiting in the V-V channel (6) is endothermic. The possibility of near resonant V→V transfer does not enhance vibrational relaxation of ions as it does for neutrals. In the case of neutrals the near resonance can give rise to dipole-dipole or dipole-quadrupole interactions that lead to efficient vibrational energy transfer. In the case of ions the already existing electrostatic long range interactions appear to dominate.

In the Birmingham experiments, the vibrational transfer

$$N_2(v=1) + O_2^+(v=0) \rightarrow O_2^+(v=1) + N_2(v=0) + 458\ cm^{-1} \quad (7)$$

was observed to be fast, $k \sim 10^{-12}\ cm^3\ s^{-1}$. In this case the maximum expected energy transfer from translation to rotation, $\sim 2kT = 400\ cm^{-1}$, is not sufficient to preclude (7).

10. ACKNOWLEDGMENT

The Defense Nuclear Agency has played a key role in supporting the Aeronomy Laboratory Ion Chemistry program over many years. Communication of unpublished results by Professors Lindinger and Lineberger is gratefully appreciated.

11. REFERENCES

Albritton, D.L., I. Dotan, W. Lindinger, M. McFarland, J. Tellinghuisen and F. C. Fehsenfeld, J. Chem. Phys. **66**, 410 (1977).

Barlow, S.E., J.M. Van Doren, C.H. DePuy, V.M. Bierbaum, I. Dotan, E.E. Ferguson, N.G. Adams, D. Smith, B. Rowe, J.B. Marquette, J.B. Dupeyrat, and M. Durup-Ferguson, J. Chem. Phys. (submitted, 1986).

Böhringer, H. and F. Arnold, J. Phys. Chem. **84**, 2097 (1986).

Böhringer, H., M. Durup-Ferguson, D.W. Fahey, F.C. Fehsenfeld and E.E. Ferguson, J. Chem. Phys. **79**, 4201 (1983).

Derai, R., P.R. Kemper and M.T. Bowers, J. Chem. Phys. **82**, 4517 (1985).

Dobler, W., W. Federer, F. Howorka, W. Lindinger, M. Durup-Ferguson and E.E. Ferguson, J. Chem. Phys. **79**, 1543 (1983).

Dotan, I., S.E. Barlow and E.E. Ferguson, Chem. Phys. Lett. **121**, 38 (1985).

Durup-Ferguson, M., H. Böhringer, D.W. Fahey, F.C. Fehsenfeld and E.E. Ferguson, J. Chem. Phys. **81**, 2657 (1984).

Durup-Ferguson, M., H. Böhringer, D.W. Fahey, and E.E. Ferguson, J. Chem. Phys. **79**, 265 (1983).

Ferguson, E.E., in Interactions Between Ions and Molecules, Pierre Ausloos, Ed., Plenum Press, New York, 1975, p. 313.

Ferguson, E.E., in Swarms of Ions and Electrons in Gases, W. Lindinger, T.D. Märk and F. Howorka, Eds., Springer-Verlag, Wien, 1984.

Ferguson, E.E., J. Phys. Chem. **90**, 731 (1986).

Ferguson, E.E., N.G. Adams, and D. Smith, Chem. Phys. Lett. (in press, 1986).

Ferguson, E.E., N.G. Adams, D. Smith and E. Alge, J. Chem. Phys. **80**, 6095 (1984).

Ferguson, E.E., F.C. Fehsenfeld and A.L. Schmeltekopf, Adv. Atomic Molec. Phys. **5**, 1 (1969).

Hamilton, C.E., V.M. Bierbaum and S.R. Leone, J. Chem. Phys. **83**, 601 (1985).

Heilweil, E.J., M.P. Cassassa, R.R. Cavanaugh and J.C. Stephenson, J. Chem. Phys. **82**, 5216 (1985).

Jones, M.E., S.E. Barlow, G. B. Ellison and E.E. Ferguson, Chem. Phys. Lett. (submitted, 1986).

Kriegl, M., R. Richter, P. Tosi, W. Federer, W. Lindinger and E.E. Ferguson, Chem. Phys. Lett. **124**, 583 (1986).

Lambert, J.D., Vibrational and Rotational Relaxation in Gases, Oxford University Press, London, 1977.

Landau, L.D. and E. Teller, Phys. Z. Sowjetunion, **10**, 34 (1936).

Lin, S.L. and J.N. Bardsley, J. Chem. Phys. **66**, 435 (1977).
Lindinger, W., F.C. Fehsenfeld, A.L. Schmeltekopf and E.E. Ferguson, J. Geophys. Res. **79**, 4753 (1974).
Lindinger, W., M. McFarland, F.C. Fehsenfeld, D.L. Albritton, A.L. Schmeltekopf and E.E. Ferguson, J. Chem. Phys. **63**, 2175 (1975).
McFarland, M., D.L. Albritton, F.C. Fehsenfeld, E.E. Ferguson and A.L. Schmeltekopf, J. Chem. Phys. **59**, 6610, 6620, 6629 (1973).
Schmeltekopf, A.L., E.E. Ferguson and F.C. Fehsenfeld, J. Chem. Phys. **48**, 2966 (1968).
Van Doren, J.M., S.E. Barlow, C.H. DePuy, V.M. Bierbaum, I. Dotan and E.E. Ferguson, J. Phys. Chem. (May 26, 1986).
Velsko, S. and R.M. Hochstrasser, J. Phys. Chem. **89**, 2240 (1985).
Viehland, L.A., S.L. Lin and E.A. Mason, Chem. Phys. **54**, 341 (1981).
Viehland, L.A. and E.A. Mason, J. Chem. Phys. **66**, 422 (1977).
Yardley, J.T., Introduction to Molecular Energy Transfer, Academic Press, New York, 1980.

KINETIC ENERGY DEPENDENCE OF ION-MOLECULE REACTIONS: FROM TRIATOMICS TO TRANSITION METALS

P. B. Armentrout
Chemistry Department
University of California
Berkeley, CA 94720, USA

ABSTRACT. The kinetic energy dependence of a variety of ion-molecule reactions are examined using guided ion beam mass spectrometry. This experimental technique is shown to provide unprecedented detail in reaction excitation functions over an extremely wide kinetic energy range. Fundamental triatomic systems ($A^+ + H_2$, HD, D_2 where A = O, N, C, Si, Ne, Ar, Kr, and Xe) are examined with an eye on understanding the details of variations in the reaction excitation functions. These include the effects of thermochemistry, electronic degeneracy, adiabatic versus diabatic potential energy surfaces, and spin-orbit coupling. This insight is then applied to the reactions of hydrogen with atomic transition metal ions in specific electronic states. Several diabatic "rules" of reactivity become evident from these studies. These rules are further examined in more complex systems, the reactions of atomic metal ions with alkanes. Finally, studies of metal dimer and cluster ions are discussed with an emphasis on the effects of internal excitation.

INTRODUCTION

One of the truly powerful features available in the study of ion-molecule chemistry is the ease with which the kinetic energy of the reactants can be altered over a very wide range. A comparable capability in neutral chemistry is achieved over much narrower energy regions with only Herculean efforts. This ability is a very useful tool for probing the potential energy surfaces of reaction systems. In theory, details of the kinetics, dynamics, and thermodynamics of the system can be elucidated. In practice, the utility of this ability has seen surprisingly limited use. In this article, we explore some of the variations one can encounter in examining the kinetic energy dependence of ion-molecule reactions. The systems considered begin with simple atom-diatom (triatomic) systems and range to fairly complex transition metal systems. It will be seen that this progression is an important one. By understanding the simple systems, more information can be garnered from the complex systems than would otherwise have been possible.

P. Ausloos and S. G. Lias (eds.), Structure/Reactivity and Thermochemistry of Ions, 97–164.

EXPERIMENTAL

All experiments described in this article were performed using our guided ion beam tandem mass spectrometer. This apparatus has been described in detail.[1] Briefly, ions are formed in one of several sources, described below, extracted and focused into a magnetic sector for mass analysis. The analyzed beam is decelerated to a desired kinetic energy and focused into an octopole ion beam guide which passes through a collision cell filled with the neutral reactant. Pressures of this gas, measured with an MKS Baratron capacitance manometer, are kept sufficiently low that single collision conditions dominate. Product and reactant ions drift from the gas cell to the end of the octopole where they are extracted and focused into a quadrupole mass filter. After mass analysis, ions are detected using a secondary electron scintillation ion detector (Daly type)[2] and counted using standard pulse counting electronics. A DEC MINC computer sweeps the kinetic energy of the ion beam while monitoring the reactant ion and all product ions so that extensive signal averaging can be performed easily.

Ion intensities are converted to absolute reaction cross sections as described previously.[1] We estimate that uncertainties in the absolute cross sections are ±20% while relative uncertainties are ±5%. It is difficult to determine whether there are systematic errors in the absolute cross section measurements made with this apparatus since these are among the most precise measurements ever made. Comparison with calculated capture rate constants (see discussion of $O^+ + H_2$ system below) suggest that the accuracy is very good. Comparisons to thermal rate constant measurements also indicate that our measurements suffer from no serious systematic errors.

The octopole ion beam guide, first developed by Teloy and Gerlich,[3] is the primary key to the unique capabilities of this apparatus when compared with more conventional tandem mass spectrometers. This device comprises eight rods held in a cylindrical array. Alternate phases of an rf potential are applied to alternate rods such that a potential well in the radial direction is established. Energies along the axis of the octopole are perturbed very little. No consequences resulting from such a perturbation have been observed. This is because the potential well, which depends on the inverse sixth power of the octopole radius, is quite flat at the bottom but has steep sides. This contrasts with a quadrupole field, such as used in triple quad analytical instruments, which varies as the inverse square power. While higher order n-pole fields have even better characteristics for chemical physics experiments, the octopole field appears to be an ideal compromise when problems associated with machining and alignment are considered.

One of the very pleasant virtues of the octopole is the ability to use retarding field analysis to measure the absolute energy of the ion beam and its distribution. Conventional retarding field analysis is beset by several problems. One, the analysis region and the interaction region are physically distinct and therefore have different energies due to contact potentials, space charge effects and focusing aberrations. Two, as ions are decelerated, they are defocused and lost even though they may have non-zero kinetic energies in the forward direction. The octopole avoids both these problems since now the analysis region and the interaction region are physically the same and the trapping field of the octopole prevents the loss of ions even at very low kinetic energies. We have verified the ability of the octopole to accurately measure kinetic energies by using time-of-flight (TOF) analysis. Within the 0.15 eV uncertainty of the TOF analysis, the measurements agree to 0.1 eV. We conservatively quote this 0.1 eV figure as the absolute uncertainty in our laboratory energy scale. However, a more severe test of the accuracy of our kinetic energy is provided by the energy dependence of ion-molecule reaction cross sections. In every case where careful measurements have been taken (see for example the reaction system $O^+ + H_2$, below), deviations from the expected energy behavior are not observed. This indicates that our energy scale is probably accurate to better than 0.05 eV in the laboratory frame.

Laboratory energies are converted to center of mass (CM) energies using the stationary target assumption. Thus, $E(CM) = E(lab) \times m/(M+m)$ where m and M are the masses of the neutral and ionic reactants, respectively. At the very lowest energies, this approximation is not adequate since the ion energy distribution is being truncated. The means of correcting for this has been discussed in detail.[1] E(CM) does not include the thermal motion of the reactant neutral. These molecules have a Maxwell-Boltzmann distribution of velocities at 305 K, the gas cell temperature. The effects of this motion can be severe and will obscure sharp features in the true excitation functions. This so-called Doppler broadening[4] and the ion beam energy distribution are explicitly considered by convoluting a trial function for the energy dependent cross sections (detailed below) with these experimentally known energy distributions.[5] Comparison to the data is made after this convolution and the trial function adjusted using a least squares analysis.

Three types of ion sources are used most routinely in our experiments: surface ionization (SI), electron impact (EI) ionization (with either a crossed or in-line[6] geometry) and EI combined with a drift cell (EI/DC). In some cases, several of these sources are used in order to vary the degree of internal energy of the ion. The SI source[7] is useful for production of low ionization potential species such as metal atoms. Ions are produced with a Maxwell-Boltzmann distribution of internal states at the temperature of the ionizing filament, usually about 2200 K. EI[7] is useful for ionization and fragmentation of volatile gases. It produces extensive amounts of internally excited species especially at elevated electron energies (Ee). Some control over the extent of excitation can be exercised by variation of Ee. The drift cell allows further control of the excitation. Here, ions produced by EI are injected into a 2 cm long cell filled with a bath gas. Drift rings establish a field which draws the ions to the exit of the cell. Under typical conditions, ions undergo several thousand collisions with the bath gas. This generally is sufficient to relax all excited states. Alternatively, a reactant gas can be introduced in the cell such that the injected ions react to produce a new ionic species or a specific state of the injected ion beam is selectively depleted.

THEORETICAL

Energy Behavior of Ion-Molecule Reactions

As discussed by Doug Ridge elsewhere in this volume, exothermic ion-molecule reactions are often observed to proceed at the capture rate. For molecules with dipoles, the calculation of the capture rate is an ongoing area of discussion. For polarizable molecules without dipole moments, the energy dependent capture cross section for a reaction proceeding from reactants (channel 1) to products (channel 2) is $\sigma_{12}(E)$. This is given fairly accurately by the Langevin-Gioumousis-Stevenson (LGS) equation,[8]

$$\sigma_{LGS} = \sigma_{12}(E) = \pi e(2\alpha_1/E_1)^{1/2} \tag{1}$$

where e is the electron charge, α_1 is the polarizability of the neutral molecule in channel 1, and E_1 is the relative kinetic energy of the species in channel 1. Deviations from this behavior abound as discussed by Henchman.[9]

One reason that deviations occur is that the reaction may be endothermic. As a first guess to the expected behavior for an endothermic ion-molecule reaction, microscopic reversibility can be applied to eq 1.[10,11,12] This yields the cross section for proceeding from channel 2 to channel 1 as

$$\sigma_{21}(E) = \sigma_0(E_2 - E_0)^{1/2}/E_2 \tag{2}$$

where E_0 is the endothermicity of the reaction, E_2 is the relative kinetic energy of the species in channel 2, and

$$\sigma_0 = \pi e(2\alpha_1)^{1/2}(\mu_1 g_1/\mu_2 g_2) \tag{3}$$

where μ is the reduced mass of the species in the specified channel and g is the electronic degeneracy of the specified channel. While it would be nice if eq 2 did accurately reflect the energy dependence of all endothermic ion-molecule reactions, experimentally this is not the case. Deviations from this form are expected for a number of reasons including those explaining deviations from the LGS form for exothermic processes.

Alternate procedures for determining the threshold behavior of ion-molecule reactions (or any reactions for that matter) include trajectory calculations, statistical theories, and empirical theories. Trajectory calculations potentially provide the most direct comparison between theory and experiment. They are not generally useful, however, because they require a very good potential energy surface which is rarely available. More useful is the application of statistical theories, e.g., phase space theory (PST)[13] or transition state theory (TST).[14] These theories require the molecular constants for all species in every reaction channel and, in the case of TST, for the intermediates as well. Dynamics are not explicitly included although conservation of angular momentum and energy is required. In our work we have made extensive use of PST and found it to be fairly useful in describing the behavior of several systems.[7,15-18] These comparisons make it clear that angular momentum effects are very important for bimolecular reactions.

The final method of describing the threshold behavior of endothermic reactions is to develop empirical theories (or more precisely, models) of this behavior. Such models can be quite general. A well-crafted model is particularly useful because it can often lead to physical insight which is easily lost in more thorough and thus complex theories. In our work, the general model used is

$$\sigma_{21}(E) = \sigma_0(E_2 - E_0)^n/E_2^{\,m} \tag{4}$$

where σ_0 need not be given by eq 3 but is merely a scaling factor. Note that this form includes eq 2 as a particular example. Indeed, most theoretical predictions of the threshold behavior can be expressed in this form.[19] Unfortunately, for even the simple atom-diatom case, there are many theoretically predicted values of n and m. No one form emerges as <u>the</u> threshold dependence. This is as it should be since experimentally, this variation is also observed. No one set of values for n and m works for all systems. Hence, in applying eq 4 to cross section data, the values of σ_0, n, m, and, if unknown, E_0 are treated as adjustable parameters. This procedure has proven to be quite adequate for reproducing the data very accurately over broad energy ranges. In most cases, the thermochemistry derived via E_0 compares favorably with any independent measures of the threshold energy.

Exceptions to this latter statement can easily arise if there is an activation barrier in excess of the endothermicity. This is, of course, equivalent to there being a barrier to reaction in the exothermic direction. Since the long-range ion-induced dipole potential is often sufficiently strong to overcome small barriers,[20] exothermic ion-molecule reactions are often (though not always) observed to proceed without an activation energy. The converse is also apparently true. Endothermic ion-molecule reactions are often (though not guaranteed) to proceed once the available energy exceeds the thermodynamic threshold.

At high kinetic energies, the consequences of energy disposal among the products need to be accounted for. In the general atom-diatom reaction,

$$A + BC \rightarrow AB + C \tag{5}$$

the excess energy available to the products must be in translation or in internal modes of AB. At sufficiently high reactant kinetic energies, the internal energy of AB can exceed the dissociation energy of this diatom such that process 6 occurs.

$$A + BC \rightarrow A + B + C \tag{6}$$

This so-called indirect collision induced dissociation has a thermodynamic threshold equal to D(BC), the bond energy of BC. As we shall see, cross sections for formation of AB in process 5 are often observed to decline rapidly above D(BC) due to the onset of process 6. This decline can be delayed in energy if the reaction dynamics tend to place much of the excess energy in product translation. In more complex systems (where A, B, or C represent chemical groups instead of atoms), this decline in the cross section for process 5 can also be delayed if the product C carries away energy in internal modes. This is often the result if C is a large fragment with low frequency vibrations and AB is a small species with only high frequency vibrations (such as a diatomic hydride). Complex systems also exhibit cross sections which decline before process 6 is thermodynamically accessible. This interesting result indicates that one of the intermediates formed during production of AB is being depleted by an alternate reaction channel. In such circumstances, the peak in the cross section for AB usually correlates with the onset of some other reaction channel.

Inter- and Intra-molecular Isotope Effects

One of the tools which we have used extensively in the studies discussed below is examination of the inter- and intra-molecular isotope effects, specifically with regards to reactions with H_2, HD, and D_2. In particular, the branching ratio between reactions 7a and 7b,

$$A^+ + HD \rightarrow AH^+ + D \quad (7a)$$

$$\rightarrow AD^+ + H \quad (7b)$$

appears to be quite sensitive to the reaction dynamics. For endothermic reactions, three general classes of behavior have been observed: 1) the branching ratio is near unity, 2) process 7a is favored by a factor of about 3 or 4, and 3) process 7b is favored by a very large factor. Coupled with these observations are the fact that for classes 1 and 2 the reaction thresholds for H_2, HD, and D_2 are all the same given the differences in zero point energy. This implies that the thermodynamic threshold is being observed. In contrast, for class 3, large shifts are observed in the thresholds for reactions 7a and 7b compared to those for H_2 and D_2. Clearly, the thermodynamic thresholds are not observed here.

The first type of intramolecular isotope effect, class 1, is approximately the effect expected for a statistically behaved system, i.e., if all degrees of freedom are in equilibrium. This can be seen by examining the density of states for the products of reactions 7a versus 7b. The following arguments presume that the mass of A greatly exceeds the mass of H and D, a reasonable approximation for everything but He. The density of internal states favors production of AD^+. In the classical limit, the density of vibrational states is $1/\hbar\omega$. Since $\omega = (k/\mu)^{1/2}$ where $\mu(AD^+) \approx 2$ and $\mu(AH^+) \approx 1$, this favors AD^+ by a factor of $2^{1/2}$. The classical density of rotational states is 1/hcB where $B \propto 1/\mu$ such that AD^+ is favored by a factor of 2. The density of translational states is proportional to $m^{3/2}$ where m is the reduced mass of the reactant or product channel. Here, $m(AD^+ + H) \approx 1$ while $m(AH^+ + D) \approx 2$ such that this favors AH^+ by a factor of $2^{3/2}$. Overall, these factors cancel such that the classical statistical isotope effect is about 1:1 formation of AH^+ and AD^+. This simple treatment ignores quantum effects but does capture the essence of a statistically behaved system. More detailed calculations using phase space theory bear this out by giving a branching ratio near 1:1.

In class 2 behavior, process 7a is favored by a factor which would appear to be a simple mass factor. In the analysis above, it is the internal density of states that favors process 7b. The obvious means to enhance AH^+ production is to neglect this contribution. The internal and translational degrees of freedom no longer equilibrate and the reaction is no longer sensitive to the density of internal energy states. In other words, the reaction is direct. Note that this favors production of AH^+ by a factor of $2^{3/2} = 2.8$. We have previously pointed out that an equivalent way of thinking about this effect is in terms of angular momentum conservation.[7,19,21] This analysis also concludes that the reaction must be direct.

Class 3 behavior is a most unusual effect. Qualitatively, the explanation for this isotope effect is that the energy relevant to reaction 5 is not the center of mass (CM) energy but a "pairwise" interaction energy. In the CM system, the energy available to cause chemical change is the relative kinetic energy between the incoming atom of mass A and the reactant molecule with mass (B + C). Hence, $E(CM) = E(lab) \times (B + C)/(A + B + C)$ where E(lab) is the energy of the ion in the laboratory frame. In a pairwise interaction, A is sensitive only to the potential between A and the atom B which is transferred in reaction 5. Hence, the pairwise energy for transfer of B from molecule BC is $E(B, BC) = E(lab) \times B/(A + B)$. In situations where $A >> B$ or C, $E(B, BC) \approx E(CM) \times B/(B + C)$ such that the energy available for chemical change in the pairwise frame is always less than in the CM frame. For reaction where BC = H_2 or D_2, the factor B/(B + C) is 1/2; while for BC = HD, it is 1/3; and for BC = DH, it is 2/3. This means that if the thermodynamic threshold occurs at E_0, the pairwise threshold will occur at a CM energy of 2 E_0 in reaction with H_2 or D_2; at 3 E_0 for reaction 7a; and at 1.5 E_0 for reaction 7b. Thus, the enhanced production of AD^+ in process 7b is due the lower apparent threshold for this reaction compared to that for reaction 7a. This pairwise scheme also readily explains the shifts in threshold observed for the H_2, HD, and D_2 systems. The extraordinary effect will become much more evident in the examples given below.

It should be noted that this pairwise energy frame may be familiar from the "spectator stripping" model (SSM).[22] The SSM is a highly specific example of a model which incorporates the pairwise energy concept. The difference is that the SSM allows no momentum transfer to the product atom C while the more general pairwise model does not forbid such a transfer. Thus, the SSM makes a specific prediction about the velocity and internal energy of the products while the pairwise model allows for distributions of these quantities. The latter, not surprisingly, corresponds most readily to observation. The point of the exercise above is that the pairwise energy scale, an extremely useful concept, can easily be derived without the severe assumptions made in the SSM.

FUNDAMENTAL SYSTEMS

In this section, we examine several chemically simple reactions,

$$A^+ + H_2 \rightarrow AH^+ + H \tag{8}$$

where A = O, N, C, Si, and the rare gases, Ne, Ar, Kr, and Xe. Most of these reactions have been studied using other techniques and there are often theoretical calculations as well. This wealth of information provides us with an opportunity to understand the dynamics of these systems in detail. In addition, we can critically evaluate the theoretical and empirical models used to analyze the kinetic energy dependence of these systems.

Several terms which will be used extensively in the following discussion are adiabatic, diabatic, and non-adiabatic. At times there seems to be a fair amount of confusion over these very useful terms. In this paper, we use the term diabatic to refer to a potential energy surface (PES) which evolves from reactants to products with an unchanged electron configuration (atomic orbital population). Diabatic PESs can and often do cross one another. If such surfaces have the same spin and the same orbital symmetry, then the crossing is avoided and the surfaces combine to form two new adiabatic PESs. Under ordinary conditions at thermal energies, reactive systems generally evolve along adiabatic surfaces. At high kinetic energies, nuclear motion may be rapid enough that electrons do not relax to the lowest energy surface. Thus, the system behaves diabatically. This is one type of non-adiabatic effect. Spin-orbit mixing and other effects may also induce non-adiabatic behavior in a reacting system.

Molecular Orbital Considerations

One of the features common to all the systems which will be discussed in this section is that the valence orbitals on the reactant ion are p orbitals. A great deal of insight into these reactions can be obtained by using simple molecular orbital arguments.[12,23] Figure 1 shows the dominant interactions which might be expected for approach of a p_x, p_y, and p_z orbital up to a hydrogen molecule in C_{2v} symmetry. The p_x and σ_u orbitals both have b_2 symmetry and thus combine into bonding and anti-bonding molecular orbitals (MOs). Since the σ_u orbital of H_2 is empty, occupation of the p_x orbital leads to occupation of the bonding b_2 orbital. This leads to an attractive bonding interaction between A^+ and H_2. The p_y orbital is out of the plane of the three reactant atoms and therefore has no interaction with H_2. Occupation of the p_y therefore is neither attractive nor repulsive but merely non-bonding. The p_z mixes with the σ_g orbital of H_2 since both have a_1 symmetry. Again bonding and anti-bonding MOs are formed. Since there are two electrons in the σ_g orbital, occupation of the p_z leads to occupation of the anti-bonding a_1* MO. Thus, this leads to a repulsive interaction between A^+ and H_2.

In $C_{\infty v}$ symmetry, the situation changes somewhat. Now, both the p_x

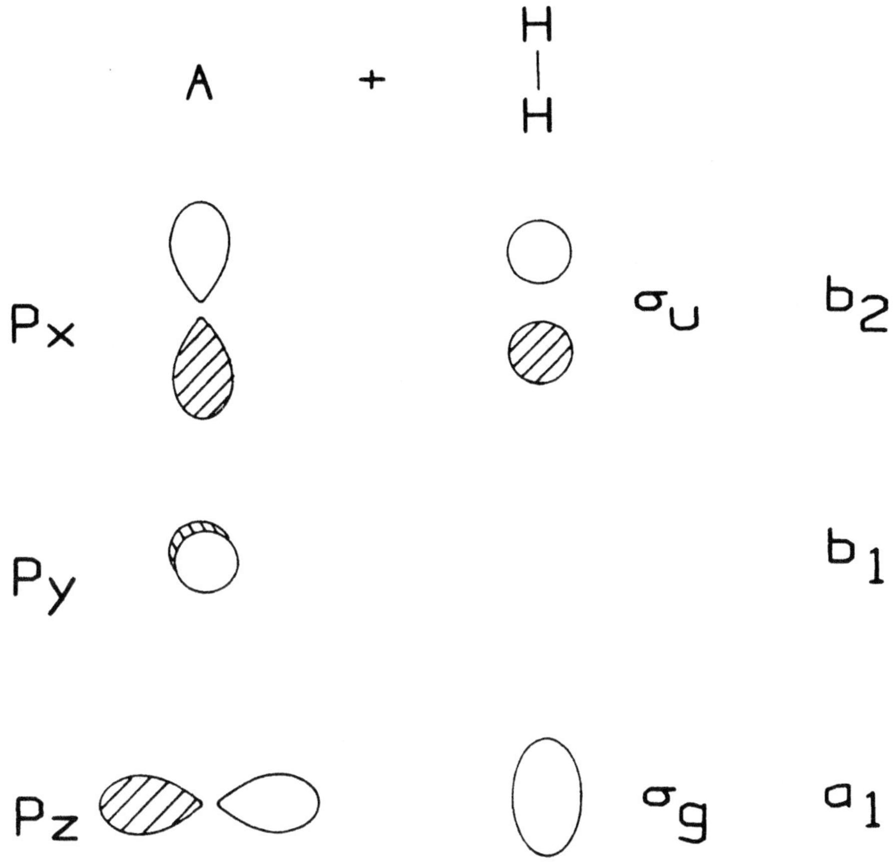

Figure 1. Orbital interactions for the p orbitals of atom A with the molecular orbitals of H_2 in C_{2v} symmetry. Symmetry groups are indicated on the right.

and p_y are non-bonding. The p_z orbital can interact with both the σ_g and σ_u MOs of H_2 such that three new MOs are formed: one is bonding, another is largely non-bonding, and the third is anti-bonding. Since there are two electrons in the σ_g orbital, the bonding orbital is always fully occupied. Occupation of the atomic p_z orbital therefore leads to occupation of the non-bonding MO. This results in an interaction which is less repulsive than that resulting from occupation of the anti-bonding a_1* MO in C_{2v} symmetry but is still less attractive than if the p_z orbital (and hence, the non-bonding MO) were empty.

The consequences of these general ideas depends on the specific occupations of all the orbitals on the reactant atom. These will be detailed for each system below. Qualitatively, we will find that the dominant effects in the entrance channel can be understood using these ideas.

$O^+(^4S) + H_2$

One of the difficulties of examining the reactions of atomic oxygen ions is that excited electronic states are easily produced. In this study, atomic oxygen ions are produced by using the EI/DC source. Ions are first generated by electron impact ionization and fragmentation of CO_2. These are passed through the drift cell filled with molecular nitrogen. Excited state O^+ rapidly reacts with N_2 by charge transfer while ground state $O^+(^4S)$ reacts very slowly.[24] The O^+ emerging from the drift cell is found to have less than 0.1% excited states as determined by examination of the charge transfer reaction with N_2 in the main reaction chamber.[25]

Reaction 8 with $A^+ = O^+(^4S)$ is exothermic by 0.6 eV.[26] The cross section for this process, shown in Figure 2, can be seen to have three

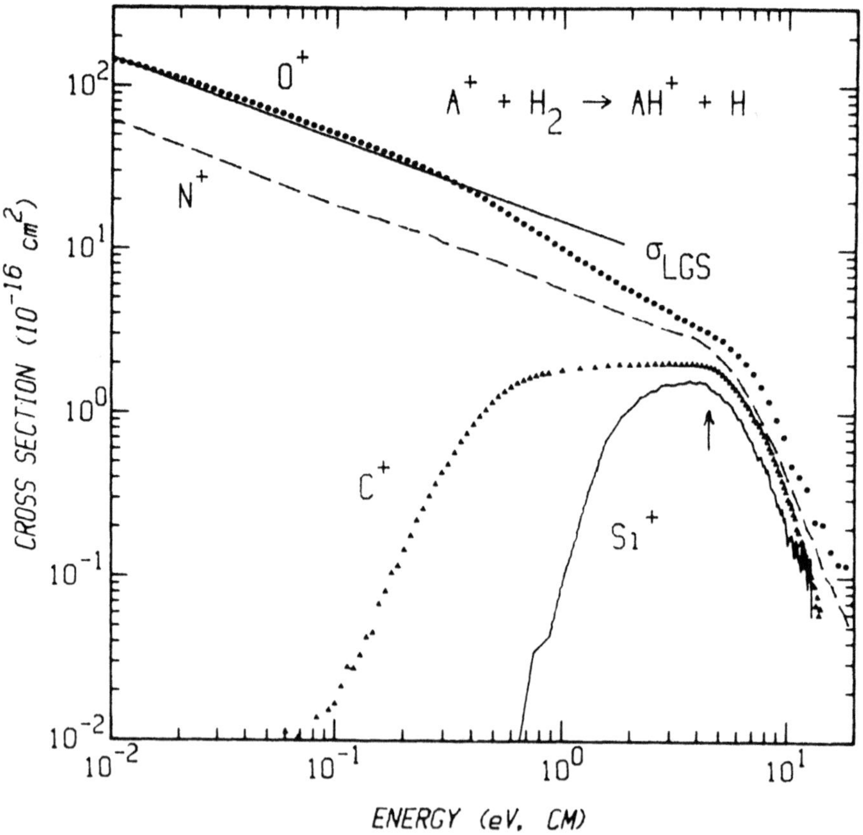

Figure 2. Cross sections for reaction of $O^+(^4S)$, $N^+(^3P)$, $C^+(^2P)$, and $Si^+(^2P)$ with H_2 as a function of relative kinetic energy. The line labeled σ_{LGS} is calculated by using eq 1. The arrow marks the bond energy of H_2 at 4.5 eV.

distinct energy regimes. At the lowest energies, <0.25 eV, the cross section decreases as $E^{-1/2}$ as predicted by eq 1. Indeed, as shown in Figure 2, the absolute agreement between the experimental data and the LGS cross section [$\alpha(H_2)$ = 0.7894 $Å^3$][27] is superb (better than 5%). While not shown, results for HD and D_2 exhibit identical behavior (within experimental error limits) and agree equally well with the absolute LGS cross section. To achieve this agreement, the comparison of the H_2, HD, and D_2 systems must be made on a center of mass (CM) energy scale. For the three systems combined, the data averages to 0.996 ± 0.055 that of the calculated LGS cross sections at low energies. At the present time, this is certainly one of the best (perhaps the only) example of an experimental reaction cross section which behaves precisely as predicted by eq 1.

At energies above 0.3 eV, the cross section deviates from the LGS prediction such that the reaction efficiency drops. In the region up to about 5 eV, the cross section falls approximately as E^{-1}. This is true for all three isotopic systems. The explanation for this behavior lies in the fact that this reaction occurs via a direct mechanism. Experimentally, this has been shown by crossed beam studies[28] but the molecular orbital ideas outlined above demonstrate this as well. First note that while H_2O^+ is a stable molecule, it has a $^2B_1$ ground state. Thus, access to the deep potential well corresponding to this state is spin-forbidden from the $O^+(^4S)$ + $H_2(^1\Sigma_g^+)$ reactants. The qualitative character of the quartet surfaces evolving from these reactants can be gleaned by examining the occupancy of the O^+ atomic orbitals. For $O^+(^4S)$, this is $(1s)^2(2s)^2(2p)^3$ such that the p_x, p_y, and p_z are all singly occupied. In C_{2v} symmetry, the potential energy surface is repulsive due to occupation of the p_z orbital. This repulsion is relieved by going to a collinear approach, $C_{\infty v}$ symmetry. _Ab initio_ calculations verify these qualitative ideas and establish that for a collinear approach there is no barrier to reaction.[29] This, of course, is consistent with the observed behavior at low energies. Apparently, at low kinetic energies, the O^+ ion and H_2 molecule have time to orient into the most favorable geometry. This ability to orient is probably enhanced by the existence of the ion-induced dipole bound $O^+ - H_2$ complex. As the kinetic energy is increased, the time available to orient decreases (or equivalently the lifetime of the $O^+ - H_2$ complex decreases) such that the reaction efficiency falls off.

At still higher energies, the cross section drops more rapidly, $\approx E^{-3}$. This is due to the dissociation of the product ion in process 6. The thermodynamic threshold for this reaction is 4.5 eV, the bond energy of H_2.[30] The observed onset of this decline is somewhat higher, about 6 eV. This suggests that some of the energy available to the products is preferentially placed in translation. This is consistent with a direct mechanism for the reaction. The HD and D_2 systems also show this high energy behavior and have similar onsets for process 6.

$N^+(^3P) + H_2$

The cross section for reaction 8 with $A^+ = N^+(^3P)$ is also shown in Figure 2. These ground state atomic nitrogen ions are formed in the drift cell by the charge transfer reaction,

$$He^+ + N_2 \rightarrow N^+(^3P) + N + He \quad (9)$$

This reaction does not have sufficient energy to form excited states of N^+.[26] Note that the behavior of this cross section is similar to that for O^+ at low energies but the magnitude is less. The high energy behavior is comparable, consistent with a similar onset for product dissociation in process 6.

The explanation for the decrease in magnitude returns to the molecular orbital ideas. The atomic orbital occupation of $N^+(^3P)$ is $(1s)^2(2s)^2(2p)^2$ such that there is one empty 2p orbital. In both C_{2v} and $C_{\infty v}$ symmetries, the most favorable geometry is to have the p_z orbital empty. This leads to an attractive surface having $^3A_2$ or $^3\Sigma^-$ symmetry. Occupation of the p_z orbital leads to repulsive surfaces of $^3B_2$ or $^3\Pi$ symmetry for an empty $2p_y$ and $^3B_1$ or $^3\Pi$ symmetry for an empty $2p_x$. Ab initio calculations[31] again verify these qualitative ideas. The final result is that of the three triplet surfaces which evolve from $N^+(^3P) + H_2(^1\Sigma_g^+)$, only one is attractive in the entrance channel and eventually leads to products. In fact, the absolute cross section for reaction 8 is close to 1/3 of the LGS cross section at low energies, Figure 2. This observation suggests that as the N^+ and H_2 approach on one of the three surfaces, they get locked onto it with a random probability. Apparently, non-adiabatic effects which would allow reactants on one the repulsive surfaces to relax to the attractive surface are inefficient.

Another difference between O^+ and N^+ systems concerns the isotope effects. Unlike the O^+ system where the total cross sections are identical for H_2, HD, and D_2, the N^+ system clearly shows differences, Figure 3. This system has seen a flurry of recent activity,[32-34] all of which establishes that reaction of $N^+ + H_2$ is slightly endothermic. Thus, the differences in reactivity observed in Figure 3 are due to differences in the zero point energies for the various reactions. Relative to reaction 8, reaction of N^+ with D_2 is more endothermic by 31 meV; reaction 7a is more endothermic by 36 meV; and reaction 7b is less endothermic by 12 meV.[30] The shapes of the cross sections at low energies are consistent with these relative energetics. Preliminary analysis finds that these four cross sections can be described reasonably well by using eq 2 with a 0 K endothermicity for the H_2 reaction of 15-35 meV. The exact result depends on the precise way in which we account for the rotational energy of the hydrogen molecule. Overall, the consensus of all experiments is that reaction between N^+ and H_2 is endothermic by 10-30 meV. It is interesting to note that until 1985, it was not even known whether or not this reaction was exothermic. Now it probably has one of the best known heats of reaction.

Several interesting questions remain to be answered before this system is completely characterized. Why do small discrepancies in the experimentally determined endothermicities exist? What is the role of H_2 rotational energy? Do the reactivities of the spin-orbit states of $N^+(^3P_0 = 0$ meV; $^3P_1 = 6$ meV; $^3P_2 = 16$ meV$)$[35] differ? What is the electronic state of the NH^+ product? (This is not known unambiguously

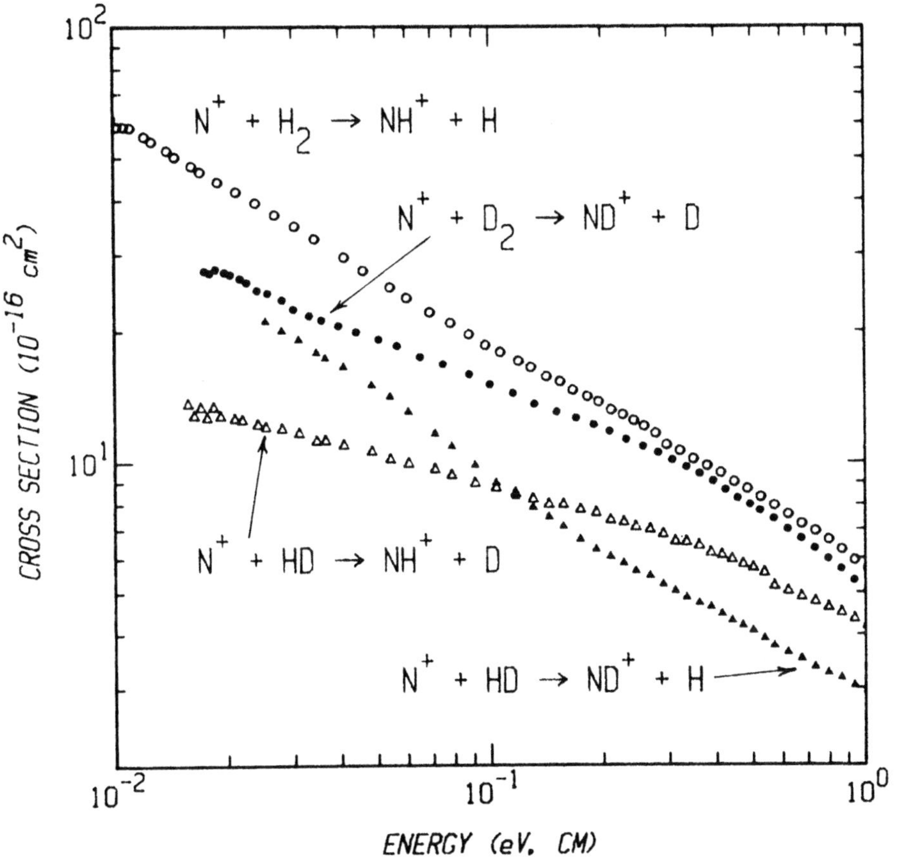

Figure 3. Cross sections for reaction of $N^+(^3P)$ with H_2, D_2, and HD at low relative kinetic energies. The data for the H_2 reaction is the same as that shown in Figure 2.

since the ground state, $^2\Pi$, and the first excited state, $^4\Sigma$, are separated by only 62 meV.[30]) Does the quadrupole moment of H_2 have any influence on the reactivity at low energies and temperatures?[1]

$C^+(^2P) + H_2$

Reaction 8 with $A^+ = C^+(^2P)$ is probably the best studied endothermic ion-molecule reaction. The endothermicity of 0.398 ± 0.003 eV is known extremely well from spectroscopic data.[36] Despite the wealth of data on this reaction and its extreme importance in interstellar chemistry, the kinetic energy dependence for this reaction was not well characterized until recently.[36] In this work, ground state $C^+(^2P)$ was generated by electron impact of a high pressure of CO in the drift cell. The cross section measured is shown in Figure 2. The reaction is clearly endothermic with an extended plateau followed by a decline beginning at $D(H_2)$. The apparent threshold is substantially lower than the known endothermicity, a result of the thermal motion of the H_2 reactant gas. The true threshold behavior is an excitation function which rises steeply from the thermodynamic threshold, Figure 4, and can be modeled using several forms of eq 4. While eq 2 clearly fails to reproduce this excitation function, a similar function, $n \approx 0.7$ and $m = 1.0$, works well. This implies, not unreasonably, that the centrifugal barrier in the exit channel is a dominant constraint in this reaction. This analysis shows unequivocally that there is no activation barrier in excess of the endothermicity for reaction of C^+ with H_2.

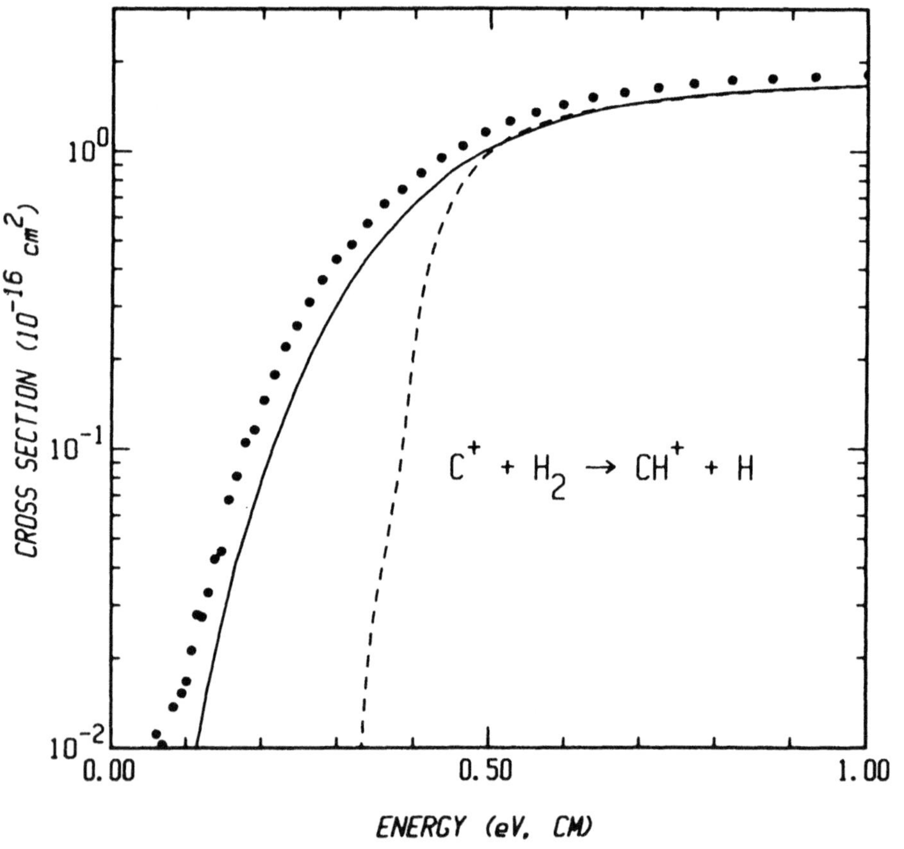

Figure 4. Cross section for reaction of $C^+(^2P)$ with H_2 at low kinetic energies. The dashed line shows a phase space calculation for this process. The full line is this calculation convoluted with the experimental energy broadening (see text).

As noted above, phase space theory (PST) provides another approach to analyzing the kinetic energy dependence of ion-molecule reactions. Indeed, an early test of PST found it to describe the major features of the $C^+ + H_2$ system quite well.[37] Comparisons to the present data, shown in Figure 4, continue to find that PST (with no adjustable parameters) describes this system exceptionally well.[16] Clearly, the shape of the cross section is reproduced very well and the absolute magnitude of the calculation is within 15% of the experimental results (which have a ±20% uncertainty). This agreement is sufficiently good that it is surprising that a statistical theory reproduces the behavior of an atom-diatom reaction at elevated energies as well as it does here. This agreement has been attributed to the very deep potential well corresponding to the CH_2^+ intermediate.

The input parameters to the PST calculation include the molecular parameters for reactants and products, all of which are well known, and the electronic degeneracies of the reactants and products. The agreement between PST and the data shown in Figure 4 is obtained only when 1/3 of the reactant $C^+ + H_2$ surfaces are presumed to lead to products. As for the N^+ system, this factor emerges from the molecular orbital analysis. $C^+(^2P)$ has a $(1s)^2(2s)^2(2p)^1$ electron configuration. If the single 2p electron resides in the p_x, an attractive surface ($^2B_2$ or $^2\Pi$) results. If the p_z is occupied, a repulsive surface of $^2A_1$ or $^2\Sigma$ symmetry results. If the p_y is occupied, a fairly flat surface ($^2B_1$ or $^2\Pi$) is anticipated. Ab initio calculations verify these ideas and show that the $^2B_1$ surface is repulsive at short internuclear distances. These calculations further show that the attractive $^2B_2$ surface crosses with a $^2A_1$ surface which leads to the deep potential well of ground state CH_2^+. This crossing is avoided in C_s symmetry (essentially all collisions). The final result is that only one of the three doublet surfaces evolving from $C^+(^2P) + H_2(^1\Sigma_g^+)$ leads to ground state products, $CH^+(^1\Sigma) + H(^2S)$, for collisions in all symmetries.[16] As in the N^+ system, this is in agreement with observation and again indicates that non-adiabatic effects are not important in this reaction.

In reaction with HD, formation of CD^+ is favored over that of CH^+ by a factor of about 1.4 from threshold until the onset of product dissociation.[36] Part of this is attributable to the fact that CD^+ has a lower zero point energy and therefore a lower threshold by 45 meV. Indeed the precision of our experimental results allows us to measure this zero point energy difference albeit with a rather poor accuracy.[36] Otherwise this isotope ratio is qualitatively comparable to the statistical prediction discussed above. However, neither the simple theory nor PST quantitatively describe the branching ratio between formation of CH^+ and CD^+.[16] This is an indication that, not unexpectedly, dynamics does play a role in this system.

$Si^+(^2P) + H_2$

As noted above, it is believed that the main reasons for the near statistical behavior of the $C^+ + H_2$ system is the deep potential well of CH_2^+, ≈4 eV, coupled with the relatively low endothermicity, 0.4 eV. To test whether these features are indeed requirements for such behavior, we have examined the isovalent reaction with silicon, $A^+ = Si^+(^2P)$ produced by SI.[38] Like $C^+(^2P)$, $Si^+(^2P)$ should easily insert into H_2 on one of three doublet surfaces without activation barrier. Unlike $C^+ + H_2$, the Si^+ reaction proceeds via a SiH_2^+ potential well which is only 0.81 ± 0.06 eV deep[39] and the overall reaction is endothermic by 1.26 ± 0.03 eV.[40]

The experimental results for $Si^+ + H_2$ are shown in Figure 2. Clearly, the endothermicity of this reaction is greater than that for C^+. The peak in the cross section again correlates with $D(H_2)$. The cross section behavior again cannot be represented by eq 2 but is reproduced extremely well by using eq 4 with n = m = 1 and E_0 = 1.245 ± 0.036 eV. This clearly establishes that the reaction proceeds without an activation barrier in excess of the endothermicity. It also demonstrates that accurate thermochemical data can be derived from beam studies of endothermic reactions.

Figure 5 shows the results of a phase space calculation again with no adjustable parameters and with only one third of the reactant surfaces active. The agreement between PST and the data is excellent both with regards to absolute magnitude and dependence on kinetic energy.

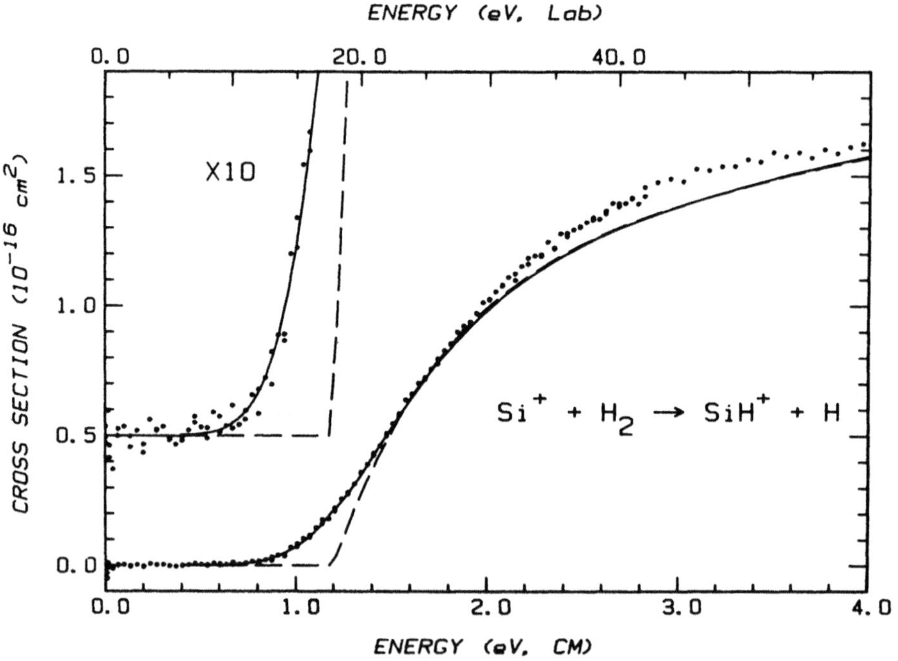

Figure 5. Cross section for reaction of $Si^+(^2P)$ with H_2 at low kinetic energies. The dashed line shows a phase space calculation for this process. The full line is this calculation convoluted with the experimental energy broadening (see text). The inset shows the data and calculated curves on a scale expanded by a factor of 10 and offset from zero.

The statistical-like behavior of this reaction is further substantiated by the reaction of Si^+ with HD.[41] The branching ratio between SiH^+ and SiD^+ is very close to 1:1 throughout the threshold region. Again PST does not quantitatively reproduce this ratio probably for reasons similar to those discussed above.

The observation that both $C^+ + H_2$ and $Si^+ + H_2$ behave "statistically" (although not without dynamic effects) indicates that vibrational periods are quite sufficient for extensive energy randomization in the transiently formed triatomic intermediates. A very deep potential well is not required nor must the intermediates be "long-lived" in the sense of isotropic distributions observed in differential cross sections. This behavior is also a reflection that total cross section measurements such as these involve substantial averaging of molecular orientations and impact parameters which probably serve to wash out dynamic behavior.

Rare Gas Ions + H_2

Rare gas ions (Rg^+) have the valence electron configuration $(ns)^2(np)^5$. Therefore the ground states are ^2P, like C^+ and Si^+. However, the reactivities of the ^2P states of Rg^+ are quite different from that of carbon and silicon ions. The MO arguments show this readily. As noted above, the less electron density there is in the p_z orbital, the more favorable the interaction between the ion and H_2 is in any geometry. Since Rg^+ must occupy the p_z, it is most favorable to singly occupy this orbital. (We will refer to this as the P_z state of Rg^+. Similarly, if the hole is in the p_x or p_y, these will be the P_x and P_y states.) This leads to a repulsive interaction in C_{2v} symmetry which is relieved in $C_{\infty v}$ symmetry. The P_x and P_y states, both having two electrons in the p_z, are much more repulsive in either geometry. Thus, we might expect that the Rg^+ reactions will occur on 1/3 of the available surfaces with a preference for a collinear geometry. Note that these MO considerations lead naturally to a predicted difference in reactivity between ions with less than half filled shells (like N^+, C^+, and Si^+) and those with more than half filled shells (like O^+ and Rg^+). The former can proceed via insertion to form strongly bound AH_2^+ intermediates while the latter react in more direct processes.

A more detailed examination of the $Rg^+ + H_2$ surfaces reveals a complication in these systems not encountered in the systems above. In the reactions of O^+, N^+, C^+, and Si^+ with H_2, the reactants in process 8 readily reach the products on a single adiabatic potential energy surface. For the rare gas ions, however, there is a second important surface in the entrance channel. This is the charge transfer channel, $Rg + H_2^+$. The coupling between these two entrance channel surfaces depends strongly on their relative energies as determined by the relative ionization potentials of H_2 and the rare gas. Thus, the effects of this coupling vary substantially as the rare gas identity changes.

$Ne^+(^2P) + H_2$: Reaction 8 where $A^+ = Ne^+(^2P)$ is exothermic by 5.6 eV.[42] It is therefore surprising to obtain the results shown in Figure 6 for ground state $Ne^+(^2P)$ (produced by low energy EI). Not only is the

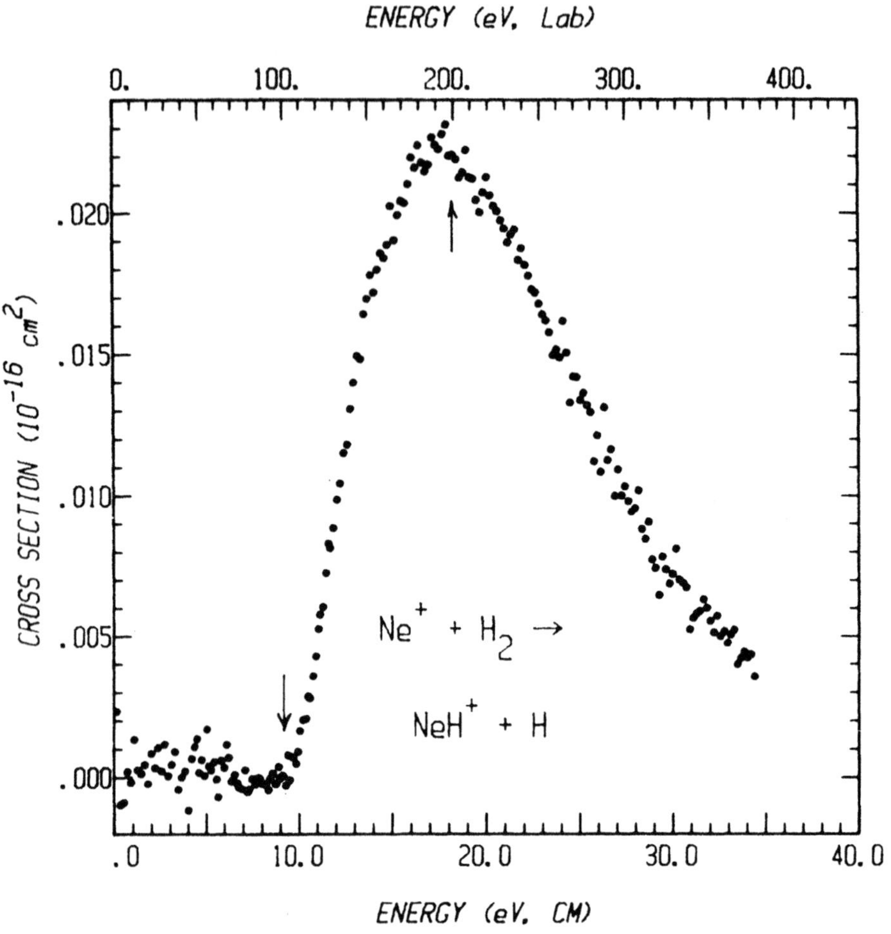

Figure 6. Cross section for reaction of $Ne^+(^2P)$ with H_2 as a function of the relative kinetic energy (lower scale) and the ion energy in the laboratory frame (upper scale). Arrows indicate the pairwise threshold and peak positions discussed in the text.

reaction very inefficient (small cross section) but it appears to be endothermic by about 9 eV. Insight into the Ne^+ reaction can be obtained from the analogous process with $He^+(^2S)$. This reaction, exothermic by 8.4 eV,[42] has been studied in detail by Tiernan and co-workers.[43] It behaves very similarly to the Ne^+ system but has a smaller cross section and an apparent threshold of about 8 eV. Mahan[23] was the first to explain this odd behavior in the He^+ case. He notes that the HeH^+ + H products correlate adiabatically (and diabatically) with He + H_2^+ not with He^+ + H_2. Indeed, the endothermic reaction,

$$He + H_2^+ \rightarrow HeH^+ + H \quad (9)$$

is efficient and easily driven by translational and vibrational energy.[44,45] The situation is comparable for the Ne system.[23] Figure 7 shows the relevant surfaces.

An important clue to the behavior of the $He^+ + H_2$ reaction comes from studies of Tiernan and co-workers[43] who examined the luminescence from this reaction. Extensive Balmer and Lyman emission was observed with an onset for Lyman α emission coincident with the onset for HeH^+ formation. Apparently, in order to form stable HeH^+, the hydrogen atom product must be formed in an excited state. The lowest of these, H(2p), is 10.2 eV above the H(1s) ground state.[35] The location of this surface in the Ne^+ system is shown in Figure 7. It is now clear why the reaction of Ne^+ and H_2 does not behave as a normal exothermic ion-molecule reaction. The process being observed at threshold is really

$$Ne^+(^2P) + H_2(^1\Sigma_g^+) \rightarrow NeH^+(^1\Sigma) + H^*(2p) \quad (10)$$

endothermic by 4.6 eV.[42]

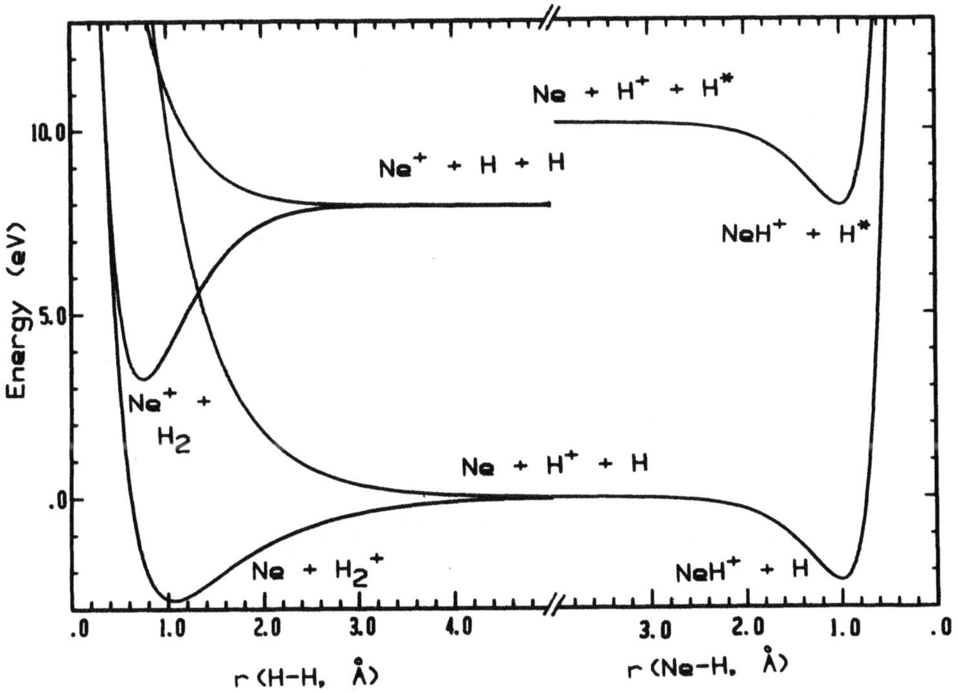

Figure 7. Schematic asymptotic potential energy curves for the [Ne + H + H]$^+$ system. The curves on the left represent the energies as a function of r(H-H) with r(Ne-H_2) = ∞. The curves on the right represent the energies as a function of r(Ne-H) with r(NeH-H) = ∞. The curves for H* are drawn for the energy of H(2p).

We now seek to explain why the observed threshold is at 9 eV and the peak occurs at about 20 eV. Examination of reactions with D_2 and HD help supply the answer. The cross section for reaction with D_2 is identical in shape but about 15% smaller than the cross section shown in Figure 6. The cross section for reaction with HD is shown in Figure 8. A very strong isotope effect is observed, quite unlike anything observed in the systems discussed above. The explanation for this isotope effect is that the energy relevant to this reaction is not the center of mass (CM) energy but the "pairwise" interaction energy discussed above.

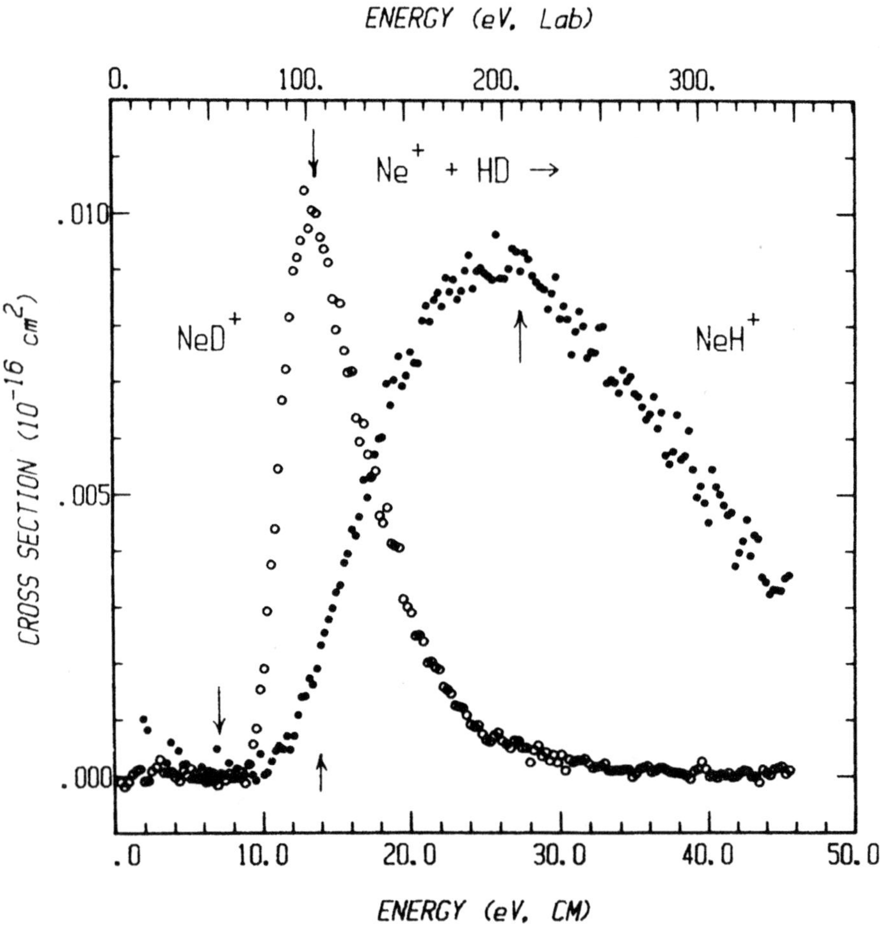

Figure 8. Cross section for reaction of $Ne^+(^2P)$ with HD to produce NeH^+ (closed symbols) and NeD^+ (open symbols) as a function of the relative kinetic energy (lower scale) and the ion energy in the laboratory frame (upper scale). Arrows indicate the pairwise threshold and peak positions discussed in the text.

Thus, the threshold for reaction 10 (or reaction with D_2) is twice the thermodynamic threshold of 4.6 eV or 9.2 eV in the CM frame. For reaction with HD to form NeH^+, the pairwise threshold is 3 × 4.6 = 13.8 eV and to form NeD^+, it is 1.5 × 4.6 = 6.9 eV. These energies are marked by arrows in Figures 6 and 8. The peaks for all products fall at a thermodynamic energy of about 9 eV: 18 eV for H_2 and D_2, 27 eV for HD, and 13.5 eV for DH, Figures 6 and 8. This energy is somewhat higher than the 6.8 eV thermodynamic onset for dissociation to $Ne^+(^2P)$ + H(1s) + H*(2p). While other factors could be involved in the isotope effects observed here, it is clear that the pairwise interaction scheme can explain much of the observed behavior. The observation that the CM energy is not useful in this case tells us that the reaction surface is repulsive. Therefore in collisions between Ne^+ and H_2, three center potential energy terms do not contribute extensively to the overall interaction potential. The collision partners behave approximately like hard-spheres.

$Ar^+(^2P) + H_2$: Reaction 8 with $A^+ = Ar^+(^2P)$ was long considered to be a classic example of an exothermic reaction which proceeded at the LGS collision limit.[9] Our recent study[1] of this system shows that this belief persisted largely because the cross section measurements were insufficiently accurate for a definitive evaluation. In fact, the cross section, shown in Figure 9, clearly deviates from the behavior predicted by eq 1 such that at thermal energies the reaction occurs with an efficiency of two-thirds. This result is in agreement with several rate constant measurements[46,47] and with other guided beam measurements.[3] Note that this disagrees with the conclusions of the simple MO treatment discussed above.

As the kinetic energy increases, the reaction efficiency increases to about 90% near 1.0 eV. Above this energy, the cross section drops as E^{-1} until 4 eV where it falls off rapidly. The cross section for H_2 is identical to this throughout the energy range examined. The total cross section for HD is also comparable except in the region where dissociation of the product becomes important, >4 eV. The onset of dissociation corresponds nicely to that predicted by the pairwise energy scheme outlined above.[1] The branching ratio between ArH^+ and ArD^+ is nearly unity until dissociation. At the highest energies, all three systems show a slight increase in the cross section which we attribute to a process like reaction 10.

These features in the energy dependent cross sections can be understood by examining the reaction surfaces. As for the He^+ and Ne^+ systems, it is the [Rg + H_2^+] surface which diabatically correlates with the [RgH^+ + H] product surface.[1,23] Unlike these systems, however, the [Ar^+ + H_2] surface and the [Ar + H_2^+] surface have nearly the same energy, Figure 10. Thus, as Ar^+ and H_2 approach, these two diabatic surfaces readily mix to form two adiabatic surfaces, Figure 10. The lower of these evolves directly to the desired products. Because the diabatic curves cross at the bottom of the H_2 well, no strong inter- or intra-molecular isotope effects are observed. The decrease in reaction efficiency observed above 1 eV corresponds nicely to a concomitant increase in the charge transfer cross section.[1] This is a reflection of

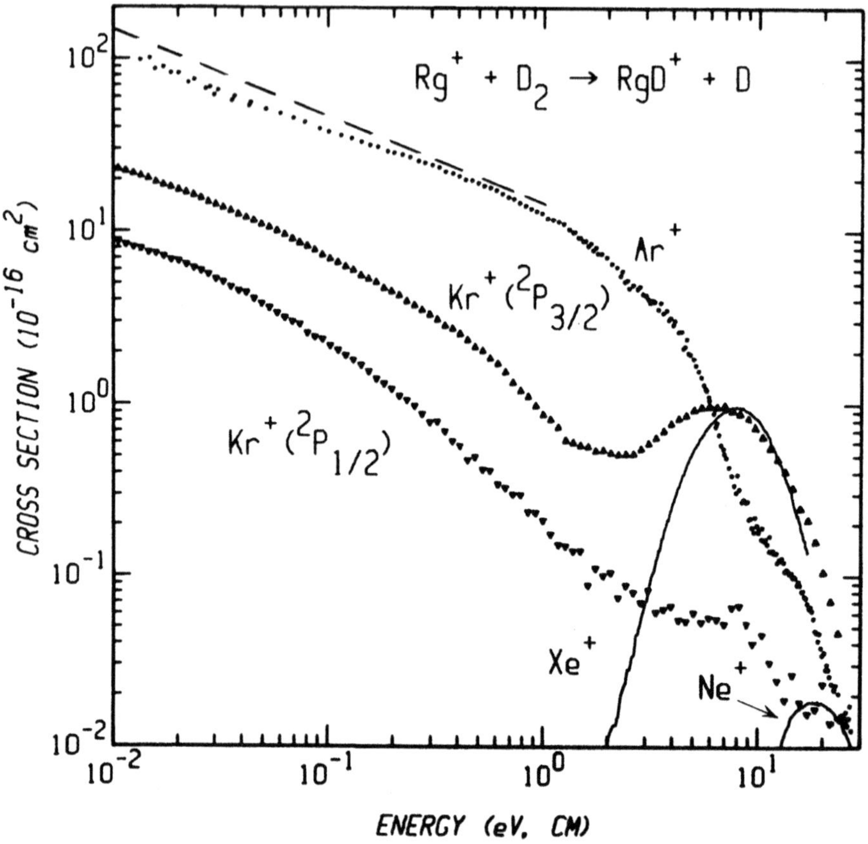

Figure 9. Cross sections for reaction of $Ne^+(^2P)$, $Ar^+(^2P)$, $Kr^+(^2P_{3/2})$, $Kr^+(^2P_{1/2})$, and $Xe^+(^2P)$ with D_2 as a function of relative kinetic energy. The dashed line is calculated by using eq 1.

increasingly diabatic behavior at these elevated kinetic energies.

The final point of interest is the reaction efficiency at low kinetic energies. In our paper on this system,[1] we discussed a number of possible explanations. Since that time we have conducted experiments, described below, on the reactions of H_2 and $Kr^+(^2P)$ in specific spin-orbit states. These results lead us to believe that spin-orbit effects are responsible for the low energy behavior in the Ar^+ system. As shown in Table I, the products, $ArH^+(^1\Sigma) + H(^2S)$, have $m_J = \pm 1/2$ and $^2A'$ symmetry. The reactant states having this quantum number and symmetry are the $m_J = \pm 1/2$ components of $Ar^+(^2P_{3/2})$ and $Ar^+(^2P_{1/2})$. Note that these spin-orbit states are mixtures of the P_x, P_y, and P_z states discussed above and that both have some P_z character. Thus, the spin-orbit coupling mixes these states such that they both react efficiently.

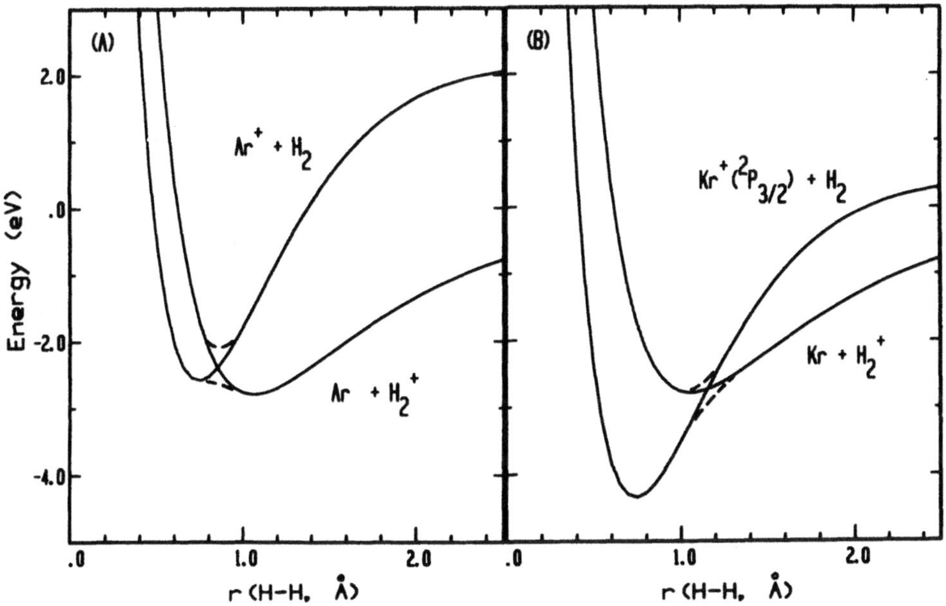

Figure 10. Schematic asymptotic potential energy curves for the $[Ar + H + H]^+$ and $[Kr + H + H]^+$ systems. The full curves represent the energies as a function of r(H-H) with $r(Rg-H_2) = \infty$. The dashed curves are approximate representations of the adiabatic curves at smaller $r(Rg-H_2)$ distances.

Table I. Asymptotic electronic states in rare gas ion - H_2 systems

Asymptotic state	J	m_J	$RgHH^+$ state	C_s sym.
$Rg^+(^2P) + H_2(^1\Sigma_g^+)$	3/2	±1/2	$\sqrt{1/6}(P_x \pm P_y) + \sqrt{2/3}P_z$	$^2A'$
$Rg^+(^2P) + H_2(^1\Sigma_g^+)$	3/2	±3/2	$\sqrt{1/2}(P_x \pm P_y)$	$^2A''$
$Rg^+(^2P) + H_2(^1\Sigma_g^+)$	1/2	±1/2	$\sqrt{1/3}(P_x \pm P_y) - \sqrt{1/3}P_z$	$^2A'$
$Rg(^1S) + H_2^+(^2\Sigma)$	1/2	±1/2		$^2A'$
$RgH^+(^1\Sigma) + H(^2S)$	1/2	±1/2		$^2A'$

In contrast, the $m_J = \pm 3/2$ components of $Ar^+(^2P_{3/2})$ have no P_z character and the wrong symmetry, $^2A''$. They therefore cannot react efficiently at low energies. In a beam with a statistical mixture of $^2P_{3/2}$ and $^2P_{1/2}$ states, this leads to the observed reaction efficiency of 2/3. As the kinetic energy is increased, mixing of the $^2A''$ and $^2A'$ surfaces can occur leading to an increase in the reaction efficiency. (The nature of this interaction is discussed below.) Further experiments are needed to verify whether this proposed explanation is valid but it is consistent with the spin-orbit state-specific experiments of Tanaka et al.[48] and Chupka and Russell.[45]

$Kr^+(^2P_J) + H_2$: Reaction 8 with $A^+ = Kr^+(^2P)$ is exothermic by 0.29 ± 0.06 eV for reaction of the $^2P_{3/2}$ ground spin-orbit state.[42] The $^2P_{1/2}$ state is 0.67 eV higher in energy such that the reaction of this state is exothermic by 0.96 eV. This spin-orbit splitting is sufficiently large that state-specific ion-molecule reactions[47,49] can be used to produce ion beams of a particular state. Here, ground state $Kr^+(^2P_{3/2})$ is produced via the resonant charge transfer reaction with CO^+ in the drift cell source.[49] Excited state $Kr^+(^2P_{1/2})$ is produced by EI ionization of Kr followed by charge transfer with N_2O which occurs rapidly with $Kr^+(^2P_{3/2})$ but slowly with $Kr^+(^2P_{1/2})$.[49] The purity of these beams is verified by examination of the reaction of Kr^+ with methane.[49] Beams of either spin-orbit state with purities exceeding 95% can be generated in this manner.[50]

The results of this study[50] are shown in Figure 9. Several features are immediately clear from this figure. Reaction with Kr^+ is much less efficient than with Ar^+. This efficiency is a strong function of the spin-orbit state. The cross section for the $^2P_{3/2}$ state shows a feature at high energies which is not found in the cross section for the $^2P_{1/2}$ state. Similar results are also obtained for reaction with HD and H_2 although the total reaction cross sections show an unusual intermolecular isotope effect such that the relative magnitudes are $\sigma(HD) > \sigma(H_2) > \sigma(D_2)$. The position of the high energy feature in the $^2P_{3/2}$ cross section shifts according to the pairwise energy scheme described above.

Many of these observations can be explained by examination of the diabatic and adiabatic surface correlations. Like the other rare gas ion systems, formation of products in this reaction involves a crossing from the reactant $[Kr^+ + H_2]$ diabatic surface to the charge transfer $[Kr + H_2^+]$ diabatic surface. As in the Ar^+ system, these diabatic curves mix as the reactants approach to form the adiabatic surfaces shown in Figure 10. This mixing is apparently much less efficient than in the case of Ar^+ probably because the bottom of the reactant well is displaced from the bottom of the charge transfer well, Figure 10. This delays charge transfer until Kr^+ and H_2 approach more closely than required for $Ar^+ + H_2$.[1] Indeed, this led early workers to hypothesize that a barrier would be observed in the Kr^+ system,[51] a result which is clearly inconsistent with experiment.[52] However, this surface topology may be evidencing itself in the unusual intermolecular isotope effect which suggests a dependence on the specific molecular constants of the reactant molecule. The true origins of this effect require further investigation.

The spin-orbit effects discussed in the Ar^+ system are now much clearer. Again it is the $m_J = \pm 1/2$ components of $Kr^+(^2P_{3/2})$ and $Kr^+(^2P_{1/2})$ which have the right symmetry to correlate with the charge transfer species, $Kr + H_2^+$, and the products, Table I. These components are responsible for the low energy reactivity of the two spin-orbit states. However, only the ground state $Kr^+(^2P_{3/2}, m_J = \pm 1/2)$ correlates adiabatically and hence it reacts much more efficiently than the $Kr^+(^2P_{1/2}, m_J = \pm 1/2)$ state. This latter state can react only via non-adiabatic transitions to the $[Kr^+(^2P_{3/2}, m_J = \pm 1/2) + H_2]$ reactive surface. Note that these same non-adiabatic transitions are apparently quite efficient in the $Ar^+ + H_2$ system presumably because the spin-orbit splitting is much smaller.

Since the high energy feature in the cross sections is observed only for reaction of $Kr^+(^2P_{3/2})$, we believe the $m_J = \pm 3/2$ components of this state are responsible. As noted above, this surface has the wrong symmetry, $^2A''$, to mix with the reactive $^2A'$ surface such that ordinary non-adiabatic transitions cannot occur. We hypothesize that rotationally non-adiabatic (coriolis) coupling[53] is the mechanism for this coupling.[50] As noted above, a similar effect occurs in the Ar^+ system although at much lower energies.

<u>$Xe^+(^2P) + H_2$:</u> Reaction 8 with $A^+ = Xe^+(^2P_J)$ has not previously been reported. It is endothermic by 0.6 eV for the $^2P_{3/2}$ ground spin-orbit state but exothermic by 0.7 eV for the $^2P_{1/2}$ excited state.[42] This system differs from the other rare gas systems in that the $Xe^+ + H_2$ reactants diabatically correlate with the products. No crossing to the charge transfer surface is required. Otherwise the adiabatic correlations are the same as those for Kr^+ and Ar^+. Only the $m_J = \pm 1/2$ components of the $^2P_{3/2}$ state correlate adiabatically with products. The result observed for a Xe^+ beam with a statistical distribution of 3/2 and 1/2 states, Figure 9 and 11, is thus not surprising. No exothermic reaction due to $Xe^+(^2P_{1/2})$ is evident presumably because the non-adiabatic coupling from this state to $Xe^+(^2P_{3/2}, m_J = \pm 1/2)$ is too inefficient. It is unclear from the present results whether this state and the $m_J = \pm 3/2$ components of the $^2P_{3/2}$ state react or not.

The threshold of the reaction is close to the thermodynamic result but rises slowly in contrast with the threshold behavior observed for $C^+(^2P)$ and $Si^+(^2P)$. This makes the determination of thermochemical data from analysis of this cross section difficult and unreliable. As noted above, MO considerations suggest that the most favorable approach of Xe^+ to H_2 is collinear. This steric constraint presumably limits the efficiency of reaction immediately above threshold. The peak in this cross section is substantially higher than the thermodynamic onset for dissociation at 4.5 eV. The peaks for reaction with HD and D_2 move as suggested by the pairwise energy scheme. Continued work in our laboratory seeks to examine the spin-orbit state-specific reaction cross sections for this system.

ATOMIC METAL IONS + H_2

The study of these simplest of all ionic transition metal reactions,

$$M^+ + H_2 \rightarrow MH^+ + H \qquad (11)$$

dates all the way back to the late 1970s when Jack Beauchamp and I first examined the reaction of Co^+ with D_2 using an ion beam apparatus.[54]

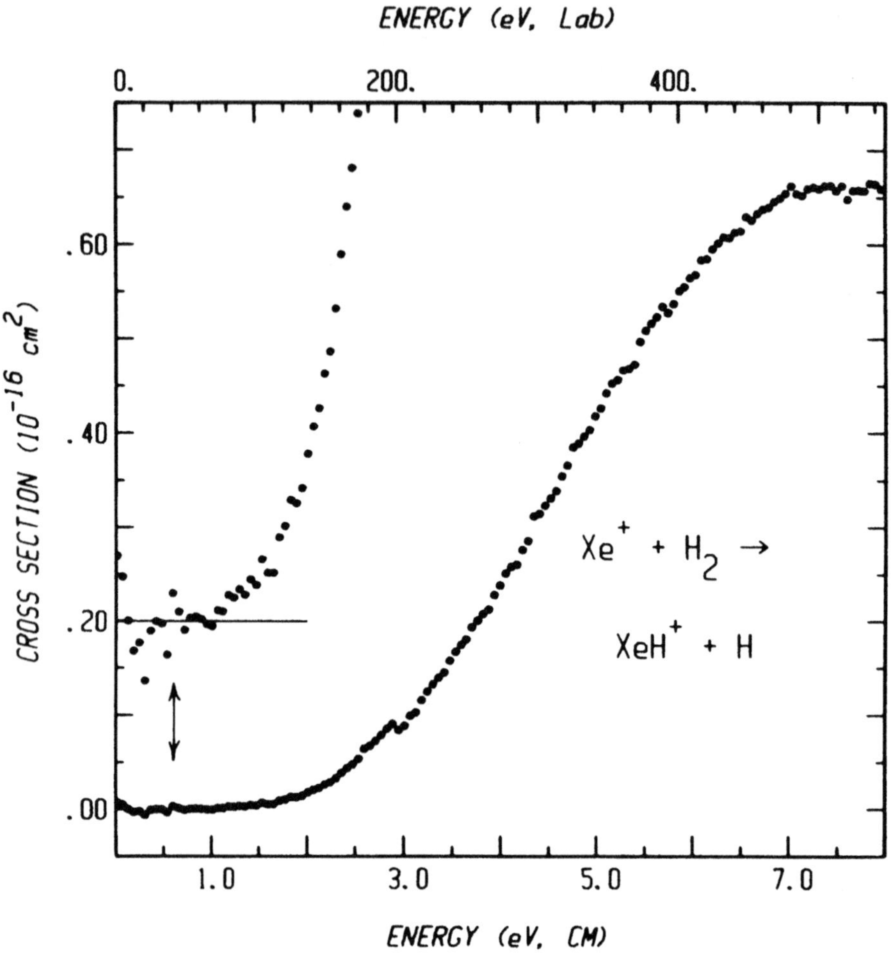

Figure 11. Cross section for reaction of $Xe^+(^2P)$ with H_2 as a function of relative kinetic energy (lower scale) and the ion energy in the laboratory frame (upper scale). The inset shows the cross section expanded by a factor of 10 and offset from zero. The arrow indicates the 0.6 eV threshold for reaction of $Xe^+(^2P_{3/2})$.

Because these reactions are endothermic, they cannot be studied using techniques like ICR or flowing afterglow. Beam studies of metal ions had been performed previously[55] but not for transition metals. One of the intimidating features of metal ions is the large number of electronic states. For instance, V^+ has eight electronic states below 2 eV (28 spin-orbit levels)[35] which will form 186 separate potential energy surfaces in the interaction with H_2. Clearly, explicit account of all these surfaces cannot be achieved easily.

A question I hoped to address in our studies at Berkeley was whether any account of the influence of electronic states could be achieved. Are there simple, unifying features which govern transition metal reactions? These studies naturally returned to the examination of reaction 11 since these systems provide the best experimental opportunity to elucidate the effects of electronic excitation. If we could not evaluate the influence of excited states in these simple processes, there seemed little hope of doing so in more complex systems.

Intrinsic to our understanding of these reactions are the guidelines established in the previous section. In essence, the "fundamental" systems have calibrated us. We now have a feeling for why the energy dependence of reaction cross sections can differ from the LGS formalism, eq 1. We have established under what circumstances reliable thermochemistry can be obtained. Simple considerations of electronic degeneracies and spin-orbit effects can be included. This base of knowledge enables us to extract more information from the following studies than would otherwise have been possible.

$V^+ + H_2$

Published results[7] for reaction 11 where M = V are shown in Figure 12. When V^+ is made by surface ionization (SI), the cross section rises from an apparent threshold of about 2 eV up to a peak between 4 and 5 eV before declining. This behavior is similar to that observed for reactions of C^+ and Si^+. As shown in Table II, this reactivity is principally due to the ^5D ground state of V^+. Analysis of this data by using eq 4 and phase space theory yields the bond energy, $D^\circ_0(V^+\text{-}H) = 2.05 \pm 0.06$ eV.[7] Examination of the reaction of V^+(SI) with HD shows nearly equal amounts of VH^+ and VD^+, again similar to C^+ and Si^+. These results are taken to indicate that $V^+(^5D, 3d^4)$ reacts via insertion into the H_2 bond.[7]

Figure 12 also shows our results[7] for V^+ produced by electron impact (EI) ionization of $VOCl_3$ at electron energies (Ee) of 30 and 50 eV. The appearance potential of V^+ from this compound is 26.8 ± 0.4 eV.[56] Thus, at Ee = 30 eV, little excess energy is available for electronic excitation. The apparent threshold for both sets of EI data has shifted to about 0.5 eV from 2 eV. In addition, the peak has shifted to lower energies by a comparable amount and the reactivity at high energies (>4 eV) has decreased. At Ee = 50 eV, a small exothermic reaction is observed. Note that the cross section magnitude does not increase appreciably under EI conditions.

These results clearly indicate that EI ionization produces electronically excited V^+. The shift in the threshold and the peak is consistent with production of the triplet states of V^+, Table II. A more detailed threshold analysis[7] indicates that under the Ee = 50 eV source conditions, the dominant component of the beam is the ^3F state (≈75%) with little contribution from ^5D or ^5F. Apparently, electronic energy and kinetic energy are comparable in promoting formation of VH^+. This is reasonable since formation of $VH^+(^4\Delta) + H(^2S)$ ground state products can occur either along a quintet surface or a triplet surface. Given this conclusion, it is at first surprising that in reaction with HD, V^+(50 eV) yields about 3-4 times as much VH^+ as VD^+. As discussed

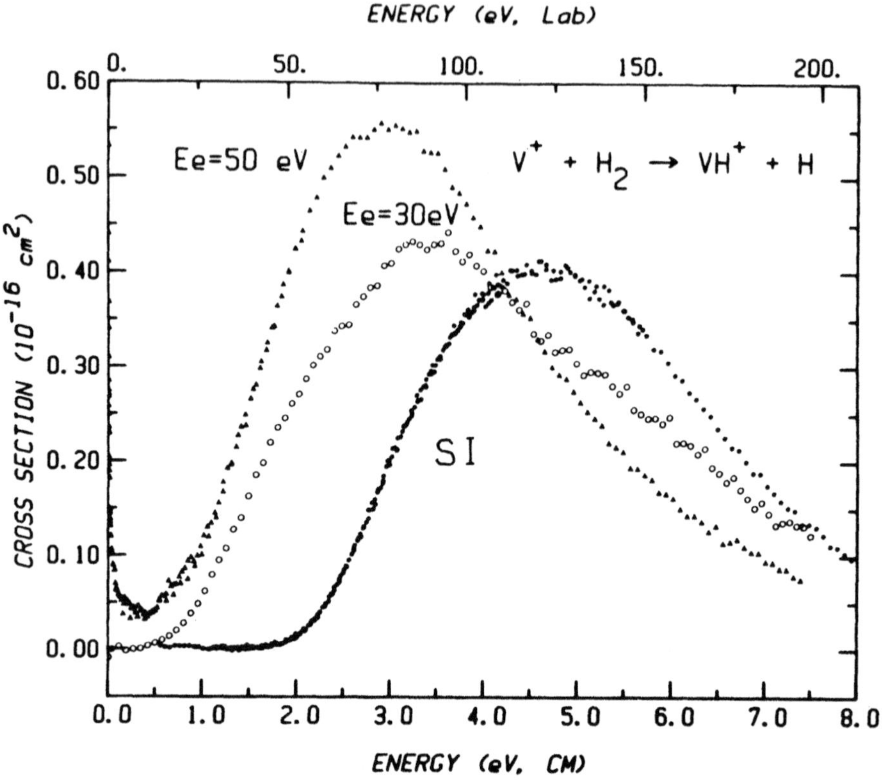

Figure 12. Cross section for reaction of V^+ with H_2 as a function of relative kinetic energy (lower scale) and the ion energy in the laboratory frame (upper scale). Data is shown for ions produced by surface ionization (SI) and by electron impact of $VOCl_3$ at electron energies (Ee) of 30 and 50 eV.

above, this type of isotope ratio is consistent with a direct reaction dominated by angular momentum constraints. Without amgiquity then we find that while the energetics of the $V^+(^5D)$ and $V^+(^3F)$ reactions are comparable, their mechanisms are not.

$Mn^+ + H_2$

Results[17] for reaction 11 where $M^+ = Mn^+$(SI) are shown in Figure 13 which also reproduces the V^+(SI) result. The behavior of Mn^+ is clearly very different than the more typical results for V^+. There are two features to the cross section. The first is at low energies and has a threshold below the thermodynamic limit of 2.4 eV.[17,57] This observation implies that this feature is due to excited states as is verified by examination of EI generated Mn^+.[17] Note that the apparent onset of

Table II. Electronic states of first row transition metal ions

Ion	State	Config.	Energy[a]	SI Pop.[b]
Sc^+	^3D	4s3d	0.01	0.88
	^1D	4s3d	0.32	0.06
	^3F	$3d^2$	0.61	0.06
V^+	^5D	$3d^4$	0.03	0.83
	^5F	$4s3d^3$	0.36	0.17
	^3F	$4s3d^3$	1.10	<0.01
	^3P	$3d^4$	1.45	
Mn^+	^7S	$4s3d^5$	0.00	1.00
	^5S	$4s3d^5$	1.17	<0.01
	^5D	$3d^6$	1.81	
Fe^+	^6D	$4s3d^6$	0.05	0.79
	^4F	$3d^7$	0.30	0.20
	^4D	$4s3d^6$	1.03	<0.01
Co^+	^3F	$3d^8$	0.09	0.85
	^5F	$4s3d^7$	0.52	0.15
	^3F	$4s3d^7$	1.30	<0.01
Ni^+	^2D	$3d^9$	0.08	0.99
	^4F	$4s3d^8$	1.16	0.01
	^2F	$4s3d^8$	1.76	<0.01
Cu^+	^1S	$3d^{10}$	0.00	1.00
	^3D	$4s3d^9$	2.81	<0.01

[a] Energies are statistically weighted averages over the J levels. Values are taken from ref 35.

[b] Surface ionization population. Maxwell-Boltzmann distribution at 2200 K.

about 1 eV is consistent with the thermodynamic threshold expected for $Mn^+(^5S)$, 1.2 eV. Analysis of the threshold for this feature indicates that $D_0^\circ(Mn^+\text{-}H) = 2.06 \pm 0.15$ eV.[17] The odd part about observing excited state reactions is that the Mn^+(SI) beam contains only 0.17% excited state, mostly ^5S. To observe such a small fraction, the excited state must be <u>much</u> more reactive than the ^7S ground state which comprises 99.83% of the SI beam.

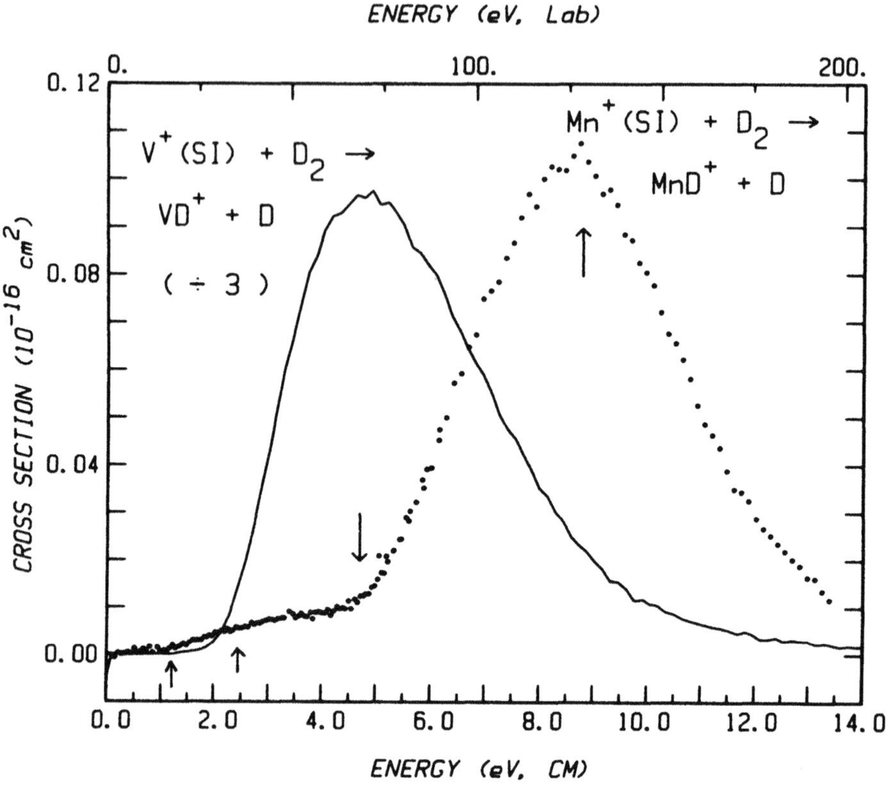

Figure 13. Cross section for reaction of Mn^+ and V^+ (both produced by surface ionization) with D_2 as a function of relative kinetic energy (lower scale) and the Mn^+ ion energy in the laboratory frame (upper scale). Arrows show the thermodynamic thresholds for reaction of $Mn^+(^5S)$ and $Mn^+(^7S)$ at 1.25 and 2.42 eV, respectively; the pairwise threshold for $Mn^+(^7S)$ at 4.7 eV; and the pairwise dissociation energy at 8.8 eV.

This conclusion is substantiated by examination of the second feature in the Mn^+(SI) cross section. This dominant feature has an apparent onset between 4 and 5 eV and peaks at about 9 eV. These values are both well above the thermodynamic values of 2.4 and 4.5 eV. Examination of EI generated Mn^+ verifies that this feature is due to ground state Mn^+.[17] The key to understanding this bizarre cross section comes from an examination of the reaction of Mn^+(SI) with HD, Figure 14. Here the main product is MnD^+ which has an apparent threshold between 3 and 4 eV and peaks at about 6 eV. Little MnH^+ is formed except at the low energies corresponding to excited state reaction. In this energy regime, MnH^+ is formed approximately 3-4 times as much as MnD^+. The threshold for both products is about 1 eV, as observed for H_2 and D_2 reactions.

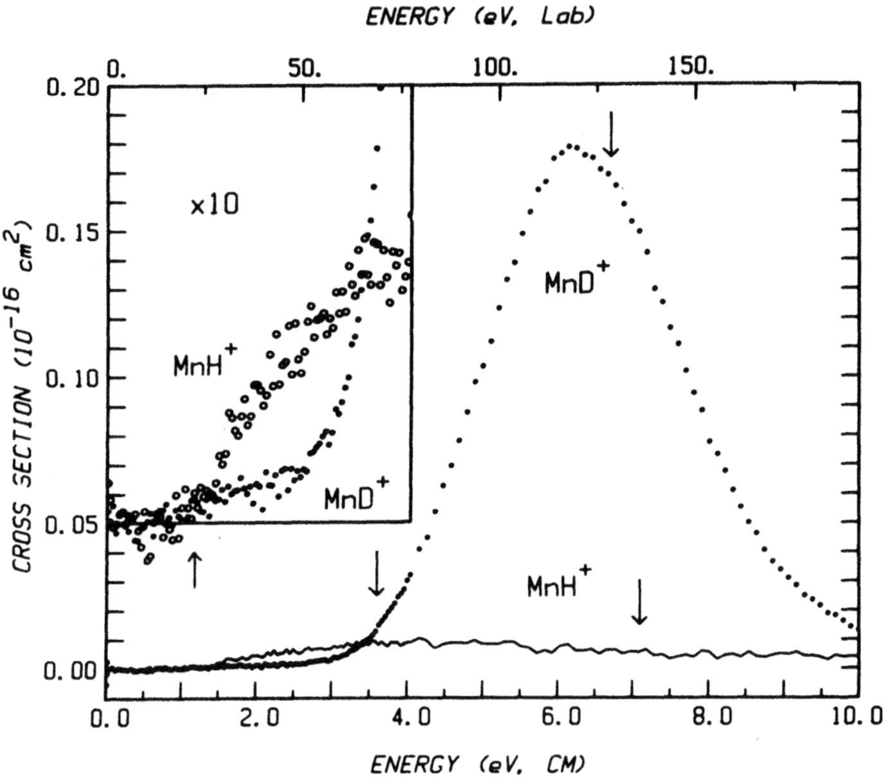

Figure 14. Cross sections for reaction of Mn^+ (produced by surface ionization) with HD to form MnH^+ (line) and MnD^+ (points) as a function of relative kinetic energy (lower scale) and ion energy in the laboratory frame (upper scale). The inset shows the threshold data for MnH^+ (open circles) and MnD^+ (points) expanded by a factor of 10. Arrows indicate the thermodynamic threshold for reaction of $Mn^+(^5S)$ at about 1.2 eV, the pairwise threshold for production of MnD^+ and MnH^+ at 3.6 eV and 7.1 eV, respectively, and the pairwise dissociation energy for MnD^+ at 6.7 eV.

The strong isotopic shifts observed in the dominant cross section feature are in quite good agreement with the pairwise energy scheme discussed above. Predictions for the apparent thresholds and peaks in the cross section are indicated by arrows in Figures 13 and 14. The fact that the pairwise scheme works indicates that the reaction surface for ground state $Mn^+(^7S)$ is repulsive and sterically hindered. As such, the reactivity of this state is quite poor.

Co^+, Ni^+, Cu^+ + H_2

Results[58] for reaction 11 where M^+ = Co^+, Ni^+, and Cu^+ (all produced by SI) are shown in Figure 15. The data for Co^+ and Ni^+ are in good agreement with previous studies of these systems.[54,59] In all cases, the reactivity observed is due almost exclusively to ground state ions. For Ni^+ and Cu^+, this is shown in Table II and has been verified by studies of EI generated ions.[58] In the case of Co^+, there is potentially a contribution from the first excited state, ^5F, but detailed studies suggest that it is not as reactive as the ^3F ground state.[58] Note that the shape of the cross section for all three systems is very similar given the differences in thresholds. These curves are straightforwardly analyzed by using eq 4 and phase space theory to give the bond energies, $D_0^\circ(Co^+{-}H)$ = 1.98 ± 0.06 eV, $D_0^\circ(Ni^+{-}H)$ = 1.68 ± 0.08 eV, and $D_0^\circ(Cu^+{-}H)$ = 0.92 ± 0.13 eV.[58]

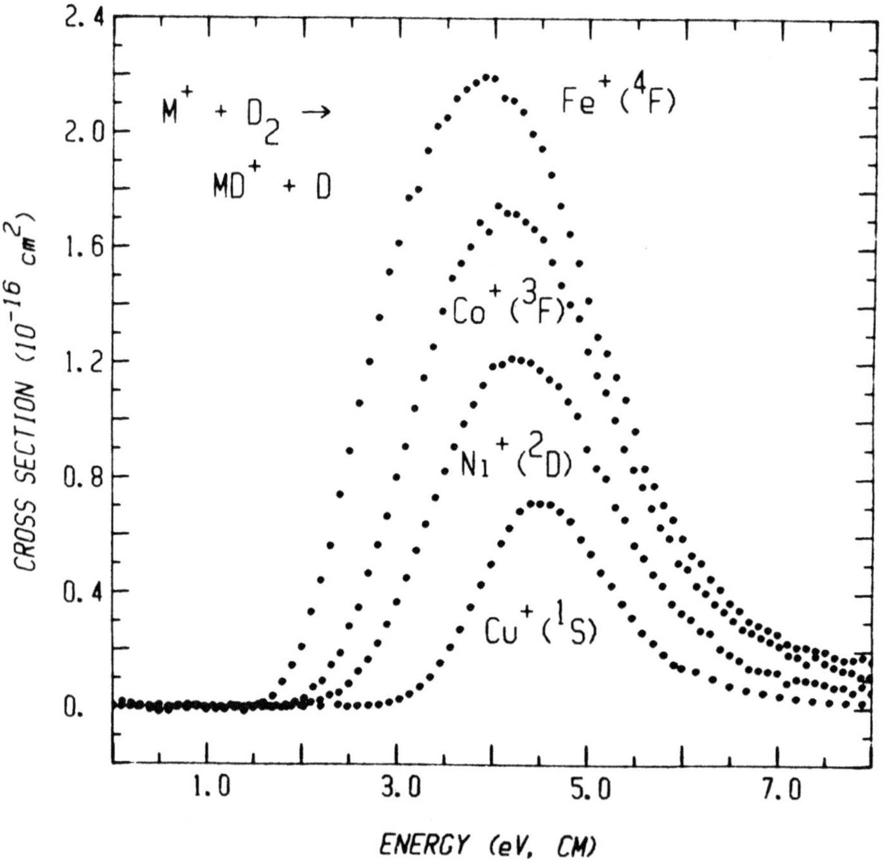

Figure 15. Cross sections for reaction of $Fe^+(^4F)$, $Co^+(^3F)$, $Ni^+(^2D)$, and $Cu^+(^1S)$ with D_2 as a function of relative kinetic energy.

Reactions of these three ions with HD also show very similar cross sections and branching ratios.[58] Figure 16 shows the example of Ni^+. The branching ratio clearly favors formation of NiH^+ at all energies. This type of isotope effect indicates a direct reaction as discussed above.

One surprising result observed in our studies of these systems is shown in Figure 17. For EI generated ions, the reactivity due to the ground state decreases, just as we observed above for V^+, Figure 12, and Mn^+. This unambiguously shows that fewer ground state ions are being produced by EI ionization. Unlike Figure 12, however, Figure 17 displays no obvious low energy feature (nor are any observed in EI results for Co^+ or Cu^+). Figure 18 shows where these excited states are hiding by comparing the EI data on an expanded scale with the SI (ground state Ni^+) data scaled down. There is, in fact, a low energy feature having a threshold which corresponds precisely with that expected for $Ni^+(^2F)$, Table II. The remaining difference in the SI and EI data curves occurs at high energies as might be expected for a process occurring via a

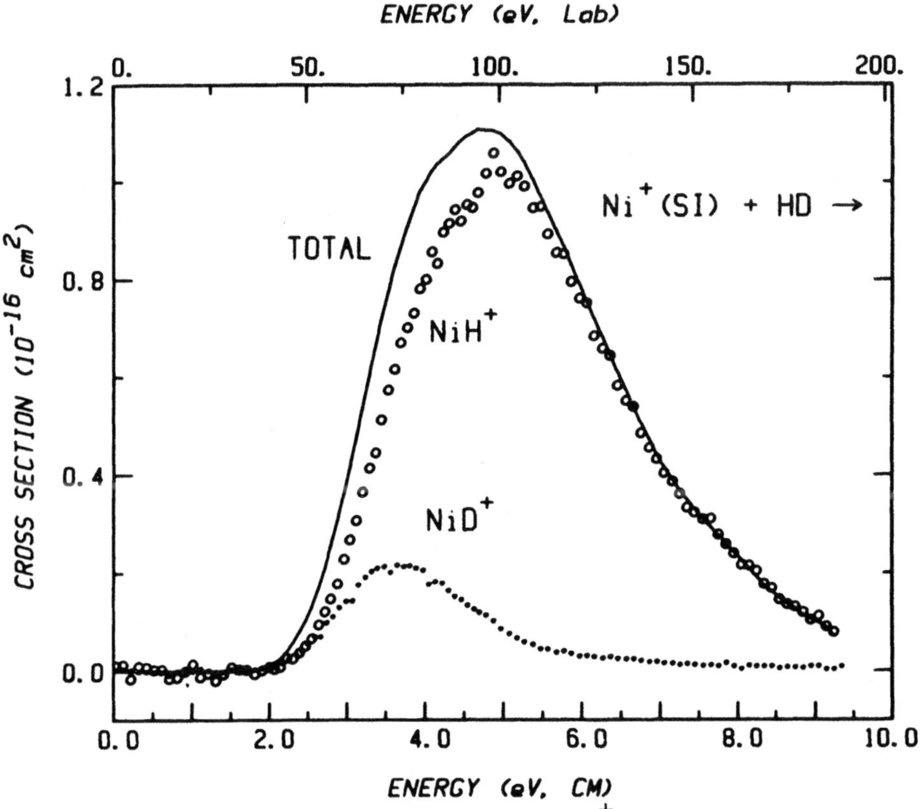

Figure 16. Cross sections for reaction of Ni^+ (produced by surface ionization) with HD to form NiH^+ (open circles) and NiD^+ (points) as a function of relative kinetic energy (lower scale) and ion energy in the laboratory frame (upper scale).

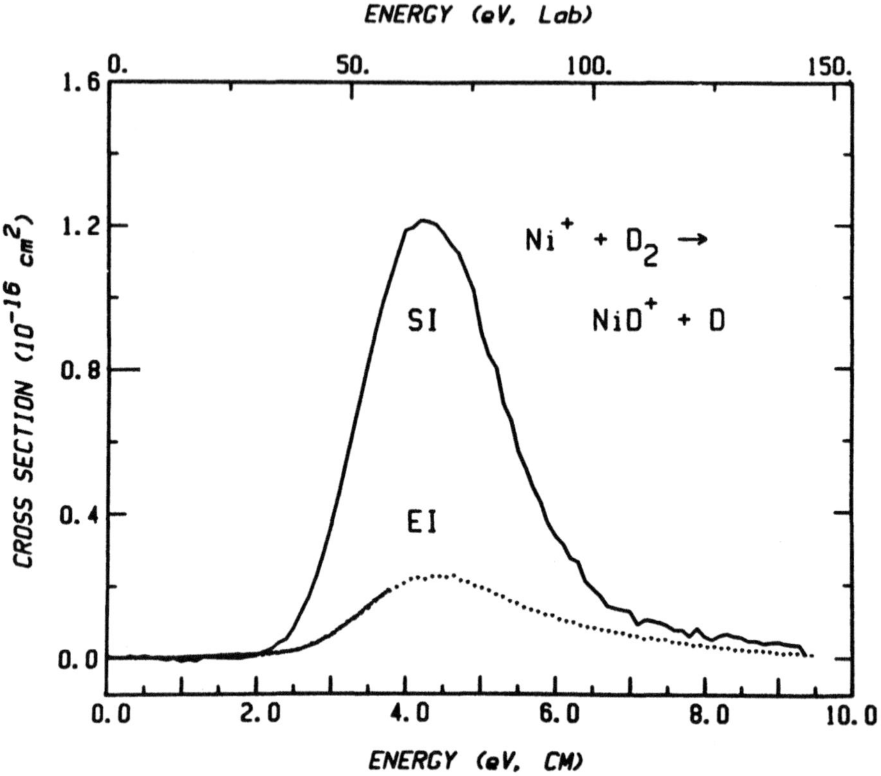

Figure 17. Cross sections for reaction of D_2 with Ni^+ produced by surface ionization (SI, line, same data as Figure 15) and electron impact at 50 eV (EI, points) as a function of relative kinetic energy (lower scale) and ion energy in the laboratory frame (upper scale).

pairwise mechanism. This is verified by the observation that much more NiD^+ than NiH^+ is formed in reaction with HD at these high energies. Based on the similarity between this observed reactivity and that for $Mn^+(^7S, 4s3d^5)$, we attribute this high energy feature to the $Ni^+(^4F, 4s3d^8)$ first excited state since both of these states have high-spin coupled $4s3d^n$ configurations.

$Fe^+ + H_2$

Results[18,60] for reaction 11 with $M^+ = Fe^+$(SI) is shown in Figure 19. This data is in good agreement with previous results for this reaction.[61] The shape of the cross section is similar to those shown in Figure 15 and shows no indication of unusual behavior. However, as can be seen in Table II, the Fe^+ produced by SI contains ^6D ground state and an appreciable amount of ^4F excited state. Since the former has a

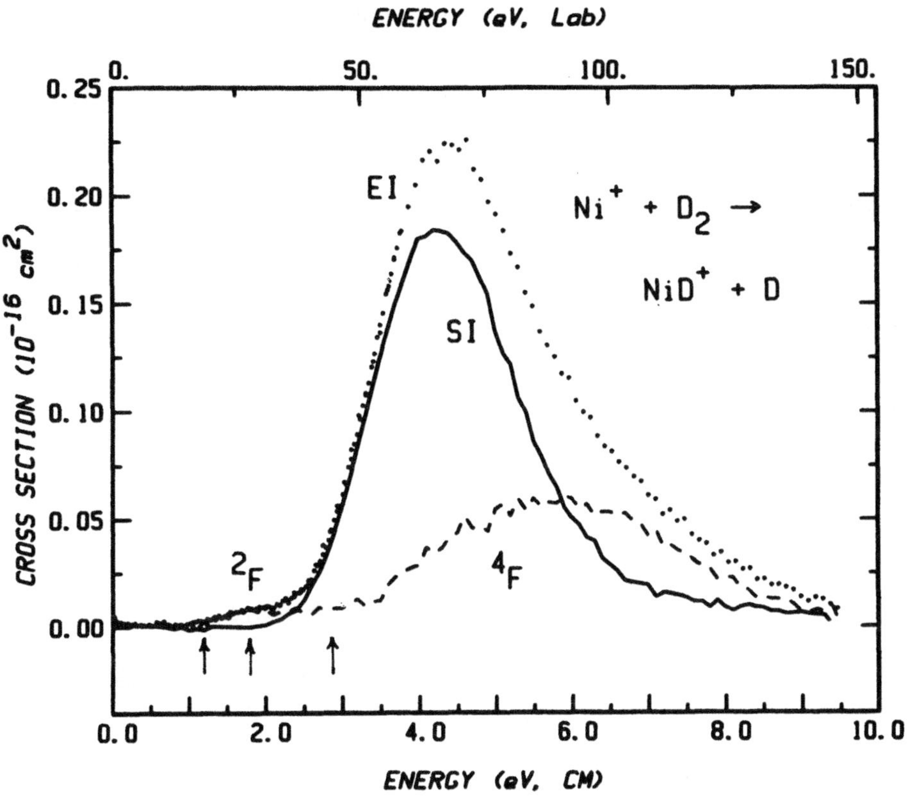

Figure 18. Cross sections for reaction of D_2 with Ni^+ produced by surface ionization (SI, line, same data as Figure 15 and 17 but reduced by a factor of 6.6) and electron impact at 50 eV (EI, points, same data as Figure 17) as a function of relative kinetic energy (lower scale) and ion energy in the laboratory frame (upper scale). The difference between these two curves (dashed line) is assigned to reactions of $Ni^+(^2F)$ at low energies and to $Ni^+(^4F)$ at high energies, see text. Arrows indicate the thermodynamic thresholds for reaction of $Ni^+(^2F)$, $Ni^+(^4F)$, and $Ni^+(^2D)$ at 1.15, 1.75, and 2.83 eV, respectively.

high-spin coupled $4s3d^n$ configuration, it seemed possible that this state, like $Mn^+(^7S)$ and $Ni^+(^4F)$, might be fairly unreactive. Experimentally, differentiating the reactivities of the ^6D and ^4F states cannot be accomplished by EI studies since they are separated by only 0.25 eV.

Our answer to this problem is the drift cell source. Figure 19 shows results for Fe^+ generated by EI but passed through the drift cell filled with Ar at the indicated pressures. At the highest pressure, Fe^+ undergoes approximately 1000 collisions. The results clearly demonstrate that these collisions alter the reactivity of the Fe^+ beam in the

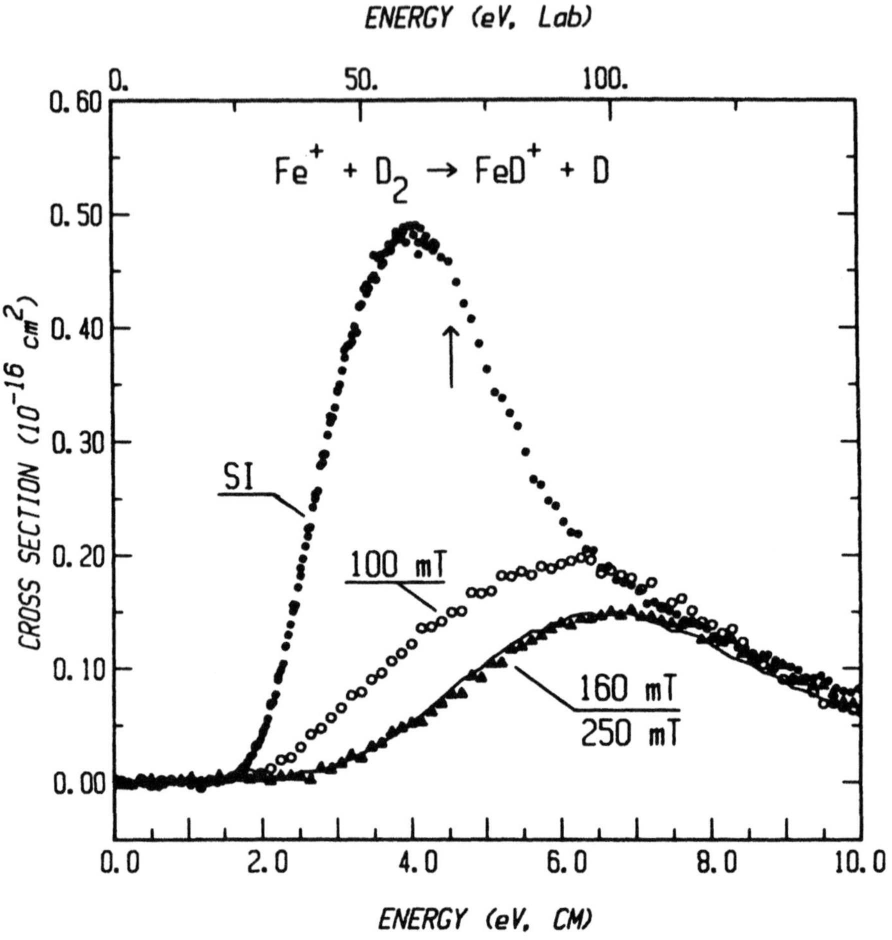

Figure 19. Cross sections for reaction of D_2 with Fe^+ produced by surface ionization (SI, closed circles, ≈80% ^6D + 20% ^4F) and in the drift cell source at Ar pressures of 100 (open circles), 160 (line) and 250 mTorr (triangles, ≈100% ^6D) as a function of relative kinetic energy (lower scale) and ion energy in the laboratory frame (upper scale). The arrow indicates the D_2 bond energy.

region of the SI peak. This change in cross section ceases above an Ar pressure of 160 mTorr indicating that the collisional process is complete by this pressure. The only explanation consistent with these observations is that the $Fe^+(^4F)$ excited state ions are quenched by collisions with Ar to the $Fe^+(^6D)$ ground state. This quenching is estimated to have a rate constant of $\approx 10^{-12}$ cm^3/s, or an efficiency of 1 in every 620 collisions.[18]

Since the ion beam emerging from the drift cell at high Ar pressures is pure $Fe^+(^6D)$ and since the SI beam has a known population of ground and excited states, the true cross sections for $Fe^+(^6D)$ and $Fe^+(^4F)$ can be extracted. These are shown in Figures 20 and 15, respectively. Note that the general behavior of these states is comparable to other metal ions having similar electron configurations, high-spin $4s3d^n$ and $3d^{n+1}$, respectively. This similarity helps confirm that the cross sections shown truly represent the reactivity of the pure states. Further evidence for this is exhibited by the results for reaction with HD, Figure 21. Note that the branching ratio for reaction of $Fe^+(^4F, 3d^7)$ is nearly identical in behavior to that shown for $Ni^+(^2D, 3d^9)$, Figure 16. Likewise reaction of $Fe^+(^6D, 4s3d^6)$ with HD produced mainly MD^+ as also observed for $Mn^+(^7S, 4s3d^5)$, Figure 14. The final point of proof is that analysis of the cross section for $Fe^+(^4F)$ shown in Figure 15 leads to $D_0^\circ(Fe^+\text{-}H)$ = 2.12 ± 0.06 eV. This differs from the value obtained by previous beam experiments,[61] 2.55 ± 0.15 eV

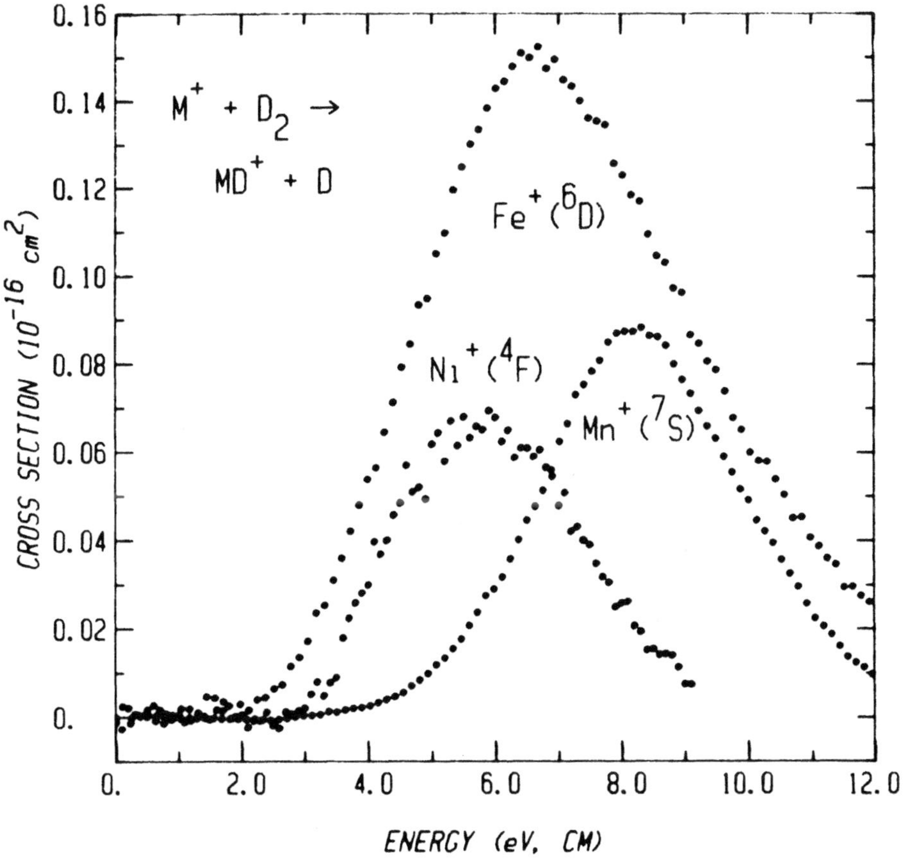

Figure 20. Cross sections for reaction of $Fe^+(^6D)$, $Ni^+(^4F)$, and $Mn^+(^7S)$ with D_2 as a function of relative kinetic energy.

largely because of the difference in identification of the reacting state. The present value is in reasonable agreement with results from ab initio calculations, 2.04 eV.[62] The small difference is within the error expected for these calculations (≈0.1 eV) and is comparable to the differences seen between these calculations and our other experimentally derived values for first row metal hydride ions.[63]

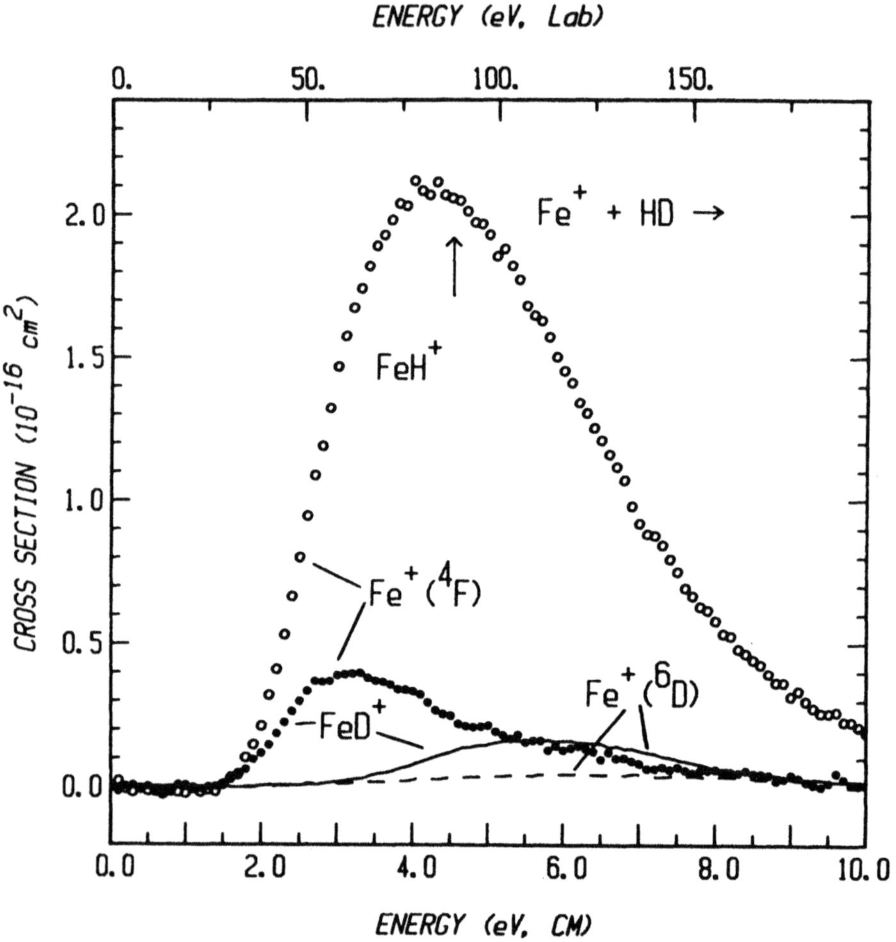

Figure 21. Cross sections for reaction of $Fe^+(^4F)$ and $Fe^+(^6D)$ with HD to form FeH^+ (open circles and broken line) and FeD^+ (points and full line) as a function of relative kinetic energy (lower scale) and ion energy in the laboratory frame (upper scale). The arrow indicates the bond energy of HD.

Periodic Trends in Thermochemistry

The bond dissociation energies (BDEs) of the diatomic metal hydride ions show a strong variation with the identity of the metal. This has been discussed several times in the past, [62,64,65] most recently by Elkind and Armentrout.[63] As shown in Figure 22, the first row metals on the left side of the periodic table, Sc and Ti, have the strongest BDEs while Cr and Cu have the weakest. These weak bond energies are easily understood by noting that the ground state configuration of Cr^+ is the very stable half-filled shell, $3d^5$. Cu^+ has a filled shell ground state, $3d^{10}$. To form bonds to either of these ions, one of the electrons must be decoupled from the others. This costs energy. This concept can be formalized in terms of a promotion energy, Ep. Here, we define Ep as the energy necessary to take a metal ion in its ground state electron configuration to a configuration where one electron is in

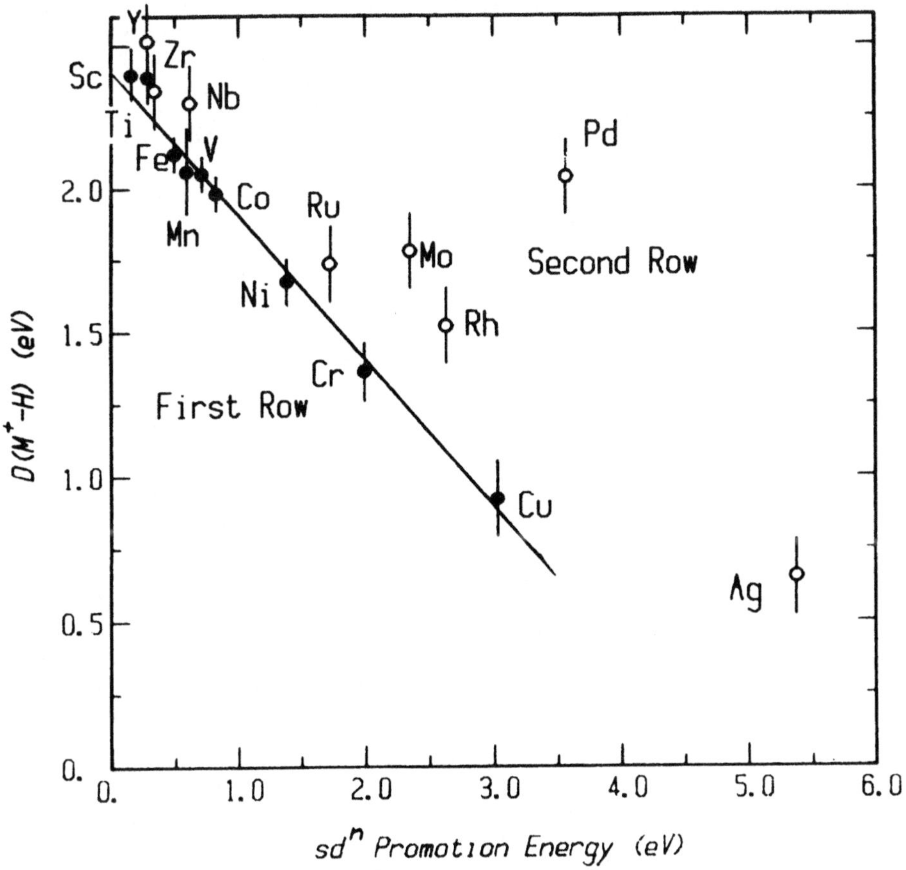

Figure 22. First row (closed circles) and second row (open circles) transition metal-hydride ion bond energies versus atomic metal ion promotion energy to an sd^n spin decoupled state, see text. The line is a linear regression fit to the first row data.

a 4s orbital and is spin decoupled from the electrons in the 3d orbitals. This energy is easily calculated as the mean energy of the electronic state which has a high-spin $4s3d^n$ configuration and that which has a low-spin $4s3d^n$ configuration.[63,65] This promotion energy is the abscissa of Figure 22. It is clear that for the first row transition metals the correlation between $D(M^+-H)$ and Ep is quite good. This implies that the dominant binding orbital on the metal is the 4s orbital. *Ab initio* calculations[62] support this idea although they establish that there is significant 3d character in the bonds, especially on the left side of the periodic table.

Molecular Orbital Considerations

As in the systems described in the FUNDAMENTAL SYSTEMS section, molecular orbital concepts are useful in discussing the reactions of atomic transition metal ions. The general ideas are much the same; only the atomic orbitals are different. For first row transition metals, the valence orbitals are the 3d, 4s, and 4p in order of increasing energy. For simplicity, we ignore the p orbitals since they are too high in energy to be very influential in the bonding.[66] The 4s orbital lies slightly above the 3d in energy (except for scandium).[67] Thus, the energy of particular electron configurations or states depends heavily on the number of electrons and the spin coupling, see Table II.

Since the 4s is the outermost valence orbital as well as the principal binding orbital in the diatomic metal hydride ions (see above), the interaction of this orbital with H_2 is expected to be dominant at long range. It plays the same role that the p_z orbital does in the fundamental reaction systems. Namely, in C_{2v} symmetry, occupation of the 4s (having a_1 symmetry) leads to a repulsive interaction with H_2 due to overlap with the doubly occupied σ_g orbital of H_2. This repulsion is relieved by going to $C_{\infty v}$ symmetry. Interactions with the 3d orbitals also have parallels in the discussion above. The 3dσ ($3dz^2$) orbital also has a_1 symmetry and therefore acts like the 4s. Indeed mixing of the 4s and 3dσ orbitals is inevitable in these systems. The 3dπ (3dxz) orbital has b_2 symmetry and therefore acts as the p_x did above. It leads to an attractive interaction with the σ_u MO of H_2 in C_{2v} symmetry and is non-bonding in $C_{\infty v}$ symmetry. The other 3dπ (3dyz) and the 3dδ orbitals [3dxy and $3d(x^2-y^2)$] do not interact with the MOs of H_2. They, like p_y above, are non-bonding in all geometries.

Now consider the electron configurations of the metals examined above. The $V^+(^5D)$ ground state has a $3d^4$ configuration such that the 4s is unoccupied, thus avoiding repulsive interactions. In addition, for 1 our of the 5 quintet surfaces evolving from $V^+(^5D) + H_2$, the 3dσ is also empty, again avoiding repulsive interactions. This leads to a $^5A_1$ surface which should be attractive since the 3dπ (b_2 symmetry) orbital of V^+ is occupied and the other 3 electrons are non-bonding. Evolution of the molecular orbitals on this surface proceed smoothly to produce $VH^+(^4\Delta)$ ground state.[7] This is consistent with the observation of statistical like behavior for this state. The other 4 surfaces should be less reactive since the 3dσ orbital is occupied. Substantiating evidence for this analysis comes from a comparison of a phase space theory (PST) calculation to the experimental results.[7] We find that PST accurately reproduces the shape of the cross section but the data has a magnitude smaller by a factor of 0.14 ± 0.03. This is comparable with our expectation of 0.20 (1/5) especially when the population of the ^5D state, 0.83 (Table II), is considered.

Now consider the ^3F excited state of V^+. This state has an electron configuration of $4s3d^3$ where the 4s electron is low-spin coupled to the three 3d electrons. As noted above, occupation of the 4s electron is expected to be repulsive in C_{2v} symmetry such that insertion into H_2 is not expected. In $C_{\infty v}$ symmetry, this repulsion is relieved. Consequently, reaction of V^+ (^3F) can occur but via a direct mechanism rather than via insertion as observed for $V^+(^5D)$. This is consistent with the change in the intramolecular isotope ratio observed for this state compared with $V^+(^5D)$.

For manganese, first note that the behavior of $Mn^+(^5S)$ is comparable to that observed for $V^+(^3F)$. This is completely consistent with the fact that both have low-spin $4s3d^n$ configurations. $Mn^+(^7S)$, on the other hand, has a high-spin $4s3d^n$ configuration. The striking pairwise behavior of this state is repeated for $Fe^+(^6D)$ and $Ni^+(^4F)$ which also have high-spin $4s3d^n$ configurations, Figure 20. At first glance, the difference between low-spin and high-spin configurations would not appear to affect the MO considerations discussed above since occupation of the 4s orbital dominates the interaction and tends to favor a collinear reaction for both cases. The difference clearly must lie in the interactions between the 3d electrons and H_2. While this difference may be viewed in several ways, one of the easier is to consider the reverse reaction, i.e., approach of MH^+ and H. The H(1s) electron can either be low- or high-spin coupled with the non-bonding 3d electrons of MH^+. If low-spin coupled, there can be favorable bonding interactions between the H atom and the metal. These are most effective if the reaction geometry differs from $C_{\infty v}$. If high-spin, a node must exist between the incoming H(1s) and the metal 3d electrons. This leads to a very repulsive interaction in anything but strict $C_{\infty v}$ symmetry. Thus, for reaction of $M^+ + H_2$, the difference between low- and high-spin states is explained as a change in the repulsiveness of the potential energy surface as the reaction deviates from a strict collinear geometry. For high-spin $4s3d^n$ states, the repulsiveness increases as this deviation occurs. For low-spin $4s3d^n$ states, the repulsion is mediated by bonding interactions between the trailing hydrogen atom and the metal 3d electrons.

Finally, consider $Fe^+(^4F)$, $Co^+(^3F)$, $Ni^+(^2D)$, and $Cu^+(^1S)$ which all have $3d^{n+1}$ configurations (n = 7, 8, 9, 10, respectively). They all behave very similarly in reaction 11, Figure 15, and in reaction with HD. Unlike the $3d^{n+1}$ configuration where $n+1 < 5$ (such as $V^+(^5D)$), these states cannot have an empty 3dσ orbital. Occupation of this orbital, like occupation of the 4s, is apparently sufficient to make these reactions occur in a direct mechanism rather than via insertion. Note that the spin state for these configurations is the same as for the low-spin $4s3d^n$ configurations of these metals. Consequently, there may be extensive interactions between the diabatic surfaces correlating with these configurations.

Summary

The experimental results and theoretical considerations discussed here lead to a reasonably comprehensive view of the reactions of atomic transition metal ions with molecular hydrogen. Four categories of reactivity appear to exist based on the electron configuration of the metal.

1) $3d^{n+1}$, $(n+1) < 5$): The systems react efficiently on at least one potential energy surface. The branching ratio in reaction with HD is nearly 1:1. Overall the behavior is near statistical. MO concepts indicate that these states should be able to insert into H_2 to form a metal-dihydride intermediate.

2) $3d^{n+1}$, $(n+1 > 5)$: The systems react efficiently via a direct mechanism. The branching ratio in reaction with HD is ≈4:1 in favor of the MH^+ product. MO concepts indicate that these states should prefer a collinear reaction geometry but that other geometries are not unfavorable.

3) $4s3d^n$, low-spin: The systems behave the same as group 2.

4) $4s3d^n$, high-spin: The systems react inefficiently via a pairwise interaction. The branching ratio in reaction with HD favors production of the MD^+ product substantially. MO concepts indicate that these states should have repulsive surfaces in all but a collinear reaction geometry.

Despite the utility of these reactivity "rules," they clearly can be broken. For instance, work in our laboratory[68] shows that ground state $Cr^+(^6S, 3d^5)$ does not conform to the behavior of groups 1 or 2 but rather shows the pairwise type of interaction characteristic of group 4. While a detailed analysis of this behavior is still underway, note that like group 2, $Cr^+(3d^5)$ must occupy the 3dσ orbital but unlike group 2, this configuration does not have the same spin as group 3. Instead, it will couple most readily with the group 4 high-spin coupled $4s3d^n$ state, $^6D(4s3d^4)$. The $Cr^+(^6S)$ state is unique in this respect.

Second Row Transition Metals

It is of obvious interest to know whether similar ideas extend beyond the first row transition metals. Some idea of this can be obtained from the results for second row metals shown in Figure 22. This shows that the metals to the left side of the periodic table, Y, Zr, and Nb, may correlate with Ep in much the same way that the first row metals do. Clearly, however, the metals to the right side deviate from this correlation. The reasons for this probably lie in the ability of the second row metals to form reasonably strong bonds by using the 4d orbital rather than the 5s. Pd^+ provides the best example. It has a $^2D(4d^9)$ ground state such that one 4d electron is spin-decoupled from the remaining electrons. Thus the promotion energy if H binds to the 4dσ is 0 eV while Ep is 3.56 eV. The intrinsic bond strength of a M(4d)-H(1s) bond is weaker than a M(5s)-H(1s) bond but, for Pd, this is compensated for by the huge difference in promotion energy. Similar qualitative arguments hold for all the right side second row transition metals.[63] Silver, Ag^+, is in a dilemma since it cannot form a covalent bond without promoting from a $4d^{10}$ ground state configuration to a $5s3d^{9}$ configuration, a very costly 5.38 eV. An interesting possibility in this case is that AgH^{+} has only a single electron binding the two atoms, i.e., a protonated silver atom.

The implications of this difference in bonding character for the mechanisms of reaction 11 with second row metal ions is still undergoing investigation. Preliminary results[68] for reactions of these species with HD show both similarities and differences with the first row analogs.

ATOMIC METAL IONS + ALKANES

In the past few years, a tremendous amount of experimental work has centered on the reactions of atomic transition metal ions with alkanes. Ever since the first observation by Allison, Freas, and Ridge[69] that M^{+} can activate C-H and C-C bonds, these processes have captured the fancy of ion-molecule chemists. Our work in this area has centered on reactions with small alkanes, particularly methane and ethane. With few exceptions, these hydrocarbons do not undergo exothermic reactions with atomic metal ions. They are therefore ideally suited to beam studies.

A question of obvious interest in these studies is whether the "rules" of transition metal ion reactivity outlined above continue to hold. In considering this, note that these rules are formulated for the diabatic character of the reaction surfaces, i.e., they pertain to a specific electron configuration of the metal ion. Such diabatic rules should work well for the H_2 system where the reactions are endothermic and the intermediate lifetimes are short. However, as the reactant neutral increases in size, the polarizability, the number of degrees of freedom, and the intermediate lifetimes all increase. As we shall see, this can permit these diabatic correlations to break down resulting in different kinds of reactivity.

$Sc^{+} + CH_4$

Results for the reaction of Sc^{+} (generated by SI) with methane are shown in Figure 23. The three products observed correspond to the reactions:

$$Sc^{+} + CH_4 \rightarrow ScH^{+} + CH_3 \quad (12)$$

$$\rightarrow ScCH_3^{+} + H \quad (13)$$

$$\rightarrow ScCH_2^{+} + H_2 \quad (14)$$

The dominant reaction at high energies is process 12. Based on the thermochemistry derived from analysis of reaction 11 ($M^{+} = Sc^{+}$),[63] the thermodynamic threshold for reaction 12 is about 2.1 eV. By using eq 4, an analysis of the cross section for this process yields a threshold in good agreement with this. The apparent threshold in Figure 23 appears much lower because of the Doppler broadening due to the thermal motion of the methane gas.

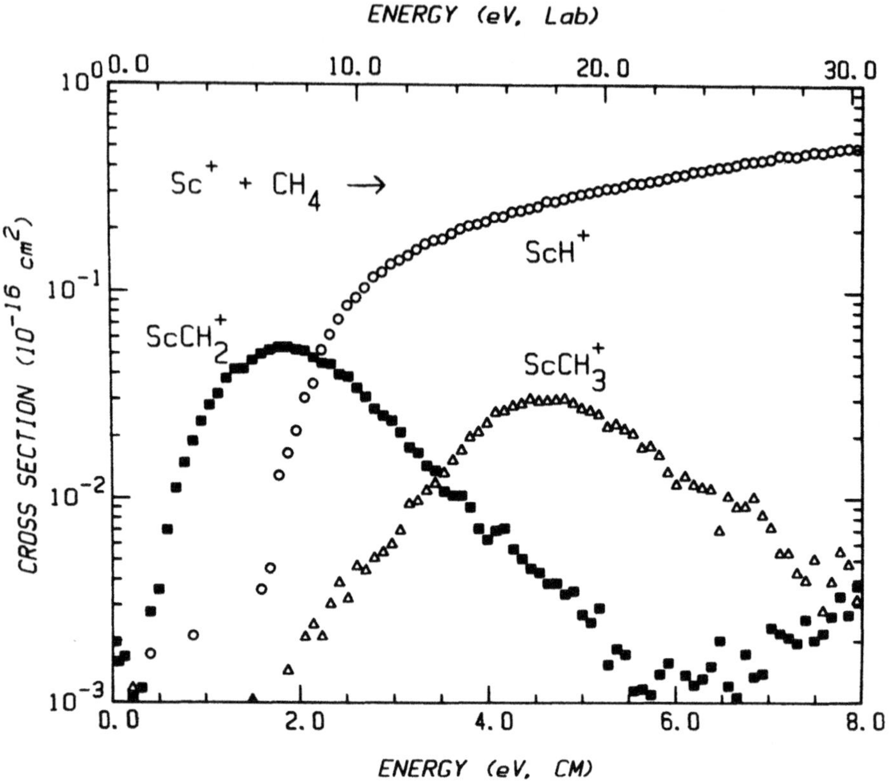

Figure 23. Cross sections for reaction of Sc^+ (produced by surface ionization) with methane as a function of relative kinetic energy (lower scale) and ion energy in the laboratory frame (upper scale).

Reaction 13 has an apparent threshold comparable to that for process 12. This is consistent with the fact that the measured bond energies are similar, D°(Sc^+-H) = 2.44 ± 0.10 eV and D°(Sc^+-CH_3) = 2.56 ± 0.13 eV.[21] Given this, it seems odd that the probability of forming $ScCH_3^+$ is much less than for forming ScH^+. The reasons for this lie in the relative ability to conserve angular momentum in reactions 12 and 13. Note that the reduced mass of the products in reaction 12 is similar to that of the reactants (11.3 amu vs 11.8 amu). In contrast, the reduced mass of the products in reaction 13 is much less (1.0 amu). Since these are endothermic reactions, this puts a severe constraint on the phase space available to the latter products and hence results in a reduction in the cross section. This argument is similar to that described above for the ratio of MH^+ to MD^+ in a direct reaction of M^+ with HD. Indeed if one were to carry the analogy to completion, the translational density of states favors the ScH^+ product by a factor of 38 (= $11.3^{3/2}$) compared to the $ScCH_3^+$ product. The fact that the observed ratio is less than this is consistent with a more statistical process where the internal density of states favors formation of $ScCH_3^+$.

The cross section for process 13 reaches a peak at about 4.5 eV. This corresponds to decomposition of $ScCH_3^+$ in reaction 15,

$$Sc^+ + CH_4 \rightarrow ScCH_3^+ + H \rightarrow Sc^+ + CH_3 + H \tag{15}$$

which has a thermodynamic threshold of 4.5 eV, D°(CH_3-H). Interestingly, the ScH^+ product, which can also begin decomposing at this energy, shows no decline up to the 8 eV energy examined. This means that the ScH^+ product carries away little of the excess energy in internal modes. Most of this must reside in translation or in internal modes of the CH_3 neutral product. A secondary decomposition pathway for $ScCH_3^+$ is loss of a hydrogen atom,

$$Sc^+ + CH_4 \rightarrow ScCH_3^+ + H \rightarrow ScCH_2^+ + H + H \tag{16}$$

This is shown by the rise in the $ScCH_2^+$ cross section beginning about 6 eV. Note that the fact that $ScCH_3^+$ decomposes primarily via process 15 is evidence that this is truly a metal-methyl ion rather than a hydrido-metal-carbene, $H-Sc^+=CH_2$.

At low energies the dominant reaction is process 14, elimination of molecular hydrogen. The only alternative reaction to form $ScCH_2^+$ is process 16 already discussed. Preliminary analysis of the cross section for reaction 14 indicates that D°(Sc^+-CH_2) ≥ 4.05 ± 0.11 eV. At higher energies, the $ScCH_2^+$ cross section reaches a peak and declines in magnitude. As seen in the previous sections, this type of behavior ordinarily occurs because the product begins to decompose. For $ScCH_2^+$, the possible decomposition products are $ScC^+ + H_2$, $ScCH^+ + H$, and $Sc^+ + CH_2$. The first two possibilities are discounted since the ionic decomposition products are not observed in these studies. (They are observed in other transition metal ion - methane systems, however, verifying that we can observe them if they are there.) The last possibility cannot be detected explicitly since the decomposition product is identical to the reactant. Overall, this reaction corresponds to

$$Sc^+ + CH_4 \rightarrow Sc^+ + CH_2 + H_2 \tag{17}$$

which is endothermic by 4.8 eV. Therefore, it too cannot account for the decline in the $ScCH_2^+$ cross section. The only remaining rationale for this behavior is that the intermediate involved in reaction 14 is being depleted by another reaction (or reactions). Formation of both ScH^+ and $ScCH_3^+$, reactions 12 and 13, are obviously such processes. This shows that these three products all have a common intermediate. Why, then, do reactions 12 and 13 dominate reaction 14 despite being thermodynamically less favorable? The answer must be because they are kinetically favored by much larger preexponential factors (equivalently, larger phase space). This indicates that reaction 14 must involve a tighter transition state than either reaction 12 or 13.

A mechanism for the interaction of Sc^+ with methane consistent with all this information is shown in Figure 24. Sc^+ inserts into the C-H bond of methane to form $H-Sc^+-CH_3$, I. At low energies, this

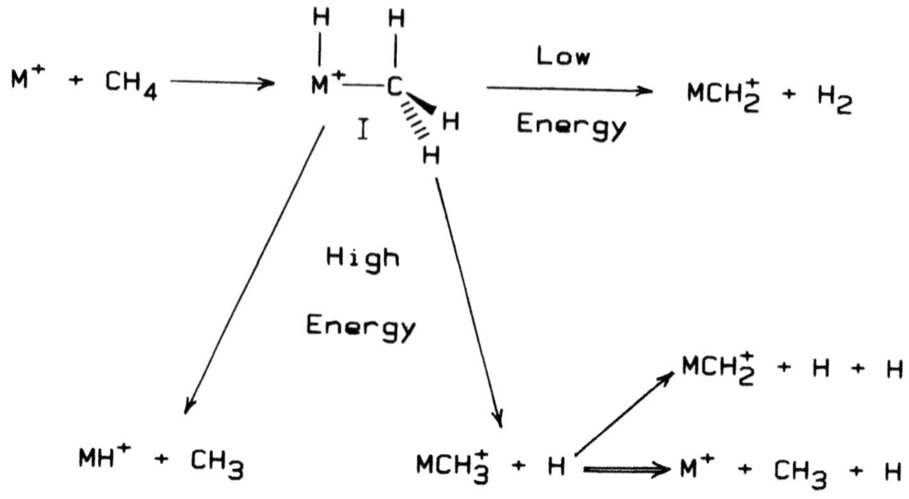

Figure 24. Mechanism for reaction of atomic metal ions with methane.

emits H_2 in a four-center elimination. While ordinarily symmetry forbidden, this process is allowed in this reaction because the metal-ligand bonds have substantial 3d character in them.[70] Note that Sc^+ has only two valence electrons and therefore cannot eliminate hydrogen via an intermediate like II,

$$\begin{matrix} H \\ H \end{matrix} \rangle Sc^+ = CH_2 \qquad II$$

At higher energies, I can decompose to form ScH^+ or $ScCH_3^+$. These simple bond fissions should have loose transition states and consequently are much more facile than the four-center elimination. At still higher energies, $ScCH_3^+$ decomposes to $Sc^+ + CH_3$ and $ScCH_2^+ + H$.

Other metals to the left side of the periodic table, Ti^+, V^+ and Cr^+, show comparable behavior to Sc^+ in their interactions with methane when in their ground states.[71,72] The primary product at low energies is MCH_2^+ and at high energies is MH^+. The mechanisms of these reactions are presumably identical to that of Sc^+. Ti^+ and V^+ both exhibit MCH^+ products as minor channels at elevated energies. These species are apparently another minor decomposition channel for MCH_3^+.

$Fe^+ + CH_4$

Results[73] for the reaction of iron ions with methane are shown in Figure 25 for Fe^+ generated by SI and in the drift cell (DC). Note that for both source conditions, no $FeCH_2^+$ is observed. As in the H_2 system, Figure 19, the FeH^+ product from Fe^+(DC) is much smaller than from Fe^+(SI). Also $FeCH_3^+$ from Fe^+(DC) is below our limit of detectability.

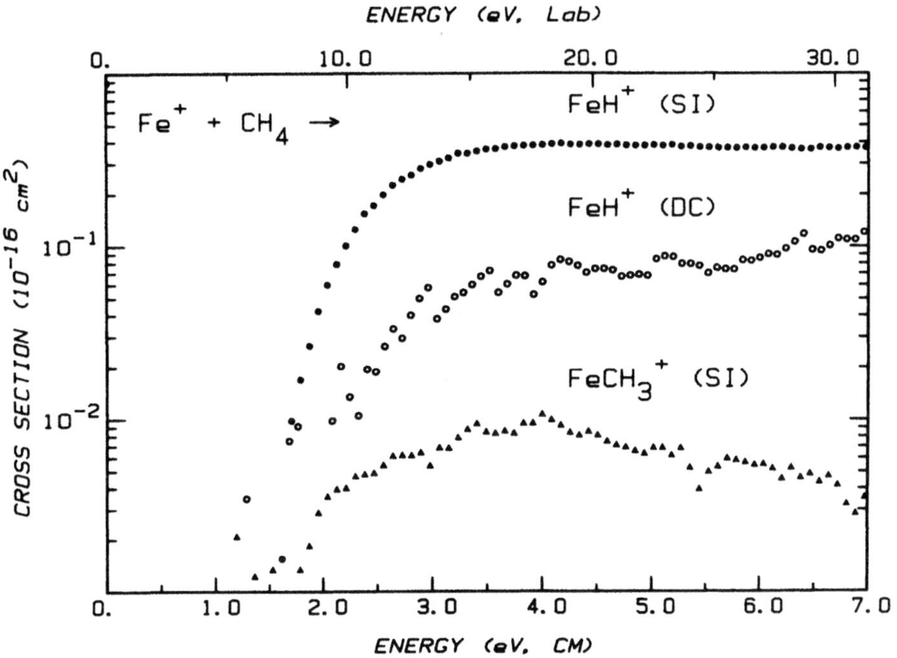

Figure 25. Cross sections for reaction of methane with Fe^+ produced by surface ionization (SI, closed symbols, ≈80% ^6D + 20% ^4F) and in the drift cell source (DC, open symbols, ≈100% ^6D) as a function of relative kinetic energy (lower scale) and ion energy in the laboratory frame (upper scale).

Clearly, as we concluded above, $Fe^+(^4F)$ is much more reactive than $Fe^+(^6D)$. To a first approximation, the true cross section for $Fe^+(^4F)$ is five times the cross section shown for Fe^+(SI).

The failure to observe $FeCH_2^+$ suggests that reaction of $Fe^+(^4F)$ does not occur via insertion. This is consistent with the conclusion that reaction of this state with H_2 is direct. It is also consistent with the observation that the ratio of MCH_3^+ to MH^+ is less in the case of Fe^+ than in the case of Sc^+. Indeed, it is about a factor of 40, close to the ratio of 42 predicted by the translational density of states.

$V^+ + C_2H_6$

Our published results[19] for the reaction of V^+ (produced by SI) with ethane are shown in Figure 26. The dominant channel is endothermic cleavage of the C-C bond to form VCH_3^+,

$$V^+ + C_2H_6 \rightarrow VCH_3^+ + CH_3 \quad (18)$$

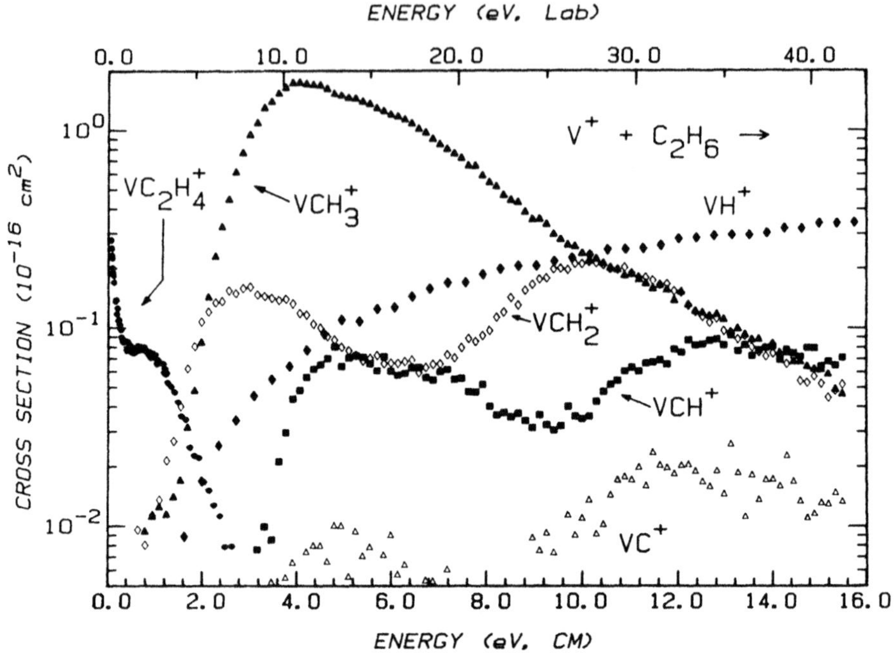

Figure 26. Cross sections for reaction of V^+ (produced by surface ionization) with ethane as a function of relative kinetic energy (lower scale) and ion energy in the laboratory frame (upper scale).

This cross section reaches a peak at 4 eV due to decomposition of this product to $V^+ + CH_3$. This is in agreement with the thermodynamic onset equal to $D°(CH_3-CH_3) = 3.9$ eV. Formation of VCH_2^+ can be seen to have two features. The lower energy feature corresponds to reaction 19,

$$V^+ + C_2H_6 \rightarrow VCH_2^+ + CH_4 \tag{19}$$

The peak in the cross section for process 19 occurs below any threshold for decomposition of VCH_2^+. This indicates that reaction 18, which has an onset at an appropriate energy, is depleting a common intermediate,

```
V+ - CH2
|     |          III
H3C   H
```

This is presumably formed by insertion of V^+ into the C-C bond of ethane. At low energies, III decomposes via a four-center elimination in process 19. At high energies, the bond fission process, reaction 18, dominates the reaction probability.

The second peak in the VCH_2^+ cross section relates to reaction 20,

$$V^+ + C_2H_6 \rightarrow VCH_2^+ + H + CH_3 \quad (20)$$

which requires 4.5 eV (= D°(H - CH_3)) more energy than process 19. Reaction 20 is a secondary decomposition of VCH_3^+ which can be seen as a break in the VCH_3^+ cross section at about 6.5 eV. The other high energy products, VCH^+ and VC^+, come from secondary decompositions of VCH_3^+ and VCH_2^+ by loss of H or H_2.

Formation of VH^+ begins at the thermodynamic limit, 2.2 eV, for process 21,

$$V^+ + C_2H_6 \rightarrow VH^+ + C_2H_5 \quad (21)$$

Competing with this is dehydrogenation of ethane,

$$V^+ + C_2H_6 \rightarrow VC_2H_4^+ + H_2 \quad (22)$$

Deuterium labeling studies (with CH_3CD_3) demonstrate that this product is a vanadium-ethene ion complex.[19] The odd part about this exothermic cross section is that while it follows the $E^{-1/2}$ energy dependence of eq 1, it is smaller than this equation predicts by a factor of ≈500. The reasons for this become apparent by varying the temperature of the surface ionization (SI) filament from 1800 to 2200 K. This changes the population of the ^5D and ^5F states very little, Table II, but the ^3F population varies by over a factor of two. The magnitude of the exothermic part of process 22 tracks with this variation. This unequivocally demonstrates that this process is due to the ^3F excited state. Not until about 0.4 eV do the quintet states begin to form $VC_2H_4^+$.

In retrospect, this result may not be so surprising. Based on the mechanisms discussed above, dehydrogenation should occur via insertion into a C-H bond of ethane to form IV, H - V^+ - C_2H_5. Since two of the four electrons on the metal are involved in binding the H and ethyl radical ligands, only two unpaired electrons remain on the metal ion. Thus, the ground state of IV (and III as well) must have triplet spin. Therefore, these intermediates correlate not with ground state reactants (a quintet and singlet) but with the lowest triplet state, $V^+(^3F)$. Apparently there is a 0.4 eV barrier to formation of IV from $V^+(^5D)$ ground state. We hypothesize that this corresponds to the curve crossing between the surface correlating with $V^+(^5D)$ and $V^+(^3F)$. For endothermic reactions, the effects of such a curve crossing are not easily observed because the endothermicity exceeds the height of the curve-crossing barrier. Presumably, such a spin-forbidden curve crossing is allowed because of spin-orbit coupling in these transition metal systems.

$Sc^+ + C_2H_6$

Figure 27 shows the results for reaction of Sc^+ (produced by SI) with ethane.[21] Variation of the SI filament temperature shows that the reactivity shown is due primarily to ground state $Sc^+(^3D)$. The cross

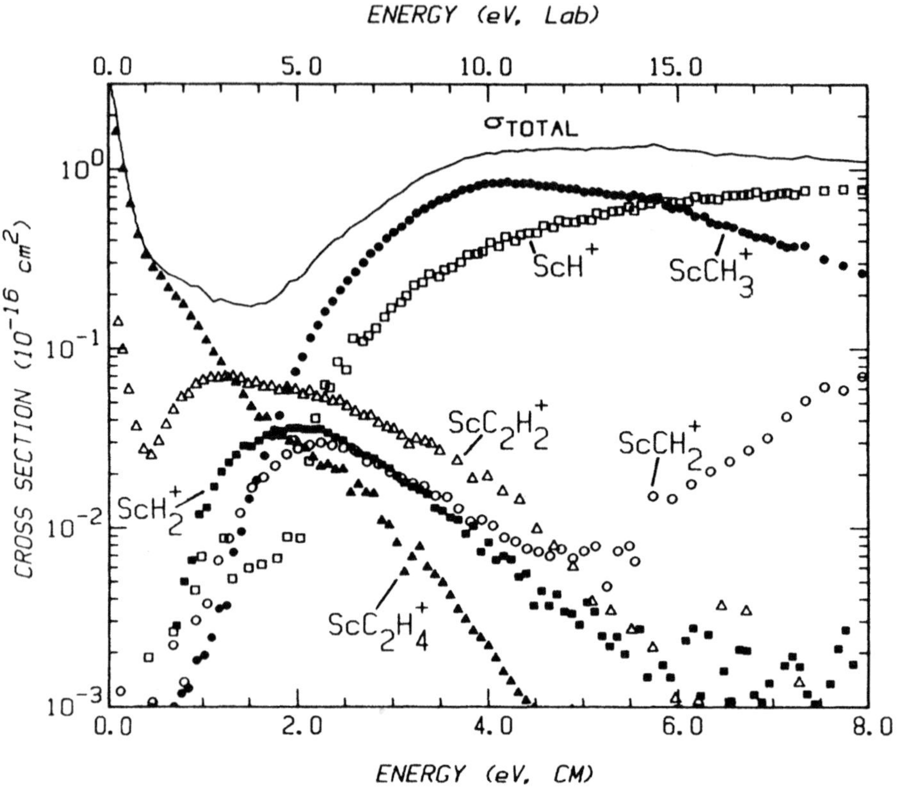

Figure 27. Cross sections for reaction of Sc^+ (produced by surface ionization) with ethane as a function of relative kinetic energy (lower scale) and ion energy in the laboratory frame (upper scale).

sections for the C-C cleavage products, $ScCH_3^+$ and $ScCH_2^+$, are quite similar in shape and only somewhat reduced in magnitude compared to those for V^+, Figure 26. We presume that a similar reaction mechanism is involved. The differences between the results shown here and those for V^+ lie in the C-H bond cleavage products. The most evident difference is that two new products are observed: $ScC_2H_2^+$ and ScH_2^+. In addition, the C-H bond cleavage products, $ScC_2H_4^+$ and ScH^+, are larger than for V^+ although production of the former still occurs at a cross section which is well below the collision limit, about 1/60.

Further insight into the C-H cleavage process can be obtained by isotopic labeling studies. In reaction of Sc^+ with CH_3CD_3, we find that the major process at low energies is

$$Sc^+ + CH_3CD_3 \rightarrow ScC_2H_2D_2^+ + HD \qquad (23)$$

As for V^+, this shows that dehydrogenation occurs mainly across the C-C bond and implies that the product is a Sc^+ - ethene complex. At the very lowest energies, reaction 24 and 25,

$$Sc^+ + CH_3CD_3 \rightarrow ScC_2HD_3^+ + H_2 \quad (24)$$

$$\rightarrow ScC_2H_3D^+ + D_2 \quad (25)$$

are also observed with equal intensities about a factor of 5 less than reaction 23. This means that scrambling can occur at the lowest energies (corresponding to the longest intermediate litetimes). A purely random result would yield a factor of 4 to 1. These results are consistent with a mechanism which involves C-H bond activation to form V followed by β-H transfer to form VI, Figure 28. Scrambling can occur via an equilibrium between V and VI. Reductive elimination of H_2 from VI yields the major product, Sc^+-ethene, while loss of ethene from VI yields ScH_2^+. This metal-dihydride ion is not observed for reaction of ethane with any other first row transition metal ion although it is observed for Y^+ and La^+ which are isovalent with Sc^+. This is presumably because these three metal ions have only two electrons both of which are involved in bonding the two H atoms in VI. Thus, the H_2Sc^+-C_2H_4 bond is a dative bond involving only donation of the ethene π electrons to the metal. Metal ions with more electrons can back-bond to the ethene ligand making elimination of C_2H_4 energetically less favorable.

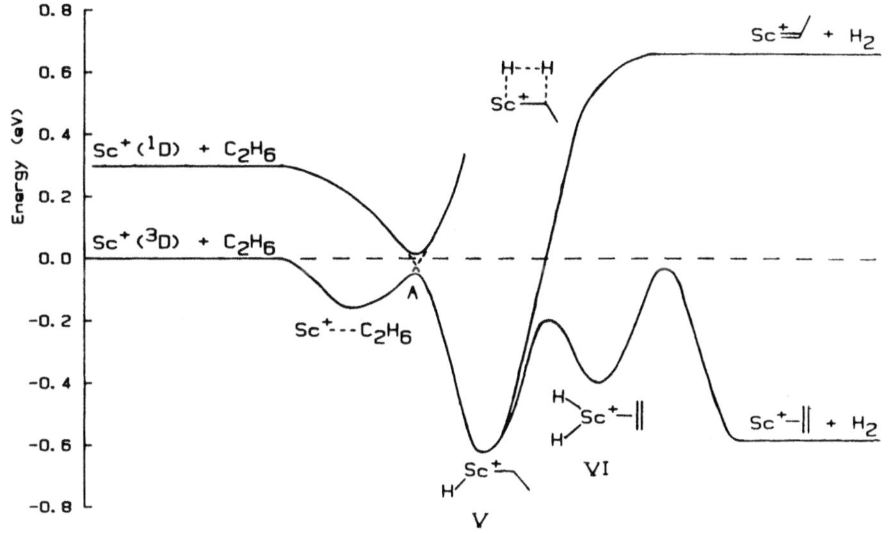

Figure 28. Potential energy surface for the dehydrogenation of ethane by Sc^+.

At higher energies, reactions 24 and 25 become more probable than reaction 23. Since these are 1,1-dihydrogen eliminations, this indicates that a scandium-ethylidene ion, $Sc^+ = CHCH_3$, is being formed, Figure 28. This 1,1-H_2 elimination is analogous to reaction 14 in the methane system.

Another effect of isotopic substitution is observed in the total cross sections for dihydrogen elimination from d_0-, 1,1,1-d_3- and d_6-ethane. A strong effect is seen such that the efficiency of dehydrogenation decreases as the degree of deuterium substitution increases. This is interpreted in terms of the need for a singlet-triplet surface crossing in formation of V. Note that this intermediate must be in a singlet state while the Sc^+ ground state is a triplet. We hypothesize that the potential energy surface (PES) looks something like Figure 28. The efficiency of reaction decreases because the position of point A on the PES depends on the relative strength of the C-H vs. C-D bond. Since the latter is stronger, point A is higher for d_6-ethane and the reaction is less efficient.

As in the case of V^+, it seems clear that a spin-forbidden surface crossing must occur if insertion to form V in its ground state is to take place. It is the need for this crossing which we believe explains the inefficiency of the reaction between Sc^+ and ethane even though the reaction is exothermic with no activation barrier. Note that this differs from the V^+ case which does exhibit a barrier. The reason for this lies in the larger splitting between the $V^+(^5D)$ and $V^+(^3F)$ states, 1.1 eV, compared to that for $Sc^+(^3D)$ and $Sc^+(^1D)$, 0.3 eV. The curve crossing thus occurs at a lower energy in the Sc^+ system.

One final point of interest in this system is the dual nature of the $ScC_2H_2^+$ product. As can be seen in Figure 27, this product is formed both at the lowest energies in an exothermic reaction with no barrier and also at higher energies with a barrier of about 0.5 eV. This process also shows[21] a very strong dependence on the degree of deuterium substitution such that the low energy feature disappears for d_6-ethane. This complex behavior is interpreted to indicate that loss of the first hydrogen molecule can occur either in the ground or first vibrational level. In the former case (v = 0), the $ScC_2H_4^+$ can further decompose by loss of a second H_2 molecule. In the latter case (v = 1), production of $ScC_2H_2^+$ is endothermic by about 0.4 eV. Zero point energy differences between H_2, HD, and D_2 can then explain the severe isotope dependences observed.

$Fe^+ + C_2H_6$

Fe^+ (produced by SI) reacts with ethane as follows:

$$Fe^+ + C_2H_6 \rightarrow FeCH_3^+ + CH_3 \quad (26)$$

$$\rightarrow FeC_2H_4^+ + H_2 \quad (27)$$

$$\rightarrow FeH^+ + C_2H_5 \quad (28)$$

$$\rightarrow FeC_2H_5^+ + H \quad (29)$$

The results[73] are shown in Figure 29. When Fe^+ is produced via the DC source, reactions 27 and 29 are no longer observed. The cross sections for processes 26 and 28 decrease by factors of about 2 and 14, respectively, at 4 eV. This behavior is similar to the changes in reactivity observed for H_2 and CH_4.

Several other features of the data shown in Figure 29 are of interest. No $FeCH_2^+$ is observed at low energies. This may imply that formation of $FeCH_3^+$ in this system occurs via a direct mechanism as postulated for reaction with H_2 and CH_4. The observation of $FeC_2H_4^+$ in contrast suggests that C-H bond activation may be occurring via an intermediate analogous to IV and V. However, this exothermic process[74] obviously has an activation barrier, the origins of which have not been fully investigated. Note that the barrier presumably cannot arise from a spin-forbidden surface crossing since the $Fe^+(^4F)$ reactive state already has the proper spin to correlate with an insertion intermediate.

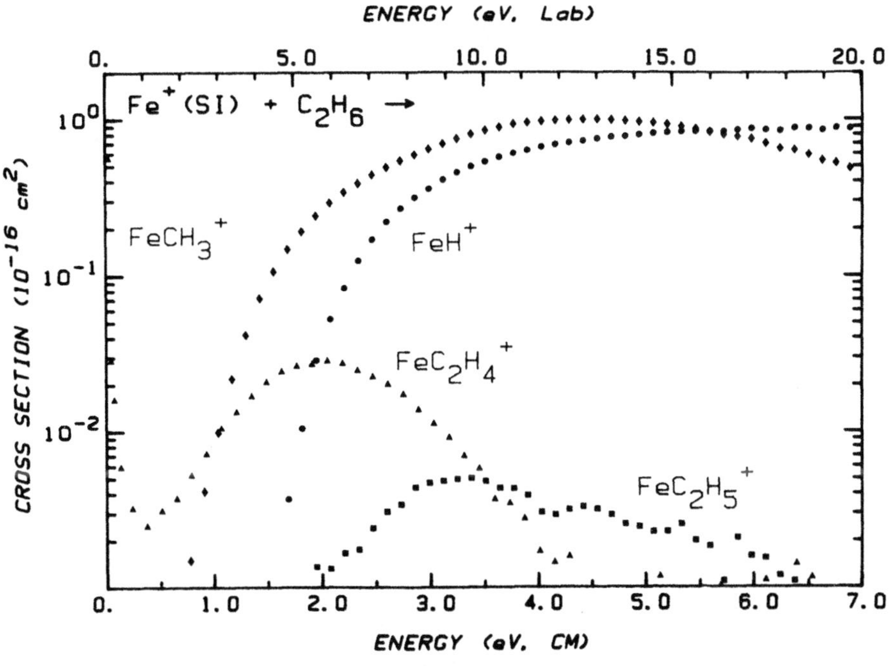

Figure 29. Cross sections for reaction of Fe^+ (produced by surface ionization, ≈80% ^6D + 20% ^4F) with ethane as a function of relative kinetic energy (lower scale) and ion energy in the laboratory frame (upper scale).

$Fe^+ + C_3H_8$

Figure 30 shows results[73] for reaction of propane with Fe^+ produced by SI. The dominant products at low energy are loss of methane to form $FeC_2H_4^+$ and loss of H_2 to form $FeC_3H_6^+$. The branching ratio at low energies is 7:3 in good agreement with ICR[75] and previous beam results.[74,76] At higher energies, the C-C cleavage products, $FeCH_3^+$ and $FeC_2H_5^+$, and the C-H bond cleavage products, FeH^+ and $C_3H_7^+$, are formed. For Fe^+(DC), cross sections for the endothermic reactions all decrease similarly to the effect observed for reaction with ethane. Surprisingly, the exothermic reactions do not show a decrease in reactivity. In fact, as shown in Figure 31, Fe^+(DC) and Fe^+(SI) show nearly identical behavior at low energies. Thus, it is the $Fe^+(^6D)$ ground state which reacts exothermically with propane! The $Fe^+(^4F)$ state does not begin to react until about 0.5 eV. The reactivity of this state peaks at about 3 eV for the $FeC_2H_4^+$ channel and about 1.7 eV for the $FeC_3H_6^+$ channel.

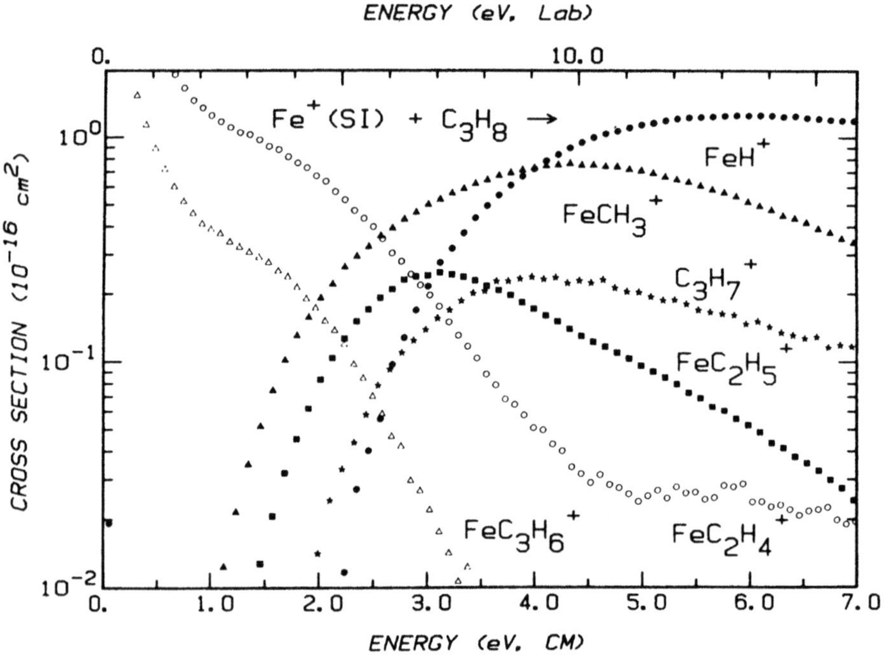

Figure 30. Cross sections for reaction of Fe^+ (produced by surface ionization, ≈80% ^6D + 20% ^4F) with propane as a function of relative kinetic energy (lower scale) and ion energy in the laboratory frame (upper scale).

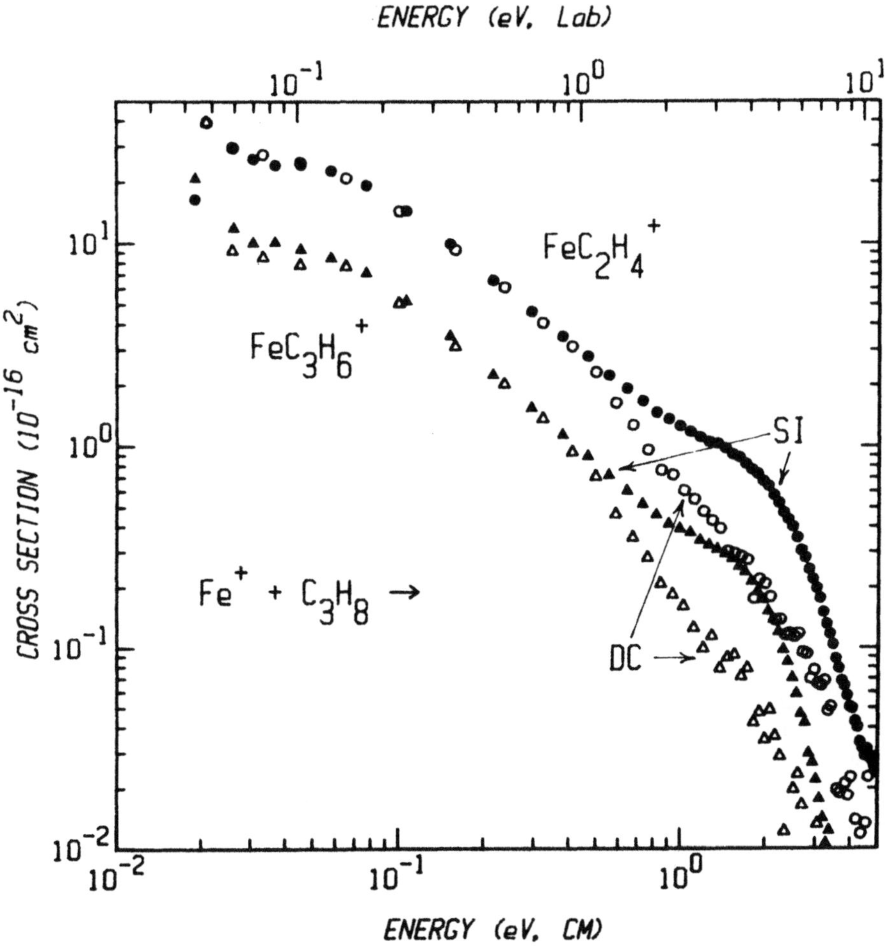

Figure 31. Cross sections for the exothermic reactions of propane with Fe^+ produced by surface ionization (SI, closed symbols, 80% ^6D + 20% ^4F) and in the drift cell source (DC, open symbols, 100% ^6D) as a function of relative kinetic energy (lower scale) and ion energy in the laboratory frame (upper scale).

The reason for this change in behavior returns to the idea of a surface crossing introduced in the V^+ + ethane system. Presume that formation of $FeC_2H_4^+$ proceeds via the C–C bond activation intermediate VII, $H_3C-Fe^+-C_2H_5$. This species must be in a state having quartet spin and thus correlates diabatically with the $Fe^+(^4F)$ state of reactants. The diabatic correlation rules established for H_2 suggest that $Fe^+(^6D)$ and alkanes should have largely repulsive interactions. Thus, the diabatic potential energy surfaces shown in Figure 32 can be drawn. The crossing can be avoided due to spin-orbit interactions to produce

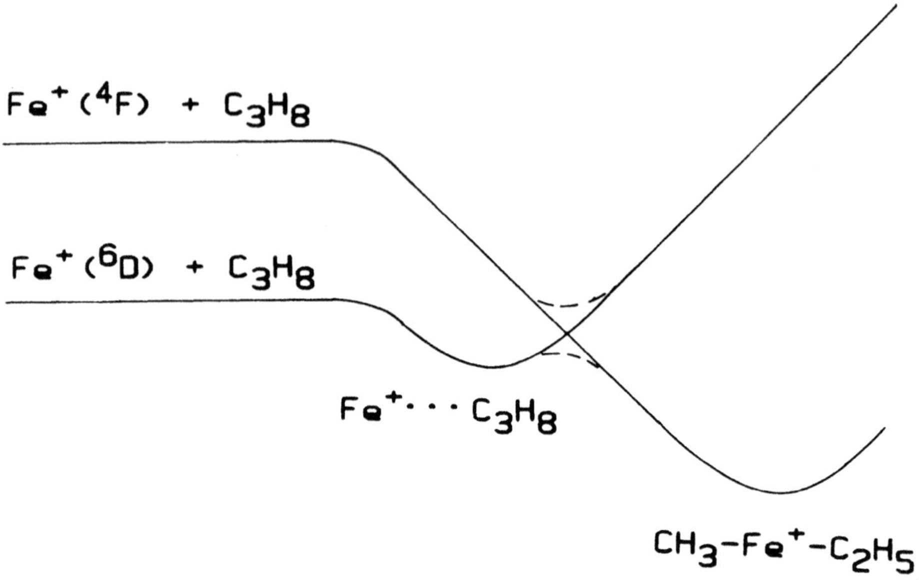

Figure 32. Potential energy surface for C-C insertion of Fe^+ in propane. Full lines represent diabatic surfaces and dashed lines are adiabatic surfaces.

two new adiabatic potential energy surfaces. This crossing can be lower in energy than the reactants due to the ion-induced dipole attraction. Since the depth of the well corresponding to the Fe^+ - alkane complex depends on the polarizability of the neutral, the larger alkanes have better access to this crossing. Also the larger alkanes will have longer lived complexes since they have more degrees of freedom and should therefore sample the crossing point more often. As the energy is raised, the reactants begin to behave diabatically. A "threshold" is thus observed for reaction of $Fe^+(^4F)$ which corresponds to the onset of diabatic behavior. By the time the endothermic products are energetically accessible, the diabatic behavior dominates the reactivity as for smaller systems (H_2, CH_4, and C_2H_6). For alkanes larger than propane (for which we have yet to perform experiments), these considerations suggest that the exothermic channels will also be due to reaction of $Fe^+(^6D)$. Indeed it is known that the efficiency of the reactions is near unity.[74-76]

METAL CLUSTER ION CHEMISTRY

A natural extension of the atomic transition metal ion reaction studies discussed above are studies of the reactivity of size-specific transition metal cluster ions. Toward this end, we have examined the reactivity of the dimer ions, Mn_2^+ and Co_2^+, and compared them to the monomer reactivity. As will be seen below, these studies demonstrate that the influence of internal energy on the reactivity of these ions can be severe. As a consequence, in order to provide quantitative information on the kinetic and thermodynamic behavior of metal cluster ions, the cluster ions must be internally cold or at least thermalized. Much of our recent work in this area has centered on the development of a source for such thermalized species.

Collision induced dissociation of Mn_2^+ and Co_2^+

In our studies of the reactions of Mn_2^+ and Co_2^+, these ions are produced by electron impact ionization and fragmentation of $Mn_2(CO)_{10}$ and $Co_2(CO)_8$, respectively. In order to limit the internal energy, the electron energy is maintained near the appearance potential of these ions, 18.8 eV for Mn_2^+ and 17.8 eV for Co_2^+.[77] Three variations of this source have been utilized in our laboratories: an in-line EI source,[6] a side-on EI source, and the EI/DC source. The first of these produces fairly hot ions while the latter two generate internally colder ions, as ascertained by collision induced dissociation (CID).

In our initial report[78] on the collision induced dissociation (CID) of Mn_2^+, reaction 30,

$$Mn_2^+ + Ar \rightarrow Mn^+ + Mn + Ar \tag{30}$$

we found that the kinetic energy threshold for CID depended strongly on the electron energy (Ee) used to form the dimer ion. A similar effect is observed for CID of Co_2^+, Figure 33. This effect is clearly due to the production of internally excited ions which therefore require less kinetic energy to dissociate. At the lowest Ee which still produced a useable beam of Mn_2^+, ~18 eV, the kinetic energy threshold for CID was determined to be 0.85 ± 0.2 eV.[78] This value is a lower limit to the bond dissociation energy of Mn_2^+, $D°(Mn_2^+)$, since this threshold is the sum of $D°(Mn_2^+)$ and the residual internal energy in the dimer ion. Since our study, Jarrold, Illies and Bowers[79] have measured this bond energy by photodissociation and obtained a lower limit of 1.39 eV. The discrepancy with our value almost certainly is a result of excited Mn_2^+ ions in our work.

To test our ability to produce a "cold" beam of ions, Mn_2^+ was produced in the side-on EI source at low Ee and in the high pressure drift cell designed to produce thermalized ions. The results for CID of Mn_2^+ ions produced using these sources give comparable results and CID thresholds of 1.1 ± 0.2 eV. Compared with the photodissociation value, this indicates that there is still a couple of tenths of an electron volt of internal excitation. While the ions are certainly not "cold" in a spectroscopic sense, the internal excitation is reasonably under control. It is somewhat mystifying that this residual excitation cannot be eliminated by any means. Similar analysis of "cold" Co_2^+, Figure 33, suggests that $D°(Co_2^+) \geq 2.8 \pm 0.3$ eV. This value is comparable to bond energies measured for Fe_2^+, 2.75 ± 0.3 eV, and $FeCo^+$, 2.86 ± 0.3 eV.[80]

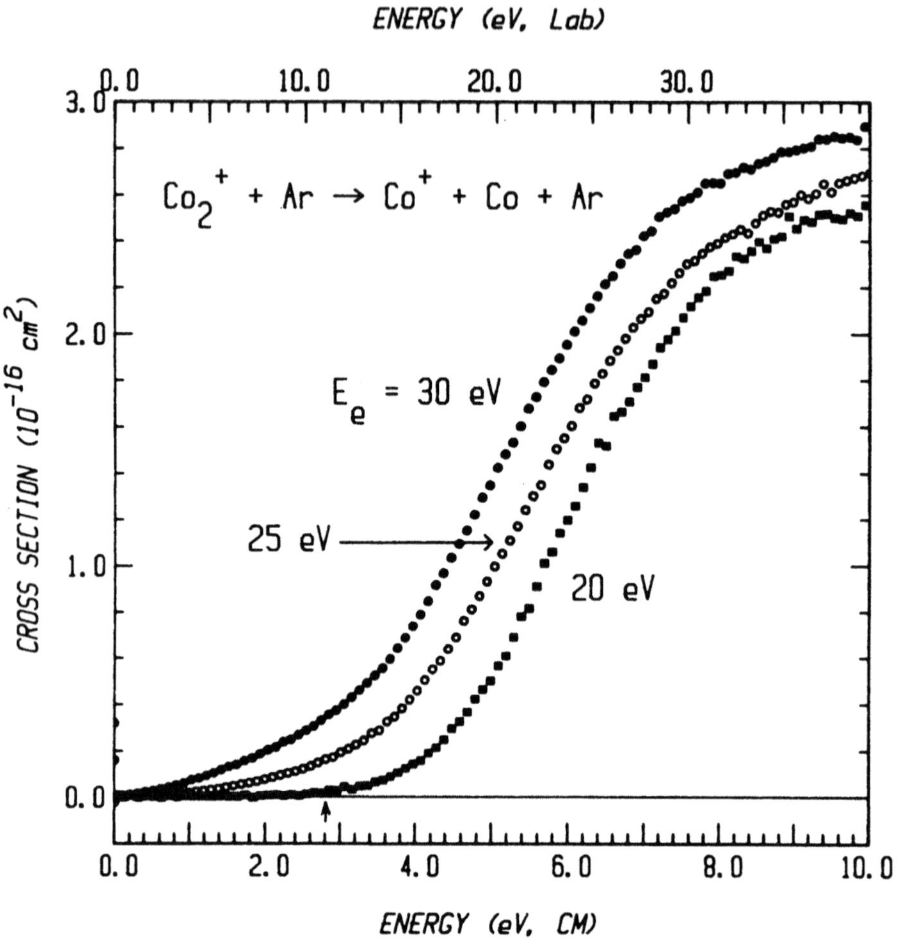

Figure 33. Collision induced dissociation of Co_2^+ by argon as a function of relative kinetic energy (lower scale) and ion energy in the laboratory frame (upper scale). Results are shown for Co_2^+ produced by electron impact of $Co_2(CO)_8$ at electron energies of 20, 25, and 30 eV. The arrow indicates the approximate threshold for the 20 eV data at 2.8 eV.

$Co_2^+ + D_2$

Results for reaction of Co_2^+ (formed at low electron energies) with D_2 are compared with results for reaction of Co^+ in Figure 34. Interestingly, both the shapes and absolute magnitudes of the hydrogen atom transfer reactions are comparable for these ionic species. In fact, the difference observed may only be due to residual internal excitation in the dimer ion. The implication of this result is that the monomer and

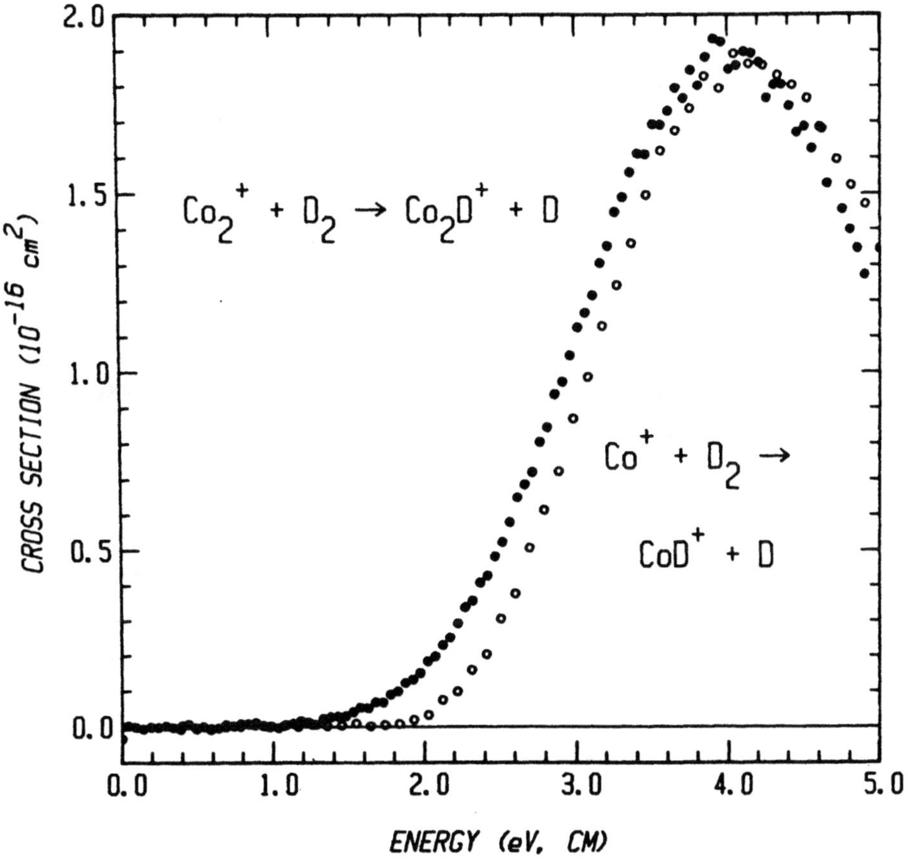

Figure 34. Cross sections for reaction of Co_2^+ (produced by electron impact at 20 eV) and Co^+ (produced by surface ionization) with D_2 as a function of relative kinetic energy.

dimer ion - deuteride bond energies are quite similar. This can be rationalized by noting that since a D atom can form a single covalent bond only, the strength of this bond may not alter appreciably when a second metal site is available for binding.

The similarity in reactivity between Co^+ and Co_2^+ observed in Figure 34 contrasts with observations of Ridge.[81] He finds that Co_2^+ is unreactive at thermal energies with large alkanes (butanes and larger) while Co^+ reacts efficiently. While much experimental work remains to be done to understand this change in behavior, one possible avenue of explanation involves the type of diabatic and adiabatic rules which control the reactivity of the monomer ions. As noted above, these "rules" change as one progresses from small to large systems.

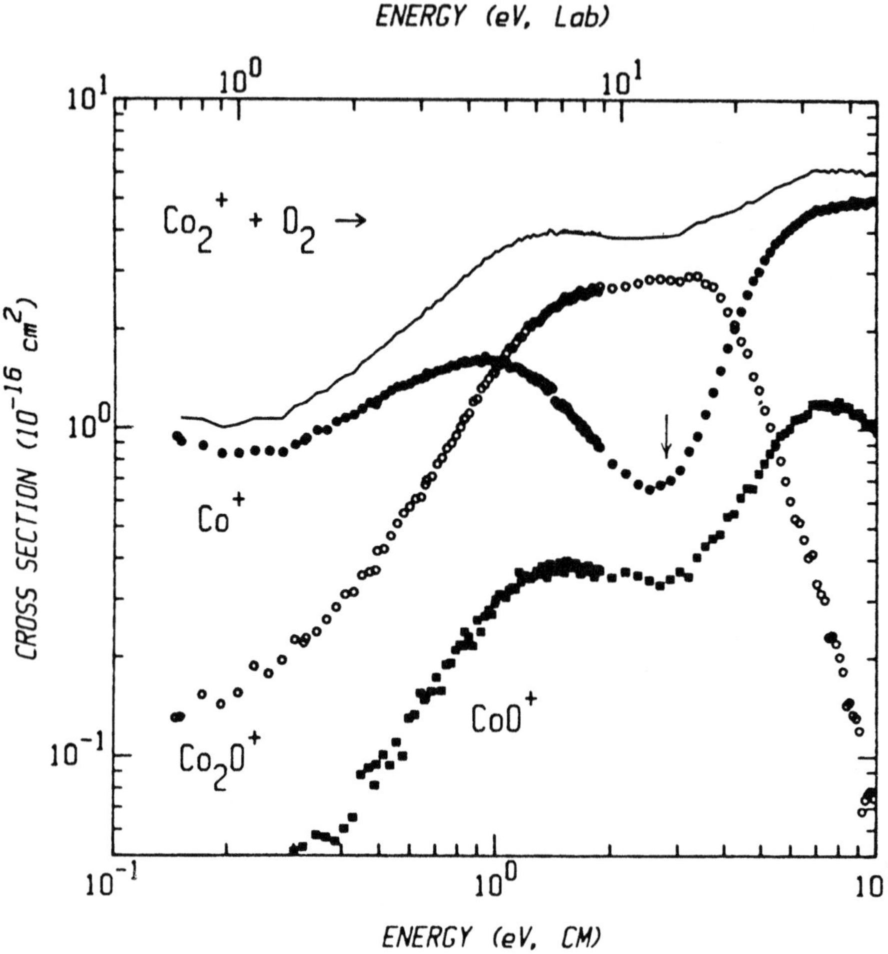

Figure 35. Cross sections for reaction of Co_2^+ (produced by electron impact at 20 eV) with oxygen as a function of relative kinetic energy (lower scale) and ion energy in the laboratory frame (upper scale). The arrow indicates the approximate onset for collision induced dissociation of Co_2^+ at 2.8 eV.

Co_2^+, Mn_2^+ + O_2

Results for reaction of molecular oxygen with "cold" Co_2^+ are shown in Figure 35. We have reported similar results for the reaction of Mn_2^+ with O_2.[82] The following reactions can be discerned:

$$Co_2^+ + O_2 \rightarrow Co^+ + CoO_2 \quad (31)$$

$$\rightarrow Co^+ + Co + O_2 \quad (32)$$

$$Co_2^+ + O_2 \rightarrow Co_2O^+ + O \quad (33)$$

$$\rightarrow CoO^+ + CoO \quad (34)$$

$$\rightarrow CoO^+ + Co + O \quad (35)$$

Reaction 31, the low energy peak in formation of Co^+, must be slightly endothermic since the cross section rises with increasing energy. The high energy peak in Co^+ corresponds to the CID process, reaction 32, beginning about 2.8 eV. Reactions 34 and 35 correspond to the two features in the cross section for CoO^+. Note that the ability to determine the multitude of reaction pathways in this system is a direct result of measuring the kinetic energy dependence of the cross sections.

The behavior shown in Figure 35 contrasts with the observed reactivity of Co^+ with O_2.[83] This ion reacts only to form $CoO^+ + O$ in a strongly endothermic reaction, 2.4 eV, which reaches a peak magnitude of about 1.5 Å^2 at about 5 eV. From this comparison, it is obvious that the dimer ion is much more reactive than the atomic ion. While this enhanced reactivity is partly due to the increase in the number of atoms in this system, it is interesting that reaction 33 has a threshold which is about 2 eV lower than the monomer reaction. Obviously, the dimer ion binds an oxygen atom much more strongly than the monomer ion, a result which is also obtained for Mn.[82] Note that this conclusion differs from that for binding a deuterium atom. This difference is presumably due to the ability of O to form multiple bonds.

The results shown in Figure 35 agree qualitatively but not quantitatively with recent FTICR results of Jacobson and Freiser.[84] They observe slow reaction of Co_2^+ with O_2 in reactions 31 and 33 with roughly a 2 to 1 ratio. Further they interpret the observation of reaction 33 to indicate that $D°(Co_2^+ - O) > D°(O_2)$. The present results show that reaction 33 is at least 10 times slower than reaction 31 and is clearly endothermic, i.e., $D°(Co_2^+ - O) < D°(O_2) = 5.2$ eV. The reason for the discrepancy is the effects of internal energy as shown in Figure 36. At the lowest kinetic energies, the cross section for reaction 33 increases by about a factor of 10 in going from production of Co_2^+ at Ee = 20 eV to 25 eV. Reaction 31, on the other hand, is not very sensitive to internal excitation. Thus, the increase in reaction 33 relative to reaction 31 observed in the FTICR studies is due to internal excitation approximately equivalent to ionization at Ee = 24 eV. This demonstrates how pervasive the effects of internal excitation are likely to be in studies of metal cluster ions.

Metal Cluster Ion Sources

As discussed above, one method for producing ionized metal clusters is the ionization and fragmentation of organometallic cluster compounds. This can be accomplished using electrons[78,79,81,82,85] as in the studies outlined above but also using photoionization.[86] An alternate route to metal cluster ion generation is the use of ion chemistry synthesis followed by CID. Freiser has pioneered this approach to form a variety of

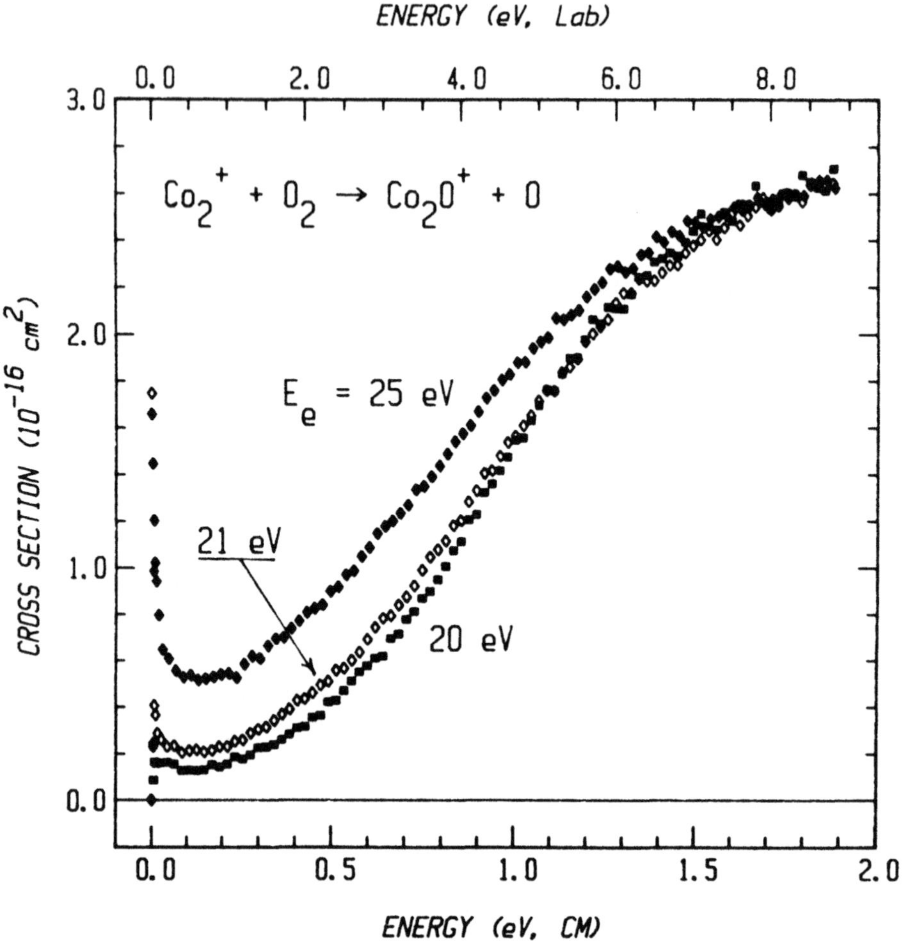

Figure 36. Cross sections for reaction 33 as a function of relative kinetic energy (lower scale) and ion energy in the laboratory frame (upper scale). Results are shown for Co_2^+ produced by electron impact at 20, 21, and 25 eV.

dimer and trimer and mixed metal cluster ions.[84,87] Particle beam sputtering has also been shown to yield intense beams of large metal cluster ions.[88-90] As detailed above, the prime difficulty associated with all of these approaches is control of the internal energy of the cluster ions (while retaining sufficient intensity) yet it is clear from the work described above that this control must be exercised if reliable, quantitative data is to be obtained.

Most of our recent efforts have been directed at developing a source of ionic metal clusters designed specifically to overcome this problem.[91] This source has four important features: 1) the metal cluster ion beam is essentially continuous; 2) clusters are internally cooled by collisions in a carrier gas flow and by supersonic expansion; 3) the initial cluster distribution is mass-selected to provide a single cluster ion; and 4) the source should be capable of generating cluster ions of most metals. Smalley's group has also shown the capability of forming cold metal cluster ions for chemical studies in a pulsed source.[92]

In our source, metal ions are produced by laser vaporization of a rotating target in a configuration similar to that of Smalley[93] using a high repetition rate (8 kHz) Cu-vapor laser (CVL). The metal vapor is entrained in a continuous flow of He through a 5.7 cm long nozzle. Here, a sputtered metal ion is rapidly thermalized and clusters with other metal atoms. Residence times are ≈170 μs such that particles entering the flow experience $\approx 10^5$ collisions with He. The clusters then undergo a mild adiabatic expansion into the initial vacuum chamber. Although we have no direct means of measuring energy distribution yet, the calculated conditions[94] of the expansion yield a translational temperature of less than 10 K. Cooling is arrested at a skimmer located approximately 2/3 the distance to the mach disk.[94] Sufficient numbers of ions are created to obviate the need for an external ionizer.[91]

Figure 37 shows a mass spectrum of an iron sample obtained with this source.[91] Cluster ions, $n < 10$, show intensities which are at least one order of magnitude higher than previously reported for transition metal clusters, although absolute intensities are frequently not reported. Ion signals are $>10^5$ counts/s to $n = 6$ and $>10^4$ counts/s to $n = 13$, an effective lower limit for ion-beam reaction studies. Fe_nO^+ and $Fe_nO_2^+$ are observed due to oxygen impurities in either the sample or the He flow. $Fe_nO_3^+$, which in neutral clusters indicate formation in a hot plasma, are not observed.[95]

We have also tested the ability of this source to produce ionic species other than pure metal clusters.[91] By introducing O_2 to the He flow upstream of the vaporization region, we are able to cleanly convert the bare metal cluster ions to metal cluster oxide ions. The absolute amount of ions does not change appreciably under these conditions. Species like $Fe_2O_y^+$ ($y = 2$-5), $Fe_3O_y^+$ ($y = 2$-6), $Fe_4O_y^+$ ($y = 4$-7) and $Fe_5O_y^+$(5-8) can easily be produced. It is clear that larger clusters show more extensive oxygenation, consistent with a preference of iron to retain an oxidation state of between 2 and 3. This indicates that the metal species undergo sufficient collisions with O_2 to reach a thermodynamically stable configuration, but not so many that the clustering process is disrupted.

These preliminary results demonstrate that intense continuous beams of cold metal cluster ions can be made by using fast laser vaporization and rare gas condensation. Further, size-specific cluster ions are easily obtained by momentum-analysis. We are nearing completion of an apparatus which includes a reaction region and mass spectrometer for analysis of reaction products. The instrument is designed to measure the energy dependences of absolute reaction cross sections. This will allow detailed characterization of the cluster energy distributions, studies of cluster binding energies and examinations of reaction thermochemistry, mechanisms, and dynamics.[78,82] The ability to create both bare and ligated clusters enables study of a variety of reaction sequences which may be useful in understanding, modeling, and designing catalytic systems.

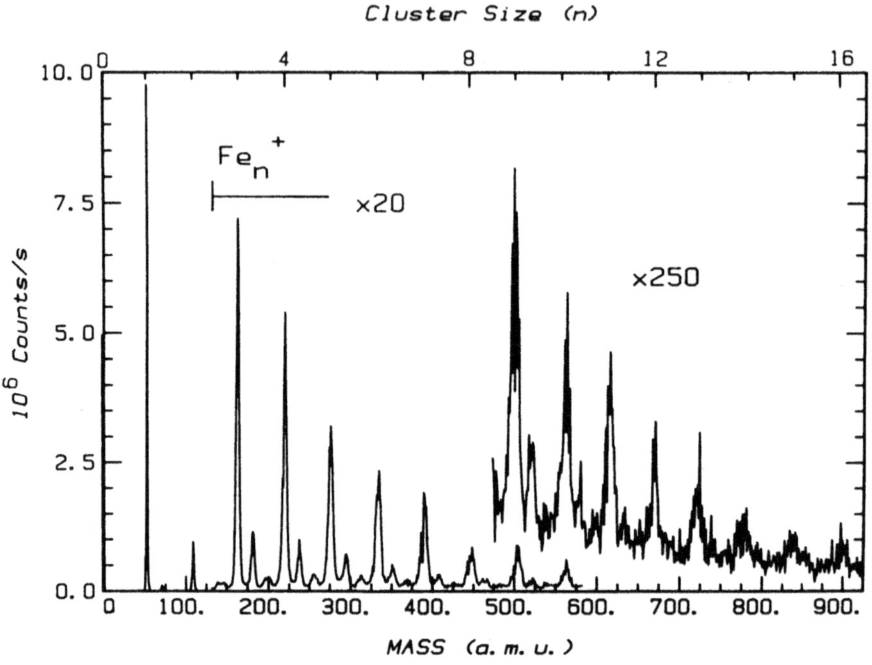

Figure 37. Mass spectrum of ionic species produced by vaporization of an iron sample with a copper vapor laser in a flow of He, see text. Positions of the bare metal clusters are indicated by the upper scale. Satellite peaks to the right of these are oxides.

ACKNOWLEDGMENT

The work described in this article was performed by my students over the past several years, Natasha Aristov, Jerry L. Elkind, Kent M. Ervin, Steve K. Loh, Rosie Georgiadis, David A. Hales, Lee Sunderlin, Mary Ellen Weber, Joel Burley, Richard Schultz and Bong Hyun Boo. I am very grateful for their dedication, hard work and insight. Early phases of this work were funded by the ACS PRF, Research Corp, and the Dreyfus Foundation. Continuing support has been provided by the National Science Foundation, the NSF Presidential Young Investigator program, and the Army Research Office.

REFERENCES

1. K. M. Ervin and P. B. Armentrout, J. Chem. Phys. 83, 166 (1985).
2. N. R. Daly, Rev. Sci. Instrum. 31, 264 (1959).
3. E. Teloy and D. Gerlich, Chem. Phys. 4, 417 (1974); D. Gerlich, Diplomarbeit, University of Freiburg, Federal Republic of Germany, 1971.
4. P. J. Chantry, J. Chem. Phys. 55, 2746 (1971).
5. C. Lifshitz, R. L. C. Wu, T. O. Tiernan, and D. T. Terwilliger, J. Chem. Phys. 68, 247 (1978).
6. H. Udseth, C. F. Giese, and W. R. Gentry, Phys. Rev. A 8, 2483 (1973).
7. J. L. Elkind and P. B. Armentrout, J. Phys. Chem. 89, 5626 (1985).
8. G. Gioumousis and D. P. Stevenson, J. Chem. Phys. 29, 294 (1958).
9. M. Henchman, "Ion-Molecule Reactions," Vol. 1, Ed. J. L. Franklin, (Plenum, New York, 1972), pg. 101.
10. B. H. Mahan, J. Chem. Ed. 52, 299 (1975).
11. R. D. Levine and R. B. Bernstein, J. Chem. Phys. 56, 281 (1972).
12. K. M. Ervin and P. B. Armentrout, J. Chem. Phys. 80, 2978 (1984).
13. J. C. Light, J. Chem. Phys. 40, 3221 (1964); P. Pechukas and J. C. Light, Ibid. 42, 3281 (1965); E. E. Nikitin, Teor. Eksp. Khim. 1, 135, 144, 248 (1965); [Theor. Exp. Chem. (Eng. Trans.) 1, 83, 90, 275 (1975)].
14. D. R. Bates, Proc. R. Soc. London A 360, 1 (1978).
15. M. E. Weber, J. L. Elkind, and P. B. Armentrout, J. Chem. Phys. 84, 1521 (1986).
16. K. M. Ervin and P. B. Armentrout, J. Chem. Phys. 84, 6750 (1986).
17. J. L. Elkind and P. B. Armentrout, J. Chem. Phys. 84, 4862 (1986).
18. J. L. Elkind and P. B. Armentrout, J. Phys. Chem., in press.
19. N. Aristov and P. B. Armentrout, J. Am. Chem. Soc. 108, 1806 (1986).
20. V. L. Talrose, P. S. Vinogradov, and I. K. Larin, "Gas Phase Ion Chemistry," Vol. 1, Ed. M. Bowers (Academic, New York, 1979), pg. 305.
21. L. Sunderlin, N. Aristov, and P. B. Armentrout, J. Am. Chem. Soc., submitted for publication.
22. A. Henglein and K. Lacmann, Adv. Mass Spectrom. 3, 331 (1966); A. Henglein, "Ion-Molecule Reactions in the Gas Phase," Ed. P. J. Ausloos (American Chemical Society, Washington, D. C., 1966), pg. 63; A. Ding, K. Lacmann, and A. Henglein, Ber. Bunsenges. Phys. Chem. 71, 596 (1967).
23. B. H. Mahan, J. Chem. Phys. 55, 1436 (1971); Accts. Chem. Res. 8, 55 (1975).
24. E. E. Ferguson, F. C. Fehsenfeld, and D. L. Albritton, "Gas Phase Ion Chemistry," Ed. M. Bowers, (Academic, New York, 1979), pg. 45.
25. J. D. Burley, K. M. Ervin, and P. B. Armentrout, J. Chem. Phys. submitted for publication.
26. D. D. Wagman, W. H. Evans, V. B. Parker, R. H. Schumm, I. Halow, S. M. Bailey, K. L. Churney, and R. L. Nuttall, J. Phys. Chem. Ref. Data 11, Supp. 2, (1982).
27. J. O. Hirshfelder, C. R. Curtiss, and R. B. Bird, "Molecular Theory of Gases and Liquids," (Wiley, New York, 1954), pg. 947.
28. K. T. Gillen, B. H. Mahan, and J. S. Winn, Chem. Phys. Lett. 22, 344 (1973); J. Chem. Phys. 58, 5373 (1973); Ibid. 59, 6380 (1973).
29. D. M. Hirst, J. Phys. B 17, L505 (1984).

30. K. P. Huber and G. Herzberg, "Molecular Spectra and Molecular Structure IV. Constants of Diatomic Molecules," (Van Nostrand, Princeton, 1979).
31. M. A. Gittins and D. M. Hirst, Chem. Phys. Lett. 35, 534 (1975); C. F. Bender, J. H. Meadows, and H. F. Schaefer, Faraday Discuss. Chem. Soc. 62, 59 (1977).
32. J. A. Luine and G. H. Dunn, Ap. J. 299, 167 (1985).
33. J. B. Marquette, B. R. Rowe, G. Dupeyrat, and E. Roueff, Astron. Astrophys. 147, 115 (1985).
34. N. G. Adams and D. Smith, Chem. Phys. Lett. 117, 67 (1985).
35. C. E. Moore, Natl. Stand. Ref. Data Ser., Natl. Bur. Stand. 1, No. 35 (1970).
36. K. M. Ervin and P. B. Armentrout, J. Chem. Phys. 84, 6738 (1986).
37. D. G. Truhlar, J. Chem. Phys. 51, 4617 (1969).
38. J. L. Elkind and P. B. Armentrout, J. Phys. Chem. 88, 5454 (1984).
39. The heat of formation of Si^+ is 294.63 ± 1 kcal/mol [JANAF Tables, see M. W. Chase, J. L. Curnutt, J. R. Downey, R. A. McDonald, A. N. Syverud, and E. A. Valenzuela, J. Phys. Chem. Ref. Data, 11, 695 (1982)] and that for SiH_2^+ is 276 ± 1 kcal/mol [A. Ding, R. A. Cassidy, L. S. Cordis, and F. W. Lampe, J. Chem. Phys. 83, 3426 (1985); B.-H. Boo and P. B. Armentrout, J. Phys. Chem., submitted for publication].
40. From spectroscopic data given in ref 38.
41. J. L. Elkind and P. B. Armentrout, work in progress.
42. Thermochemical data for RgH^+ is taken from S. G. Lias, J. F. Liebman, and R. D. Levin, J. Phys. Chem. Ref. Data, 13, 695 (1984). Supplementary data is from ref 26.
43. E. G. Jones, R. L. C. Wu, B. M. Hughes, T. O. Tiernan, and D. G. Hopper, J. Chem. Phys. 73, 5631 (1980).
44. H. von Kock and L. Friedman, J. Chem. Phys. 38, 1115 (1963).
45. W. A. Chupka and M. E. Russell, J. Chem. Phys. 49, 5426 (1968).
46. P. R. Kemper and M. T. Bowers, Int. J. Mass Spectrom. Ion Phys. 52, 1 (1983); I. Dotan and W. Lindinger, J. Chem. Phys. 76, 4972 (1982); W. Lindinger, E. Alge, H. Stori, M. Pahl, and R. N. Varney, Ibid. 67, 3495 (1977); N. G. Adams, D. K. Bohme, D. B. Dunkin, and F. C. Fehsenfeld, Ibid. 52, 1951 (1970).
47. R. D. Smith, D. L. Smith, and J. H. Futrell, Chem. Phys. Lett. 32, 513 (1975); Int. J. Mass Spectrom. Ion Proc. 19, 395 (1976).
48. K. Tanaka, J. Durup, T. Dato, and I. Koyano, J. Chem. Phys. 74, 5561 (1981).
49. N. G. Adams, D. Smith, and E. Alge, J. Phys. B: Atom. Molec. Phys. 13, 3235 (1980).
50. K. M. Ervin and P. B. Armentrout, J. Chem. Phys. in press.
51. P. J. Kuntz and A. C. Roach, J. Chem. Soc. Faraday Trans. 2 68, 259 (1972).
52. P. F. Fennelly, J. D. Payzant, R. S. Hemsworth, and D. K. Bohme, J. Chem. Phys. 60, 5115 (1974).
53. H. Laue, J. Chem. Phys. 46, 3034 (1967).
54. P. B. Armentrout and J. L. Beauchamp, J. Am. Chem. Soc. 103, 784 (1981).
55. S. A. Safron, G. D. Miller, F. A. Rideout and R. C. Horvat, J. Chem. Phys. 64, 5051 (1976); G. D. Miller and S. A. Safron, Ibid. 64, 5065

(1976); J. A. Rutherford and D. A. Vroom, Ibid. 65, 4445 (1976); P. B. Armentrout, R. V. Hodges, and J. L. Beauchamp, Ibid. 66, 4683 (1977); J. Am. Chem. Soc. 99, 3162 (1977).
56. G. D. Flesch and H. J. Svec, Inorg. Chem. 14, 1817 (1975).
57. A. E. Stevens and J. L. Beauchamp, Chem. Phys. Lett. 78, 291 (1981).
58. J. L. Elkind and P. B. Armentrout, J. Phys. Chem. submitted for publication.
59. P. B. Armentrout and J. L. Beauchamp. Chem. Phys. 50, 37 (1980).
60. J. L. Elkind and P. B. Armentrout, J. Am. Chem. Soc. 108, 2765 (1986).
61. L. F. Halle, F. S. Klein, and J. L. Beauchamp, J. Am. Chem. Soc. 106, 2543 (1984).
62. J. B. Schilling, W. A. Goddard, and J. L. Beauchamp, J. Am. Chem. Soc. 108, 582 (1986).
63. J. L. Elkind and P. B. Armentrout, Inorg. Chem. 25, 1078 (1986).
64. P. B. Armentrout, L. F. Halle, and J. L. Beauchamp, J. Am. Chem. Soc. 103, 6501 (1981).
65. M. L. Mandich, L. F. Halle, and J. L. Beauchamp, J. Am. Chem. Soc. 106, 4403 (1984).
66. Ab initio calculations (ref 62) indicate that the MH^+ species contain only ≈10% 4p character.
67. C. J. Ballhausen and H. B. Gray, "Molecular Orbital Theory," (Benjamin/Cummings, Reading, 1964).
68. J. L. Elkind and P. B. Armentrout, work in progress.
69. J. Allison, R. B. Freas, and D. P. Ridge, J. Am. Chem. Soc. 101, 1332 (1979).
70. M. L. Steigerwald and W. A. Goddard, J. Am. Chem. Soc. 106, 308 (1984).
71. N. Aristov, L. Sunderlin, R. Georgiadis, and P. B. Armentrout, work in progress.
72. Because of increased sensitivity, the results of ref 70 differ from those reported in L. F. Halle, P. B. Armentrout, and J. L. Beauchamp, J. Am. Chem. Soc. 103, 962 (1981).
73. R. H. Schultz, J,L. Elkind, and P. B. Armentrout, J. Am. Chem. Soc., submitted for publication.
74. L. F. Halle, P. B. Armentrout, and J. L. Beauchamp, Organometallics 1, 963 (1982).
75. G. D. Byrd, R. C. Burnier, and B. S. Freiser, J. Am. Chem. Soc. 104, 3565 (1982); D. B. Jacobson and B. S. Freiser, Ibid. 105, 5197 (1983).
76. R. Houriet, L. F. Halle, and J. L. Beauchamp, Organometallics 2, 1818 (1983).
77. R. E. Winters and R. W. Kiser, J. Phys. Chem. 69, 1618 (1965).
78. K. Ervin, S. K. Loh, N. Aristov, and P. B. Armentrout, J. Phys. Chem. 87, 3593 (1983).
79. M. F. Jarrold, A. J. Illies, and M. T. Bowers, J. Am. Chem. Soc. 107, 7339 (1985).
80. D. B. Jacobson and B. S. Freiser, J. Am. Chem. Soc. 106, 4623 (1984).
81. R. B. Freas and D. P. Ridge, J. Am. Chem. Soc. 102, 7129 (1980); D. P. Ridge, "Ion Cyclotron Resonance Spectrometry," Ed. H. Hartman and

K. P. Wanczek, (Springer-Verlag, New York, 1982).
82. P. B. Armentrout, S. K. Loh, and K. M. Ervin, J. Am. Chem. Soc. 106, 1161 (1984).
83. P. B. Armentrout, L. F. Halle, and J. L. Beauchamp, J. Chem. Phys. 76, 2449 (1982).
84. D. B. Jacobson and B. S. Freiser, J. Am. Chem. Soc. 108, 27 (1986).
85. K. G. Leopold, T. M. Miller, and W. C. Lineberger, J. Am. Chem. Soc. 108, 178 (1986).
86. V. Vaida, N. J. Cooper, R. J. Hemley, and D. G. Leopold, J. Am. Chem. Soc. 103, 7022 (1981); D. G. Leopold and V. Vaida, Ibid. 105, 6809 (1983).
87. D. B. Jacobson and B. S. Freiser, J. Am. Chem. Soc. 106, 4623, 5351 (1984); 107, 1581 (1985); R. L. Hettich and B. S. Freiser, Ibid. 107, 6222 (1985).
88. G. Delacretaz, P. Fayet, and L. Woste, Ber. Bunsenges. Phys. Chem. 88, 284 (1984); P. Fayet and L. Woste, Surf. Sci. 156, 134 (1985).
89. L. Hanley and S. L. Anderson, Chem. Phys. Lett. 122, 410 (1985); Proc. SPIE 669 (1986).
90. R. B. Freas and J. E. Campana, J. Am. Chem. Soc. 107, 6202 (1985).
91. S. K. Loh, D. A. Hales, and P. B. Armentrout, Chem. Phys. Lett., in press.
92. L. S. Zheng, P. J. Brucat, C. L. Pettiette, S. Yang, and R. E. Smalley, J. Chem. Phys. 83, 4273 (1985); P. J. Brucat, L. S. Zheng, C. L. Pettiette, S. Yang, and R. E. Smalley, Ibid. 84, 3078 (1986).
93. T. G. Dietz, M. A. Duncan, D. E. Powers, and R. E. Smalley, J. Chem. Phys. 74, 6511 (1981).
94. R. Campargue, J. Phys. Chem. 88, 4466 (1984); J. P. Toennies and K. Winkelmann, J. Chem. Phys. 66, 3965 (1977).
95. R. L. Whetten, D. M. Cox, D. J. Trevor, and A. Kaldor, J. Phys. Chem. 89, 566 (1985).

REACTIONS OF TRANSITION METAL IONS WITH CYCLOALKANES AND METAL CARBONYLS

Douglas P. Ridge
Department of Chemistry
University of Delaware
Newark, DE 19716

INTRODUCTION

It has long been known that metal containing ions formed by electron impact on metal compounds undergo a variety of interesting ion molecule reactions. The formation of "triple-decker sandwiches" in the ion molecule reactions of ferrocene and nickelocene was reported in 1966 (Schumacher and Taubenest, 1966). Cationic clustering reactions in metal carbonyls were reported in 1971 (Kraihanzel, Conville and Sturm, 1971 and Foster and Beauchamp, 1971). Clustering reactions of transition metal containing anions were reported in 1973 (Dunbar, Ennever and Fakler, 1973). Reactions of metal containing ions with ligand molecules were described in 1971 (Muller and Fenderl, 1971, and Foster and Beauchamp, 1971). Proton affinities of transition metal complexes were reported in 1975 (Foster and Beauchamp, 1975a and 1975b). Observation of alkane activation by a transition metal containing ion was first reported in 1973 (Muller and Goll, 1973). This early work led to a substantial body of research on organometallic ion molecule reactions. Rather than attempt to review that work, this discussion will emphasize (1) some recent developments in the study of reactions involving the formation of metal to carbon bonds and (2) clustering reactions in transition metal carbonyls. In doing this, the significance of the unique properties of transition metals and the use of a variety of mass spectrometric techniques to solve problems in transition metal ion chemistry will be discussed.

OXIDATIVE ADDITION AND THE FORMATION OF METAL TO CARBON BONDS

One of the characteristics of transition metals is, in the terminology of formal oxidation states, their ability to change oxidation states. A particularly interesting example is the reaction of Fe^+ with methyl iodide first observed using ion cyclotron resonance (icr) techniques (Allison and Ridge, 1975):

P. Ausloos and S. G. Lias (eds.), Structure/Reactivity and Thermochemistry of Ions, 165–175.

$$Fe^+ + CH_3I \longrightarrow \begin{cases} \xrightarrow{48\%} FeI^+ + CH_3 \\ \xrightarrow{52\%} FeCH_3^+ + I \end{cases} \qquad (1)$$

This is interesting for two reasons. First, the formal oxidation state of the metal changes in both channels. Second, a metal to carbon sigma bond is formed. This suggested that the metal forms a strong bond to carbon. It was possible to form a complex of Fe^+ and methyl iodide. The chemistry of that complex suggested that it had the structure of an oxidative addition product: $I\text{-}Fe\text{-}^+\text{-}CH_3$ (Allison and Ridge, 1976). The reactions of alkyl halides and alcohols with group 8 atomic transition metal ions were found to be generally consistent with an oxidative addition mechanism (Allison and Ridge, 1976 and 1979a). The reaction of larger alkyl halides was found to involve the shift of an H atom to the metal from a carbon β to the metal. This is followed by loss of the halide and the hydrogen atom as a hydrogen halide molecule. An elimination of this kind is referred to as a reductive elimination (Allison and Ridge, 1976 and 1979a).

In 1979 Allison, Freas and Ridge (1979b) described icr experiments in which the reactions of 2-methyl propane with Fe^+ were observed:

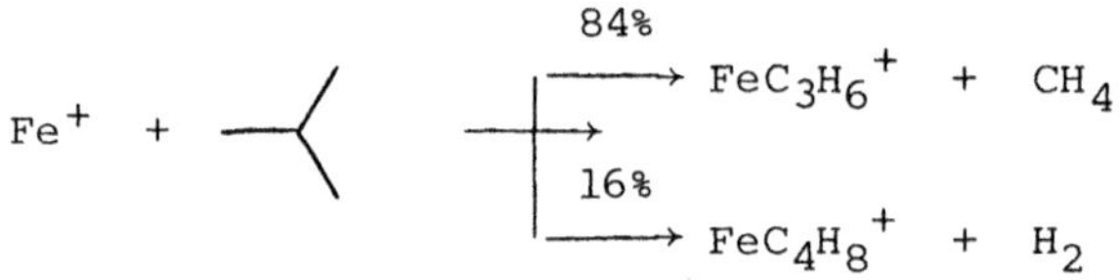

(2)

The mechanism suggested for the reaction resembled that postulated for the alkyl halides. It involves oxidative addition of the C-C bond to the metal followed by a β H atom shift and reductive elimination of methane:

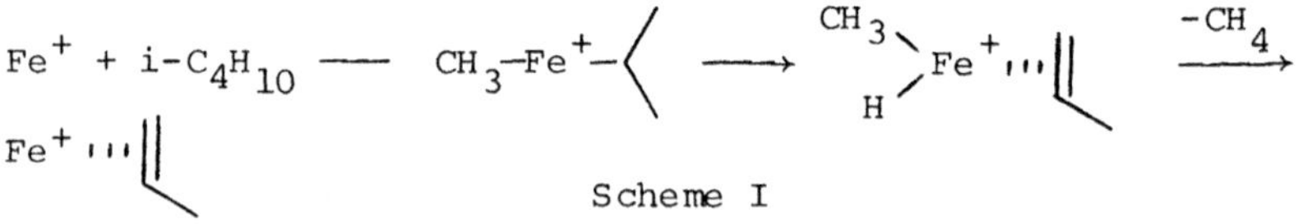

Scheme I

The H_2 loss channel was proposed to proceed in an analogous fashion:

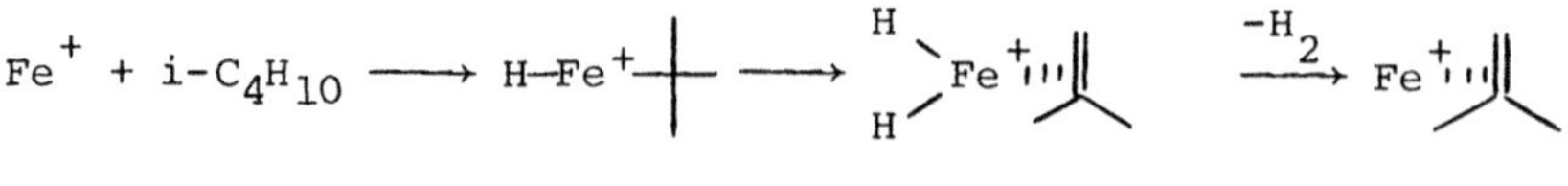

Scheme II

Oxidative addition of alkane bonds, particularly C-C bonds, to transition metal species is very unusual so observation of reaction (2) and the analogous reactions for Co^+ (Armentrout and Beauchamp, 1980) and Ni^+ (Freas and Ridge 1980) led to a number of studies of similar processes (see for example Houriet, Halle and Beauchamp, 1983, Larsen and Ridge, 1984 and Jacobsen and Freiser 1983a). These studies suggest that alkanes react readily with a number of atomic transition metal ions. The reactions result in cleavage of both C-C and C-H bonds. In general, the observed chemistry is consistent with the oxidative addition, β - H atom shift, reductive elimination mechanism illustrated in Schemes I and II. Interesting questions remain, however, about the details of the mechanism in a number of specific instances (see particularly Houriet, Halle and Beauchamp, 1983).

The failure of Fe^+ to cleave C-C bonds in cyclohexane, for example, (Allison, 1978, Kalmbach and Ridge, 1983, Peake, Gross and Ridge, 1984 and Jacobsen and Freiser, 1983b) seemed to contradict the preference for C-C bond cleavage in the chemistry of Fe^+ with straight chain alkanes (Allison, Freas and Ridge, 1979, Freas and Ridge, 1980, Houriet, Halle and Beauchamp, 1983, Jacobsen and Freiser, 1983a and Larsen and Ridge, 1984). This led to the suggestion (Kalmbach and Ridge, 1983) that the low energy structure for the product of oxidative addition of a C-C bond to Fe^+ is linear (1):

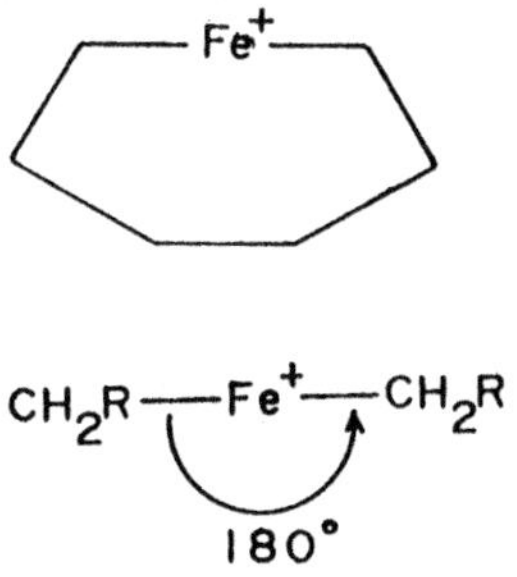

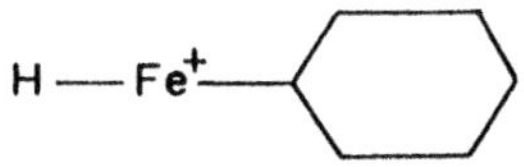

(1) (2)

A cyclohexane or cyclopentane ring could not accommodate a linear $C\text{-}Fe^+\text{-}C$ moiety without ring strain. This would explain the failure of the C-C bonds in cyclohexane to add oxidatively to Fe^+. The C-H bonds add to the metal instead (2), and molecular hydrogen is eliminated from the cycloalkane by a mechanism analogous to Scheme II. To test this hypothesis, the reactions of larger cycloalkanes were examined. It was expected that for larger rings C-C bond cleavage would occur. This was, in fact, observed. All of the products of reaction of cyclooctane with Fe^+ involve cleavage of C-C bonds:

$$Fe^+ + \text{cyclooctane} \longrightarrow \begin{cases} \xrightarrow{73\%} FeC_6H_{12}^+ + C_2H_4 \\ \xrightarrow{12\%} FeC_2H_4^+ + C_6H_{12} \\ \xrightarrow{9\%} FeC_4H_8^+ + C_4H_8 \\ \xrightarrow{6\%} FeC_4H_6^+ + C_4H_8 + H_2 \end{cases} \qquad (3)$$

This is a dramatic chemical effect. The fraction of overall product corresponding to loss of molecular hydrogen without loss of carbon containing fragments is plotted as a function of ring size in Fig. 1.

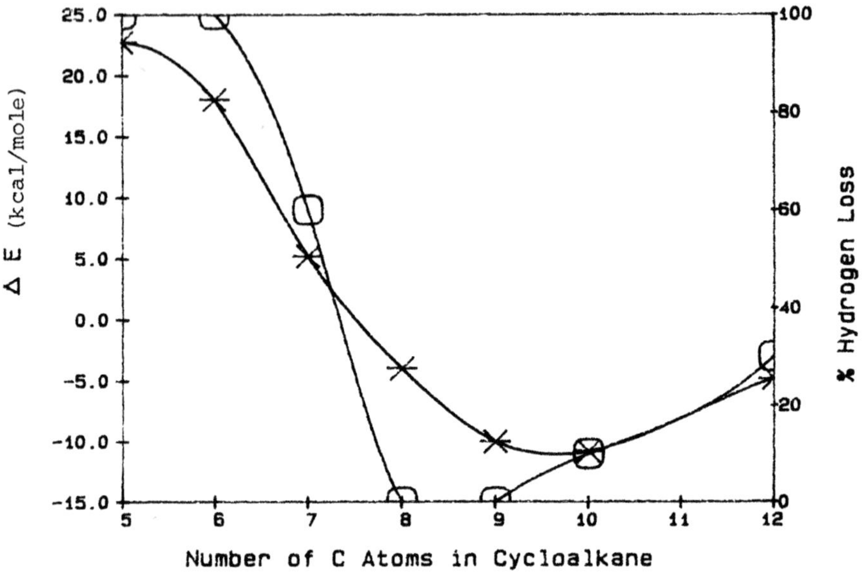

Fig. 1. ΔE, steric energy of inserting acetylenic group into alkanes, and H_2 loss as a percent of total reaction with Fe^+ as a function of number of carbons in the cycloalkane. The circles represent percent H_2 loss and the asterisks, ΔE.

Cyclopentane and cyclohexane react only to lose molecular hydrogen. There is no C-C bond cleavage. Cyclooctane and Cyclononane, on the other hand, react only by C-C bond cleavage. This tends to support the hypothesis of a preference for a linear structure for the product of oxidative addition of a C-C bond to Fe^+.

Molecular mechanics calculations further verify this "ring strain" effect. Parameters for transition metals are lacking in standard programs, but the effect of inserting a long, stiff group into a cycloalkane can be modelled by comparing cycloalkanes and cycloalkynes (Kalmbach and Ridge). Introduction of a triple bond requires a stiff, linear four carbon segment in an alkyl chain. As suggested in Fig. 2, the length of such a segment is comparable to the expected length of a linear C-Fe^+-C segment. The difference between the "steric energy" of a cycloalkane with n+2 carbons and that of a cycloalkyne with n+4 carbons is thus taken as a measure of the steric consequence of inserting Fe^+ in a linear fashion into a ring with n+2 carbons. In terms of Fig. 2, ΔE steric and ΔE' steric are assumed to vary in the same way with n. The results of such steric energy calculations done with the

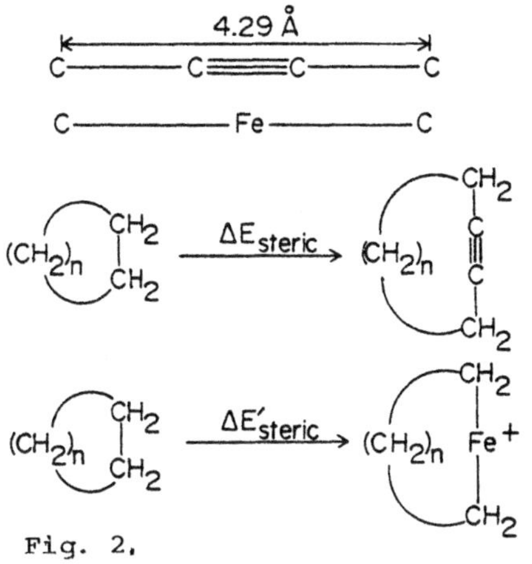

Fig. 2.

standard MM2 program are shown in Fig. 1. Note that steric energy includes only the energy of bending, torquing, compressing and stretching bonds as well as van der Waals interactions between atoms not bonded to each other.

The results in Fig. 1 show that introducing a linear group into cyclopentane or cyclohexane produces substantial steric strain. Introducing a linear group into cyclooctane or cyclononane actually reduces the steric strain. The eight and nine membered rings are large enough to accommodate the linear group without bending the linear bonds. In addition, introducing the linear group reduces the transannular repulsive interaction between H atoms found in intermediate rings. This difference in steric energy, of course, coincides with the change in chemistry. When the steric energy is unfavorable for insertion of a linear group, only hydrogen loss occurs. When the steric energy is favorable for insertion of a linear group, only C-C bond cleavage occurs. This further substantiates the hypothesized linear structure for the oxidative addition product (1).

An interesting result of the molecular mechanics calculations is illustrated in Fig. 3.

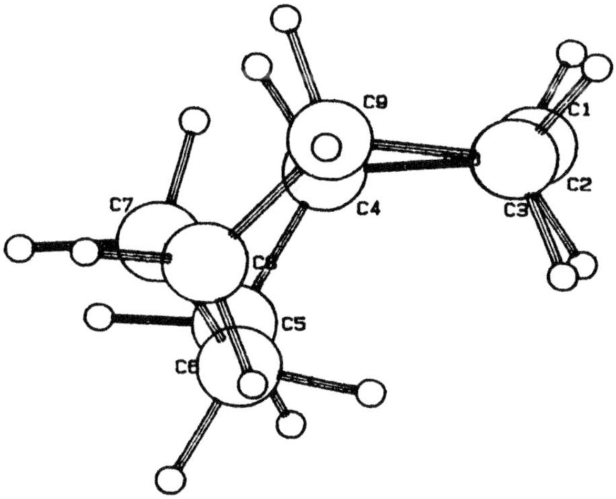

Fig. 3. Structure of cyclodecyne looking down the axis of the triple bond between C_1 and C_2.

This shows the calculated low energy configuration of cyclodecene. Cyclodecene is the model for the product of insertion of Fe^+ into cyclooctane. The viewpoint for the diagram is looking down the axis of the triple bond between carbon 1 and carbon 2. An interesting feature is the proximity of the H atom on carbon 6 to the triple bond. Were the triple bond replaced with an Fe^+ (as suggested in Fig. 2) the H atom on carbon 6 would be the one closest to the metal. This suggests the possibility of the mechanism shown in Scheme III:

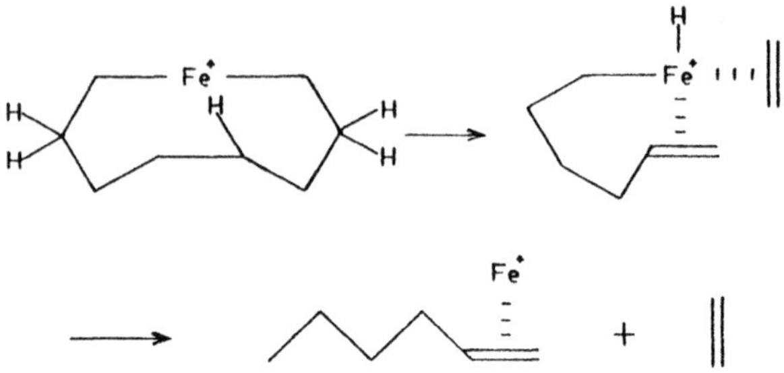

Scheme III

This predicts that the product of reaction of cyclooctane with Fe^+ would be a complex of 1 hexene with Fe^+. As indicated in reaction (3), the major product has that stoichiometry. Characterization of that product by collision induced dissociation verifies that the structure is that of a 1 hexene complex (Kalmbach and Ridge). This again substantiates the hypothesis of a linear oxidative addition product.

It is difficult to construct a bonding scheme for a linear $C\text{-}Fe^+\text{-}C$ moiety without invoking the participation of a 4p orbital on the metal. The promotion energy of the 4p orbitals is quite high (4.77 eV), but if the charge is delocalized to the alkyl groups, the promotion energy will drop. Preliminary results suggest that the chemistry of Co^+ and Ni^+ with the cycloalkanes is consistent with a bent oxidative addition product (Kalmbach and Ridge). This is consistent with the fact that the 4p promotion energies of Co^+ (5.60 eV) and Ni^+ (6.39 eV) are higher than that for Fe^+.

Study of the chemistry of Cr^+ provided insight into the relationship between electronic structure of a metal ion and its reactivity (Freas and Ridge, 1980, Armentrout, Halle and Beauchamp, 1981 and Reents, Strobel, Freas, Wronka and Ridge, 1985). Kinetic studies indicated that two long lived states of Cr^+ are formed by electron impact on chromium hexacarbonyl. The excited state was found to have a lifetime in excess of 2 seconds. This state reacts with methane to form a chromium methylene cation. Analysis of the kinetic data indicated, however, that the primary product of the interaction between methane and the excited state of Cr^+ is ground state Cr^+:

$$Cr^{+*} + CH_4 \longrightarrow H\text{—}Cr^+\text{—}CH_3 \longrightarrow CrCH_2^{\ +} + H_2$$

$$\qquad\qquad\qquad\qquad \lfloor\!\longrightarrow Cr^+ + CH_4^{\ *}$$

Scheme IV

Appearance potential measurements suggest the excited state to be about 2.5 eV above the ground state (Armentrout, Halle and Beauchamp, 1981). The candidate states are the a ^4D ($3d^4$ 4s) at 2.421 eV and the a ^4G ($3d^5$) at 2.543 eV. Methane has no excited electronic states at that energy. This clearly indicates that the usual rules regarding collisional quenching are not valid for systems such as this one. Spin and parity forbidden transitions seem to occur quite readily. This must be considered when applying Armentrout's rules to reactive systems involving large molecules. Armentrout's rules, described elsewhere in this volume, account for the reactivity of atomic transition metal ions with molecular hydrogen. These rules, derived from simple molecular orbital theory, specify the relationship between the electronic configuration of the metal ion and the mechanism of its reaction with hydrogen. In applying these rules to reactions of larger molecules, the possibility must be considered that the reaction does not occur on the surface on which the initial approach occurs. Even nominally forbidden transitions may readily occur.

CLUSTERING REACTIONS OF METAL CARBONYLS

The next topic illustrates the importance of the ligand environment in the chemistry of transition metals. As mentioned in the introduction, metal carbonyls react with their electron impact produced fragment ions to produce clusters with several metal atoms and a number of ligand molecules. Shown in Fig. 4 is the evolution with time following an ionizing pulse of a series of such cluster ions in decacarbonyl dirhenium (Meckstroth, Reents, and Ridge). Data are plotted for clusters containing six metal atoms and various numbers of carbonyls. The ions initially are formed in sequential reactions of smaller ions with the neutral metal carbonyl. Most of them then begin to disappear as they react with the metal carbonyl to form clusters with eight metal atoms. It is clear from the curves that the more carbonyls a cluster has the more stable it is. The clusters with 14 and 15 carbonyls react rapidly while the clusters with 20 or more carbonyls react slowly if at all.

These results illustrate the general relationship between reactivity and coordinative unsaturation that obtains in coordination chemistry. Each metal center can accommodate 18 valence electrons. A ligand such as CO donates two electrons to a metal. In a metal carbonyl cluster, each metal-metal bond increases the total number of valence electrons by two. If the ionic rhenium clusters are considered to have closed polyhedral metal cores with metal to metal bonds along the edges, then the number of valence electrons in each cluster can be determined. If that number is less than 18 per metal atom, then an electron deficiency can be assigned. An electron deficiency of two indicates that on average each metal atom in the cluster needs an additional two electrons to fill its valence shell to 18. Such a cluster should be very reactive. Clusters with smaller electron deficiencies should be less reactive. Fig. 5 shows a plot of the log of the rate constants derived from the data in Fig. 4 and related data

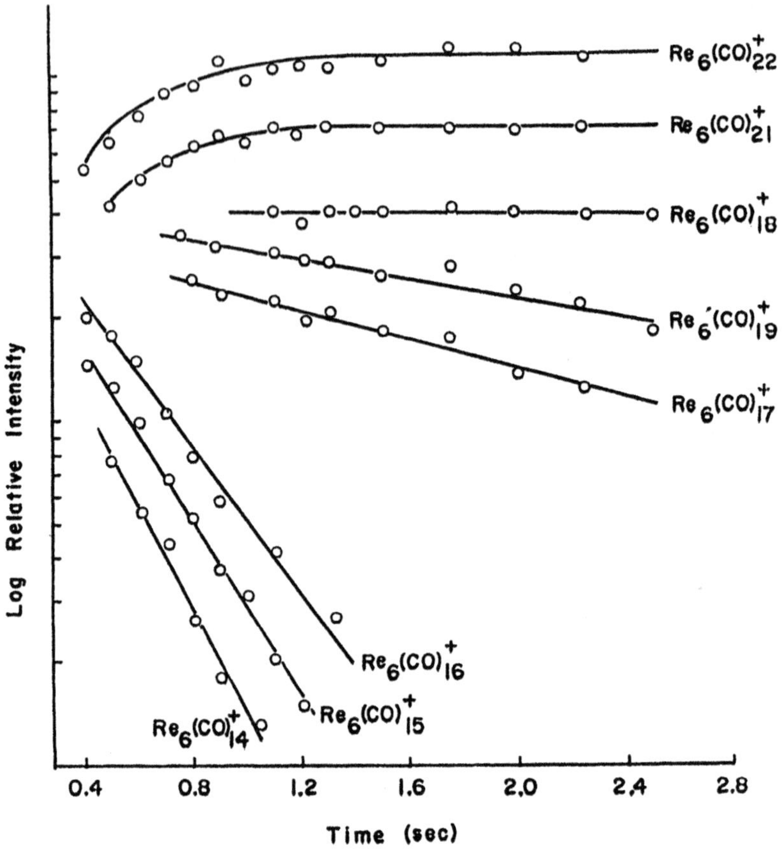

Fig. 4. Time dependence of relative concentrations of $Re_6(CO)_m{}^+$ species in 9.5×10^{-8} torr of $Re_2(CO)_{10}$.

on other rhenium carbonyl cluster ions (Meckstroth, Ridge and Reents, 1985). For large clusters molecular orbital theory predicts that coordinative saturation sometimes requires more than 18 electrons per metal atom. This has been accounted for in assigning the electron deficiencies plotted in Fig. 5. As expected, the reactivity drops off as the electron deficiency drops below two.

There have now been several studies of clustering reactions which have used analyses of this type to suggest structural information about these cluster ions (Wronka and Ridge, 1984, Meckstroth, Freas, Reents and Ridge, 1985, Freeden and Russell, 1985 and 1986). That is, the relationship between electron deficiency and reactivity can be used to deduce electron deficiency from reactivity. From the electron deficiency, the number of metal-metal bonds can be deduced. From the number of metal-metal bonds, possible structures can be deduced.

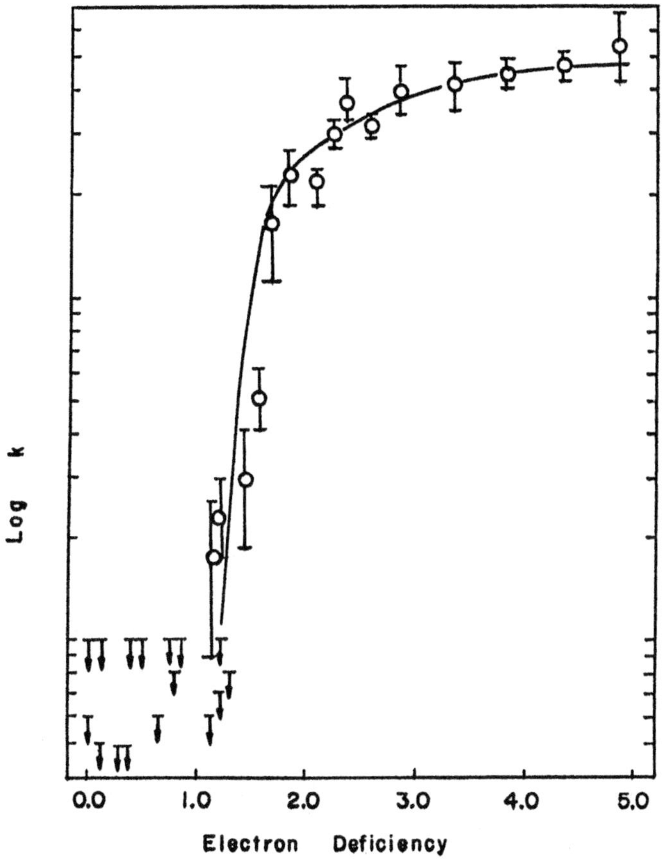

Fig. 5. Relative rate constants for reaction of $Re_n(CO)_m^+$ ions with $Re_2(CO)_{10}$ vs electron deficiency.

OTHER PROBLEMS AND FUTURE PROSPECTS

The clustering results emphasize the importance of a proper understanding of metal-ligand interactions. Unfortunately, quantitative data on the strengths of metal-ligand bonds are few. There are several scales of relative transition metal ligand bond affinities (Kappes and Staley 1982a, b, Jones and Staley 1982a, b), but only two scales of absolute ligand affinities for transition metal ions (Corderman and Beauchamp, 1976, Larsen, Freas and Ridge, 1984). This is an area in need of attention.

There have been few studies of atomic transition metal anions. Such studies can produce valuable thermochemical data as well as interesting chemistry (Squires, 1985). There is a recent very interesting study of the chemistry of a dipositive transition metal ion

(Weisshaar, 1986). Both of these areas deserve much more study.

There have been relatively few attempts to model condensed phase processes in which transition metal containing ions are thought to play an important role. Particularly elegant exceptions are a study of the mechanism of the water gas shift reaction (Lane, Lee, Sallans and Squires, 1985) a study of charge transfer reactions of cyclopentadienyl complexes (Eyler, 1985). This is another area of opportunity in need of attention.

Recent work on bare metal cluster ions is too extensive to review here, but it promises to be a rich and productive area for some years to come.

REFERENCES

Allison, J., Ph.D. Dissertation, University of Delaware, 1977.

Allison, J.; Freas, R. B., III and Ridge, D. P. (1979) J. Amer. Chem. Soc. 101, 1332.

Allison, J. and Ridge, D. P. (1976) J. Amer. Chem. Soc. 98, 7445.

Allison, J. and Ridge, D. P. (1979) J. Amer. Chem. Soc. 101, 4998.

Allison, J. and Ridge, D. P. (1975) J. Organomet. Chem. 99, C11.

Armentrout, P. B. and Beauchamp, J. L. (1980) J. Amer. Chem. Soc. 102, 1736.

Armentrout, P. B.; Halle, L. F. and Beauchamp, J. L. (1981) J. Amer. Chem. Soc. 103, 6501.

Corderman, R. R. and Beauchamp, J. L. (1976) J. Amer. Chem. Soc. 98, 3998.

Dunbar, R. C.; Ennever, J. F. and Fackler, J. P., Jr. (1973) Inorg. Chem. 12, 2735.

Eyler, J. R. and Richardson, D. E. (1985) J. Amer. Chem. Soc. 107, 6130.

Foster, M. S. and Beauchamp, J. L. (1971) J. Amer. Chem. Soc. 93, 4924.

Foster, M. S. and Beauchamp, J. L. (1975a) J. Amer. Chem. Soc. 97, 4808.

Foster, M. S. and Beauchamp, J. L. (1975b) J. Amer. Chem. Soc. 97, 4814.

Freas, R. B., III and Ridge, D. P. (1980) J. Amer. Chem. Soc. 102, 7129.

Fredeen, D. A. and Russell, D. H. (1985) J. Amer. Chem. Soc. 107, 3762.

Fredeen, D. A. and Russell, D. H. (1986) J. Amer. Chem. Soc. 108, 1860.

Houriet, R.; Halle, L. F. and Beauchamp, J. L. (1983) Organomet. 2, 1818.

Jacobsen, D. B. and Freiser, B. S. (1983) J. Amer. Chem. Soc. 105, 5197.

Jacobsen, D. B. and Freiser, B. S. (1983) J. Amer. Chem. Soc. 105, 7484.

Jones, R. W. and Staley, R. H. (1982a) J. Phys. Chem. 86, 1387.

Jones, R. W. and Staley, R. H. (1982b) J. Amer. Chem. Soc. 104, 2296.

Kalmbach, K. A. and Ridge, D. P. (1983) presented at the 31st Annual Conference on Mass Spectrometry and Allied Topics; Boston, MA; May 8-13, 1983.

Kalmbach, K. A. and Ridge, D. P. unpublished results.

Kappes, M. M. and Staley, R. H. (1982a) J. Amer. Chem. Soc. 104, 1813.

Kappes, M. M. and Staley, R. H. (1982b) J. Amer. Chem. Soc. 104, 1819.

Kraihanzel, C. S.; Conville, J. J. and Sturm, J. E. (1971) Chem. Comm. 159.

Lane, K. R.; Sallans, L. and Squires, R. R. (1985) J. Amer. Chem. Soc. 106, 5767.
Larsen, B. S.; Freas, R. B., III and Ridge, D. P. (1984) J. Phys. Chem. 88, 6014.
Larsen, B. S. and Ridge, D. P. (1984) J. Amer. Chem. Soc. 106, 1912.
Meckstroth, W. K.; Freas, R. B.; Reents, W. D., Jr. and Ridge, D. P. (1985) Inorg. Chem. 24, 3139.
Meckstroth, W. K.; Ridge, D. P. and Reents, W. D., Jr. (1984) J. Phys. Chem. 89, 612.
Meckstroth, W. K.; Ridge, D. P. and Reents, W. D., Jr. (1985) J. Phys. Chem. 89, 612.
Muller, J. and Fenderl, K. (1971) Chem. Ber. 104, 2207.
Muller, J. and Goll, W. (1973) Chem. Ber. 106, 1129.
Peake, D. A.; Gross, M. L. and Ridge, D. P. (1984) J. Amer. Chem. Soc. 106, 4307.
Reents, W. D., Jr.; Strobel, F.; Freas, R. B., III; Wronka, J. and Ridge, D. P. (1985) J. Phys. Chem. 89, 5666.
Schumacher, E. and Taubenest, R. (1966) Helv. Chim. Acta. 47, 1525.
Squires, R. R. (1985) J. Amer. Chem. Soc. 107, 4385.
Tonkyn, R. and Weisshaar, J. C. (1986) J. Amer. Chem. Soc., submitted.
Wronka, J. and Ridge, D. P. (1984) J. Amer. Chem. Soc. 106, 67.

GAS PHASE METAL ION CHEMISTRY:

SUMMARY OF THE PANEL DISCUSSION

Robert R. Squires
Department of Chemistry
Purdue University
West Lafayette, Indiana 47907

Investigation of the properties and reactions of transition metal ions and metal-containing compounds continues to represent one of the most active research areas in contemporary gas phase ion chemistry. The rapid growth of this field has been inspired by an increased need for basic physical data for inorganic and organometallic species, and by the rich and fascinating reactivity which bare atomic metal ions and metal-ion fragments have shown with simple organic compounds. The family of experimental methods used to study gas phase metal ion chemistry has grown steadily over the last decade, and now includes guided-ion beam instruments, flowing afterglows, high pressure mass spectrometers, and multiple sector instruments, in addition to the ICR and FT-ICR methods which fostered the early development of the field. Notable recent advances in gas phase metal ion chemistry include the measurement of highly precise absolute bond energies for metal-hydrogen, metal-alkyl and metal-oxygen bonds using ion-beams, thermal energy ion-molecule reactions and photodissociation; increased activity involving transition metal negative ion chemistry; the development of means to study novel metal cluster ions, including homonuclear and heteronuclear clusters ranging in size from two to twenty metal atoms; and formation and study of reactive metal ion complexes with intermediate ranges of coordinative unsaturation (12-16 electrons).

In addition to the specific participant contributions which follow, the group assembled for this workshop briefly discussed several outstanding problems in gas phase metal ion chemistry. A general concern was expressed regarding the role of excited states in atomic metal ion and metal ion fragment reactions. Useful probes for the presence of excited states include the use of multiple neutral precursors for the metal ion reactants, careful attention to the occurrence of endothermic reaction channels, and testing for possible pressure effects and/or electron energy effects on the occurrence or the kinetics of reactions. The need for further development of methods for ionizing high molecular weight, involatile, and thermally labile (fragile) metal complexes was cited. Experiments designed to systematically evaluate the effects of coordinative unsaturation on metal ion reactivity would be extremely useful for developing a relationship between the atomic metal ion

P. Ausloos and S. G. Lias (eds.), Structure/Reactivity and Thermochemistry of Ions, 177–184.

chemistry which has become so prevalent, and the chemistry of organometallic complexes in solution. More transition metal thermochemistry is needed; especially useful would be means to relate the growing database for metal ions to neutral metal species. Additional studies of second and third-row metal ions was also called for in order to evaluate columnar trends in reactivity. Finally, continued work with transition metal cluster ions and neutral metal clusters was recommended in order to further evaluate the possible relationship between these species and bulk metals and surfaces.

STEREOSELECTIVE REACTIONS OF TRANSITION METAL ATOMIC IONS (D. P. Ridge, Department of Chemistry, University of Delaware, Newark, Delaware 19716

We describe elsewhere in this volume evidence that Fe^+ does not attack C-C bonds in 5-6 membered rings because of a ring strain effect.[1] We suggested that this is because the preferred geometry of the product of oxidative addition of a C-C bond to Fe^+ is linear. One consequence of this is that the chemistry of Fe^+ with cyclopentanes and cyclohexanes is dominated by elimination of molecular hydrogen from the cycloalkane and formation of a metal cycloalkene complex. We have postulated that Fe^+ induces hydrogen elimination by oxidative addition of a C-H bond to the metal, shift of a beta H atom to the metal and finally reductive elimination of molecular hydrogen from the metal.[2] This would effect a syn 1-2 elimination of hydrogen.

This picture of the reactions of Fe^+ with cyclohexanes leads us to expect that the reactions of fused ring compounds will show stereospecificity. In particular it should not be possible for Fe^+ to eliminate hydrogen across the bond between the bridgehead carbons of a trans fused ring compound made up of six-membered rings. On the other hand, the metal should readily eliminate hydrogen across the bond between the bridgehead carbons in a cis ring fused compound. As a result, the expected reactions for cis and trans decalin are given in reactions (1) and (2).

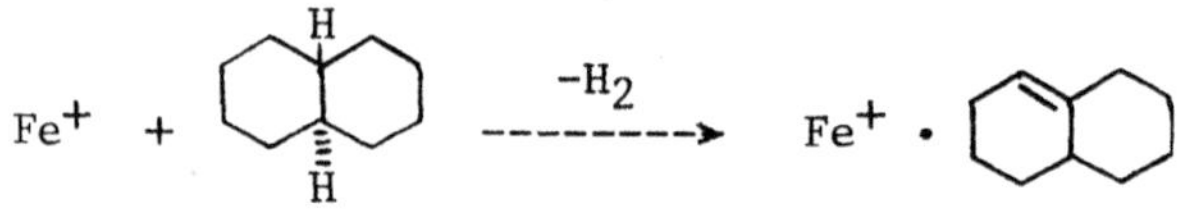

(1)

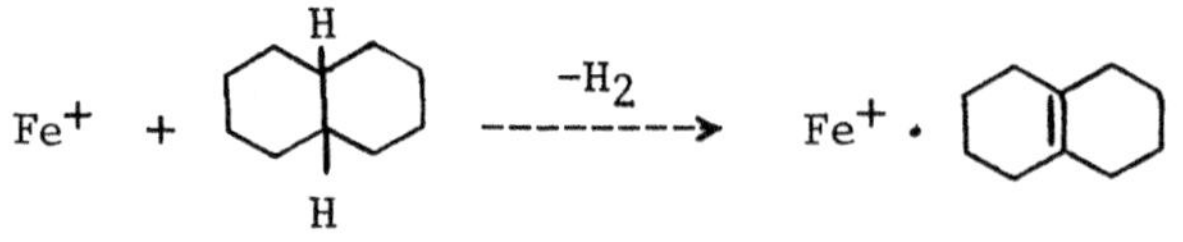

(2)

The stoichiometries of the observed products of reaction of cis and trans decalin are given in reactions (3) - (10).

$$Fe^{+} + \text{trans-decalin} \xrightarrow{36\%} Fe(C_{10}H_{16})^{+} + H_2 \quad (3)$$

$$\xrightarrow{20\%} Fe(C_{10}H_{14})^{+} + 2H_2 \quad (4)$$

$$\xrightarrow{25\%} Fe(C_7H_8)^{+} + C_3H_6 + H_2 \quad (5)$$

$$\xrightarrow{19\%} Fe(C_{10}H_{12})^{+} + 3H_2 \quad (6)$$

$$Fe^{+} + \text{cis-decalin} \xrightarrow{24\%} Fe(C_{10}H_{16})^{+} + H_2 \quad (7)$$

$$\xrightarrow{38\%} Fe(C_{10}H_{14})^{+} + 2H_2 \quad (8)$$

$$\xrightarrow{2\%} Fe(C_7H_8)^{+} + C_3H_6 + H_2 \quad (9)$$

$$\xrightarrow{47\%} Fe(C_{10}H_{12})^{+} + 3H_2 \quad (10)$$

CID experiments were performed to test whether the products of reactions (3) and (7) have the structures shown in reactions (1) and (2). Ions which we took to have the indicated structures were prepared in reactions (11) and (12).

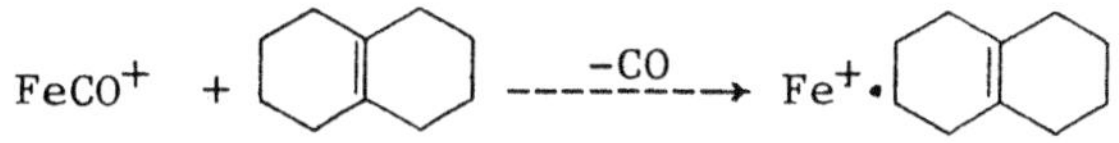

(11)

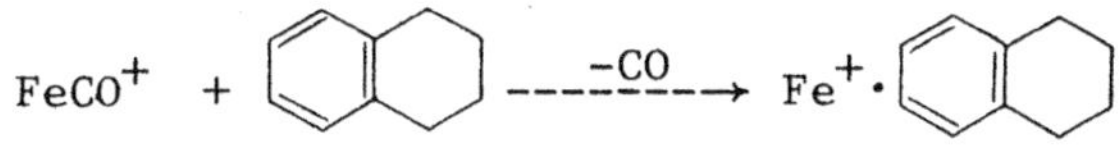

(12)

The products of reactions (7) and (11) were found to have an identical CID spectrum which differed substantially from the CID spectrum of the product of reaction (3). This verifies our initial expectation for the product of the cis decalin reaction.

We note also that the products of Reactions (10) and (12) have an identical CID spectrum which differs from that of the product of reaction (10). This suggests that successive hydrogen eliminations of hydrogen molecules all occur on the same side of the ring. Hence the cis and trans decalin on elimination of three hydrogen molecules give different tetralin isomers. Formation of a bond between the bridgehead carbons is only possible when those carbons bear hydrogens on the same side of the ring, i.e. in the cis decalin.

The product of reaction (5) is particularly interesting. It is an important product in the trans decalin and almost absent in the cis compound. The mechanism outlined in Scheme I rationalizes the reaction in the trans case. Elimination of two hydrogen molecules from one side of the first ring proceeds first. Another syn elimination is not possible. A beta alkyl shift rather than a beta hydrogen shift, however, may be particularly favorable. There is a cis beta alkyl

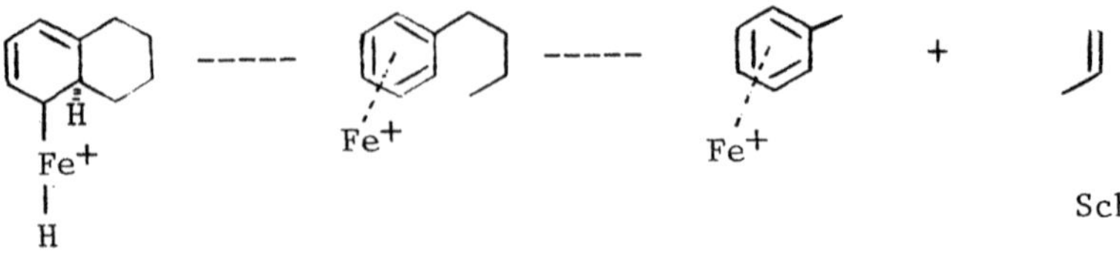

Scheme I

group and the pertinent C-C bond is allylic and therefore activated. Reductive elimination of a C-H bond followed by attack of a benzylic C-C bond gives the observed product. This suggests that the multiple elimination reactions involve sequential steps in which the metal moves from one carbon to its neighbor. Presumably it moves in an energy well resulting from favorable interactions of the metal with one lobe of a pi system formed by the eliminations.

These results tend to verify our initial picture of the reactivity of the cycloalkanes with Fe^+. They are in agreement with the predictions implied by that picture. The results also suggest that atomic transition metal ion chemistry can be used to characterize subtle structural molecular features.

THE STUDY OF VIC-DIOXIMATES OF NICKEL, PALLADIUM AND PLATINUM BY METASTABLE MAPPING (M. G. Santana-Marques and A. J. Ferrer-Correia, CEQMA, Department of Chemistry, University of Aveiro, 3800 Aveiro, Portugal)

The vic-dioximes and vic-dioximates shown in Figure 1 were studied by metastable mapping, using a reverse geometry mass spectrometer.

Computer storage of the detector output while both the magnetic field (B) and the ESA field (E) strengths are scanned (to cover the whole of the B/E plane) allows a series of peak intensities superimposed on a bidimensional surface to be obtained. With this type of data acquisition, all the existing information can be examined simultaneously, either in a "tridimensional" representation or using color coded maps. The second method is to be preferred as the information gathered is more obvious to the user.

By means of the appropriate software, different types of scans can be simulated, i.e. parents in the 1st and 2nd FFR, constant neutral losses, etc., and the fragmentations in the different regions of the mass spectrometer can then be identified. Simultaneously, as the origins and nature of all the ion currents in the whole of the mass spectrum can be understood, the so called "artifact" peaks acquire a new meaning and can be studied as well as the "normal" peaks, and consecutive fragmentations along the flight path can be easily detected. When appropriate, the mass and energy of some of the metastable peaks were confirmed using a high resolution triple sector mass spectrometer.

Based on these maps the metastable supported fragmentation pattern shown in Figure 2 was proposed for the dimethylglyoximates of Ni, Pd and Pt. As it can be observed, there are two main branches which eventually intercross, originated in ions $MetL_2$ - OH and $MetL_2$ - NO. The fragmentations proceed preferentially by loss of small mass fragments and no metastable peak for the loss of a ligand and/or a ligand minus

vic-dioximes

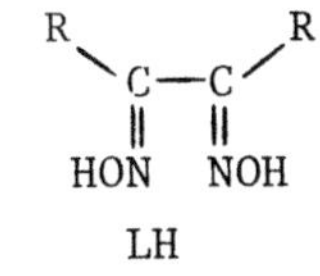

R=R =H	glyoxime
R=H, R =CH_3	methylglyoxime
R=R =CH_3	dimethylglyoxime
R=R =C_6H_5	diphenylglyoxime
R=R =C_4H_3O	α-furildioxime

vic-dioximates

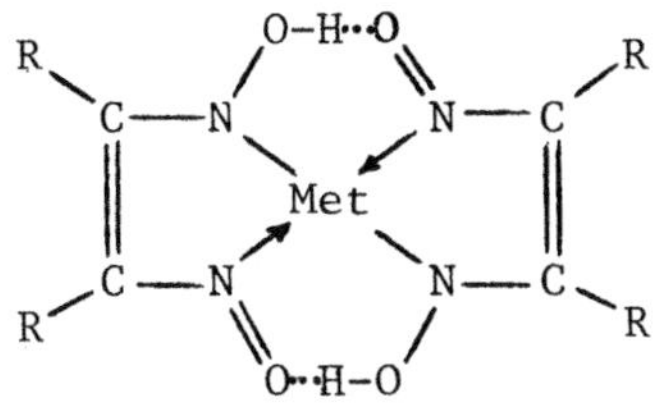

$MetL_2$
Met = Ni, Pd, Pt

Figure 1. Metal complexes studied by metastable mapping.

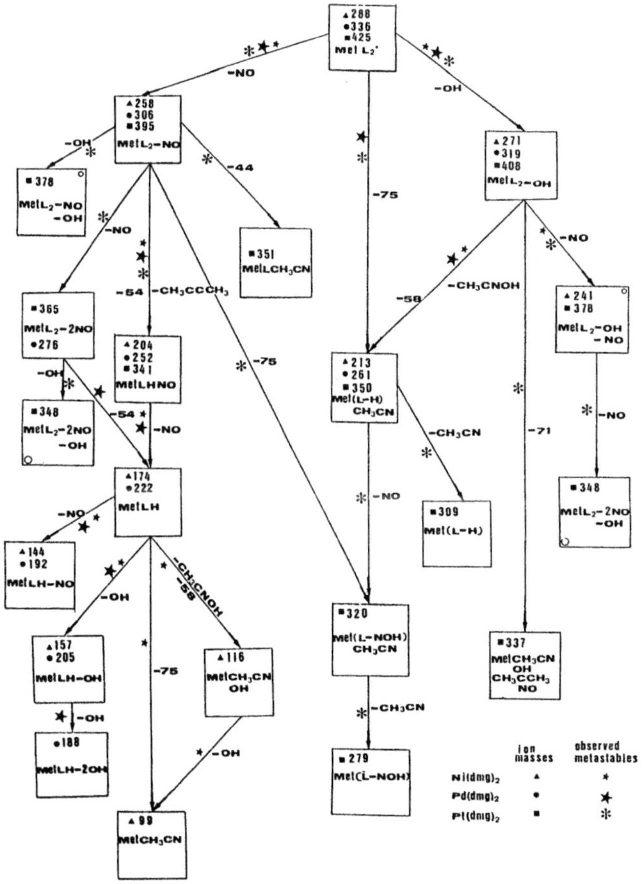

Figure 2. Fragmentation pathways for Ni, Pd and Pt dimethylglyoximates.

a hydrogen, [suggested by some authors[3,4]] were observed. Some of the structures indicated in the figure have already been proposed[4,5], but metastable evidence for the fragmentation pathways has been so far very scarce.

In general, we can observe a difference in the behavior of platinum complexes as compared with those of nickel and palladium (more similar with each other). This gas phase behavior is different from what is more generally known about the solution chemistry of the same elements, and has to be attributed to the greater ionic radius of the former element and consequently a larger "size" for the orbitals involved in the bonding.

METAL-LIGAND BOND ENERGIES (P. B. Armentrout, Department of Chemistry, University of California, Berkeley, California 94720)

During the last several years, we have measured a number of metal-ligand bond dissociation energies (BDEs) using our guided ion beam mass spectrometer (discussed in detail elsewhere in this volume). Table I summarizes much of the data we have obtained, some of which is not yet available in the literature. Some of this information (e.g. periodic trends in $D(M^+-H)$) is discussed elsewhere in this volume. Several comparisons in this data are of particular interest. For example, $D(M^+-CH_3)$ are comparable to $D(M^+-H)$ on the left side of the periodic table, but are noticably higher on the right side (Fe^+ and Ni^+). Previously, this had been attributed to the larger polarizability of the methyl radical compared with a hydrogen atom.[6] However, if this were the reason, the effect should be the same for both left and right side metal ions. One intriguing explanation for this difference is that MCH_3^+ species on the right side have agostic interactions which strengthen the observed bonding compared to the hydrides.[7]

In a recent paper, we noted a correlation between the bond order and the bond energy of V^+-L species in the series L = H, CH_3, CH_2, and CH.[8] Examinations of the data in Table I shows that a similar correlation also occurs for the left side metals - Sc^+, Ti^+ and Cr^+.

Table I also includes results for the <u>second</u> bond energies of H and CH_3 to Sc^+. We find that the first and second hydride and methyl BDEs to Sc^+ are comparable and all are strong. This is in accord with the promotion energy idea since promotion for the first bond is small for Sc^+ (0.15 eV).[9] For the second bond, the promotion energy is approximately zero since there is only a single non-bonding electron remaining on ScR^+.[10]

Finally Table I includes preliminary results for the <u>neutral</u> thermochemistry, $D(M-CH_3)$. Such data are obtained by a straightforward extension of the techniques used to obtain ionic thermochemistry and has been used before to estimate neutral metal-ligand BDEs.[6] For any reaction like process (1), there is a competing reaction (2). Just

$$M^+ + RX \rightarrow MX^+ + R \qquad (1)$$

$$\rightarrow MX + R^+ \qquad (2)$$

Table 1. Bond Dissociation Energies (eV) at 298 K [a]

M	$D(M^+-H)$[b]	$D(M^+-CH)$ ±0.2	$D(M^+-CH_2)$ ±0.2	$D(M^+-CH_3)$ ±0.2	$D(M-CH_3)$ ±0.3
Ca	1.95(0.2)				
Sc	2.44(0.10)[c]		≥4.05[c]	2.56[c]	1.41
ScH	2.57(0.16)[c]			2.8	
$ScCH_3$	2.68(0.2)			2.61	
Ti	2.43(0.10)[d]	5.27	3.69	2.45	2.0
V	2.09(0.06)[e]	4.90[f]	3.30[f]	2.17[f]	2.05
Cr	1.40(0.11)[g]		2.21	1.28	1.76
Mn	2.10(0.15)[h]			2.2[i]	<1.3(≈0.5)[i]
Mn_2	2.1 (0.3)[i]			2.4[i]	
Fe	2.12(0.06)[j]			2.56	
Co	2.02(0.06)[k]				
Ni	1.72(0.08)[k]			2.13	
Cu	0.96(0.13)[k]				
Zn				3.03[m]	0.86[m]
Y	2.52(0.15)[d]				
Zr	2.34(0.15)[d]				
Nb	2.31(0.15)[d]				
Mo	1.80(0.15)[d]				
Rh	1.56(0.15)				
Pd	2.06(0.15)				
Ag	0.65(0.15)[d]				

[a] Unless otherwise noted values are from J. L. Elkind, N. Aristov, R. Georgiadis, L. Sunderlin, and P. B. Armentrout, to be published. These values have not yet been critically evaluated and are subject to change.

[b] Uncertainties in parentheses.

[c] L. Sunderlin, N. Aristov and P. B. Armentrout, J. Am. Chem. Soc., submitted.

[d] J. L. Elkind and P. B. Armentrout, Inorg. Chem. 25, 1078 (1986).

[e] J. L. Elkind and P. B. Armentrout, J. Phys. Chem. 89, 5626 (1985).

[f] N. Aristov and P. B. Armentrout, J. Am. Chem. Soc. 108, 1806 (1986).

[g] J. L. Elkind and P. B. Armentrout, J. Phys. Chem. submitted.

[h] J. L. Elkind and P. B. Armentrout, J. Chem. Phys. 84, 4862 (1986).

[i] P. B. Armentrout, "Laser Applications in Chemistry and Biophysics," M. A. El-Sayed, Ed., Proc. SPIE 620, 38 (1986).

[j] J. L. Elkind and P. B. Armentrout, J. Phys. Chem. in press.

[k] J. L. Elkind and P. B. Armentrout, J. Phys. Chem. submitted.

[m] R. Georgiadis and P. B. Armentrout, J. Am. Chem. Soc. 108, 2119 (1986).

as threshold measurement of reaction (1) can be used to measure $D(M^+-X)$, measurement of the threshold for reaction (2) can yield $D(M-X)$. Indeed, the likelihood of reaction (2) compared with process (1) can be enhanced by choosing R groups with low ionization potentials, e.g. t-butyl. The ability of the guided ion beam apparatus to collect the slow R^+ ions produced in reaction (2) makes quantitative measurements feasible. We have now demonstrated the technique for X = H, O and CH_3 and tested its accuracy in studies of $ZnCH_3$ and $ZnCH_3^+$.[11] Insufficient data is available for a definitive analysis of the periodic trends in these values. Qualitatively, however, the neutral BDEs are usually less than the ionic counterpart except for Cr. We believe this is due to the fact that the ground state of $Cr(^7S, 4s3d^5)$ has a much lower promotion energy than $Cr^+(^6S, 3d^5)$. Likewise, the promotion energies of the other neutral atoms is larger than the ionic counterparts. The arguments are similar to those discussed by Squires for neutral diatomic metal hydrides.[12]

REFERENCES

1. Ridge, D. P., 'Transition Metal Ion Chemistry,' this volume.

2. Allison, J.; Freas, R. B.; Ridge, D. P., J. Am. Chem. Soc., 1979, 101, 1332.

3. Ablov, A. V.; Kharitonov, Kh. Sh.; Vaisbein, Zh. Yu. Dok I. Chem., 1979, 201, 916-919.

4. Charalambous, J.; Soobramanien, G.; Stylianou, A. D.; Manini, G.; Operti, L.; Vaglio, G. A. Org. Mass Spec., 1983, 18(9), 406-409.

5. Westmore, J. B.; Fung, D. K. Inorg. Chem, 1983, 2(6), 902-907

6. Armentrout, P. B.; Beauchamp, J. L. J. Am. Chem. Soc., 1981, 103, 784.

7. Calhorda, M. J.; Martinho Simoes, J. A. J. Am. Chem. Soc., submitted.

8. Aristov, N.; Armentrout, P. B. J. Am. Chem. Soc., 1984, 106, 4065.

9. Elkind, J. L.; Armentrout, P. B. Inorg. Chem., 1986, 25, 1078.

10. Sunderlin, L.; Aristov, N.; Armentrout, P. B. J. Am. Chem. Soc., submitted.

11. Georgiadis, R.; Armentrout, P. B. J. Am. Chem. Soc., 1986, 108, 2119.

12. Squires, R. R. J. Am. Chem. Soc., 1985, 107, 4385.

DYNAMICS OF DISSOCIATION AND REACTIONS OF CLUSTER IONS

A. W. Castleman, Jr. and R. G. Keesee
Department of Chemistry
The Pennsylvania State University
152 Davey Laboratory
University Park, PA 16802

ABSTRACT. Under certain conditions following the ionization of a cluster, reactions proceed between various moieties with the cluster ion that, in terms of products, parallel those resulting from bimolecular gas-phase ion-molecule reactions. In addition, the exothermicity of these reactions and the dielectric relaxation that occurs in newly ionized clusters lead to metastability and dissociation of the clusters. Studies with time-of-flight techniques that have been made on clusters of ammonia, methanol, xenon, and p-xylene•trimethylamine exemplify some of these processes. Fast-flow reaction techniques are used to study the association kinetics of metal ions with neutral ligands. A model for angular momentum coupling which involves vibrational relaxation is consistent with the results. Work on the photodissociation of CO_3^- hydrates and $(SO_2)_2^-$ has revealed the channels and processes of energy redistribution in clusters.

1. INTRODUCTION

During the last decade or so, there has been a nearly explosive growth of activity in the field of cluster research. Reference to clusters permeates the literature of an amazingly broad cross-section of basic and applied science from that dealing with condensed matter and surfaces to interstellar space. Clusters provide a way of investigating properties of matter at different degrees of aggregation. One of the primary thrusts of current basic work stems from the realization that the results of such research will serve to elucidate the behavior of condensed matter and surfaces at the microscopic molecular level. The present NATO conference deals largely with the field of gas-phase ion-molecule reactions, and hence the primary examples of the implications and applications of research on clusters discussed here are drawn from work on ionized systems in the gas phase.

Table I lists some of the more important fundamental and applied areas to which cluster research have a bearing. The listing

P. Ausloos and S. G. Lias (eds.), Structure/Reactivity and Thermochemistry of Ions, 185–217.

Table I. Some Motivations for Studies of Clusters

Fundamental Aspects	Applications
interaction potentials	atmospheric chemistry/ physics (8)
structure	interstellar chemistry (9)
reaction rate theory (1)	obscuration
energy transfer and redistribution (2)	combustion (10)
electron/charge transfer	catalysis (11)
nucleation and phase transitions (3)	corrosion (12)
solvation (4)	radiolysis (13)
polarons (5)	nuclear fusion technology (14)
effects of solvation on reactivity (6)	laser technology (15)
development of bulk properties (7)	photography
	surface science (16)
	thin film deposition (17)
	microelectronics

is not meant to be exhaustive, nor are the references; a few are shown in parentheses which will serve as a guide to introduce the reader to the literature. In addition, the diversity and scope of cluster research is also evident from recent journal issues devoted to the subject (18-21).

This paper is devoted to three major subjects: Photoionization of clusters, metal ion-ligand association reactions, and dynamics of cluster ion photodissociation. No attempt is made to survey all aspects of the literature, and most of the results come directly from the authors' own laboratory.

2. PHOTOIONIZATION OF CLUSTERS

Neutral clusters are often formed by supersonic expansion techniques and investigated by mass spectrometry as cluster ions. Although chemical ionization is sometimes utilized in their study, neutral clusters are most commonly ionized by either electron impact or photoionization methods. A subject of considerable interest is the relative abundance of the ensuing cluster ions as a function of the degree of aggregation. Many systems display anomalous abundances at certain cluster sizes which are invariant with the expansion conditions used to produce the neutral clusters. These features are commonly called magic numbers. In many cases these have been attributed to special features in the size distribution of the neutral clusters, but mounting evidence in most cases points to the importance of the processes which follow ionization and the properties of the cluster ions. Interest in the reactions in clusters following ionization also derives from the belief that the results will further elucidate the relationship of ion-molecule reactions in the gas phase to those in condensed media. Ionization of clusters also provides a method to investigate the spectroscopic solvent shifts of chromophores as a function of the extent of solvation. Finally, studies of ionization thresholds provide information that can be directly compared to the work function for the bulk system under consideration. Several examples from our laboratory are discussed herein.

2.1 Experimental Techniques

2.1.1 Supersonic Expansions. Supersonic expansion techniques are commonly used to produce beams of neutral clusters (22). In our own laboratory, we employ both pulsed jets and continuous expansion sources. In both cases cooling of the beam is accomplished through the conversion of thermal energy in the high pressure source gas into a directed beam velocity u. Taking C_p to represent the heat capacity of the expanding gas and T_s the absolute temperature in the expansion source, a simple energy balance , Equation (1), for an ideal gas of molecular weight m displays why the temperature T_b in the beam is colder.

$$C_pT_s = 1/2\ mu^2 + C_pT_b \qquad (1)$$

This cooling promotes condensation. This simple relationship, however, neglects the effect of the latent heat of condensation resulting from the formation of clusters.

The significant cooling which can be accomplished by an expansion is apparent by comparing Figures 1 through 3. In all cases ammonia is excited by a two-photon absorption to the designated state with a third photon leading to ionizaton. Figure 1 shows the multiphoton ionization spectrum obtained during a weak expansion where the vibrational quanta in the $\tilde{B}$ and $\tilde{C}'$ states are discernible but substantial congestion fails to reveal a well-defined rotational structure. Such structure becomes easily seen in Figures 2 and 3 for the rotational distribution in the $\tilde{C}'(0)$ and $\tilde{B}(5)$ states where the first is accomplished by expansion of neat ammonia at 200 torr and the second demonstrates the additional cooling which obtains through dilution of ammonia in argon. Clusters do not generally attain temperatures as low as unclustered species, since the latent heat of condensation is released during the aggregation process leads to internal vibrational and rotational heating of the aggregate. Of course, cooling collisions with an inert gas help to reduce the internal temperature of the cluster.

2.1.2. Time-of-Flight Mass Spectrometry. The time-of-flight (TOF) mass spectrometer has been in use for many years. The instrument is experiencing a resurgence in popularity due to the advent of pulsed lasers which supply efficient and short durations of ionization in a

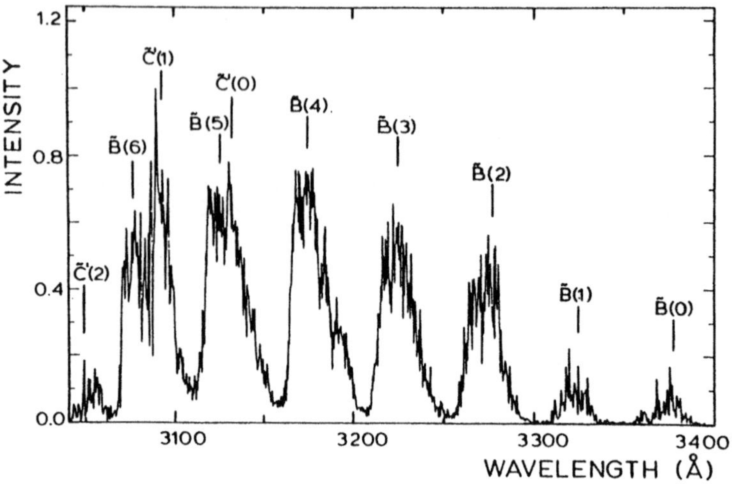

Figure 1. Multiphoton ionization (2+1) spectrum of ammonia under mild expansion conditions. Resonances with the $\tilde{B}$ and $\tilde{C}'$ manifolds are observed.

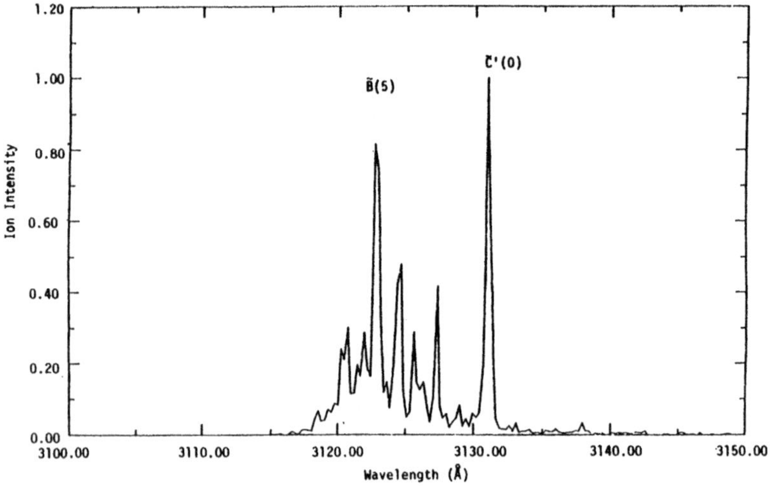

Figure 2. Multiphoton ionization ionization spectrum of ammonia under stronger expansion conditions (200 torr of ammonia behind nozzle) where rotational congestion between the $\tilde{B}(5)$ and $\tilde{C}'(0)$ states is decreased.

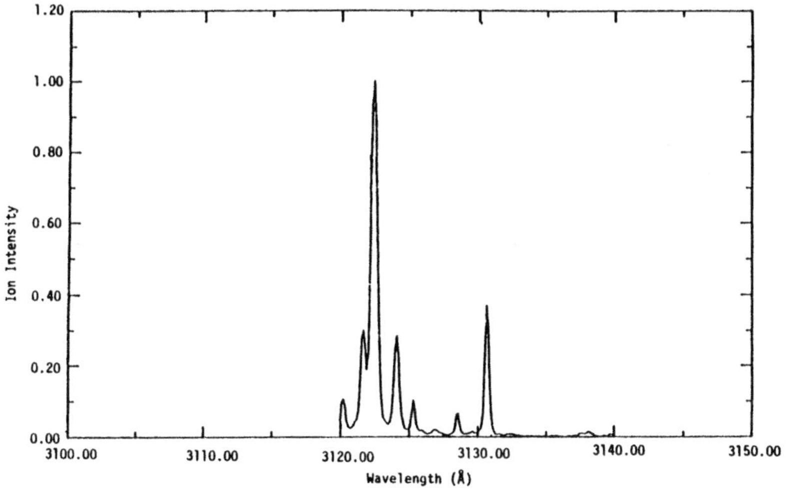

Figure 3. Multiphoton ionization spectra of ammonia obtained with 200 torr behind nozzle, but 5% NH_3 diluted in argon.

small volume. In a typical time-of-flight mass spectrometer, either a two element or alternately a single element accelerating field may be used in the region of ionization. This is followed by a field-free drift region at the end of which the arrival of the ions is detected. Using the conventional TOF method, dissociation which occurs with rates in the neighborhood of 10^6 to 10^8 sec^{-1} can be investigated by either of two methods. One involves analyzing the peak shape (arrival spectrum) of the ions under dual field accelerating conditions, in which cases a knee is observed since the ions spend far more time in the first low-field region where ionization is initiated than in the second high-field region where the bulk of the acceleration occurs (23). An alternate method is to operate under single field conditions and deduce rates from the shape of the late arriving tail of the peak (24).

A third method of studying dissociation employs a reflecting field (reflectron). Although originally designed to enhance the resolution of the TOF method (25), a reflectron can also be employed to investigate dissociation in the field-free drift region, so that slower processes may be observed. These experiments are performed by subjecting the cluster beam to multiphoton ionization using a Nd:YAG pumped dye laser with wavelength extension components that provide frequency doubling and mixing capabilities. Typical photon fluxes are about 10^{16} photons per pulse of approximately 6 ns duration. The ions are accelerated in the accelerating field to about 2 KeV whereafter they enter a field-free region and then are electrically reflected and detected as shown in Figure 4. With appropriate potentials applied to the the reflectron grids, non-dissociating parent ions can be separated from those that dissociate while in the field-free region. A unique identification of these daughter ions can be accomplished by the time separation or by an energy analysis with the reflectron. The separation of the parent and daughter ions occurs as a result of the loss in kinetic energy with essentially no change in velocity of the ion upon dissociation. Hence species with greater kinetic energy (parents) have a longer path to the detector than do the daughter (dissociation) products.

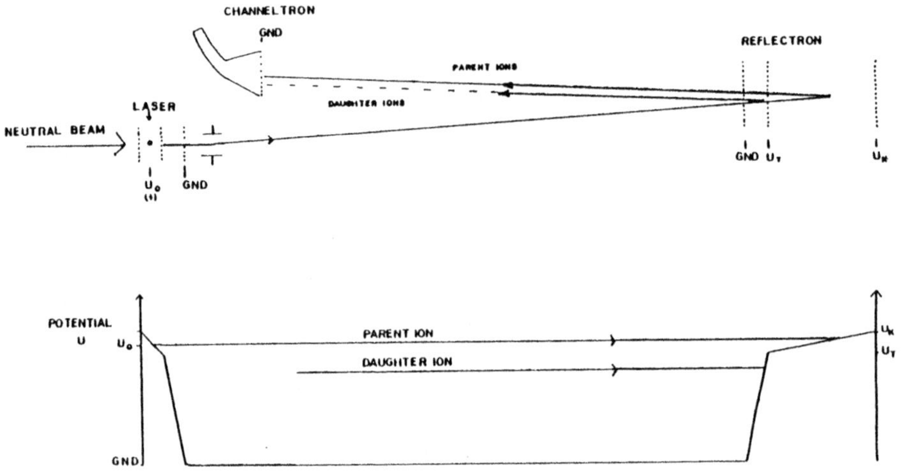

Figure 4. Schematic of time-of-flight mass spectrometer ith reflectron where daughter ions are reflected before parent ions. Lower figure depicts the electrostatic potentials.

2.2. Internal Ion-Molecule Reactions and Dissociation

2.2.1. Ammonia Clusters. We have conducted extensive investigations of the dissociation of ammonia clusters following multiphoton ionization at 266 nm with these techniques (26). Interaction with the laser beam leads to ionization of one of the molecules in the hydrogen bonded neutral ammonia cluster and ejection of an electron from the cluster. Due to the large proton affinity of NH_3 (8.85 eV) and analogous to the well-known gas phase ion-molecule reaction between NH_3^+ and NH_3, an internal (intracluster) ion-molecule reaction rapidly forms NH_4^+ with the ejection of NH_2. Thereafter the ammonia molecules reorient to accommodate the proton. Unimolecular (unicluster) dissociation may occur as a result of the excess vibrational energy created during the conversion from a neutral to ionized system. In practice, collision-induced dissociation may also be observed as a result of the interaction of the cluster ions with residual molecules in the vacuum chamber.

$$(NH_3)_n + h\nu \rightarrow NH_3^+(NH_3)_{n-1} + e \qquad (2)$$

$$NH_3^+(NH_3)_{n-1} \rightarrow [NH_4^+(NH_3)_{n-2}]^* + NH_2 \qquad (3)$$

$$[NH_4^+(NH_3)_{n-2}]^* \rightarrow NH_4^+(NH_3)_{n-2-m} + mNH_3 \qquad (4)$$

Figure 5 shows a conventional time-of-flight spectrum. Figure 6 (upper) shows a spectrum obtained with the reflectron where a daughter ion (from the loss of one ammonia molecule during the flight in the field-free region prior to the reflectron) precedes each of the corresponding parents. By reducing the applied potential at the end of the reflectron (i.e., $U_K < U_T$, see Figure 4),

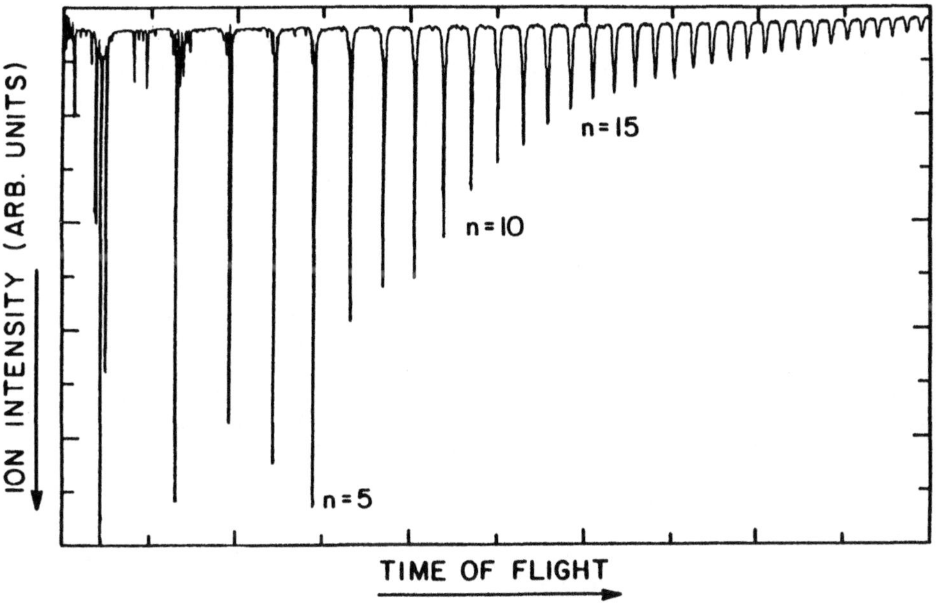

Figure 5. A conventional TOF mass spectrum of $H^+(NH_3)_n$ clusters from the multiphoton ionization of ammonia clusters at 266 nm.

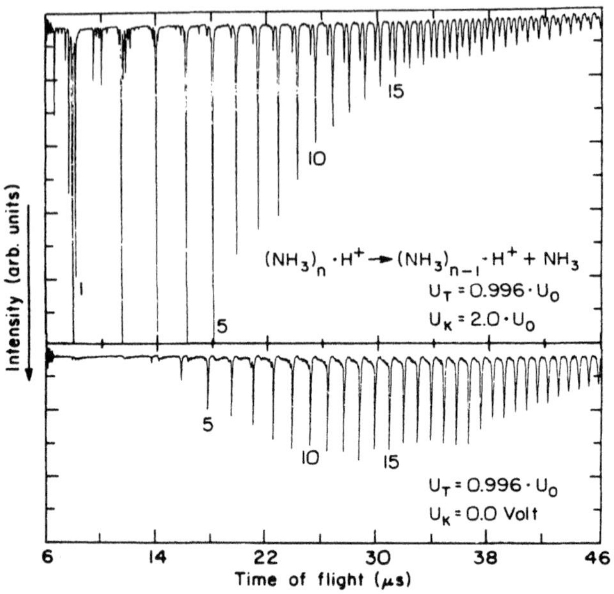

Figure 6. Time-of-flight mass spectrum taken using the reflectron. In the upper mass spectrum, the potentials U_T and U_K are chosen such that all ions are reflected but with the daughter ions arriving at the particle detector earlier than their corresponding parent ions. In the lower mass spectrum, U_K is lowered to eliminate the parent ions. Both spectra are accumulated ion signals over 2460 laser shots.

only the lower kinetic energy products are reflected. The non-dissociating ions are eliminated from the spectrum as shown in the lower part of Figure 6. Further reduction of the reflecting potential U_T (with $U_K < U_T$) improves the ability to discern small contributions from more extensive dissociation. An example is shown in Figure 7 for the clusters $NH_4^+(NH_3)_n$ with sizes n ranging from 4 to 9. The lowest spectrum shows that we have observed the loss of up to six ammonia molecules from $NH_4^+(NH_3)_9$ during its flight through the field-free region.

Both unimolecular and collision-induced dissociation processes are possible. The contribution of the latter is determined through experiments in which the pressure in the field-free region is varied. An extrapolation of the data to a zero of pressure can be made in a linear fashion under the thin collision approximation as shown in Figure 8. Within the uncertainty of the true zero of pressure, the ordinate gives a direct measure of the fraction of

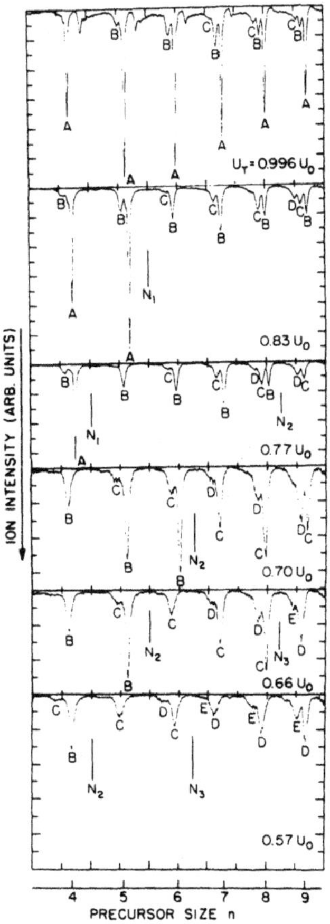

Figure 7. Energy analysis of the "reflected" mass spectra of daughter ion signals. The potential of the first barrier of the mass reflectron, U_T, is varied with respect to the kinetic energy of parent ions, $q \cdot U_o$. All daughter ions lighter than their precursor ions by ix17 amu belong to class i; in this study we observe five classes corresponding to i=1, 2, ..., 5 (labeled A-E). For each mass spectrum taken at its particular potential, U_T, we also show the calculated cluster cutoff size N_i for a particular class. The vertical scale is changed arbitrarily to show low-intensity peaks of higher classes of daughter ions.

dissociation by unimolecular (evaporative) loss. The component from unimolecular decay increases steadily with cluster size from n=4 to 25 as seen in Figure 8. Although insufficient data are available for a full RRKM analysis, simple RRK considerations show that the rate constant for unimolecular dissociation k_u should vary with the number of molecules N in the cluster and the energy difference between the internal energy E of the cluster and the dissociation energy E_o by the relationship

$$k_u \propto N^{2/3} \left[\frac{E-E_o}{E} \right]^{6(N-1)-1} \qquad (5)$$

The factor $N^{2/3}$ represents the pathway degeneracy and 6(N-1) is the number of intermolecular vibrational degrees of freedom in the cluster. The implication of our results is that each larger cluster progressively has more excess internal energy due to factors such as

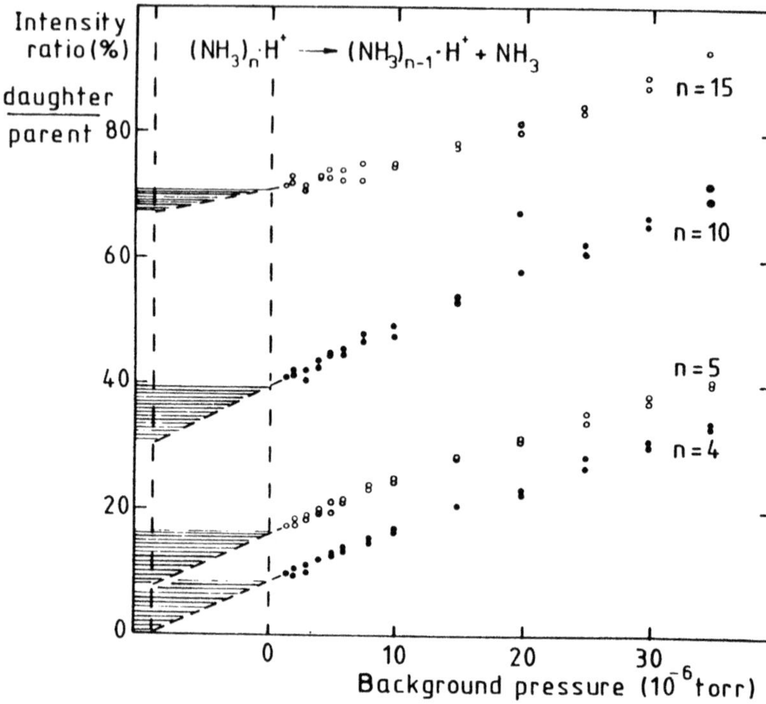

Figure 8. Pressure dependence of daughter to parent ion intensity ratios. The plotted ratios correspond to the accumulated daughter ion (via loss of one NH_3 unit) signal due to dissociation in the field free region before reflectron. The horizontal axis represents the gauge pressure. The actual effective point of zero pressure is bracketed between the dashed lines.

dielectric relaxation after ionization, latent heat from the formation of the neutral cluster, and the lower ionization threshold of larger clusters. Confirmation of this picture follows from the observation that the rate for the loss of a second ammonia molecule decreases with increasing cluster size in accordance with expectations for clusters having little internal energy as a result of the loss of the first ammonia molecule.

The ammonia system is of further interest since $H^+(NH_3)_5$ is distinctive in its relative intensity (see Figure 5). This ion can be considered to be an NH_4^+ ion solvated by one ammonia molecule at each hydrogen. The role of laser fluence for multiphoton ionization at 266 nm is seen by comparing the data in Figures 5 and 9 where the ionization was accomplished under nearly identical expansion

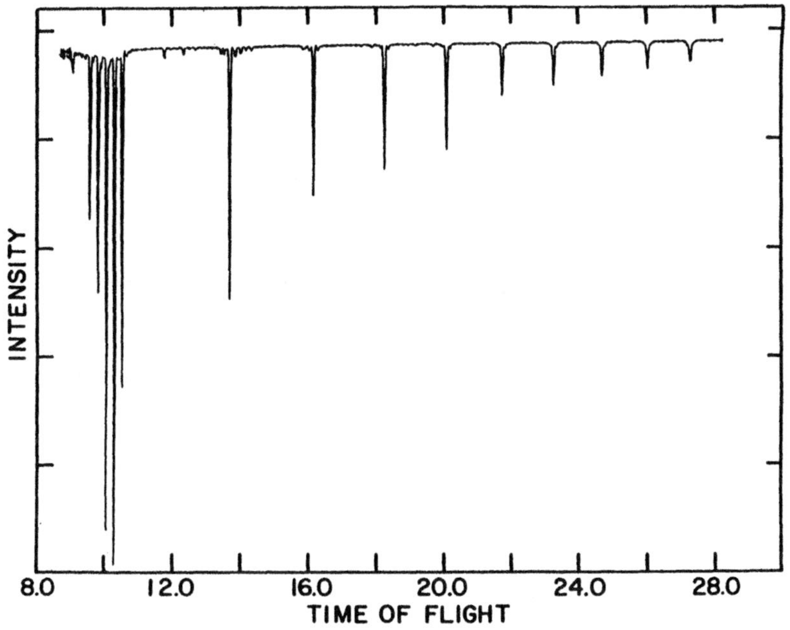

Figure 9. TOF mass spectrum of $H^+(NH_3)_n$ from the multiphoton ionization of ammonia clusters at 266 nm. Laser fluence is about 100 times that used to obtain Figure 5. The $H^+(NH_3)_5$ peak is at 20 μsec.

conditions but at much different laser fluences. The prominence of the protonated pentamer decreases substantially at higher laser fluences providing further evidence for the importance of fragmentation and dissociation in effecting magic numbers in hydrogen bonded clusters (27).

2.2.2. Methanol Clusters. Similar studies to the foregoing were undertaken using 266 nm light to ionize $(CH_3OH)_n$ clusters (28,29). Several processes were observed, the first of which has a direct analogy to the ammonia studies. In particular, following ionization, methanol clusters undergo proton transfer with a concomitant loss of CH_3O. The exothermicity of the proton transfer reaction and subsequent relaxation of the alcohols around the protonated moiety also leads to excess internal energy and dissociation processes. The dissociation processes lead to unimolecular evaporation rates in a manner similar to the ammonia system. Rate data for selected cluster sizes are shown in Table II.

Table II. Unimolecular Dissociation (Evaporative) Rate Constants (in 10^5 sec^{-1}) for $H^+(CH_3OH)_n$

n	rate	n	rate	n	rate
4	1.8	9	6.5	14	4.7
5	3.9	10	4.1	15	3.6
6	8.9	11	5.1	16	3.6
7	2.3	12	6.0	17	2.6
8	3.5	13	---	18	2.4

In contrast to the ammonia system, a nonevaporative loss process was also observed; namely the elimination of H_2O from the protonated dimer. This process, however, was not detected for the larger cluster ions. This observation is in direct correspondence with observations (30) of gas-phase ion-molecule reactions

$$CH_3OH_2^+ + CH_3OH \longrightarrow (CH_3)_2OH^+ + H_2O \qquad (6)$$

$$\xrightarrow{M} CH_3OH_2^+ \cdot CH_3OH \qquad (7)$$

and

$$CH_3CH_2^+ \cdot CH_3OH + CH_3OH \not\longrightarrow (CH_3)_2OH^+ \cdot CH_3OH + H_2O \qquad (8)$$

$$\xrightarrow{M} CH_3OH_2^+(CH_3OH)_2 \qquad (9)$$

The rate for the H_2O loss from $CH_3OH_2^+ \cdot CH_3OH$ has been measured to be $5.5x10^5$ s^{-1}, in good agreement with the range of 10^5 to 10^6 s^{-1} estimated for the gas-phase ion-molecule process.

Based on measurements (31) and estimates for the thermodynamics of various dissociation channels in $H^+(CH_3OH)_3$ one would expect the loss of H_2O from this and larger methanol clusters to be at least as energetically favorable as that for the CH_3OH. Nibbering and co-workers (32) have suggested that intermediate A is important in the rearrangement process which leads to the loss of H_2O. Bowers and his colleagues (30) have suggested that the major product ion in the ion-molecule reaction is the formation of the symmetrical, proton-bound dimer as shown for structure B.

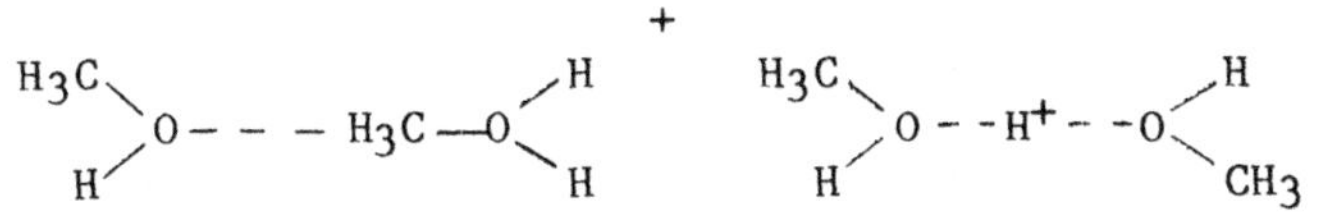

A B

It may be that rearrangement from structure B to structure A is necessary for the water loss process and that this process occurs readily in the protonated dimer but is impeded in higher order clusters. Hence, we think these data reflect the importance of structural rearrangements necessary for certain unimolecular processes to be operative, and hence give further insight into differences which may be seen in the condensed phase compared to the gas for ion molecule reactions.

2.2.3. Xenon Clusters. Recently, we conducted a multiphoton ionization study of xenon clusters. Figure 10 shows the cluster abundances for the system Xe_n^+. Mass spectra with high-energy electron impact ionization are virtually identical (33). This observation indicates the importance of the energy liberated following ionization of the cluster in effecting the magic numbers and size distribution of weakly bound clusters as compared to the energy deposited to initiate ionization.

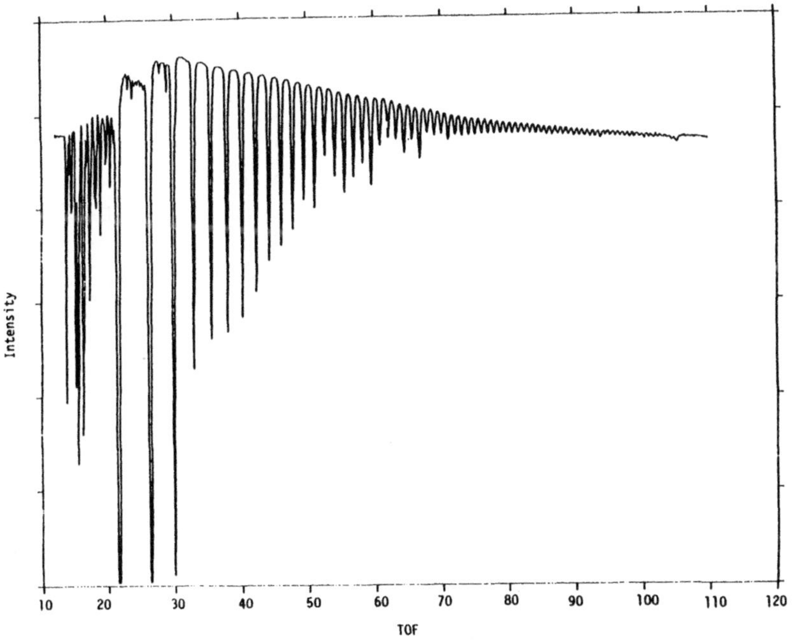

Figure 10. TOF mass spectra of Xe_n^+ obtained by multiphoton ionization of xenon clusters.

Figure 11 shows an example of the peak shapes of the Xe^+, Xe_2^+, and Xe_3^+ in the time-of-flight spectra under single field acceleration. The tail toward longer times is the result of rapid dissociation of larger clusters to the indicated ion. Note that no tail is apparent in the Xe^+ peak. By analysis of the other peak shape the dissociation rates for Xe_5^+, Xe_4^+, and Xe_3^+ are found to lie in the neighborhood of $5x10^7$ sec^{-1}. These are evidently the only species which are metastable in the time domain accessible by this method. The dissociation rates for larger clusters (ten or more atoms) have been reported to be much slower (34).

2.2.4. Internal Penning Ionization. During recent years we have conducted extensive studies of the spectral shifts in clusters of substituted benzenes through resonance enhanced multiphoton ionization (35,36). These studies have revealed that weakly bound clusters can remain intact with ionization near threshold by using two-color methods. An example is p-xylene clustered by argon where the first color is tuned into resonance with the excited state (S_1) of the clustered p-xylene and the second promotes the excited state into the ionization continuum.

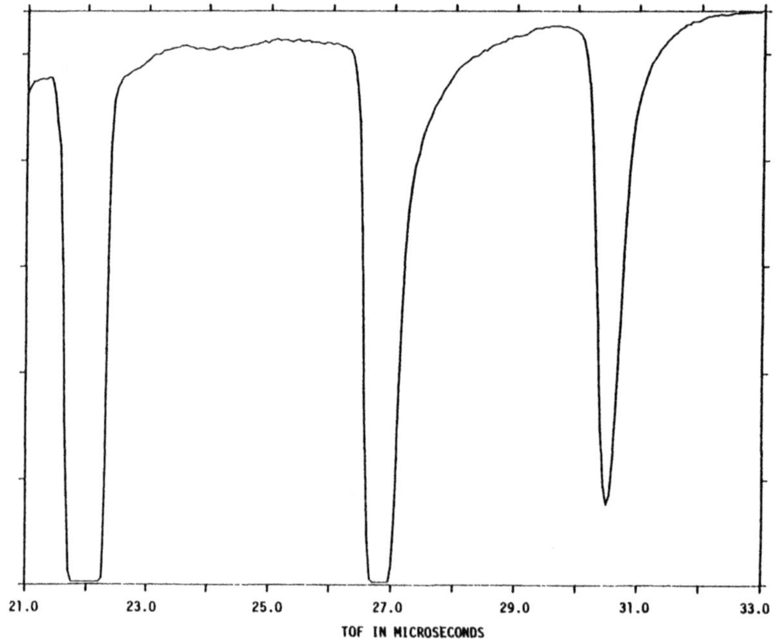

Figure 11. Detail of the TOF peak shapes for Xe^+, Xe_2^+, and Xe_3^+ under single field acceleration. Note the pronounced tails on Xe_2^+ and Xe_3^+ which indicate that Xe_3^+ and Xe_4^+ are undergoing dissociation in the acceleration field.

It occurred to us that a rather interesting possibility existed for charge transfer to an optically transparent partner via an internal Penning ionization process if the ionization threshold of p-xylene is higher than that of the partner (37). The ionization potential of p-xylene is 8.445 eV, with the S_1 state lying 4.45 eV above the ground state. Trimethylamine (TMA) has an ionization potential of 7.82 eV and was chosen as the partner. Ammonia, which has a higher ionization potential than p-xylene, was also studied for comparison. The channels corresponding to NH_4^+ and NH_3^+ become observable, respectively, at 0.1 and 1.8 eV above the ionization threshold of p-xylene. In our studies of ammonia bound to p-xylene, we found that the cluster ion is detected at threshold ionization. When the second ionizing photon is tuned to the appropriate energies the NH_3^+ and NH_4^+ channels are obtainable and observed in the mass spectra. This is also the case at high fluences where three-photon absorption is significant.

For the case of trimethylamine, the onset of ionization produces mainly TMA^+ and some protonated TMA (from an internal proton transfer reaction). None of the cluster ion (p-xylene•TMA)$^+$ is seen. We believe ionization occurs through Rydberg states with a transfer of an electron from trimethylamine to the excited p-xylene and subsequent loss of the excited electron. That this process is taking place through the cluster rather than by collisions in the gas phase has been established by tuning the wavelength of the first photon and observing that a spectral shift of the S_1 state expected for the bound complex is necessary for the observation of this process. Figure 12 shows the peak shape of TMA^+ and $TMAH^+$ under dual field acceleration where a smooth tail can only result from a slow ionization process. Interestingly, the time constant is longer when the photon absorption is lower in the Rydberg manifold. The fact that the tail corresponds largely to the channel for the production of trimethylamine ions rather than the protonated channel was established through isotopic substitution experiments which yielded $TMAD^+$ with no corresponding shift in the position of the tail.

The data reveal an amazingly slow electron transfer after excitation. In analogy to the ammonia system, it is believed that the excited states of trimethylamine would also be planar and that a large geometry change is required for electron transfer and ensuing ionization. These results compare with observations (38) of slow ionization processes in the condensed phase.

2.3. Ionization Thresholds

2.3.1. Paraxylene•Ar_n. We have investigated the changing ionization potential of paraxylene-argon clusters as a function of the number of attached argon atoms through two-color resonance enhanced ionization (36). With this method, we were able to tune the first photon to the shifted resonance as done in the p-xylene•TMA experiments discussed in the preceding section and to use the second photon to probe the ionization threshold. As a result of the Stark

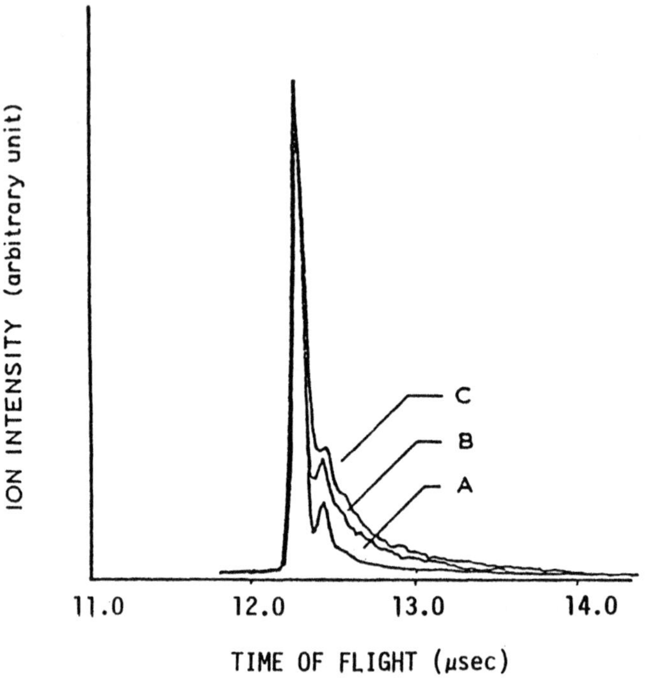

Figure 12. TOF peak shapes from a two-field acceleration of trimethylamine$^+$ and protonated trimethylamine following two-color resonant (1+1) ionization of p-xylene•trimethylamine with the ionizing photon at (A) 3.88 eV, (B) 3.688 eV and (C) 3.607 eV. The smooth tails indicate a slow ionization process.

effect, the clusters (and other molecules as well) ionize at a slightly lower energy within an electric field. The variation is small and is proportional to the square root of the field strength. The zero-field ionization potentials of the clusters are given in Figure 13. A cumulative shift of only about 720 cm^{-1} is seen at n=6 compared to a shift in the bulk condensed state of about 6000 cm^{-1}. Clearly for the case of dielectrics the shift is quite far from that expected for the bulk condensed state.

2.3.2. Sodium Clusters. Due to their simple one-electron nature, and the ease with which they can be compared with theoretical calculations, systems comprised of alkali metals are particularly interesting for study. Photoionization spectra for Na_x, with x ranging from 1 to 8, have been obtained in our laboratory using a molecular beam coupled with a UV light source and quadrupole mass spectrometer detection system (39). Sodium clusters are produced by

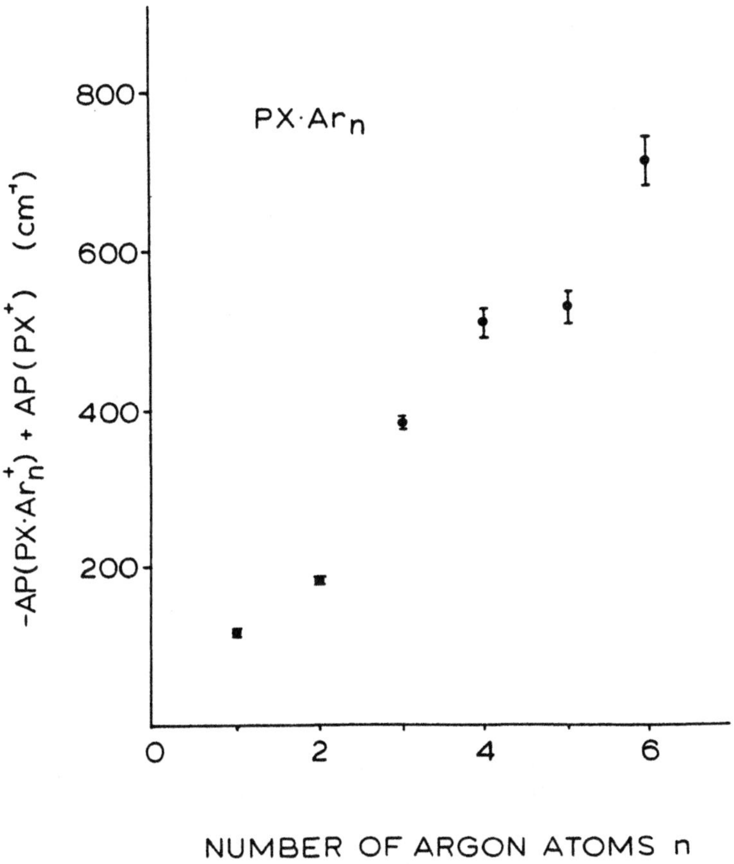

Figure 13. The reduction in ionization potential of p-xylene (PX) as a function of the number of clustering argon atoms.

vaporizing the metal and expanding it with argon gas at pressures ranging between 70 and 100 torr through a 300 μm diameter nozzle. Once formed, the clusters pass through a skimmer into a detection chamber where they are photoionized and mass analyzed. Photoionization is accomplished by use of a 500 watt xenon arc lamp. The desired wavelength is selected with a monochromator. The light intensity on the monochromator is continually monitored using a thermopile detector and the collected spectra are corrected for variations in power and the monochromator slit function.

Data from our laboratory (39) agree extremely well with measurements reported by Schumacher and coworkers (40) for all clusters with the exception of x = 5, 6, and 8 (see Table III). This is particularly evident in view of the somewhat different source and source conditions employed in the experiments. Even the measured value for Na_5 is essentially within experimental error, but Na_6 and Na_8 are unexpectedly different. Recent measurements for Na_8 reported by Buttet (41) are in good agreement with our findings.

Table III. Ionization Potentials (in eV) for Na_x

x	Present Work (±0.04) Ref. 39	Schumacher et al. (±0.05) Ref. 40
1	5.14	5.14
2	4.91±0.04	4.934±0.011
3	3.98	3.97±0.05
4	4.28	4.27
5	3.95	4.05
6	3.97	4.12
7	4.06	4.04
8	3.9	4.1

A closer inspection of our data for Na_6 (see Figure 14) reveals two breaks in the photoionization efficiency curve. One appears at 3.97 eV, which is the assigned appearance potential from our measurements and another at 4.12 eV, which is identical to the appearance potential assigned by Schumacher and coworkers. In fact, it is important to note that the data from the two laboratories are essentially identical and it is only the assigned appearance potential which is different.

At the present time, there is ambiguity in the proper assignment since several possible complicating factors may exist. In addition to possible influences due to Franck-Condon factors or hot bands, neither of which are expected to relate to the problem at hand (39), questions exist concerning the possible existence of isomers or the possibility that the results can be influenced by fragmentation of larger clusters.

We have calculated the bond energies for atoms bound to metal cluster ions using our appearance potential measurements and three sets of calculations (42-44) for the bonding of sodium atoms to neutral clusters. Based on the values derived from the calculations of Martins et al. (42) and Lindsay (43), the atoms would be sufficiently bound that fragmentation of Na_7^+ is unlikely; but, use of the results of Flad et al. (44) to calculate bond energies indicates

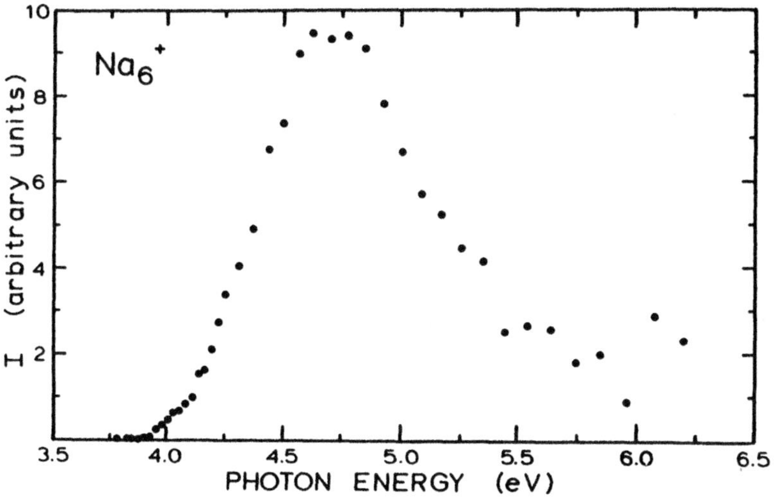

Figure 14. The photoionization efficiency curve for Na_6.

that this ion could fragment easily into Na_6^+. Hence the question of fragmentation remains open.

Another possibility is that the data provide evidence of isomers. Martins et al. (42) suggest that Na_6 should have two isomers separated in energy by only 0.06 eV. The observed difference in ionization potentials is about 0.15 eV if the difference for the two steps in the ionization efficiency curve is attributed to the onset of two different isomers.

For comparison with the bulk metal, the experimental data correlate resonably well with the classical electrostatic expression relating the work function of a system of radius r to the bulk work function, W_∞ (see Figure 15). A number of other metallic systems also correlate well with this classical theory. These results indicate that the intrinsic part of the ionization potential is independent of size (45).

2.3.3. Na_xO and Na_xCl. Subchlorides and suboxides of sodium clusters have been generated through reactions with Cl_2, HCl, N_2O, and O_2. From tabulated thermochemical data and the appearance potentials measured in our laboratory (46), bond energies have been derived for Na_2^+ with X and X^- and for Na_2 with X, where X = O or Cl (see Table IV). The bond energies for O bound to Na_2^+ and Na_2 differ by only 0.2 eV. By contrast, the bond energy for Na_2^+ with Cl lies between that of Na_2 with Cl and Na_2^+ with Cl^-. The similarity in the bond energies of O with Na_2^+ and Na_2 and the fact that the ionization potential of Na_2O is close to that of the metal is suggestive that the electron removed upon photoionization comes

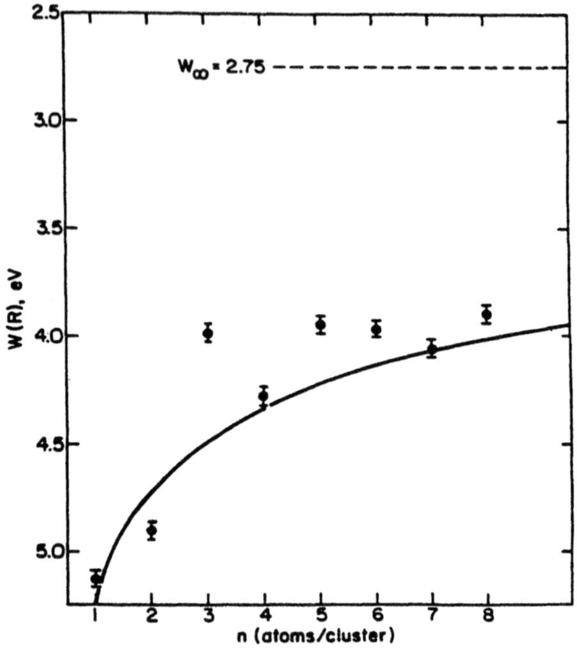

Figure 15. The ionization potential or work function of sodium clusters as a function of size. The curve is based on the classical relationship $W(R)=W_\infty + 3e^2/8R$ where radius R is related to the numnber of atoms by assuming the bulk density for the clusters.

Table IV. Bond Energies of Sodium Suboxide and Subchloride Species

Suboxides	Bond Energy (eV)	Subchlorides	Bond Energy (eV)
Na_2^+ - O	4.08	Na_2^+ - Cl	5.01
Na_2 - O	4.28	Na_2 - Cl	4.29
Na_2^+ - O^-	7.68	Na_2^+ - Cl^-	5.54

largely from a nonbinding orbital associated with the Na(3s). In the case of the chlorine containing species, the ionization potential of Na_2Cl (4.15 eV) lies between the electron affinity of chlorine (3.617 eV) and the ionization potential of sodium (5.13 eV). This result suggests that the highest occupied level may be spatially localized around Cl and influenced by the electron affinity of Cl.

Table V lists the difference in ionization potential between Na_xO and Na_x, where measurements are available from our studies for x up to 4. The ionization potential for NaO is much greater than that for Na. The difference between the oxide and pure species becomes diminishingly small for x=3 and the trend reverses for x=4. The ionization potential for the larger cluster moves into accord with the observation that the presence of adsorbed impurities on polycrystalline sodium surfaces lowers the bulk work function. The potassium system (47), while not displaying this reversal in ionization potential between oxide and pure metal, does show a diminishing trend with cluster size. It is tempting to speculate that there would be a reversal for a larger cluster (see results in Table V).

Table V. Difference in Ionization Potential[(a)] for M_x and M_xO

$\Delta \equiv IP(M_x) - IP(M_xO)$

x	Na Δ, (eV)	K Δ, (eV)
1	- 1.36	----
2	- 0.20	- 0.65
3	0.08	- 0.35
4	0.33	- 0.02

[(a)]results calculated from data given in ref. 47.

3. METAL ION-LIGAND ASSOCIATION REACTIONS

Another stimulus for research on metal clusters has been derived from recent findings (48-50) which show that certain cluster sizes are more reactive than others. Of particular interest are trends reported by Kaldor and coworkers (51) which suggest a correlation between the ionization potential of a metal and its reactivity. In view of the difficulty in selecting specific neutral metal clusters for investigation, and hence in deriving accurate rate coefficients, this interest is also promoting studies of ionized systems where specific cluster sizes can be mass selected and subjected to subsequent examination.

In order to investigate the individual reaction kinetics for a metal ion (or metal cluster ion) with a specific ligand, a fast-flow reaction apparatus has been assembled in our laboratory. Primary ions are produced by thermionic emission or laser vaporization. Reactant gases can be added at any inlet position in the flow, so kinetic rates can be determined by monitoring the reactant ion signal as either the position of reactant injection, the flow rate of reactant into the flow tube, or the bulk flow velocity is varied. The reactant ion and product ions are sampled from the flow tube into a quadrupole mass spectrometer.

Recently, we investigated the association of a variety of ligands including NH_3, CH_4, CH_3F, CH_3Cl, CH_3Br, and CO with a variety of metal ions including Pb^+, Al^+, Li^+, Ag^+, Cu^+, and the dimer ion Al_2^+. In the low pressure limit, the association reactions for the attachment of a ligand to an ion (or cluster ion) is related to the collision rate k_1 between the ion and ligand, the stabilization rate k_2 which with unit efficiency is equal to the collision rate for the complex with a third body M, and the lifetime τ of the collision complex against unimolecular dissociation. The forward rates, k_1 and k_2, are calculated from Langevin theory (52) for collisions with molecules having no permanent dipole moment and from ADO theory (53) or the method of Su and Chesnavich (54) for polar ligands. Lifetimes of the collision complexes are deduced from the experimental measurements of the overall forward rate ($k_1k_2\tau$). In the case of ammonia, these values are found to range from 21 picoseconds for Pb^+ to 1.9 ns for Cu^+. For the case of CH_3Cl, the complexes have lifetimes from 140 picoseconds (Pb^+) to 11 ns (Cu^+). By contrast, the methane systems have much shorter lifetimes. The trends in lifetimes are in rough accord with expectations based on the angular momentum coupling model (55).

Of particular significance are comparisons of reactions for the aluminum dimer ion Al_2^+ compared to the atomic ion Al^+. The dimer ion is found to react from 10 to 50 times faster with CO, CH_3Br, and NH_3, which indicates corresponding increases in lifetimes of the collison complex. Such variations in lifetimes and reaction rates are expected to reflect on the trends observed for the reactivity of metal clusters. Such work is currently in progress in our laboratory.

4. DYNAMICS OF CLUSTER ION PHOTODISSOCIATION

Study of the dynamics of photodissociation of clusters provides a way of investigating intramolecular energy transfer for a series of nearly identical molecules having varying degrees of freedom. Furthermore, the interesting question of coupling between modes of widely differing frequency can be addressed. Effects of solvation on the dynamics can also be directly investigated. Studies of ion clusters are particularly interesting since investigations can then be done on mass selected systems, and energy release upon dissociation can be determined in a relatively straightforward manner.

Vibrational predissociation can occur when one of the molecular subunits absorbs an amount of energy larger than a bond dissociation energy of the cluster. Many photodissociation experiments have involved neutral van der Waals molecules where vibrational predissociation proceeds by the transfer of a small amount of energy from the initially excited mode of a molecule to the intermolecular bonds of the cluster. The energy associated with one vibrational quantum in a molecular entity of a neutral cluster generally exceeds that required to dissociate the cluster. In the case of cluster ions, an energy corresponding to a change of several vibrational quanta is required for dissociation to occur.

This section deals with results from our laboratory of studies of the dissociation dynamics of a homomolecular system, the anion $(SO_2)_2^-$, and a heteromolecular system, CO_3^- and its hydrates. The bonding of all of these cluster anions exceeds the energy associated with one quantum of vibration in the core ion.

4.1 Experimental Technique and Data Analysis

The apparatus used for energy resolved photodissociation spectroscopy (56) consists of an ion source, a system of electrostatic ion optics to control the position and energy of the ion beam, a Wien velocity filter to mass select a particular ionic species, a tunable dye laser (with the ion beam traversing intracavity) pumped by an argon ion

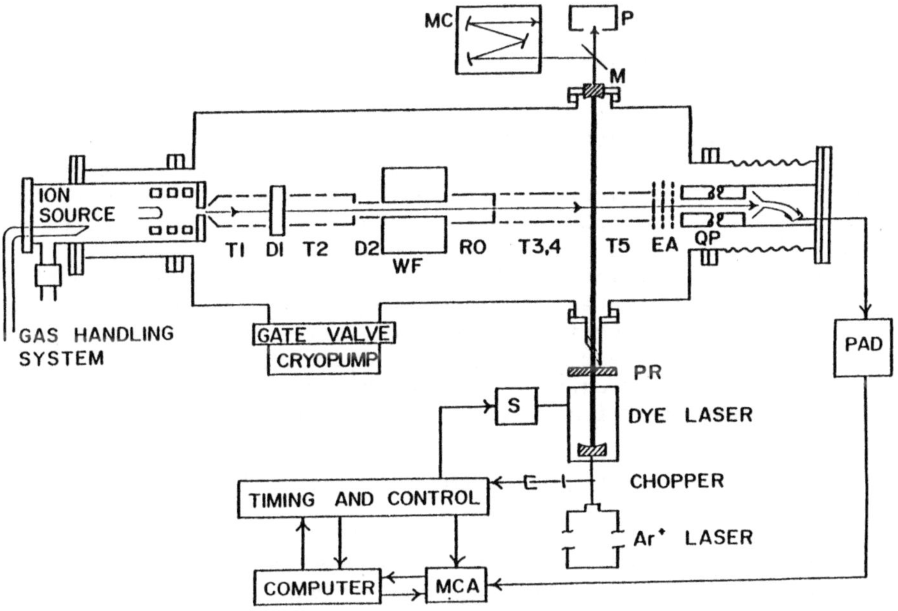

Figure 16. Schematic diagram of photodissociation spectrometer. T1-T5: Three-cylinder electrostatic tube lenses. D1, D2: Deflector pairs. WF: Wien filter. RO: Resolving orifice. EA: Energy analyzer. QP: Quadrupole mass spectrometer. M: Mirror of extended dye laser cavity. P: Power meter. MC: Monochromator. PAD: Preamplifier/discriminator. MCA: Multichannel analyzer. S: Stepping motor. PR: Polarization rotator.

laser, a triple screen retarding field energy analyzer, a quadrupole mass spectrometer, and associated signal acquisition and control electronics (see Figure 16). The ion source is of a type used for high pressure mass spectrometry in our laboratory (57). Photodissociation occurs in the interaction region which is surrounded by a cylindrical screen held at a well defined electrical potential. A chopper positioned between the argon and dye lasers is synchronized with the data collection electronics and permits continuous collection of a signal and background channel from which corrections are made for the contribution of collision-induced dissociation to the measured ion current. The intensities of parent ions and all photofragment ions are monitored by the quadrupole mass spectrometer as a function of both laser frequency and ion kinetic energy in order to determine the spectroscopic structure of the parent ions and the dynamics of the dissociation process.

Energy analysis is performed with the two outer screens of the energy analyzer held at the same potential as the surrounding ion optical elements, while the center screen is scanned over a range of repulsive potentials. Only ions whose energies are equal to or greater than the center potential are able to pass through the analyzer and be detected; all other ions are repelled. This experiment yields the integral of the lab frame energy distribution. The actual energy distribution is generated by numerically differentiating the raw data using least squares polynomial smoothing (58). If no energy is released upon dissociation and the parent ion beam is monoenergetic, the parent and fragment ions would appear as single peaks. Because the fragment has a lower mass than the parent, the fragment peak is expected to appear at a less repulsive value of the retarding field. The amount of energy released by the fragmentation process into relative translation is calculated from the location and the width of both parent and fragment peaks (56).

Briefly, an analysis is made as follows: The raw data are first smoothed and differentiated, and the zero point of the energy scale is shifted to the interaction region potential. This procedure gives values for the parent beam energy, E_b, and the projection of fragment energies along the beam axis, E_{1x}. The energy released into relative translational energy of the fragments, E_r, is then obtained from

$$E_r = \frac{1}{m_2 \cos^2 \Theta_1} \left[E_b m_1 + E_{1x} M \pm 2(m_1 M E_b E_{1x})^{1/2} \right] \qquad (10)$$

where Θ_1 is the angle that the fragment ion makes with the beam axis, M the parent mass, m_1 the fragment ion mass, and m_2 the neutral fragment mass. In carrying out calculations, we have combined all electrostatic potential calibration factors into a single correction factor which we attribute largely to contact potential effects. This correction factor is treated as an adjustable parameter and is assigned the value that satisfies the equality $E_{1x} = m_1/ME_b$ at $E_r = 0$. This value of E_{1x} corresponds to the center of the measured fragment energy distribution.

An isotropic distribution is obtained by summing together distributions measured with different laser beam polarizations with respect to the ion beam axis

$$I(E) = \frac{1}{N} \Sigma\, I(E,\chi)\Delta\chi \qquad (11)$$

where I(E) is the isotropic distribution, $I(E,\chi)$ a distribution measured at laser polarization angle, χ, and N is a normalizing factor.

The lab frame energy distribution of the photofragment can be calculated by first computing the lab energy from

$$E_{1x} = \frac{m_1}{M} E_b + \frac{m_2}{M} E_r \cos^2\Theta + \frac{2}{M} (E_b E_r m_1 m_2)^{1/2} \cos\Theta \qquad (12)$$

where M, m_1, and m_2 are the masses of the parent ion, ionic fragment and the neutral fragment, respectively, E_b is the measured parent energy, E_r is the relative kinetic energy release, and Θ is the angle the departing fragment makes with the ion beam axis. The probability of a given E_{1x} is then given by

$$P(E_{1x}) = P(E_b)P(E_r)P(\Theta) \qquad (13)$$

where $P(E_b)$ is the probability of a given parent energy (measured in an experiment), $P(E_r)$ is the probability of a given kinetic energy release (calculated from either the isotropic distribution or phase space), and $P(\Theta)$ is the probability of a given recoil angle.

The probability of product recoil per solid unit angle is (59)

$$P(\phi) = (\frac{1}{4\pi})\ [1 + \beta P_2(\cos \qquad (14)$$

where ϕ is the recoil angle with respect to laser polarization, P_2 $(\cos\phi)$ is the second Legendre polynomial, and β is the anisotropy parameter. The value of β ranges from 2 for a pure parallel transition, 0 for an isotropic transition, to -1 for a pure perpendicular transition. Physically, the anisotropy parameter can be thought of as the product of two angle dependencies; one relating the angle between the internuclear axis and transition dipole moment and the second between the internuclear and dissociation axes (60). The second arises as a result of molecular rotation which occurs after photon absorption but before dissociation.

Using the relationship between laser beam polarization, ion beam direction, and product relative velocity direction, Equation (14) can be transformed to the more useful relationship (61)

$$P(\Theta) = \frac{\sin\Theta}{2} [1 + \beta P_2(\cos\Theta)P_2(\cos\chi)] \qquad (15)$$

4.2. $(SO_2)_2^-$

The SO_2 dimer anion has been observed in our laboratory to photodissociate to SO_2^- as the sole ionic photofragment; its photodissociation cross section has been reported (62) to be a smooth peak displaying a maximum near 2.1 eV and having a total cross section of 1.9×10^{-17} cm^2. Power studies made in our laboratory at 600 nm (2.07 eV) indicate that the dissociation is a one-photon process. We have measured the bond energy for $SO_2^- \cdot SO_2$ to be 1.04 eV (63) and that for $SO_2 \cdot SO_2$ to be 0.19 eV (64). Based on the reported electron affinity for SO_2 of 1.1 eV (65), the results lead to 1.95 eV as the electron affinity of the dimer ion.

Energy analysis experiments of $SO_2^-(SO_2)$ performed at different laser polarizations show that the measured laboratory frame energy distribution of the fragment ions is widest under conditions where the laser polarization is oriented along the ion beam axis. A typical energy analysis set of data for vertical polarization is shown in Figure 17 along with the smooth derivative obtained from the computer program.

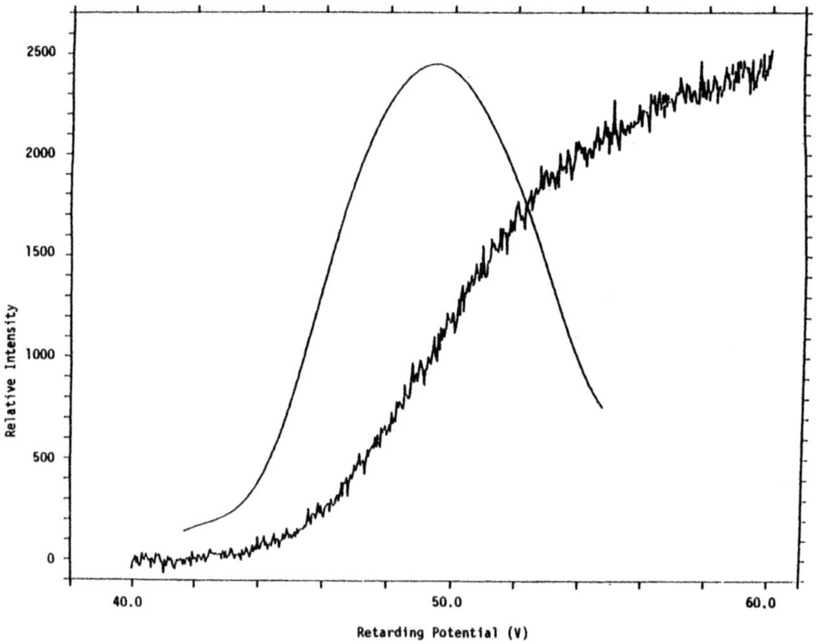

Figure 17. A typical retarding potential energy analysis of the SO_2^- photofragment ion from $(SO_2)_2^-$. The laboratory frame energy distribution is the curve obtained by differentiating and smoothing of the raw data.

With the above relationships and the isotopic distribution, computations of relative intensity as a function of retarding potential can be made (66) and compared with the original experimental data. Our experimental laboratory energy distributions of the SO_2^- photofragments are best fit by an anisotropy parameter of 1.2. This result suggests that the transition is more parallel than isotropic and that the dissociation is rapid with respect to rotation.

Figure 18 shows the probability of having a kinetic energy release exceeding a certain energy. The experimental results indicate that 50% of the fragments have a kinetic energy release of 0.2 eV or greater. The integrated probability expected if the process were statistical is also shown in Figure 17. This calculation is based on a statistical phase space theory developed by Bowers and coworkers (67). Although the calculated result would vary slightly depending upon the vibrational frequencies employed in the calculations, the difference between phase space theory and the experiment is significant and further points to dissociation on a repulsive surface. A potential energy curve which is consistent with the experimental observations is shown in Figure 19.

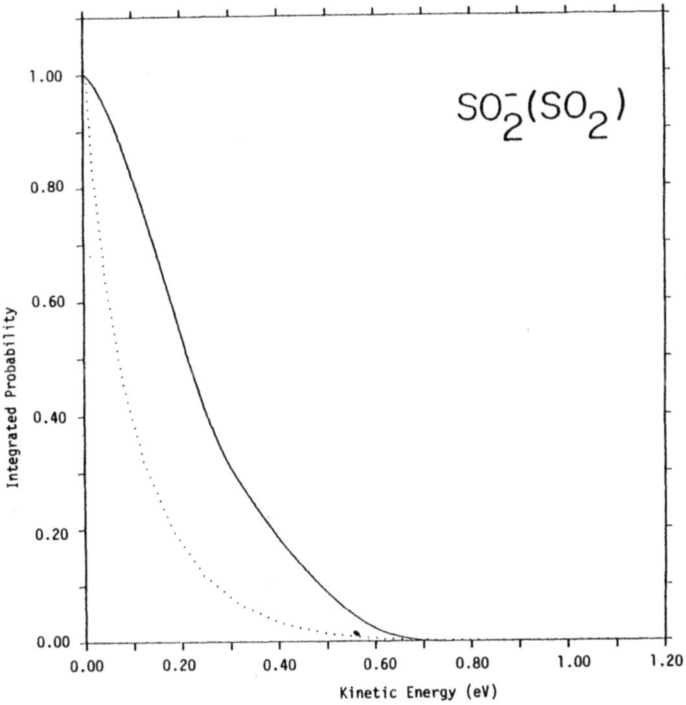

Figure 18. The experimentally determined (solid line) and statistical (dotted line) integrated probability of kinetic energy release for photodissociation of $SO_2^- \cdot SO_2$.

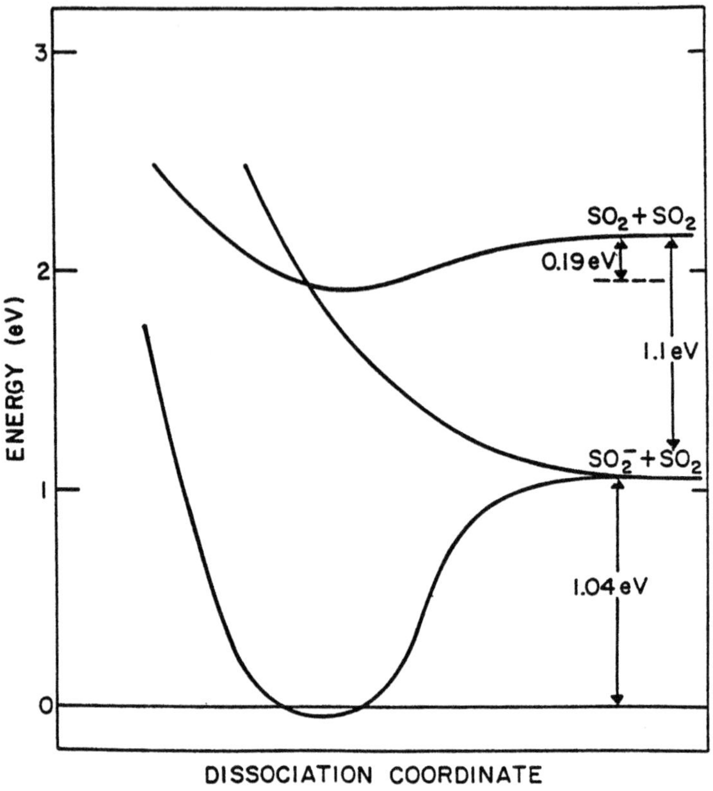

Figure 19. Potential energy surfaces for the $(SO_2)_2^-$ system.

4.3. CO_3^- and Its Hydrates

Studies of the photodissociation of CO_3^- and its hydrates have also been conducted in our laboratory (56,68). In the photon energy range 1.95 to 2.2 eV, photodissociation of CO_3^- leads to O^- and CO_2 fragments. The photodissociation spectrum measured in our laboratory shows a well-defined, sharp structure which reproduces the spectra obtained in earlier experiments (69). Excitation to a bound intermediate excited state ion has been confirmed by our work. Dissociation of the core anion occurs by two mechanisms. One involves two-photon excitation to a repulsive surface and the other results from collisional dissociation of the excited anion which has undergone internal conversion from the excited intermediate state to high vibrational levels of the ground state. Photodissociation is initiated by a $^2A_1 \leftarrow {}^2B_1$ bound-bound transition that promotes the ion from the ground state to the intermediate weakly bound excited state ion (see Figure 20). The sharp features observed in the photodissociation spectra arise from the vibrational structure of

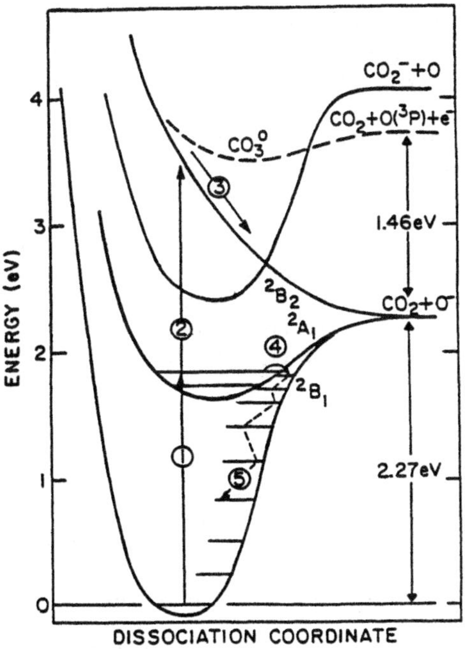

Figure 20. Model of CO_3^- electronic surfaces and transitions of the two-photon direct dissociation (TPDD) mechanism: (1) First photon excitation, $^3A_1 \leftarrow {}^2B_1$. (2) Second photon excitation, $^2B_2 \leftarrow {}^2A_1$. (3) Dissociation along the $^2B_2$ repulsive surface to the observed CO_2 + O^- photofragments. Model of collision-assisted photodissociation (CAPD) mechanism: (1) Photon excitation is by the same $^2A_1 \leftarrow {}^2B_1$ transition that occurred in TPDD. (4) Internal conversion returns the ion to the ground electronic surface. (5) Redistribution of the resulting vibrational energy to all modes of the ion leaves the ion vibrationally hot.

the intermediate $^2A_1$ state. Analysis of the kinetic energy for the two-photon mechanism indicates that a large fraction (about 20%) of the available energy is deposited into relative translation of the photofragments.

The bond energies for water with CO_3^- have been measured in our laboratory and found to be approximately 0.6 eV each (57). The only photodissociation channel observed for the CO_3^- hydrates is loss of all attached water molecules (see Table VI). It is an interesting observation that, in the larger clusters where partial dehydration is possible, the loss of all ligands is the only observed channel. The kinetic energy analysis of the CO_3^- photofragment ions from all three hydrates indicates that more energy is partitioned into translation than is predicted by phase space. The overall percentage of available energy that is channeled into relative translation is 27%, 18%, and 27% for the first, second, and third hydrates, respectively.

In the case of the hydrates, no O^- photofragment ion is detected. Conversion of energy from the core ion into the vibration of ion-water bonds leads to dissociation of the cluster. In the case of the CO_3^- hydrates neither a $\Delta v = -1$ nor -2 transition would be sufficient to result in bond dissociation. Although vibrational predissociation would be expected to pertain in these systems, a surprisingly large amount of energy is partitioned into relative translation and we cannot rule out the possibility that dissociation to some repulsive surface is operative in this cluster system.

References

1. M. Meot-Ner, in Gas Phase Ion Chemistry, ed. M. T. Bowers (Academic, New York, 1979), pp. 197-271.
2. A. Amirav, U. Even, and J. Jortner, J. Phys. Chem. 85, 309 (1981).
3. A. W. Castleman, Jr., Adv. Colloid Interface Sci. 10, 73 (1979).
4. P. Kebarle, Mod. Aspect Electrochem. 9, 1 (1974).
5. G. A. Kenney-Wallace, Acc. Chem. Res. 11, 433 (1978).
6. D. K. Bohme, in Ionic Processes in the Gas Phase, ed. M. A. Almoster-Ferreira, NATO ASI Series C 118, 111 (1984).
7. A. W. Castleman, Jr. and R. G. Keesee, Acc. Chem. Res., in press.
8. D. Smith and N. G. Adams, Top. Curr. Chem. 89, 1 (1980); E. E. Ferguson, F. C. Fehsenfeld, and D. L. Albritton, in Gas Phase Ion Chemistry, Vol. 1, ed. M. T. Bowers (Academic, New York, 1974), p. 45; F. Arnold and S. Qiu, Plant. Space Sci. 32, 169 (1984); R. G. Keesee, R. Sievert, and A. W. Castleman, Jr., in Chemistry of Multiphase Atmospheric Systems, ed. W. Jaeschke, NATO ASI Series G 6, 501 (1986).
9. D. Smith and N. G. Adams, Int. Rev. Phys. Chem. 1, 271 (1981); D. K. Bohme, this volume.
10. J. M. Goodings, S. D. Tanner, D. K. Bohme, Can. J. Chem. 60, 2766 (1982).
11. R. C. Baetzold and J. F. Hamilton, Prog. Solid State Chem. 15, 1 (1983).
12. K. D. Pitzer, J. Phys. Chem. 87, 1120 (1983).
13. G. G. Meisels, Radiat. Phys. Chem. 20, 1 (1982).
14. E. W. Becker, R. Klingelhofer, and P. Lahse, J. Naturforsch. 15a, 644 (1960); R. Beuhler and L. Friedman, Chem. Rev. 86, 54 (1986).

15. J. B. Laudenslager, in Kinetics of Ion-Molecule Reactions, ed. P. Ausloos, NATO ASI Series B 40, 405 (1979).
16. E. L. Meutterties, T. N. Rhodin, E. Bend, C. F. Brucker, and W. R. Pretzer, Chem. Rev. 79, 91 (1979); N. H. Turner, B. L. Dunlap, R. J. Colton, Anal. Chem. 56, 373R (1984).
17. I. Yamada and T. Takagi, Thin Solid Films 80, 105 (1981).
18. Ber. Bunsenges. Phys. Chem. 88, 187-322 (1984).
19. Surf. Sci. 156, 1-1043 (1985).
20. Chem. Rev. 86, 491-657 (1986).
21. Z. Phys. D 2 (1986) in press.
22. J. B. Anderson, R. P. Andres, and J. B. Fenn, Adv. Chem. Phys. 10, 275 (1966); O. F. Hagena, Surf. Sci. 106, 101 (1981).
23. J. L. Durant, D. M. Rider, S. L. Anderson, F. D. Proch, and R. N. Zare, J. Chem. Phys. 80, 1817 (1984).
24. H. Kuhlewind, U. Boesl, R. Weinkauf, H. J. Neusser, and E. W. Schlag, Laser Chem. 3, 3 (1983).
25. V. I. Karataev, B. A. Mamyrin, and D. V. Shmikk, Sov. Phys. Tech. Phys. 16, 1177 (1972); V. A. Mamyrin, V. I. Karataev, D. V. Shmikk, and V. A. Zauglin, Sov. Phys. JETP 37, 45 (1973).
26. O. Echt, P. D. Dao, S. Morgan, and A. W. Castleman, Jr., J. Chem. Phys. 82, 4076 (1985).
27. A. W. Castleman, Jr., S. Morgan, O. Echt, and P. D. Dao, "Considerations of the Origin of Magic Numbers in Hydrogen Bonded Clusters," to be submitted.
28. S. Morgan and A. W. Castleman, Jr., "Evidence of Delayed 'Internal' Ion Molecule Reactions Following the Multiphoton Ionization of Clusters: Variation in Reaction Channels in Methanol with Degree of Solvation," submitted to J. Am. Chem. Soc.
29. S. Morgan and A. W. Castleman, Jr., "Unimolecular Dissociation of Protonated Methanol Clusters," to be submitted.
30. L. M. Bass, R. D. Cates, M. F. Jarrold, N. J. Kirchner, and M. T. Bowers, J. Am. Chem. Soc. 105, 7024 (1983).
31. R. G. Keesee and A. W. Castleman, Jr., J. Phys. Chem. Ref. Data, 15 (1986) in press.
32. J. C. Kleingeld and N. M. M. Nibbering, Org. Mass Spectrom. 17, 136 (1982).
33. O. Echt, K. Sattler, and E. Recknagel, Phys. Rev. Lett 47, 1121 (1981).
34. D. Kreisle, O. Echt, M. Knapp, and E. Recknagel, Phys. Rev. A 33, 768 (1986).
35. P. D. Dao, S. Morgan, and A. W. Castleman, Jr., Chem. Phys. Lett. 111, 38 (1984).
36. P. D. Dao, S. Morgan, and A. W. Castleman, Jr., Chem. Phys. Lett. 113, 219 (1985).
37. P. D. Dao and A. W. Castleman, Jr., J. Chem. Phys. 84, 1435 (1986).
38. Y. Hatano, personal communication; see also T. Wada, K. Shinsaka, H. Namba, and Y. Hatano, Can. J. Chem. 55, 2144 (1977).

39. K. I. Peterson, P. D. Dao, R. W. Farley, and A. W. Castleman, Jr., J. Chem. Phys. 80, 1780 (1984).
40. A. Herrmann, S. Leutwyler, E. Schumacher, and L. Woste, Helv. Chim. Acta 61, 453 (1978).
41. J. Buttet, Proc. Int. Symp. on Metal Clusters, Heidelberg, Apr. 7-11, 1986, p. 12.
42. J. L. Martins, J. Buttet, and R. Car, Phys. Rev. B 31, 1804 (1985).
43. D. Lindsay, personal communication.
44. J. Flad, H. Stoll, and H. Preuss, J. Chem. Phys. 71, 3042 (1979).
45. M. M. Kappes, M. Schar, P. Radi, and E. Schumacher, J. Chem. Phys. 84, 1863 (1986).
46. K. I. Peterson, P. D. Dao, and A. W. Castleman, Jr., J. Chem. Phys. 79, 777 (1983).
47. P. D. Dao, K. I. Peterson, and A. W. Castleman, Jr., J. Chem. Phys. 80, 563 (1984).
48. K. Liu, E. K. Parks, S. C. Richtsmeier, L. G. Pobo, and S. J. Riley, J. Chem. Phys. 83, 2882 (1985).
49. M. E. Geusic, M. D. Morse, and R. E. Smalley, J. Chem. Phys. 82, 590 (1985); M. D. Morse, M. E. Geusic, J. R. Heath, and R. E. Smalley, J. Chem. Phys. 83, 2293 (1985).
50. R. L. Whetten, D. M. Cox, D. J. Trevor, and A. Kaldor, J. Phys. Chem. 89, 566 (1985).
51. R. L. Whetten, D. M. Cox, D. J. Trevor, and A. Kaldor, Phys. Rev. Lett. 54, 1494 (1985).
52. E. W. McDaniel, V. Cermak, A. Dalgarno, E. E. Ferguson, and L. Friedman, Ion Molecule Reactions, eds. C. F. Barnett and D. M. Cobble (Wiley Interscience, New York, 1970) p. 16.
53. T. Su and M. T. Bowers, Int. J. Mass Spect. Ion Phys. 12, 347 (1973).
54. T. Su and W. J. Chesnavich, J. Chem. Phys. 76, 5183 (1982).
55. F. J. Schelling and A. W. Castleman, Jr., Chem. Phys. Lett. 111, 47 (1984); E. E. Ferguson, J. Phys. Chem. 90, 731 (1986).
56. D. E. Hunton, M. Hofmann, T. G. Lindeman, and A. W. Castleman, Jr., J. Chem. Phys. 82, 134 (1985).
57. R. G. Keesee, N. Lee, and A. W. Castleman, Jr., J. Am. Chem. Soc. 101, 2599 (1979).
58. A. Savitzky and M. J. Golay, Anal. Chem. 36, 1627 (1964).
59. R. N. Zare, Mol. Photochem. 4, 1 (1972); G. E. Busch and K. R. Wilson, J. Chem. Phys. 56, 3638 (1972).
60. S.-C. Yang and R. Bersohn, J. Chem. Phys. 61, 4400 (1974).
61. M. F. Jarrold, A. J. Illies, and M. T. Bowers, J. Chem. Phys. 81, 214 (1984).
62. R. V. Hodges and J. A. Vanderhoff, J. Chem. Phys. 72, 3517 (1980).
63. R. G. Keesee, N. Lee, and A. W. Castleman, Jr., J. Chem. Phys. 73, 2195 (1983).
64. J. J. Breen, K. Kilgore, K. Stephan, R. Sievert, B. D. Kay, R. G. Keesee, T. D. Mark, J. van Doren, and A. W. Castleman, jr., Chem. Phys. 91, 305 (1984).

65. R. J. Celotta, R. A. Bennett, and J. L. Hall, J. Chem. Phys. 60, 1740 (1974).
66. M. F. Jarrold, A. J. Illies, and M. T. Bowers, J. Chem. Phys. 79, 6086 (1983).
67. W. J. Chesnavich and M. T. Bowers, Prog. React. Kin. 11, 137 (1982); W. J. Chesnavich and M. T. Bowers, in Gas Phase Ion Chemistry, Vol. 1, ed. M. T. Bowers (Academic, New York, 1979) pp. 119-151; computational program supplied by M. T. Bowers.
68. D. E. Hunton, M. Hofmann, T. G. Lindeman, C. R. Albertoni, and A. W. Castleman, Jr., J. Chem. Phys. 82, 2884 (1985).
69. J. T. Moseley, P. C. Cosby, and J. R. Peterson, J. Chem. Phys. 65, 2512 (1976); G. P. Smith, L. C. Lee, and J. T. Moseley, J. Chem. Phys. 71, 4034 (1979); J. F. Hiller and M. L. Vestal, J. Chem. Phys. 72, 4713 (1980).

Acknowledgments

Support by the National Science Foundation, Grant No. ATM-82-04010, the Department of Energy, Grant No. DE-AC02-82-ER60055, and the Army Research Office, Grant No. DAAG29-85-K-0215, is gratefully acknowledged. The authors are indebted to various colleagues whose work is referenced and to current group members Cathy Albertoni and Harry Sarkas who did the work of the photodissociation of $(SO_2)_2^-$, Matt Cook for his work on Xe clusters, and Sterling Sigsworth, Robert Leuchtner, Robert Farley, and Hideyuki Funasaka who are engaged in studies of the lifetimes and reactivity of metal ions and cluster ions. Also, we must thank Professor K. Weil (on leave from the University of Darmstadt) for stimulating discussions and suggestions on metal ion reactivity, and Dr. Olof Echt (University of Konstanz) for collaboration on the Xe work.

GROWING MOLECULES WITH ION/MOLECULE REACTIONS

Diethard K. Bohme
Department of Chemistry and Centre for Research in Experimental Space Science
York University
Downsview, Ontario, Canada M3J 1P3

ABSTRACT. Ion/molecule reactions are viewed as agents of chemical synthesis as it may proceed in partially ionized gases. Partially ionized interstellar gas clouds, the largest known chemical factories in the universe, are chosen as exemplary environments. Emphasis is given to the synthesis of more complex interstellar molecules. Target molecules include some which have been identified in interstellar gas clouds and others which have not (yet). Hydrogenation reactions of ions are viewed as starting reactions for molecule formation and ion/molecule reactions leading to C-C bond formation are regarded in the context of the ionic growth of carbon skeletons. Particular attention is given to the synthesis of polyacetylenes, cyanoacetylene, methylcyanoacetylene, cyanopolyacetylenes, cyclopropenylidene, silacyclopropyne, cumulated carbene chains, disubstituted carbenes and ring carbenes, benzene, polycyclic aromatic hydrocarbons, and amino acids.

1. INTRODUCTION

The kinetics, dynamics, thermodynamics and mechanisms of ion/molecule reactions in the gas phase have been actively explored for several decades and continue to be rewarding subjects of research in ion chemistry. In contrast, and somewhat curiously, the synthetic aspects of ion/molecule reactions have received relatively little attention over the years. Only now is this situation beginning to change significantly. With the advent of space chemistry ion/molecule reactions have become attractive agents for the synthesis of the simple and exotic molecules recently identified by radioastronomers to be present in cool interstellar gas clouds. Current gas phase models of the cloud chemistry rely heavily on ion/molecule reactions and are beginning to account for the abundance of the ions and simple molecules observed in these clouds. Less success has been possible with more complicated molecules because of a more limited chemical data base. The synthesis of these complicated molecules provides an exciting challenge to the laboratory ion chemist and should stimulate

P. Ausloos and S. G. Lias (eds.), Structure/Reactivity and Thermochemistry of Ions, 219–246.

increasing research activity in this area.

The interest in synthetic aspects of ion/molecule reactions is by no means restricted to interstellar gas clouds. Ionic synthesis also operates in other partially ionized environments such as electrical discharges, hydrocarbon flames, irradiated gases, planetary atmospheres etc. So there is now a real need to begin to develop an understanding of the synthetic aspects of ion/molecule reactions. This article is a first attempt to do so.

2. ION/MOLECULE REACTIONS COMPARED TO NEUTRAL REACTIONS

2.1. Molecular Growth

In any partially ionized environment molecules may grow in one of two ways as is illustrated in Fig. 1. They may grow directly from atoms to large molecules by reactions between neutral species. Reactions between free radical species would be particularly effective in this regard. Alternatively, if some energy is deposited in ionization, growth may proceed by ion/molecule reactions. At the low pressures of interstellar gas clouds these reactions will be bimolecular but termolecular reactions also will be important in higher pressure environments. New large molecules will result if the large ions produced in this fashion are neutralized without disturbing the newly formed chemical bonds. Neutralization may be accomplished in a number of different ways including charge transfer, proton transfer and ion/electron recombination. Alternatively the new molecule may appear directly as the neutral product of the ion/molecule reaction. The overall relative importance of the neutral and ionic paths will clearly depend on the degree of ionization. But, as will shortly become apparent, a strong dependence of the relative importance of the neutral and ionic pathways on the temperature of the environment is also to be expected.

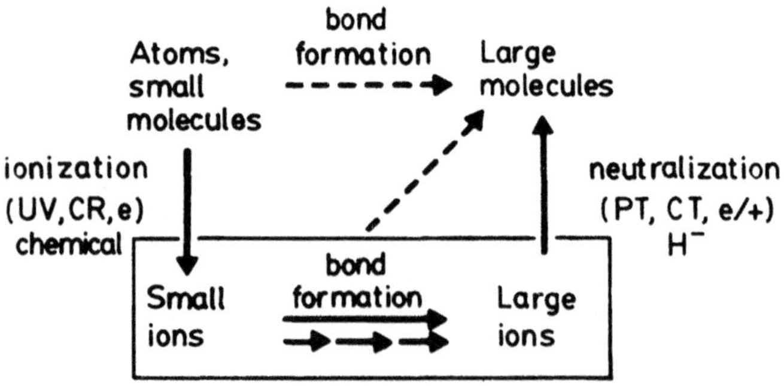

Fig. 1. Two pathways for molecular growth in partially ionized environments. Ionization may be induced by ultraviolet light (UV), cosmic rays (CR), electron impact (e) or chemical ionization. Neutralization may proceed by proton transfer (PT), charge transfer

(CT), electron/ion recombination (e/+) or hydride transfer (H^-).

2.2. The Advantages

Ion/molecule reactions have two important intrinsic advantages for chemical synthesis compared to neutral reactions. Firstly, because ions communicate with molecules at long range by electrostatic interaction such as ion-induced dipole and ion-permanent dipole interaction, they collide with molecules with high specific rates typically in the range from 10^{-8} to 10^{-10} cm^3 $molecule^{-1}$ s^{-1}. These specific rates are several orders of magnitude larger than the corresponding specific rates for neutral reactions for which the long range electrostatic interaction is absent. Furthermore the high collision rates for ion/molecule reactions persist or increase at lower temperatures. Indeed, they increase substantially for collisions with polar molecules (1).

Secondly, the electrostatic interaction between an ion and molecule also leads to energies at close separations which are sufficient to overcome most activation energies associated with the redisposition of chemical bonds. This means that many ion/molecule reactions proceed with nearly unit reaction efficiency.

The combination of high collision rate and high reaction efficiency results in a specific rate for chemical bond formation by ion/molecule reaction which can easily be several orders of magnitude larger than for the corresponding neutral reaction. Also the specific rate for ion/molecule reaction is likely to be relatively insensitive to the temperature or to have a negative temperature dependence.

2.3. The Competition

Fig. 2 presents reaction energy versus reaction coordinate diagrams for neutral and ion/molecule reactions. The profile for the neutral reaction is drawn in the usual fashion in terms of an activation energy. The ion/molecule reaction is described with a profile having a double minimum which arises due to the electrostatic attraction between the ion and the neutral reactants as well as between the ion and neutral products. For many ion/molecule reactions the initial energy of the reactants exceeds the intermediate activation energy.

Some measure of the competition between neutral and ion/molecule reactions can be derived from a comparison of the rates of the two reactions represented in Fig. 2. If an Arrhenius behaviour is assumed for the specific rate of the neutral reaction and the specific rate of the ion/molecule reaction is taken to be temperature independent with unit efficiency (or at least to approach unit efficiency at low temperatures), the relative rate of the two types of reaction in a partially ionized environment can be reduced to the following expression:

$$\frac{-d[A^+]/dt}{-d[A]/dt} = \frac{k_{A^+/B}\,[A^+][B]}{k_{A/B}\,[A][B]} \simeq 100\, f_{ion}\, e^{E_a/RT}$$

where E_a is the assumed activation energy of the neutral reaction, f_{ion} is the fractional ionization in the environment, and the collision rate constant of the ion/molecule reaction has been taken to be 100 times that for the corresponding neutral reaction.

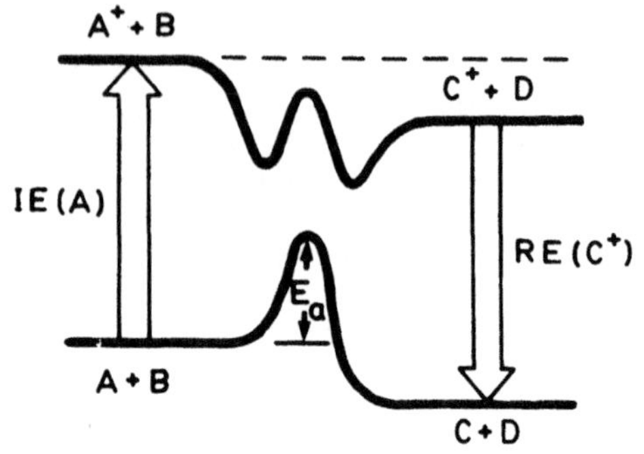

Fig. 2. Generalized profiles for the dependence of reaction energy on the reaction coordinate for ion/molecule and neutral reactions. IE and RE refer to ionization energy and recombination energy, respectively.

It is apparent from the expression above that chemical reactions between ions and molecules become increasingly competitive as the temperature is lowered. This has been recognized in space chemistry in models which attempt to account for the synthesis of molecules in cold partially ionized interstellar clouds. Calculations indicate that ion/molecule reactions without activation energies will compete effectively with neutral reactions having activation energies as small as only 1 kcal mol^{-1} at the temperature of 50 K and 1 ppm ionization typical of such clouds.

3. NEUTRALIZATION OF IONS

Unless the desired molecule is produced directly as the neutral product of an ion/molecule reaction, the ion/molecule reaction route towards molecular growth will involve ion neutralization as a last step. The neutralization of ions may take place in a number of different ways. Some important neutralization reactions are indicated below in terms of the target molecule AH. Also a measure of the excess energy of the reaction is given in parentheses.

a. Charge Transfer (Excess energy = Δ IE)

$$AH^+ + B \longrightarrow AH + B^+$$

b. Proton Transfer (Excess energy = Δ PA)

$$AH_2^+ + B \longrightarrow AH + BH^+$$

c. Hydride Transfer (Excess energy = Δ H^- A)

$$A^+ + HB \longrightarrow AH + B^+$$

d. Electron/Ion Recombination (Excess energy = IE(H) – PA(AH))

$$AH_2^+ + e \longrightarrow AH + H$$

Here IE, PA and H^-A refer to ionization energy, proton affinity and hydride affinity, respectively.

For molecular growth by the ion/molecule route to be a success the eventual neutralization of the precursor ion for a particular target molecule must preserve the desired chemical bonds. Bond rupture will be minimized if the excess energy is low or if it is distributed among as many of the bonds as possible. In practise, the favoured precursor ion for a particular AH will be determined by the rate of the neutralization reaction. The four neutralization reactions all have high specific rates at room temperature which tend to have a negative temperature dependence. The specific rate for electron/ion recombination is about 100 times higher than the others but electron/ion recombination also involves the largest excess energy. Little is known about the neutral products of such reactions and thus the degree of dissociation of the recombining ion. For the recombination of an ion of the type AH_2^+ it is commonly assumed that the dominant product involves the elimination of H which preserves the bonding in AH.

The neutralization reactions listed above indicate that a particular target molecule has a number of possible ionic parents. This notion is expressed in the reaction grid shown in Fig. 3 for the synthesis of the diatomic hydride molecule AH from the atom A.

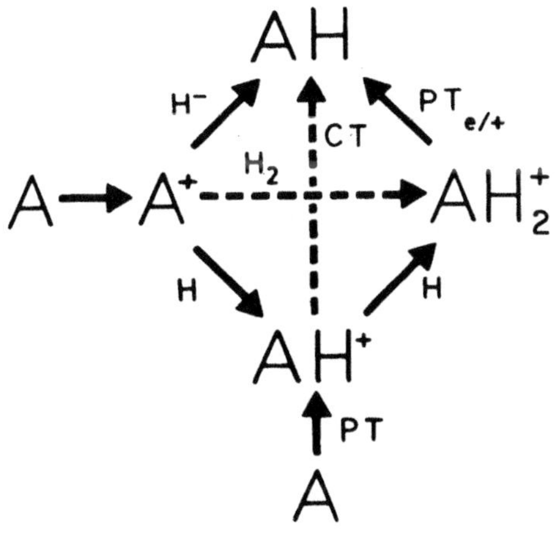

Fig. 3. Reaction grid for the synthesis of the diatomic hydride molecule AH from the atom A. Here H^-, H, H_2, CT and PT denote hydride, hydrogen atom, hydrogen molecule, charge and proton transfer respectively, and e/+ denotes electron/ion recombination.

4. FROM ATOMS TO MOLECULES/HYDROGENATION REACTIONS

4.1. Hydrogen Atom Transfer

Hydrogenation reactions are important as starting reactions for molecule formation in interstellar gas chemistry, especially those with H_2 because of its high relative abundance in certain clouds. Sequential hydrogen atom transfer reactions of the type indicated below appear to

$$AH_n^+ + H_2\ (XH) \longrightarrow AH_{n+1}^+ + H\ (X)$$

provide the most effective means for the hydrogenation of ions. Reactions of this type with molecular hydrogen are particularly important in dense interstellar clouds and the atmospheres of the outer planets. They may operate in competition with neutralization of the molecular ions to lead to the conversion of atoms to diatomic and polyatomic molecules as shown below.

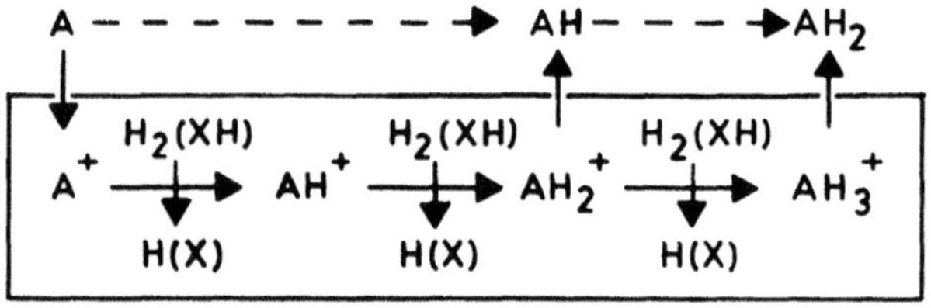

4.2. Full Saturation by Hydrogen Atom Transfer

Under favourable kinetic and thermodynamic circumstances sequential hydrogen atom transfer reactions can rapidly lead to full saturation as in, for example, the following two hydrogenation sequences:

$$O^+ \xrightarrow{H_2} OH^+ \xrightarrow{H_2} H_2O^+ \xrightarrow{H_2} H_3O^+$$

$$Cl^+ \xrightarrow{H_2} HCl^+ \xrightarrow{H_2} H_2Cl^+$$

Neutralization of the terminal ions in these sequences may lead to the stable molecules H_2O and HCl, respectively. However, full saturation in this manner is not always guaranteed for other ions.

4.3. Thermodynamic Bottlenecks

Firstly, hydrogenation sequences may be interrupted by endothermicity. This is the case, for example, in hydrogenation of carbon ions to CH_5^+ as shown in the following sequence:

$$C^+ \overset{H_2}{\not\longrightarrow} CH^+ \overset{H_2}{\longrightarrow} CH_2^+ \overset{H_2}{\longrightarrow} CH_3^+ \overset{H_2}{\not\longrightarrow} CH_4^+ \overset{H_2}{\longrightarrow} CH_5^+$$

Hydrogen-atom transfer from hydrogen to either C^+ or CH_3^+ is endothermic. In the hydrogenation of C_2^+ to $C_2H_7^+$.

$$C_2 \overset{+H_2}{\longrightarrow} C_2H \overset{+H_2}{\longrightarrow} C_2H_2 \overset{+H_2}{\not\longrightarrow} C_2H_3 \overset{+H_2}{\not\longrightarrow} C_2H_4 \overset{+H_2}{\not\longrightarrow} C_2H_5 \overset{+H_2}{\not\longrightarrow} C_2H_6 \overset{+H_2}{\not\longrightarrow} C_2H_7^+$$

only the first two H atom transfers are exothermic with molecular hydrogen so that the hydrogenation stops at $C_2H_2^+$.

4.4. Kinetic Bottlenecks

Secondly, there are instances in which exothermic hydrogen transfer reactions are kinetically unfavourable. One example important in the synthesis of cyanoacetylene is the following hydrogenation of the cyanoacetylene radical cation which is exothermic but slow at 300 K(2):

$$HC_3N^+ + H_2 \longrightarrow H_2C_3N^+ + H$$

Hydrogenation bottlenecks of this kind may be circumvented by radiative or collisional association with molecular hydrogen, but these may be quite inefficient. Slightly endothermic reactions may be driven by ion kinetic energy (3).

Another interesting example involves the hydrogenation in hydrogen of N^+ to NH_4^+ which is important to the formation of ammonia in the interstellar medium. The rate constants of both the first and last reactions in this sequence have been shown to be quite sensitive to temperature (4). The former reaction is quite rapid at room temperature but exhibits a small activation energy which can be attributed to endothermicity. The activation energy of the hydrogenation reaction of N^+ with H_2 has been measured to be 7.4 ± 0.8 meV. The rate constant of the latter exothermic reaction of NH_3^+ with H_2 drops from 5×10^{-13} cm^3 molecule^{-1} s^{-1} at room temperature to about 2×10^{-13} cm^3 molecule^{-1} s^{-1} before it recovers and increases again below about 30 K.

4.5. Importance of Bottlenecks

Kinetic or thermodynamic bottlenecks can have important implications for further synthesis. Generally speaking, failure to react with one kind of molecule will make the ion available to react with another, less abundant molecule. For example, in interstellar clouds rich in molecular hydrogen the bottlenecks in the hydrogen atom transfer sequences identified above impart a special chemical potential to the 'terminal' ions C^+, N^+, CH_3^+, $C_2H_2^+$ and HC_3N^+. Nevertheless, because of the high abundance of H_2, the chemical reactions of these ions may have to compete with much less efficient radiative association reactions with H_2 which act to 'shortcircuit' the bottlenecks in the hydrogen atom transfer sequences. One such radiative association reaction has now been characterized in the laboratory (5):

$$CH_3^+ + H_2 \longrightarrow CH_5^+ + h\nu$$

The rate constant for this reaction has been reported to be 1.1×10^{-13} cm^3 $molecule^{-1}$ s^{-1} at 13 K. For the corresponding reaction with C^+:

$$C^+ + H_2 \longrightarrow CH_2^+ + h\nu$$

the most recent measurements indicate an upper limit to the rate constant at 10 K of 1.5×10^{-15} cm^3 $molecule^{-1}s^{-1}$ (5). The latter result implies that C^+ is available to molecules in the ppm range or below in interstellar hydrogen clouds.

In higher pressure environments collisional association reactions may act to short circuit bottlenecks in bimolecular reaction sequences.

$$AH_n^+ + H_2 + M \longrightarrow AH_{n+2}^+ + M$$

We note here that C^+, CH_3^+ and $C_2H_2^+$ have been observed in the laboratory to add H_2 in termolecular association reactions (6,7,8). The derivative ions CH_5^+ and $C_2H_4^+$ might therefore become more likely ions to initiate other chemistry in higher pressure H_2 environments.

5. GROWING CARBON CHAINS/FORMATION OF CARBON-CARBON BONDS

The essential steps in molecular growth involve formation of chemical bonds between two heavy atoms and formation of larger skeletons of heavy atoms. Many ion/molecule reactions are known which lead to such growth but no attempt is made here to produce a systematic survey. Instead we restrict ourselves to a consideration of ion/molecule reactions which form bonds between carbon atoms. In the next section which deals with the synthesis of specific target molecules, ion/molecule reactions will be invoked which lead to formation of C-O, C-S and C-N bonds.

5.1. With Atomic Carbon Ions

Conceptually the most direct mode of C-C bond formation by ion/molecule reactions would be the association of an atomic carbon ion with a carbon substrate (or a carbon atom with an ionized carbon substrate). Such a scheme has been proposed by Suzuki (9) for the growth of carbon chain molecules in interstellar regions of low optical depth where carbon is mostly ionized. The essential features of this scheme may be presented as follows:

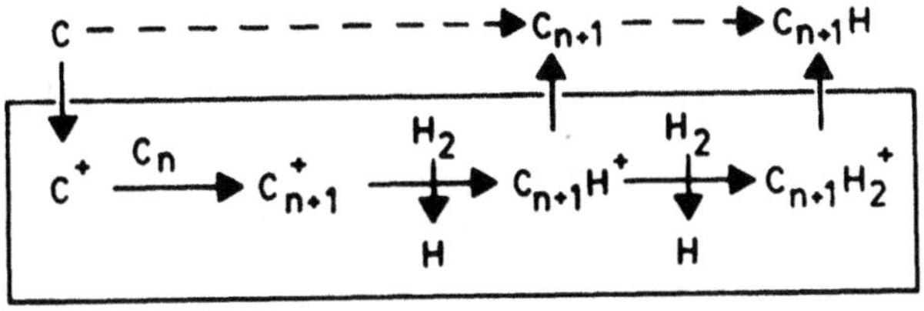

The scheme involves the radiative association of C^+ and C_n followed by the hydrogenation of the carbon cluster ions C_n^+ and neutralization of the hydrogenated carbon cluster ions. Rate constants for the radiative recombination appear to become substantial for n ≥ 4. For example, Bates has calculated rate constants of 2 x 10^{-13} and 2 x 10^{-10} cm^3 $molecule^{-1}$ s^{-1} for the radiative association at 30K for C^+ with C_3 and C_4, respectively (10). We have recently shown in our laboratory that hydrogenation of carbon cluster ions by hydrogen atom transfer with H_2 is feasible at 300K for n = 2,3,4, and 5. For n = 2 and 4 a sequential hydrogen atom transfer reaction was also observed (11). The hydrogen atom transfer reaction of C_4^+ with H_2 has also been observed to proceed rapidly at 80 K (12).

Thermochemical considerations indicate that formation of bimolecular products in the reaction of C^+ with neutral carbon clusters is endothermic and so will not compete with radiative association. This is no longer the case for reactions of C^+ with hydrocarbon molecules for which we may distinguish between the following two bimolecular reaction channels:

$$C^+ + C_nH_m \longrightarrow C_{n+1}H_p^+ + H_q \quad (a)$$

$$\longrightarrow C_{n+1-x}H_p^+ + C_xH_q \quad (b)$$

Channel (a) corresponds to molecular growth while channel (b) leads to the dissociation of the neutral hydrocarbon substrate. Both channels are often exothermic. We see that, in contrast to the reactions of C^+ with carbon clusters discussed earlier, reactions of C^+ with hydrocarbons present a competition between constructive and dissociative bimolecular reaction channels. Systematic SIFT measurements in our laboratory with homologous series of hydrocarbons have provided some insight into the nature of this competition. The results of these measurements suggest that dissociative channels become increasingly preferred as the length of the hydrocarbon 'backbone' increases (12). This is evident from the results listed below in which the percentage of the reaction with C^+ which leads to molecular growth is given in parentheses.

CH_4 (100)	C_2H_6 (30)	C_3H_8 (≤3)
C_2H_4 (70)	CH_2CCH_2 (40)	CH_3CHCH_2 (10)

C_2H_2 (100) C_4H_2 (50) CH_3C_2H (30)
HCN (100) HC_2CN (20)

Inspection of these results indicates that as the number of carbon atoms in the hydrocarbon increases more of the excess energy of the reaction of C^+ appears in C-C rather than C-H bonds! There appears to be a limit to molecular growth in this fashion.

5.2. With Molecular Ions

The carbon skeleton may also be extended by reactions of molecular hydrocarbon ions with hydrocarbon molecules. This may be accomplished by condensation or association. Condensation results when the adduct of the hydrocarbon ion and the hydrocarbon molecule dissociates into products before it is stabilized, while association results if the adduct is stabilized by collision with a third molecule or by the emission of radiation before separation into reactants or products. The various options are indicated below:

$$C_nH_x^+ + C_mH_y \longrightarrow C_{n+m}H_{x+y}^+$$
$$\longrightarrow C_{n+m}H_{x+y-1}^+ + H$$
$$\longrightarrow C_{n+m}H_{x+y-2}^+ + H_2$$
$$\longrightarrow C_{n+m-k}H_p^+ + C_kH_q$$

Association or condensation with elimination of H, H_2 or a small hydrocarbon fragment ($k<n$) will lead to growth of the ionic carbon skeleton, while elimination of small ionic fragments will lead to growth of the neutral carbon skeleton.

No data appear to be available for reactions of hydrocarbon ions with carbon cluster molecules ($y=0$). A few measurements have been made for reactions of carbon cluster ions ($x=0$) with hydrocarbon molecules and these indicate clearly that carbon cluster ions are highly suited for ionic growth. For example, 70% of a rapid reaction of C_2^+ with methane has been reported to lead to ionic growth by condensation with elimination of H, H_2, and H + H_2 (14). 100% growth by condensation with elimination of H has been reported for the reactions of C_3^+ and C_4^+ with acetylene (15) and of C_4^+ with diacetylene (16) proceeding at the low pressures of an ICR. The reactions of C_3^+ and C_4^+ with acetylene to form C_5H^+ and C_6H^+ have also been observed to proceed in our SIFT apparatus but in competition with the addition of acetylene to form $C_5H_2^+$ and $C_6H_2^+$, respectively (11). The highly unsaturated C_5H^+, C_6H^+ and C_8H^+ cations are suitable sources for the carbon cluster molecules C_5, C_6, C_8, respectively.

There have been numerous measurements of reactions of hydrocarbon ions with hydrocarbon molecules, too numerous to address in any comprehensive fashion in this article. However, a few comments are appropriate concerning the reactions of the 'terminal' interstellar hydrocarbon ions CH_3^+ and $C_2H_2^+$. The reactions of both of these ions have been investigated with methane, acetylene, ethylene, ethane (17) and diacetylene(18). 100% of the reactions with methane and acetylene lead to growth in the carbon backbone with elimination of H_2 (with CH_3^+) and H or H_2 (with $C_2H_2^+$). The ions which are spawned are $C_2H_5^+$, $C_3H_3^+$, $C_3H_4^+$, $C_3H_5^+$, $C_4H_2^+$, and $C_4H_3^+$ and these are possible precursors for a range of interesting neutral hydrocarbon molecules: C_2H_4 (ethylene), C_3H_2, C_3H_3, C_3H_4 (methylacetylene), C_4H, and C_4H_2 (diacetylene). For the reactions with ethylene, ethane and diacetylene several non-productive channels compete with ionic growth. For example, hydride transfer comprises 38% and 85% of the reactions of CH_3^+ with ethylene and ethane while 90% of the reaction with diacetylene produces $C_3H_3^+/C_2H_2$. Charge transfer is the main competitor with ionic growth in the reactions of $C_2H_2^+$ with ethylene (34%) and diacetylene (90%) while H_2 and hydride transfer (together 27%) compete with ionic growth in the reaction with ethane. These various modes of competition with molecular growth are likely to be representative of the more general situation. Little has been done to explore systematically for homologous series of molecules the dependence of the extent of competition on the size of the hydrocarbon molecule as was illustrated for the reactions of C^+. Such an experimental campaign would probably be quite instructive.

It is known that radical cations of olefins and acetylenes in general, or of polyolefinic and polyacetylenic molecules, undergo condensation and/or association reactions with their parent molecules (n=m, x=y). This often initiates consecutive additions of the parent molecule which leads to high molecular weight polymeric ions, albeit the rate constant for addition tends to decrease with the number of steps. Fig. 4 shows the rapid growth of hydrocarbon ions in acetylene initiated by ionized acetylene. Three steps are seen to proceed in succession and to lead to the connection of four C_2 carbon units with overall complete retention of the acetylenic hydrogen ($C_8H_8^+$) and with loss of H ($C_8H_7^+$) and H_2 ($C_8H_6^+$). Other examples of condensation/association reactions initiated by the reactions of radical cations of olefins and acetylenes with their parent molecules include the sequences initiated in ethylene (19) and diacetylene (20):

$$C_2H_4^+ \rightarrow C_4H_8^+ \rightarrow C_6H_{12}^+ \rightarrow C_8H_{16}^+ \rightarrow C_{10}H_{20}^+ \rightarrow C_{12}H_{24}^+$$

$$C_4H_2^+ \rightarrow C_8H_4^+ \rightarrow C_{12}H_6^+ \rightarrow C_{16}H_8^+ \rightarrow C_{20}H_{10}^+$$

$$C_4H_2^+ \rightarrow C_6H_2^+ \rightarrow C_{10}H_4^+ \rightarrow C_{14}H_6^+ \rightarrow C_{18}H_8^+$$

$$C_4H_2^+ \rightarrow C_8H_2^+ \rightarrow C_{12}H_4^+ \rightarrow C_{16}H_6^+$$

$$C_4H_2^+ \rightarrow C_6H_2^+ \rightarrow C_8H_2^+ \rightarrow C_{12}H_4^+ \rightarrow C_{16}H_6^+$$

The ion/molecule reaction sequences initiated in diacetylene have been investigated in detail in an ICR spectrometer at ca. 10^{-6} Torr and in a high-pressure photoionization mass spectrometer at ca. 10^{-2} Torr. The sequence which predominates is dependent on the degree of collisional stabilization in the initiating step. The ions produced in the sequence initiated by $C_8H_4^+$ all showed two populations, one reactive and one unreactive with diacetylene. The interesting proposal has been put forward that these two populations correspond to straight-chain and cyclic isomers, respectively (20).

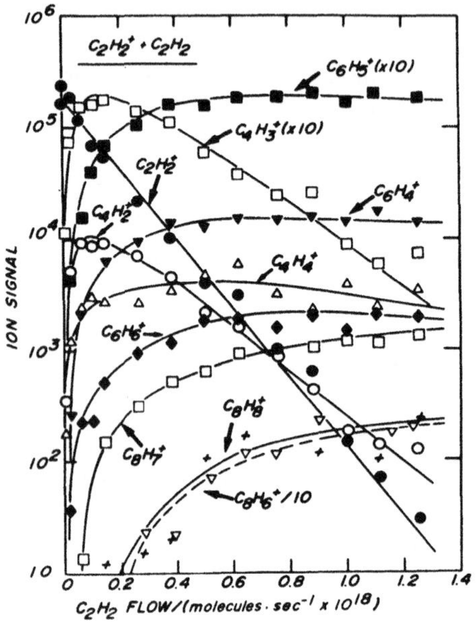

Fig. 4. Observations of the sequential acetylene chemistry initiated by $C_2H_2^+$ in a SIFT apparatus at 296 K in helium buffer gas at 0.34 Torr. The $C_2H_2^+$ is derived from acetylene by electron impact at 20 eV (11).

6. GROWING MOLECULES

The total synthesis of many heteroatomic molecules is likely to involve several interlocking chemistries. For example, in the synthesis of hydrocarbon molecules a carbon skeleton will have to be constructed and heteroatoms or functional groups may need to be added to the skeleton. Also, a desired synthesis may require transformations from acyclic into cyclic structures, either aliphatic or aromatic. This section illustrates the general approach with a detailed consideration of specific ionic syntheses of target molecules chosen primarily, but not exclusively, from those identified in interstellar gas clouds.

6.1. Growing Polyacetylenes

Plausible ionic parents for neutral polyacetylenes include the acetylenic ions $C_{2n}H_3^+$ and $C_{2n}H_2^+$ which may neutralize by electron/ion recombination, proton transfer or charge transfer to form polyacetylenes in the following manner:

$$C_{2n}H_3^+ + e(X) \longrightarrow C_{2n}H_2 + H\ (XH^+)$$

$$C_{2n}H_2^+ + X \longrightarrow C_{2n}H_2 + X^+$$

Two obvious schemes for the ionic growth of polyacetylenes involve the sequential addition of C_2 units either directly to the ion which may subsequently neutralize as follows:

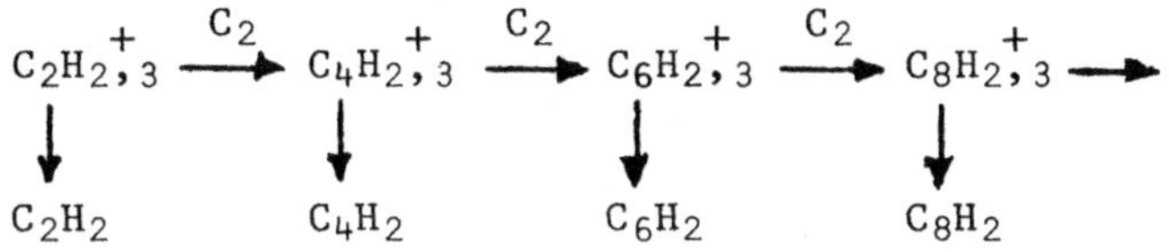

or indirectly after neutralization by repeated reaction with an ionic C_2 unit as follows:

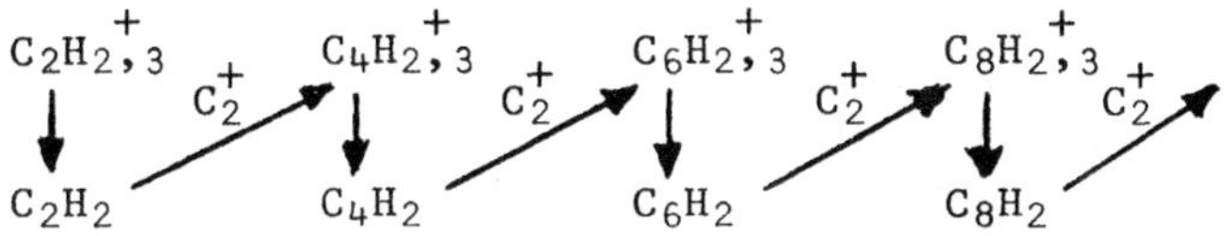

Both schemes may of course also operate with higher carbon units with even numbered carbon atoms. The addition of the carbon unit in each step may occur at low pressures by radiative association or condensation in the following general manner:

$$C_{2n}H_{2,3}^+ + C_2 \longrightarrow C_{2n+2}H_{2,3}^+ + h\nu$$

$$C_{2n}H_{2,3}^+ + C_2H_{1,2} \longrightarrow C_{2n+2}H_{2,3}^+ + H, H_2$$

or

$$C_{2n}H_2 + C_2H_{0,1}^+ \longrightarrow C_{2n+2}H_{2,3}^+ + h\nu$$

$$C_{2n}H_2 + C_2H_{2,3}^+ \longrightarrow C_{2n+2}H_{2,3}^+ + H, H_2$$

Although many of these reactions remain to be explored in the laboratory, some measurements are available which provide insight at least into the initiating steps.

The ions $C_2H_2^+$ and $C_2H_3^+$ are known to react with acetylene to establish $C_4H_2^+$ and $C_4H_3^+$. Fig. 4 shows results of SIFT studies which demonstrate how the reaction of $C_2H_2^+$ with C_2H_2 establishes $C_4H_2^+$ and $C_4H_3^+$ as well as the adduct $C_4H_4^+$ which is likely to be collisionally stabilized at the pressure of the SIFT experiments. The adduct is not observed at low pressures (17). In contrast, the $C_4H_2^+$ and $C_4H_3^+$ ions form adducts with acetylene under SIFT conditions (18) and at the low pressures of the ICR (15). There has been one report of a minor channel at low pressures leading to $C_6H_3^+$ from $C_4H_2^+$ and C_2H_2 (15). The observation of adducts at low pressures implies the occurrence of radiative association but both collisional and radiative association have been advanced as the mechanism of association (15, 21). This uncertainty in the mechanism of stabilization at low pressures makes the suggestion that the higher $C_{2n}H_{2,3}^+$ acetylenic ions react with acetylene by radiative association perhaps somewhat premature. There is no information available on the reactions of the acetylenic ions with the molecules C_2 and C_2H.

The reactions of acetylene with $C_2H_2^+$ and $C_2H_3^+$ also initiate the scheme which involves neutralization followed by reaction with ionic units and it is known that C_2H^+ reacts rapidly with C_2H_2 to produce $C_4H_2^+$ exclusively (15). As regards the second step in this scheme, SIFT measurements have shown that the chain lengthening reaction of diacetylene with $C_2H_2^+$ to form $C_6H_3^+$ must compete with a predominant charge transfer channel (18). This will also be the case for reactions of $C_2H_2^+$ with higher polyacetylenes as the ionization potential decreases. With $C_2H_3^+$ the competition will be provided by proton transfer. The proton affinity of C_2H_2 is expected to be lower than for the polyacetylenes.

Some information is also available for reactions between neutral and ionic C_4 units. Recent SIFT measurements have indicated that the reactions of $C_4H_2^+$ and $C_4H_3^+$ with diacetylene can establish $C_6H_2^+$ and $C_6H_3^+$ but that they are dominated by adduct formation as are the reactions of $C_6H_2^+$ and $C_6H_3^+$ with diacetylene. The situation is different at low pressures where collisional stabilization is less influencial. Recent ICR measurements have indicated the following product distribution for the reaction of $C_4H_2^+$ with C_4H_2 (16):

$$C_4H_2^+ + C_4H_2 \xrightarrow{0.83} C_6H_2^+ + C_2H_2$$
$$\xrightarrow{0.17} C_8H_2^+ + H_2$$
$$\xrightarrow{0.01} C_8H_3^+ + H$$

Also the reaction of $C_4H_3^+$ was observed to produce $C_6H_3^+$ exclusively (16). Other ICR measurements have shown that the $C_6H_2^+$ produced from $C_4H_2^+$ and C_4H_2 reacts with diacetylene to produce $C_8H_2^+$ and $C_{10}H_2^+$ which are possible parents for tetra and penta-acetylene, respectively (20). Here again only the adducts are observed at higher pressures. Finally, it has also been shown that C_4H^+ reacts with diacetylene to produce $C_8H_2^+$ exclusively (16).

It seems clear from these experimental results that polyacetylenes are readily generated by sequential condensation reactions involving C_2 or C_4 units at low pressures. Collisional stabilization of the adduct can dominate already at the moderate pressures of the SIFT technique and, at least for the reactions of $C_4H_{2,3}^+$ with acetylene, may operate already at quite low pressures.

6.2. Growing Polycarbon Monoxides and Sulphides

Carbene cations appear to be well suited to linking heteroatoms to carbon chains due to their ability to coordinate with non-bonded electron pairs. Recent SIFT experiments have shown that:C_3H^+ reacts rapidly with H_2O, CH_3OH, OCS, CO_2, O_2 and N_2O to produce some HC_3O^+ (22,23). This latter ion may neutralize by electron/ion recombination or proton transfer to yield tricarbon monoxide or by hydride transfer to yield ketene. The carbene cation:C_3H^+ has also been observed to add directly to CO to yield HC_4O^+ (23) as has:C_5H^+ to yield HC_6O^+ (24). The product ions of these latter two reactions may neutralize to form tetracarbon monoxide or $H_2C{=}C{=}C{=}C{=}O$ and hexacarbon monoxide or $H_2C{=}C{=}C{=}C{=}C{=}C{=}O$ respectively. Tricarbon monoxide is the latest molecule to be identified in the interstellar cloud TMC-1 (25) where :C_3H^+ may be produced from the reaction of C^+ with acetylene.

Laboratory measurements have indicated an analogous sulphur chemistry. C_3H^+ has been observed to react with H_2S and OCS to yield the HC_3S^+ ion which may neutralize to form tricarbon sulphide or $H_2C{=}C{=}C{=}S$ (23). These latter two molecules are predicted to be probably interstellar molecules since both H_2S and OCS have been identified in interstellar clouds. They have not yet been detected. The major portions of the laboratory reactions of C_3H^+ with H_2S and H_2O lead to CS and CO, respectively. This occurs either by the direct elimination of CO and CS or by the formation of CHO^+ and CHS^+ which may neutralize to yield the same molecules.

6.3. Growing Cyanoacetylene

Nitrogen can also be entrained by carbene cations. For example, we have observed that a substantial fraction of the reaction of C_3H^+ with N_2O apparently produces cyanoacetylene directly by nitride abstraction:

$$C_3H^+ + N_2O \longrightarrow HC_3N + NO^+$$

Nitrous oxide appears not to be an interstellar molecule (26) so that this reaction is not a likely source of cyanoacetylene in interstellar gas clouds. Current chemical models for these environments instead invoke reactions which generate $H_2C_3N^+$ which may neutralize to produce HC_3N. For example, $H_2C_3N^+$ is assumed to be formed directly from the reaction of the $C_3H_3^+$ hydrocarbon ion with free nitrogen atoms as follows (27,28):

$$C_3H_3^+ + N \longrightarrow H_2C_3N^+ + H$$

Alternatively, it is assumed that the $H_2C_3N^+$ originates from the hydrogenation of HC_3N^+ by H_2. The HC_3N^+ radical cation may be derived in a number of different ways including the reaction of $C_3H_2^+$ with N, the protonation of C_3N by ions such as H_3^+ and HCO^+, and the reaction of $C_2H_2^+$ with CN or of CN^+ with C_2H_2 (27 - 31). However, recent SIFT experiments suggest that the hydrogenation of HC_3N^+ by H_2 may have a positive activation energy and that other hydrogen atom donors may be required to effect sufficient hydrogenation (2).

Other, more complex condensation reactions also have been shown to qualify as possible sources of cyanoacetylene. For example, SIFT experiments have indicated that C_3H^+ reacts with acetonitrile as follows (23):

$$C_3H^+ + CH_3CN \longrightarrow C_2H_3^+ + HC_3N$$

$$\longrightarrow H_2C_3N^+ + C_2H_2$$

It is also interesting to note here that both protonated and neutral cyanoacetylene are products of the same reaction. This may be attributed to the occurrence of intramolecular proton transfer before the products become separated.

6.4. Growing Methylcyanoacetylene

The synthesis of cyanoacetylene allows growth to methylcyanoacetylene. SIFT experiments have shown that CH_3^+ adds directly to cyanoacetylene presumably to form the cyanopropenylium ion $CH_3\text{-}CH{=}C^+\text{-}CN$ which may neutralize to form methylcyanoacetylene (32).

$$CH_3^+ + HC_3N \longrightarrow C_4H_4N^+$$

Also 5% of the reaction of HC_3N^+ with methane leads to a $C_4H_4N^+$ ion in a bimolecular fashion (2):

$$HC_3N^+ + CH_4 \longrightarrow C_4H_4N^+ + H$$

Other routes leading to the formation of $C_4H_4N^+$ have been adopted in a recent molecular modelling of the molecular cloud TMC-1 (33). The acetylenic ions $C_4H_4^+$ and $C_4H_3^+$ are assumed to react with atomic nitrogen in the following manner:

$$C_4H_2^+ + N \longrightarrow HC_4N^+ + H$$

$$C_4H_3^+ + N \longrightarrow H_2C_4N^+ + H$$

The product ions HC_4N^+ and $H_2C_4N^+$ are then assumed to hydrogenate by sequential hydrogen atom transfer reactions with molecular hydrogen to form $H_4C_4N^+$. All of these reactions have yet to be established in the laboratory.

6.5. Growing Cyanopolyacetylenes

One remarkable aspect of the chemical composition of the interstellar medium is the abundance of the homologous series of cyanopolyacetylene molecules which have been identified by radioastronomers (34). These long chain organic molecules have been detected through their rotational transitions at radio frequencies in all types of molecular clouds as well as in circumstellar envelopes. One mechanism which has been proposed for their formation in these environments involves homogeneous gas-phase reactions of ions. A number of different schemes have been suggested. In all of them formation of the cyanopolyacetylene is achieved ultimately by the neutralization of $H_2C_{2n+1}^+$ through electron/ion recombination or proton transfer to a molecule with a proton affinity higher than the cyanopolyacetylene molecule:

$$H_2C_{2n+1}N^+ + e\ (X) \longrightarrow HC_{2n+1}N + H\ (XH^+)$$

The various schemes leading to the formation of $H_2C_{2n+1}^+$ differ in the manner of chain growth and nitrogen entrainment.

The earliest models put forward allowed for the buildup of the cyanopolyacetylenes in two ways (35,26). In one scheme the acetylenic chain was conceived to grow two carbon atoms at a time by reactions between acetylene and its ions and then to be substituted by reaction with HCN or CN as indicated in the following equation:

$$C_{2n}H_{2,3}^+ + CN,\ HCN \longrightarrow H_2C_{2n+1}N^+ + h\nu,\ H,\ H_2$$

Alternatively, substitution was conceived to be achieved first, followed by growth of the carbon backbone by reactions with acetylene ions:

$$C_2H_{2,3}^+ + HC_{2n+1}N \longrightarrow H_2C_{2n+3}N^+ + H,\ H_2$$

Growth by one carbon atom at a time is the distinguishing feature of the scheme proposed by Woods (37). The essential steps in this scheme are as follows:

$$C^+ + HC_{2n+1}N \longrightarrow C_{2n+2}N^+ + H$$

$$C_{2n+2}N^+ + CH_4 \longrightarrow H_2C_{2n+3}N^+ + H_2$$

Also, Millar and Freeman (33) have recently postulated a mechanism in which build-up of the carbon chain is followed by reactions with N atoms:

$$C_{2n+1}H_3^+ + N \longrightarrow H_2C_{2n+1}N^+ + H$$

Reaction of N atoms with an ionic hydrocarbon chain which is less hydrogenated, such as $C_{2n+1}H^+$, will result in an ion, such as $C_{2n+1}N^+$, which needs to be hydrogenated before neutralization.

Recent measurements have shown that the early schemes may fail to build up cyanopolyacetylenes. The ions $C_4H_2^+$ and $C_4H_3^+$ have been reported not to react with HCN at low pressures, $k \leq 10^{-11}\,cm^3\,molecule^{-1}s^{-1}$ (15). Also the ions $C_2H_2^+$ and $C_2H_3^+$ react with HC_3N by elimination of HCN and proton transfer, respectively, and so do not lead to the growth of the carbon backbone (15,32).

The reactions of ionic hydrocarbon chains with atomic nitrogen remain to be tested in the laboratory and it is not clear whether hydrogenation can be assured. For example, HC_3N^+ has been shown not to be rapidly hydrogenated by reaction with molecular hydrogen (2). However, reactions with N atoms are attractive in the chemistry of dark clouds because of the relatively high abundance of these atoms.

The plausibility of the scheme of Woods can be judged on the basis of recent SIFT measurements of the essential steps with n = 1. The experiments have shown that atomic carbon ions react with cyanoacetylene to produce C_4N^+ about 20% of the time (32):

$$C^+ + HC_3N \xrightarrow{0.8} C_3H^+ + CN$$
$$\xrightarrow{0.2} C_4N^+ + H$$

The reaction is extremely fast; the rate constant has been measured to be 6.1×10^{-9} cm^3 $molecule^{-1}$ s^{-1} . Also, C_4N^+ has been observed to react with methane with a rate constant of 5.7×10^{-10} cm^3 $molecule^{-1}$ s^{-1} (11). A variety of ionic products was evident which indicated that this reaction is a potential source of diacetylene and cyanodiacetylene. 35% of the reaction either leads to C_4H_2 directly or to its protonated form. 5% of the reaction leads to $C_5H_2N^+$ which is likely to be protonated cyanodiacetylene. The remaining 60% appears to generate neutral or protonated cyanoacetylene. In the interstellar chemistry these latter two products lead to chemical feedback in the sense that the cyanoacetylene will become available for reaction with C^+ which regenerates the C_4N^+. This is illustrated in Fig. 5 in which the chemistry established for n = 1 has been projected to higher values of n.

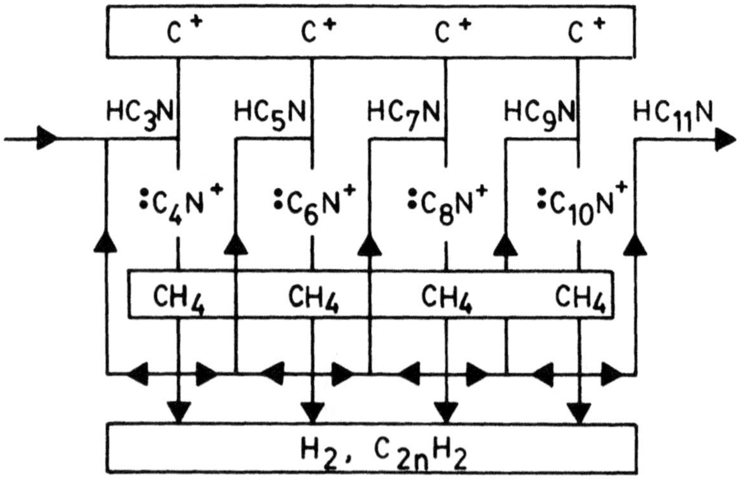

Fig. 5. Hypothetical scheme for the synthesis of cyanopolyacetylenes. The chemistry established for the synthesis of cyanodiacetylene has been projected to the higher members of the homologous series.

6.6. Growing Cyclopropenylidene

Generating molecular rings from acyclic units provides a special synthetic challenge. Two cyclic molecules have been identified in the interstellar medium so far. In particular, it has been suggested that cyclopropenylidene may be one of the most abundant molecules in many diffuse molecular clouds (38, 39). Millimeter-wave transitions of cyclopropenylidene, a planar singlet carbene with C_{2v} symmetry, have recently been identified in emission in several cold molecular dust clouds and in absorption in the direction of the galactic center (38, 39). Cyclopropenylidene may be generated by ion/molecule reactions in just three steps (40).

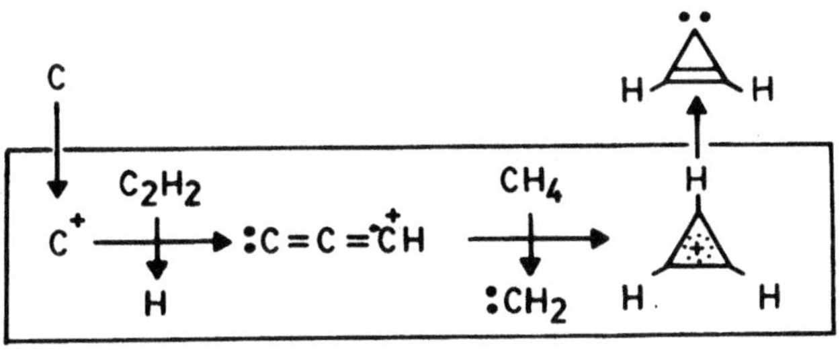

The kinetics of the first two steps have been characterized in the laboratory (23,40). Available thermochemical data for the reaction of $:C_3H^+$ with methane indicates that only formation of the cyclopropenium ion is exothermic with methylene as the required neutral product. It is the neutralization of this cyclopropenium ion by electron/ion recombination or proton transfer which may lead to cyclopropenylidene. We can therefore anticipate the formation of this ring carbene as well as methylene in ionized interstellar regions containing acetylene and methane. The radiative association of $:C_3H^+$ and H_2 has been proposed as an additional source of $C_3H_3^+$ although the efficiency of this process is uncertain and both the cyclopropenium and the linear propargyl cations are energetically possible products (41).

Interstellar spectral transitions also have been observed for the ring molecule silacyclopropyne, SiC_2. This molecule has long been of astrochemical importance. It has been known for some time now to exist in stellar atmospheres but has only recently been shown to have a cyclic ground state geometry. One ion/molecule reaction route towards the formation of SiC_2 is shown below. This particular route appears plausible in view of recent SIFT measurements at York which indicate a fast reaction, $k = 3.9 \times 10^{-10}$ cm^3 $molecule^{-1}$ s^{-1}, between Si^+ and acetylene to produce exclusively the SiC_2H^+ ion (11). The geometry of the SiC_2H^+ is not known but is likely to be cyclic in view of the cyclic ground-state geometry of SiC_2.

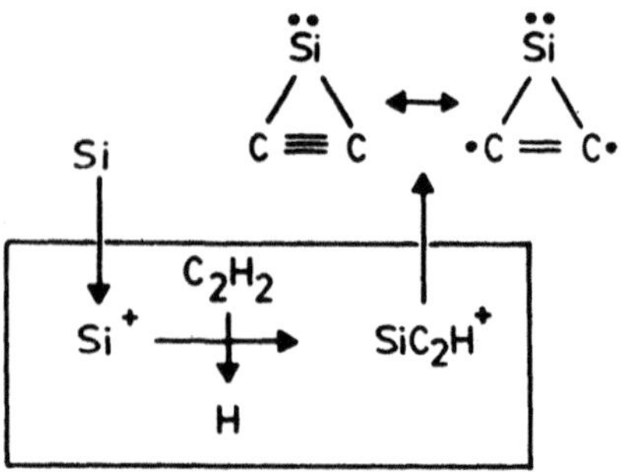

6.7. Growing Carbenes

Carbene molecules occur widely in the universe: in cometary atmospheres, stellar atmospheres, planetary atmospheres and interstellar gas clouds. One possible origin of these molecules involves ionizing radiation which can drive reactions between ions and molecules in these environments towards the formation of ionic carbene precursors. Recent laboratory measurements suggest that many ion/molecule reactions, in particular those involving atomic carbon ions or their derivatives, can direct ion chemistry towards the formation of carbene molecules of various types and complexity.

The simplest carbene, methylene, will be produced in the neutralization of CH_3^+ by proton transfer or electron/ion recombination. The CH_3^+ ion has been shown in Section 4 to be a 'terminal' ion in the hydrogenation of C^+ so that we may write the overall synthesis of methylene as follows:

$$:C: \xrightarrow{h\nu,\ H_3^+} :\dot{C}^+,\ :CH^+ \xrightarrow{H_2} CH_2^+ \xrightarrow{H_2} CH_3^+ \xrightarrow[PT]{e/+} :CH_2$$

Alternatively, as indicated earlier, $:CH_2$ may be formed directly as a neutral product, as, for example, from the reaction of $:C_3H^+$ with methane:

$$:C_3H^+ + CH_4 \longrightarrow C_3H_3^+ + :CH_2$$

$:C_3H^+$ itself is a product of the reaction of C^+ with acetylene. It may neutralize to form the dicarbene :C=C=C:. Other reactions of C^+ can lead to higher members of the homologous series of carbene cations $H(C_2)_n\text{-}C:^+$ as well as substituted carbene cations of the type $R(C_2)_n\text{-}C:^+$. For example, we have seen that C^+ reacts very rapidly with cyanoacetylene to form $:C_3H^+$ and $:C_4N^+$ in a proportion of about 4 to 1 (32). Similar reactions with higher polyacetylenes and the known interstellar polycyanoacetylenes should produce the corresponding higher carbene cations $:C_5H^+$, $:C_7H^+$ .... $:(C_2)_n\text{=}CH^+$ and $:C_6N^+$, $:C_8N^+$ .... $:C^+\text{-}(C_2)_n\text{-}CN$. The former series of ions in turn should spawn the cumulated dicarbenes $:(C\text{=}C)_n\text{=}C:$ directly by recombination with electrons or by proton transfer. Reactions with oxygen or sulphur bearing molecules will establish higher member of the unsaturated heteratomic carbenes $:(C_2)_n{=}C{=}O$ and $:(C_2)_n{=}C{=}S$ in analogy with the reaction of the $:C_3H^+$ ion discussed earlier (40).

Carbene cations are expected to react by sigma bond insertion with molecules of the type XH, possible via a mechanism involving a cyclic intermediate, and lead to the ionic carbene product $:CX^+$. The generic reaction is as follows:

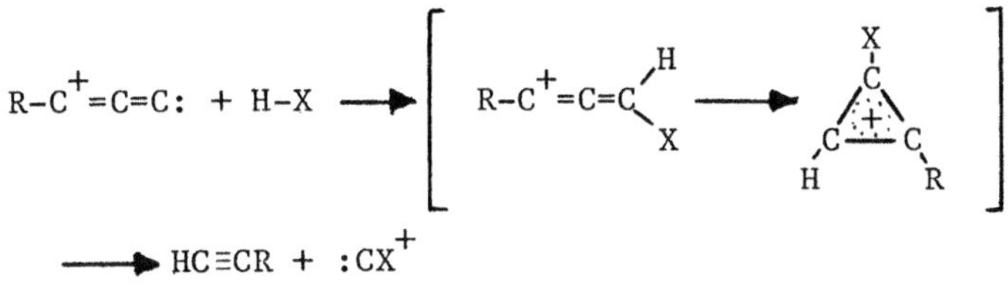

The overall reaction has been observed for R = H and CN and X = NH_2 and CH_3. Neutralization of the product carbene cation for these two X substituents may lead to the carbenes :C=NH and $:C{=}CH_2$. Indeed the laboratory experiments also have shown that a fraction of the ionic products observed for these reactions may correspond to the direct formation of :CNH and $:C_2H_2$ through neutralization by intramolecular proton transfer as the products separate.

For R = H and CN and X = H, CN and C_3N the overall reaction is endothermic and, with the exception of the reaction of $:C_4N^+$ with H_2, the product observed in the laboratory was the adduct ion, $RC_3^+.HX$ (24). The measured specific rates for these association reactions are large and consistent with covalent bond formation in the adduct ion by sigma bond insertion. The structures of the adduct ions are uncertain and may be cyclic or acyclic. For example, the product $C_3H_3^+$ in the reaction of C_3H^+ with H_2 may be the linear propargyl cation or the cyclopropenyl cation. Recent calculations for this system indicate that the initial energy of the reactants is just sufficient to allow the conversion of the propargyl cation (which is presumed to be formed in the initial insertion) to the cyclopropenyl cation (42). The acyclic adduct ions may neutralize to form the acyclic disubstituted carbenes $:C(C_2H)H$, $:C(C_2H)CN$, $:C(C_2H)C_2CN$ and $:C(C_2CN)CN$ and $:C(C_2CN)_2$. The cyclic adduct ions may neutralize to form the cyclic carbene indicated below with X = H, CN or C_2CN when R = H and X = CN or C_2CN when R = CN.

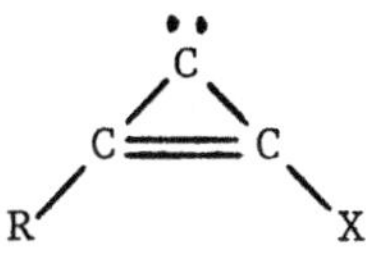

The adduct ions are likely to be stabilized by collision at the total pressures of the SIFT experiments in which they were observed. In interstellar gas clouds stabilization must occur by the emission of radiation. This may not be improbable. Recent theoretical studies by Bates suggest that radiative association is favoured by high association energies and high numbers of nuclei in the adduct ion (43).

With CH_3CN the observed products of the reactions with C_3H^+ and C_4N^+ are more consistent with insertion into the sigma bond between the two carbon atoms. Observed products with C_3H^+ include $C_2H_3^+/HC_3N$ and $C_2H_2/H_2C_3N^+$ (23) which are likely sources of the carbene $:C{=}CH_2$ and cyanoacetylene, and the adduct ion which is a potential source of the disubstituted carbene $:C{=}C{=}C(CH_3)\,CN$ or the cyclic carbene shown above with R = CH_3 and X = CN. The products observed with C_4N^+ are $C_2H_3^+/C_4N_2$ and $C_2H_2/C_4N_2H^+$ (11) which are likely sources of $:C_2H_2$ and dicyanoacetylene.

Formation of cumulated carbene chains, disubstituted carbenes and ring carbenes should provide opportunities for still further chemical growth in partially ionized gaseous environments. In terrestrial chemistry carbenes are known to be among the most versatile and synthetically useful reactive intermediates. They should have a similar function in extraterrestrial chemistry. For example, there are intriguing possibilities for the growth of graphite-like molecules as shown below:

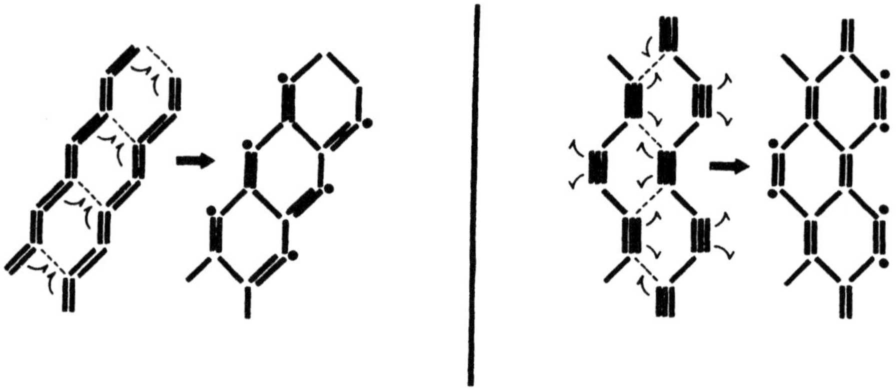

Fig. 6. Possible cross-bonding initiated by the side-on approach between extended cumulenes (left) or polyacetylenes (right).

Side-on attack between cumulenes or polyacetylenes (where one may be ionic) can lead to carbon ring formation through cross-bonding. This is the inverse of the mechanism proposed for the transformation of graphite at high temperatures (44,45) which is analogous to the well-known transformation of benzene to acetylene. Even without cross-bonding, the stacking of carbon chains of this type can lead to polymorphic forms of carbon known as carbynes. The existence of carbynes appears to have been established in highly purified carbonaceous fractions of the Allende meteorite (46).

6.8. Growing PAH

Partially hydrogenated, positively charged polycyclic aromatic hydrocarbons (PAH's) have been invoked recently to account for some of the unidentified IR emission features of the interstellar medium (diffuse interstellar bands) (47). One may speculate on the growth of PAH's from a benzene-like substrate by successive additions of C_4H_2 or C_2 units as indicated in Fig. 7. The scheme in Fig. 7 has been proposed for the high temperature polymerization of PAH's and also provides a basis for discussion of the growth of large aromatic ions which are believed to be precursors in the ionic formation of soot in flames (48,49). But the mechanism of such ambitious growth remains to be elucidated in the laboratory.

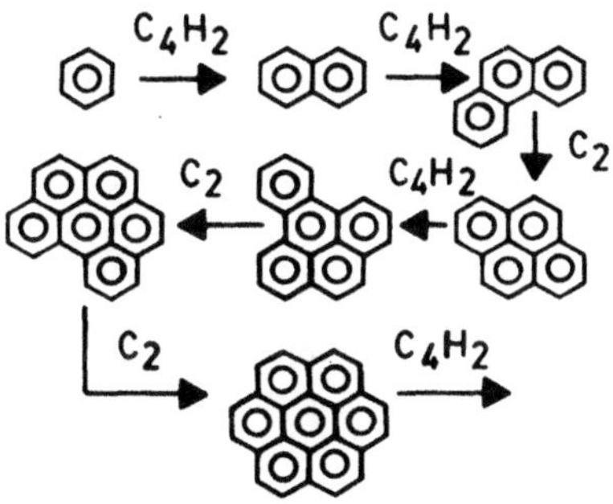

Fig. 7. Possible route toward the growth of polycyclic aromatic hydrocarbon ions.

The first step in the mechanism in Fig. 7 has recently been investigated in the York SIFT apparatus (11). The $C_6H_6^+$ cation derived from benzene has been found to rapidly add diacetylene at room temperature as follows:

$$C_6H_6^+ + C_4H_2 \longrightarrow C_{10}H_8^+$$

The effective bimolecular rate constant was measured to be 5.0 x 10^{-10} cm^3 $molecule^{-1}$ s^{-1} in helium buffer at 0.34 Torr and a helium density of 1.1 x 10^{16} atoms cm^{-3}. The product $C_{10}H_8^+$ ion may neutralize by charge transfer to produce naphthalene. The strong bonding in the adduct suggested by the large effective bimolecular rate constant for association implies that radiative association may be a viable option at the low pressures of the interstellar medium. The counter-reaction of ionized diacetylene with neutral benzene was observed to proceed rapidly in a bimolecular fashion predominantly by charge transfer with a rate constant of 9.2 x 10^{-10} cm^3 $molecule^{-1}$ s^{-1}.

6.9. Growing Benzene

Laboratory studies are now becoming available which demonstrate the growth of cyclic aromatic hydrocarbons from ion/molecule reactions involving small acyclic units. For example, several routes toward the synthesis of benzene have now been characterized. The benzenium ion (protonated benzene) is one suitable precursor for benzene. It may neutralize to yield benzene by proton transfer or recombination with electrons. The benzenium ion has been shown to be formed from three-carbon units in allene and propyne by the following bimolecular reactions (50,51):

$$H_2C{=}C{=}CH_2^+ + H_2C{=}C{=}CH_2 \longrightarrow C_6H_7^+ + H$$

$$HC{\equiv}CCH_3^+ + HC{\equiv}CCH_3 \longrightarrow C_6H_7^+ + H$$

The structures of the $C_6H_7^+$ ions produced in these two reactions have been probed with collisional activation and chemical reactivity studies. Both types of studies have indicated that the $C_6H_7^+$ ions to a large extent have the cyclic benzenium structure. The chemical reactivity studies which have probed the proton affinity of the $C_6H_7^+$ product ions with proton transfer reactions have established that 70% of the ions produced in allene and 44% of the ions produced in propyne have the benzenium structure. The structures of the remaining $C_6H_7^+$ ions appear to correspond to protonated fulvenes or protonated dimethylenecyclobutenes (51).

Another possible precursor ion for benzene is the benzene cation which may undergo a charge transfer reaction to form benzene. SIFT studies have shown that a $C_6H_6^+$ ion may be grown from three two-carbon units: acetylene cation will add two molecules of acetylene to produce $C_6H_6^+$. The data has already been presented in Fig. 4. The structure of the $C_6H_6^+$ cation produced in this fashion has not yet been elucidated.

6.10. Growing Amino Acids

It has been noted for some time that several of the organic molecules which have been identified in interstellar clouds may be only one chemical step removed from amino acids (52). For example, methanimine, CH_2NH, or methylamine, CH_3NH_2, and formic acid, HCOOH, are plausible building blocks for glycine, NH_2CH_2COOH, which is the simplest amino acid. One inconclusive search for interstellar glycine has been reported in the literature (53). Also, there appears not to have been a previous laboratory characterization of reactions between neutral molecules in the gas phase which produce the simple amino acids.

The challenge to the ion chemist is to bring about the synthesis of amino acids through ion/molecule reactions. Observations discussed earlier in this article suggest that protonated amino acids are likely to be suitable precursors for amino acids. The formation of protonated amino acids by the association of ammonium ions and carbon dioxide has recently been investigated by observing the following two reactions in a pulsed ionization high-pressure ion source (54):

$$CH_3NH_3^+ + CO_2 \longrightarrow CH_3NH_3^+.CO_2$$

$$C_2H_5NH_3^+ + CO_2 \longrightarrow C_2H_5NH_3^+.CO_2$$

The empirical formulas of the product ions of these two reactions are identical with those of protonated glycine and protonated alanine, respectively, which could result from the formation of new covalent bonds in the adduct ions. However, studies of the equilibrium constant of the two association reactions as a function of temperature indicate that the desired covalent bond formation does not take place. The enthalpies of formation of the adduct ions $CH_3NH_3^+.CO_2$ (42 kcal mol^{-1}) and $C_2H_5NH_3^+.CO_2$ (35 kcal mol^{-1}) derived from the van't Hoff plots for these reactions do not agree with the known heats of formation of protonated glycine (53 kcal mol^{-1}) and protonated alanine (43 kcal mol^{-1}), respectively (54).

A more suitable association reaction might involve protonated imines and formic acid as follows:

$$CH_2NH_2^+ + HCOOH \longrightarrow (\text{glycine})H^+$$

$$CH_3CHNH_2^+ + HCOOH \longrightarrow (\text{alanine})H^+$$

The association reaction between $CH_2NH_2^+$ (derived from CH_3NH_2 by electron impact) and HCOOH can be observed to occur rapidly in helium buffer gas at 0.35 Torr in a SIFT apparatus (11). Experiments are now in progress to deduce the structure of the adduct ion by proton-transfer bracketing experiments (the proton affinity of glycine is known) and other possible reactivity measurements. Stabilization of the adduct ion in helium at 0.35 Torr may occur by collision. In the interstellar medium stabilization must occur radiatively. No information is available on the efficiency of the radiative association of protonated imines and formic acid. Exothermic reactions with bimolecular products might be preferable as sources for protonated glycine. A search for reactions of this type has now been initiated in the Ion Chemistry laboratory at York University.

Acknowledgements: The author wishes to recognize the financial support of the Natural Sciences and Engineering Research Council of Canada and the contributions of his co-workers who are listed in the references. Also thanks go to G. Peden for typing the manuscript and C. Sammer for preparing the figures.

REFERENCES

(1) Clary, D.C., Smith, D., and Adams, N.G.: 1985, Chem. Phys. Lett. **119**, p.320.

(2) Fox, A., Raksit, A.B., Dheandhanoo, S., and Bohme, D.K.: 1986, Can. J. Chem. **64**, p.399.

(3) Adams, N.G., Smith, D., and Millar, T.J.: 1984, Mon. Not. R. Astr. Soc. **211**, p.857.

(4) Luine, J.A. and Dunn, G.H.: 1985, Ap. J. **299**, p.L67.

(5) Barlow, S.E., Dunn, G.H., and Schauer, M.: 1984, Phys. Rev. Letters **52**, p.902.

(6) Fehsenfeld, F.C., Dunkin, D.B., and Ferguson, E.E.: 1974, Ap. J. **188**, p.43.

(7) Adams, N.G. and Smith, D.: 1981, Chem. Phys. Letters **79**, p.563.

(8) Adams, N.G. and Smith, D.: 1977, Chem. Phys. Letters **47**, p.383.

(9) Suzuki, H.: 1983, Ap. J. **272**, p.579.

(10) Bates, D.R.: 1983, Ap. J. **267**, p.L121.

(11) Unpublished results from the Ion Chemistry Laboratory, York University.

(12) Herbst, E., Adams, N.G., and Smith, D.: 1983, Ap. J. **269**, p.329.

(13) Bohme, D.K., Rakshit, A.B., and Schiff, H.I.: 1982, Chem. Phys. Letters **93**, p.592.

(14) Smith, D. and Adams, N.G.: 1977, Chem. Phys. Letters **47**, p.383.

(15) Anicich, V.G., Huntress, W.T., and McEwan, M.J.: 1986, J. Phys. Chem. **90**, p.2446.

(16) Anicich, V.G., Blake, G.A., Kim, J.K., McEwan, M.J., and Huntress, W.T.: 1984, J. Phys. Chem. **88**, p.4608.

(17) Anicich, V.G. and Huntress, W.T.: 1984, Survey of Bimolecular ion-Molecule Reactions for Use in Modeling the Chemistry of Planetary Atmosphere, Cometary Comae, and Interstellar Clouds, Astrophys. J. Supplement, submitted for publication.

(18) Dheandhanoo, S., Forte, L., Fox, A., and Bohme, D.K.: 1986, Can. J. Chem. **64**, p.641.

(19) Kebarle, P. and Haynes, R.M.: 1967, J. Chem. Phys. **47**. p.1676.

(20) Buckley, T.J., Sieck, L.W., Metz, R. Lias, S.G., and Liebman, J.F.: 1985, Int. J. Mass Spectrom. Ion Processes **65**, p.181.

(21) Brill, F.W. and Eyler, J.R.: 1981, J. Phys. Chem. **85**, p.1091.

(22) Bohme, D.K., Raksit, A.B., and Fox, A.: 1983, J. Amer. Chem. Soc. **105**, p.5481.

(23) Raksit, A.B. and Bohme, D.K.: 1983/84, Int. J. Mass Spectrom. Ion Processes **55**, p.69.

(24) Bohme, D.K., Dheandhanoo, S., Wlodek, S., and Raksit, A.B.: 1986, J. Phys. Chem., submitted for publication.

(25) Matthews, H.E., Irvine, W.M., Friberg, P., Brown, R.D., and Godfrey, P.D.: 1984, Nature **310**, p.125.

(26) Wilson, W.J. and Snyder, L.E.: 1981, Ap. J. **246**, p.86.

(27) Millar, T.J. and Freeman, A.: 1984, Mon. Not. R. astr. Soc. **207**, p.425.

(28) Leung, C.M., Herbst, E., and Huebner, W.F.: 1984, Ap. J. Suppl. **56**, p.231.

(29) Prasad, S.S. and Huntress, W.T.: 1980, Ap. J. **239**, p. 151.

(30) Raksit, A.B., Schiff., H.I., and Bohme, D.K.: 1984, Int. J. Mass Spectrom. Ion Processes **56**, p.321.

(31) Schiff, H.I., Mackay, G.I., Vlachos, G.D., and Bohme, D.K.: 1981, in The Proceedings of the I.A.U. Symposium on Interstellar Molecules, edited by B.H. Andrews, D. Reidel Publ. Co., Dordrecht, p.307.

(32) Raksit, A.B. and Bohme, D.K.: 1985, Can. J. Chem. **63**, p.854.

(33) Millar, T.J. and Freeman, A.: 1984, Mon. Not. R. astr. Soc. **207**, p.405.

(34) Irvine, W.M. and Hjalmarson, A.: 1983, in Cosmochemistry and the Origin of Life, edited by C. Ponnamperuma, D. Reidel Publ. Co., Dordrecht, p.113.

(35) Winnewiser, G. and Wamsley, C.M: 1979, Astrophys. Space Sci. **65**, p.83.

(36) Schiff, H.I. and Bohme, D.K.: 1979, Ap. J. **232**, p.740.

(37) Woods, R.C.: 1983, in Molecular Ions, edited by J. Berkowitz and K.-O. Groeneveld, Plenum Publ. Co., p.511.

(38) Vrtilek, J.M., Thaddeus, P., and Gottlieb, C.A.: 1985, Bull. Am. Astr. Soc. **17**, p.568.

(39) Matthews, H.E., Irvine, W.M., Madden, S.C., and Swade, D.A.: 1985, Bull. Am. Astr. Soc. **17**, p.568.

(40) Bohme, D.K.: 1986, Nature **319**, p.473.

(41) Herbst, E., Adams, N.G., and Smith, D.: 1984, Ap. J. **285**, p.618.

(42) Hopkinson, A.C. and Lien, M.H.: 1986, J. Amer. Chem. Soc. **108**, p.2843.

(43) Bates, D.R.: 1983, Ap. J. **270**, p.564.

(44) Whittaker, A.G.: 1978, Science **200**, p.763.

(45) Heimann, R.B., Kleiman, J. and Salansky, N.M.: Carbon **22**, p.147.

(46) Hayatsu, R., Scott, R.G., Studier, M.H., Lewis, R.G., and Anders, E.: 1980, Science **209**, p.1515.

(47) Allamandola, L.J., Tielens, A.G.G.M., and Barker, J.R.: 1985, Ap. J. **290**, p.L25.

(48) Stein, S.E.: 1978, J. Phys. Chem. **82**, p.566.

(49) Calcote, H.F.: 1981, Combustion and Flame **42**, p.215.

(50) Lifshitz, C. and Gleitman, Y.: 1981, Int. J. Mass Spectrom. Ion Phys. **40**, p.17.

(51) Lias, S.G. and Ausloos, P.: 1985, J. Chem. Phys. **82**, p.3613.

(52) Hoyle, F. and Wickramasinghe, C.: 1984, From Grains to Bacteria, University College Cardiff Press.

(53) Hollis, J.M., Snyder, L.E., Suenram, R.D. and Lovas, F.J.: 1980. Ap. J. **241**, p. 1001.

(54) Meot-Ner, M., Hunter, E.P., and Field, F.H.: 1979, J. Amer. Chem. Soc. **101**, p.686..

AB INITIO STUDIES OF INTERSTELLAR MOLECULAR IONS

W. P. Kraemer
Max-Planck-Institute of Astrophysics
Karl-Schwarzschild-Str. 1
8046 Garching b. München
FRG

I. INTRODUCTION

Matter in the universe exists in a highly non-uniform distribution. It is concentrated in the galaxies which contain apart from the stars a large amount of low density interstellar matter. In our Galaxy up to about 30% of the total matter is estimated to be interstellar. This matter is contained in clouds of irregular geometric shape ranging from 0.1 to 50 light years in extent. The clouds have gaseous densities in the range of $10-10^7$ atoms or molecules per cm^3 and contain in addition dust grains of essentially yet unknown composition. Kinetic temperatures in the clouds are between 10 and 100° K. The cool and dense clouds are expected to be those regions where star formation can take place and a detailed knowledge of their physical conditions is therefore important for astrophysics. Information about the physical conditions in dense interstellar clouds cannot be obtained by optical observation techniques but rather from molecular lines in the microwave and with some restrictions also in the infrared frequency range.

Within the past 20 years the development of highly powerful radio astronomical observation techniques has led to the detection of a rapidly increasing number of absorption and emission lines in the radio frequency range and to the identification of a large variety of molecular species in the interstellar space. These discoveries have opened a new field of research which is concerned with the study of molecular reaction processes in interstellar clouds. Chemical reaction network models have been developed to investigate the formation and destruction of interstellar molecules in a steady state situation as well as in a time dependent framework. The lifetimes of any reactive neutral or ionic molecular species in interstellar clouds are short compared to typical cloud lifetimes which are estimated to be of the order of 10^5 years assuming free-fall gravitational collapse.

The extremely low temperature and low density conditions in interstellar clouds restrict the molecular formation and destruction processes to exothermic reactions that originate in binary collisions mostly between molecules in their ground electronic and vibrational

P. Ausloos and S. G. Lias (eds.), Structure/Reactivity and Thermochemistry of Ions, 247–260.

states. Exothermic binary ion-molecule reactions usually do not require any activation energies and therefore dominate interstellar chemistry. Photoionization does not contribute to the formation of interstellar ions to any larger extent since penetration of interstellar UV radiation into dense clouds is very low. However, throughout the Galaxy there exists a roughly constant flux of high-energy cosmic rays with energies up to about 10^{10} eV and the cosmic ray ionization rate per atom or molecule is estimated to be of the order of 10^{-17} s^{-1} in dense interstellar clouds. Interstellar chemistry in dense clouds is thus initiated by cosmic ray induced ionization of the most abundant species H_2, H, He. Production of these primal ions is then followed by a large variety of ion-molecule and ion-electron reactions in the gas phase or on grain surfaces. A useful introduction into the basic concepts of interstellar chemistry has been published by Duley and Williams [1]. This introduction also provides a number of references to the most important original studies related to this new field of research.

The reliability of the present chemical reaction network models and the detailed understanding of interstellar chemistry is still suffering from the lack of reliable data for most of the relevant chemical reactions. In many cases an extrapolation of laboratory measured reaction data to the extreme conditions in interstellar space is very uncertain. In this situation theoretical studies and especially quantum chemical ab initio calculations can be useful in providing missing information. Accurate ab initio calculations have been able in the past to make valuable contributions to the identification of a number of observed interstellar radio lines that could not be related to any existing laboratory spectroscopic data. Among these calculations theoretical predictions of isotopic shifts of molecular rotational transition frequencies have been particularly helpful to facilitate laboratory measurements and radio astronomical observations. Ab initio studies have also been successful in making predictions about relative stabilities of molecular species that are highly reactive and therefore difficult to study in the laboratory.

II. CALCULATION OF ROTATION-VIBRATION ENERGIES

Rotation-vibration energies are calculated according to the following computational scheme:

- Pointwise calculation of the molecular electronic potential hypersurface;
- Fitting of an analytical potential function (Dunham-type expansion or Simon-Parr-Finlan expansion) to the individual potential points;
- Treatment of nuclear motions in the previously determined electronic potential.

Among these three steps the first one represents the accuracy determining and by far the most time consuming part of this computational scheme. Much effort has to be put into the calculation of an accurate electronic potential surface because the vibrational and rotational energy levels are extremely sensitive to the shape of the potential surface around the minimum. Therefore Gaussian basis sets of at least triple-zeta quality augmented by single or multipole sets of polarization functions on each nuclear center are needed to achieve a reasonably accurate description of the molecular orbitals. Inclusion of electron correlation effects in further absolutely essential in order to obtain potential surfaces which can compete with those deduced from experimental data. Single or multi-reference configuration interaction calculations taking into account all possible single and double excitations that can be generated from the set of reference configurations have therefore to be performed at each selected point on the molecular potential surface in order to determine the main contributions of the electron correlation energy.

Since the electronic potential does not depend on the nuclear masses, the spectroscopic data of all possible isotopic forms can easily be determined without much extra work from the same potential surface. This means that isotopic shifts of rotational and vibrational frequencies are usually obtained with very good accuracy which appears to be a great advantage of theoretical calculations. Spectroscopists have utilized this fact in the past in searching for isotopic spectra.

In the remainder of this section a few more recent theoretical rotation-vibration energy calculations will be briefly discussed.

(a) The isoformyl ion, HOC^+

Over the last 10 years the isoformyl ion has received much attention among experimentalists and theoreticians. Its energetically more stable isomer, the formyl ion HCO^+, has been identified [2,3] as the carrier of the interstellar X-ogen line at 89.189 GHz [4], and is now known to be rather abundant in several interstellar sources. Many of its isotopic rotational transitions have also been detected in interstellar space and the intensities of these lines have been used to estimate atomic isotope abundances in the Galaxy.

HCO^+ and HOC^+ are isoelectronic with another isomer pair: HCN and HNC, where the HCN is one of the first molecules that have been detected by radioastronomical observation techniques [5] and is since then known to be a rather abundant constituent in interstellar clouds. In 1971 the emission line, then called "X_2", at 90.665 GHz was assigned as the J = 1-0 rotational transition in HNC [6], and this isomer has been found to be almost as abundant as the more stable HCN in a number of molecular clouds. In view of this high HNC abundance there was reason to expect that HOC^+ would also be produced and exist under dense interstellar cloud conditions.

Due to its interstellar importance a large number of ab initio studies have been undertaken to investigate the isoformyl ion [7-13]. These calculation predicted a linear geometry for HOC^+ with a rather small bending force constant and an isomerization barrier of about 60 kcal/mol, which under interstellar cloud conditions prevents the HOC^+ from isomerizing into the more stable HCO^+. A complete geometry optimization of HOC^+ including bent structures and a calculation of its spectroscopic constants has been performed by Hennig et al. [14] as part of an extensive configuration interaction study of the isoelectronic series of triatomic molecules HCN, HNC, HCO^+, HOC^+, HN_2^+ [15].

Guided by the ab initio results of Ref. [15], Gudeman and

Woods [16] have finally been able to make the first laboratory observations of the J = 1-0 rotational transitions of HOC^+ and its ^{13}C and ^{18}O isotopic forms. The experimental frequencies differ from the early theoretical predictions [15] by about 250 MHz, but the theoretical isotopic shifts of the $HO^{13}C^+$ and $H^{13}OC^+$ lines are in good agreement with the laboratory results.

Gudemann and Woods [16] suspected in their paper that the bending potential of HOC^+ "may possibly exhibit the phenomenon of quasilinearity". This idea has then initiated further refined ab initio calculations [17-19]. DeFrees et al. [19] have investigated the effect of systematically augmented basis sets to describe the bending potential of HOC^+ as accurately as possible. They find that at the correlated wave function level the HOC^+ ion is indeed linear in agreement with other extended calculations [17,18] but that the inclusion of f-type polarization functions lowers the bending potential significantly compared to that obtained in Ref. [17].

Most recently [20] a combination of the stretching part of the HOC^+ potential of Ref. [18] with the bending part determined by DeFrees et al. [19] has been used to calculate highly accurate rotation-vibration energies of HOC^+. The results are summarized in Table I, and Table II shows a comparison with observed rotational transition frequencies [21,22] for 4 different isotopic forms of HOC^+.

Table I: Rotation-Vibration Parameters (in cm^{-1}) for HOC^+ and DOC^+

	HOC^+	DOC^+
ν_1	3266	2502
ν_2	268	198
$2\nu_2$	591	439
ν_3	1935	1868
B_{000}	1.493	1.275
B_{100}	1.482	1.263
B_{010}	1.499	1.285
B_{001}	1.481	1.266

Table II: Comparison with Observed Rotational Transition Frequencies (cm^{-1})

			obs.	obs.-calc.
HOC^+	J'=4	J"=3	11.938992	-0.004
	3	2	8.954565	-0.003
	2	1	5.969865	-0.002
	1	0	2.984979	-0.001
$HO^{13}C^+$	J'=1	J"=0	2.860403	-0.001
$H^{18}OC^+$	J'=1	J"=0	2.889051	-0.001
DOC^+	J'=5	J"=4	12.738316	-0.006
	4	3	10.191103	-0.005
	3	2	7.643590	-0.004
	2	1	5.095852	-0.002

(b) CNC^+/CCN^+

Recent chemical reaction network systems [24,25] include the C_2N^+ molecular ion in describing the chemistry of interstellar carbon-nitrogen containing compounds in dense molecular clouds. The most effective formation of C_2N^+ is assumed to occur through the main destruction reaction of HCN in dense cloud regions

$$C^+ + HCN \rightarrow C_2N^+ + H \tag{1}$$

For this rapid reaction Schiff and Bohme [26] have measured a rate constant of 3.5×10^{-9} cm^3 s^{-1}. These authors also found in laboratory studies that the reaction of C_2N^+ with H_2 molecules is immeasurably slow. It is therefore mostly assumed that C_2N^+ is predominantly destroyed by dissociative recombination with electrons with an estimated rate constant of about 3×10^{-7} cm^3 s^{-1}. Details of this reaction are as yet unknown.

The finding reported in [26] that C_2N^+ does not effectively react with H_2 led Hartquist and Dalgarno [27] to the conclusion that C_2N^+ should be the most abundant interstellar molecular ion other than HCO^+. However, the C_2N^+ ion has never been observed in interstellar space.

Considering only the above formation and destruction reactions for C_2N^+, a simple steady state relation for the relative abundance of C_2N^+ is obtained

$$\frac{[C_2N^+]}{[HCN]} = \frac{k_F}{k_D} \cdot \frac{[C^+]}{[e^-]} \tag{2}$$

where [X] denotes the abundance of species X, and where k_F and k_D are the rate constants of the formation and destruction reactions. With the above values for k_F and k_D and using an estimate for $[C^+]/[e^-] \approx 10^{-1}$, the abundance of C_2N^+ is obtained as about 10^{-3} times smaller than that of HCN. This small value would explain the fact that C_2N^+ ions have not been detected in interstellar sources. However, due to the uncertainties in the estimates for k_D and $[C^+]/[e^-]$ the abundance ratio (2) could also be considerably larger.

Another interesting point about C_2N^+ was brought into the discussion by Haese and Woods [28]: By analogy to the fast reaction of C^+ with HCN there could be a similar reaction of C^+ with the HNC isomer. This would open the possibility that the two different isomers of C_2N^+ could be selectively formed by the two reactions

$$C^+ + HCN \rightarrow CNC^+ + H \tag{3a}$$

$$C^+ + HNC \rightarrow CCN^+ + H \tag{3b}$$

A recent crossed beam experiment reported by Daniel et al. [29] does in fact indicate that the symmetrical CNC^+ isomer can exclusively be formed by reaction (3a).

A complete geometry optimization of both C_2N^+ isomers including bent structures and a determination of spectroscopic data has been performed on the correlated wave function level [31,32]. Both isomers are obtained to be linear in their electronic ground state, but the bending potentials are fairly shallow. In order to assess the reliability of these calculations, identical calculations have been carried out for the isoelectronic C_3 radical. The theoretical results for all 3 molecular species are summarized in Table III with experimental data for C_3 added in parentheses.

Table III: Rotation-Vibration Energies and Constants (in cm^{-1}) for CNC^+, CCN^+ and C_3 in their electronic ground states.

	CNC^+	CCN^+	C_3	C_3(expt.)
ν_1	1340	1121	1277	(1224.5)
ν_2	106	87	68	(63.1)
$2\nu_2$	180	153	160	(132.7)
ν_3	2128	2349	2108	(2040.0)
B_{000}	0.458	0.402	0.433	(0.4305)
B_{100}	0.456	0.400	0.428	–
B_{010}	0.464	0.410	0.442	(0.4422)
B_{001}	0.454	0.399	0.436	–

(c) The ethynyl radical, CCH

Because of its astrophysical importance the ethynyl radical CCH has been subject of a number of experimental and theoretical studies over the last few years. It is well established that the radical has a linear geometry its electronic ground state being $\tilde{X}^2\Sigma^+$ and its first low lying excited state $\tilde{A}^2\Pi$. There is however still controversy about the bending frequency in the ground state and about the excitation energy $T_e(\tilde{A}^2\Pi)$. A recent theoretical study by Fogarasi et al. [33] has not been able to solve these problems.

Extensive ab initio calculations have very recently been performed [34] to determine accurate potential surfaces for the two lowest electronic states of the CCH radical. At each point on the surfaces the molecular orbitals have been obtained by a CASSCF calculation [35] and the energy has been determined by the contracted CI (CCI) method [36] using a selection of the most important configurations of the CASSCF calculation as reference configurations in the CCI. The excitation energy has been obtained as $T_e(\tilde{A}^2\Pi)$-3650 cm^{-1}. The spectroscopic parameters obtained from these calculations are summarized in Table IV.

Experimentally Jacox [36], in a matrix isolation study, has determined ν_1 = 3612 cm^{-1} and ν_3 = 1848 cm^{-1} where ν_1 is very uncertain because of the presence of the $\tilde{A}^2\Pi$-$\tilde{X}^2\Sigma^+$ band system in that region [37]. The B_{000} value has been determined by Gottlieb et al. [30] to be 1.4568256 cm^{-1}. Very recently Kanamori and Hirota [39] have determined ν_3 = 1841 cm^{-1} and B_{001} = 1.4385 cm^{-1}, and ν_2+ν_3 has been found to be 2089 cm^{-1} [40] in the gas phase.

Table IV: Rotation-Vibration Energies and Constants (in cm^{-1}) calculated for CCH ($\tilde{X}^2\Sigma^+$)

	CCH	^{13}CCH	C^{13}CH	^{13}C^{13}CH	CCD
ν_1	3497	3496	3482	3482	2589
ν_2	298	299	295	296	223
$2\nu_2$	649	650	642	644	493
ν_3	1883	1846	1854	1817	1812
$\nu_1+\nu_2$	3784	3784	3768	3768	2790
$\nu_2+\nu_3$	2107	2073	2077	2043	1985
B_{000}	1.446	1.393	1.412	1.358	1.195
B_{100}	1.440	1.387	1.405	1.352	1.188
B_{010}	1.447	1.394	1.411	1.358	1.200
B_{001}	1.436	1.383	1.401	1.348	1.190

III. THEORETICAL DETERMINATION OF PROTON AFFINITIES

Ion molecule reactions play an important role in many chemical systems and are especially important in the chemistry governing molecular formation in dense interstellar clouds. Because of the ubiquitous presence of hydrogen in the interstellar medium, many of these reactions involve proton transfer between various bases. The simple energetics for such reactions, i.e. the exo- and endothermicity of proton transfer reactions, can be determined if the proton affinities (PA) of the various bases (B_i) are known

$$B_1H^+ + B_2 \rightarrow B_1 + B_2H^+ \qquad (4)$$

with $\Delta H = PA(2) - PA(1)$, where PA of a base is defined as $-\Delta H$ for the reaction

$$B + H^+ \rightarrow BH^+. \qquad (5)$$

There exist many experimental measurements of relative proton affinities, especially for compounds with proton affinities larger than PA(H_2O) Absolute proton affinities however are less well known due to the difficulty to measure accurate heats of formation for molecular cations. Molecular orbital theory on the other hand is particularly well suited for the study of molecular proton affinities because there is no change in the number of electrons in the protonation reaction (5). This has the consequence that in general the protonation reaction (5) does not involve any dramatic change of the electronic structure and that the effect of the electronic correlation energy on the reaction energetics is usually rather moderate. Therefore in many cases simple Hartree-Fock SCF calculations are able to produce fairly reasonable proton affinities. A number of ab initio determinations of proton affinities have been performed and excellent agreement with experiment has mostly been found. With very extended calculations it is even possible to predict absolute proton affinities with lower error limits than those obtained from experiment.

Using ab initio calculated data, the proton affinity can be determined from the following expression

$$PA(B) = -\Delta E^{elec} - \Delta ZPE + \frac{5}{2} RT \quad (6)$$

where the ΔE^{elec} term includes all electronic effects, the term ΔZPE represents the difference in zero-point energies for the base B and its protonated species BH^+, and where the final term is required for the conversion from ΔE to ΔH at the temperature T (in the present case T = 300 K). If there are no low-lying excited electronic states involved in reaction (5), then the electronic term represents just the difference in energy between the ground electronic states: $\Delta E^{elec} = \Delta E_0^{elec'}$. It is assumed in expression (6) that the internal vibrational and rotational levels of base B and the protonated BH^+ are populated in the same way.

Protonation of rather abundant CO molecules in dense interstellar clouds proceeds according to the two reactions [42]

$$CO + H_3^+ \rightarrow HCO^+ + H_2 \quad (7a)$$

$$CO + H_3^+ \rightarrow HOC^+ + H_2 \ . \quad (7b)$$

Woods et al. [43] have suggested that both reactions occur at essentially the same rate in interstellar clouds. However, Bowers et al. [44] in a laboratory experiment have found that only 6±5% of the reactions of $CO+H_3^+$ lead to the formation of HOC^+. There are two different possibilities to explain this experimental finding: either the energetics of the above protonation reactions of CO are very different such that reaction (7b) is much less favorable than reaction (7a), or initially formed HOC^+ ions are immediately removed from the reaction vessel by a subsequent reaction of HOC^+ with CO or H_2. This last possibility will be discussed later in this contribution.

Extensive ab initio calculations have been performed [45] to calculate the two different proton affinities of CO and to investigate the relative energetics of the two protonation reactions of CO. Detailed calculations show that both reactions proceed via intermediate complexes:

$$CO + H_3^+ \rightarrow H_2 \cdot HCO^+ \rightarrow HCO^+ + H_2 \quad (8a)$$

$$CO + H_3^+ \rightarrow H_2 \cdot HOC^+ \rightarrow HOC^+ + H_2 \ . \quad (8b)$$

The stabilities of these intermediate complexes relative to the reactants and the reaction products are shown in the energy level diagram of Fig. 1. From this Figure and from the proton affinity values collected in Table V it follows that $PA(H_2)$ and PA(OC) at oxygen are comparable and probably within 1 kcal/mol of each other. In other words, there appears to be no strong thermodynamic force in the reaction of H_3^+ with CO leading to the formation of HOC^+; the calculations show that reaction (7b) is essentially thermoneutral. Formation of HCO^+ on the other hand is strongly exothermic and collisions of H_3^+ with CO should

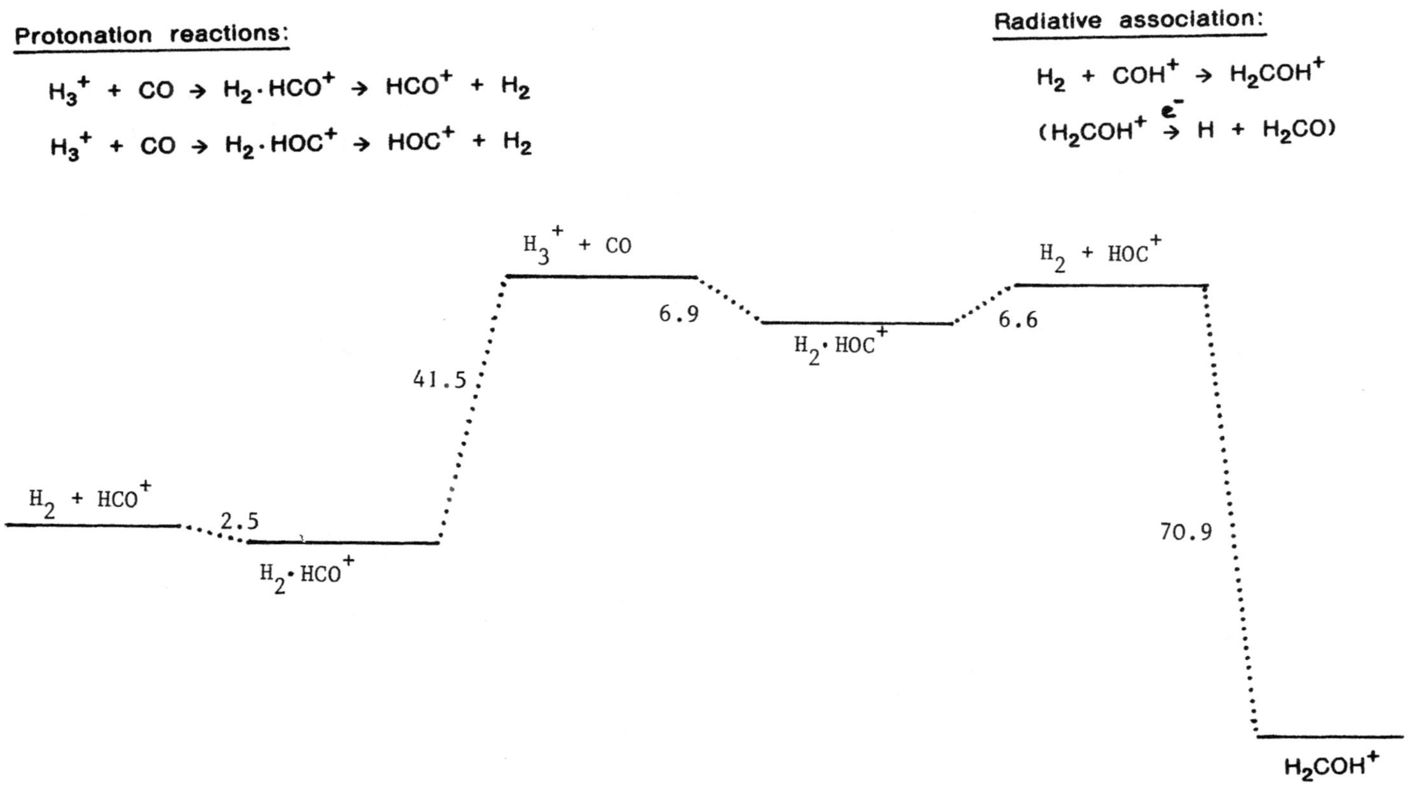

Figure 1. Schematic reaction profile for protonation reactions of CO and the radiative association reaction of HOC^+ with H_2 (energy values in kcal/mol).

lead predominantly to the formation of HCO^+. This analysis is consistent with the indirect mass spectrometric measurement of Bowers et al. [44].

Table V: Theoretical proton affinities (in kcal/mol)

	PA(H_2)	PA(CO)	PA(OC)
ΔE	106.6	147.1	106.1
ΔZPE	6.8	7.2	5.7
PA	101.3	141.4	101.9
PA(expt.)	101.3	141.9	-

Very recently McMahon and Kebarle have published experimentally determined proton affinity values for CO which are significantly different from the present theoretical results: PA(CO) = 145.6 kcal/mol and PA(OC) = 109 kcal/mol [46].

Another important species in various ionic processes and in interstellar chemistry is the protonated nitrogen molecular ion HN_2^+. This ion has been extensively studied in the laboratory [47-51] and has also been subject of a number of theoretical studies [52-55]. The ion is formed in interstellar clouds by protonation of N_2 in analogy to the protonation reaction of CO

$$N_2 + H_3^+ \rightarrow H_2 \cdot HN_2^+ \rightarrow HN_2^+ + H_2 \ . \qquad (9)$$

A recent ab initio study of the proton affinity of N_2 and the spectroscopic properties of HN_2^+ has obtained PA(N_2) = 119.7 kcal/mol. This number is 1.5 kcal/mol larger than the currently accepted experimental value of 118.2 kcal/mol [57]. Again McMahon and Kebarle have reported a value that is significantly higher: PA(N_2) = 122 kcal/mol [58].

IV. INTERSTELLAR HCO^+/HOC^+ ABUNDANCE RATIO

The abundance of the metastable isoformyl ion HOC^+ in dense molecular clouds poses still a problem in interstellar chemistry. After the first laboratory measurement of the lowest rotational transition frequency of HOC^+ and its isotopic forms [16], an extensive search for this rotational line has been performed in 14 different interstellar sources [43]. In this search only a single feature has been detected toward Sgr B2 that is consistent in frequency and shape with the laboratory measurement. Identifying this feature provisionally as being due to HOC^+ and comparing it with the 1-0 transition of $HC^{18}O^+$ in the same source, an isomeric abundance ratio of $[HCO^+]/[HOC^+] \approx 330$ has been obtained. This is in contrast to the isoelectronic isomer pair HCN/HNC where both isomers have been found to have comparable abundancies in various interstellar sources. Several attempts have been made since then to explain the low abundance of the interstellar HOC^+ ion.

According to Woods et al. [43] both isomers HCO^+ and HOC^+

are formed at the same rate via the reaction between H_3^+ and CO, and the metastable HOC^+ ion is then rapidly removed by the reaction with atomic hydrogen

$$HOC^+ + H \rightarrow HCO^+ + H \qquad (10)$$

while HCO^+ is depleted more slowly by electron dissociative recombination. However, in order to reproduce the large abundance ratio value, this suggestion would require a large abundance of atomic hydrogen much in excess of what is actually present in dense cloud regions. It has further been shown by Green [59] that reaction (10) processes a small reaction barrier which makes the reaction impossible at interstellar cloud conditions.

Another possibility of depleting the HOC^+ isomer has been pointed out by Nobes and Radom [60] investigating the isomerization reaction

$$HOC^+ + CO \rightarrow HCO^+ + CO \ . \qquad (11)$$

These authors report that their calculations at the RHF/4-31G level have indicated that reaction (11) should proceed without an activation energy. The entire reaction is exothermic and its rate coefficient has been measured to be 5×10^{-10} cm^3 s^{-1} [61]. Including this reaction in some recent model calculations, Herbst and Leung [62] have determined the $[HCO^+]/[HOC^+]$ abundance ratio to be in the range of 10–70 depending on the physical conditions and assuming that the Illies et al. [44] value of 6±5% for the probability of HOC^+ formation via reactions (7a) and (7b) is correct and not just a lower limit.

DeFrees et al. [63] have finally investigated another isomerization channel for HOC^+, namely the exothermic reaction of HOC^+ with molecular hydrogen

$$HOC^+ + H_2 \rightarrow HCO^+ + H_2 \qquad (12)$$

for which they have obtained an activation energy of 10 kcal/mol. They have concluded therefore that this reaction cannot take place under interstellar conditions. In a more recent study DeFrees et al. [64] have reinvestigated this reaction. In their recent paper they report that "they present new theoretical evidence in form of a new and improved quantum chemical calculation which results in a dramatically reduced activation energy" for reaction (12). Their current value for the activation energy is now 2 kcal/mol "with an uncertainty large enough to allow the possibility that no activation energy exists". These statements in their paper appear to be rather questionable. The activation barrier in reaction (12) is probably of physical nature and not just a basis set deficiency or the effect of any other shortcomings in the calculations.

There is however another possibility of a reaction of HOC^+ with molecular hydrogen

$$HOC^+ + H_2 \rightarrow H_2COH^+ \ . \qquad (13)$$

The protonated formaldehyde ion produced in this reaction would recombine with electrons under interstellar cloud conditions and dissociate to H_2CO and H. Preliminary CASSCF/CCI calculations using a double-zeta Gaussian basis set have recently been performed [65] which indicate that there is no activation energy required for this reaction. More extended calculations are underway. Similar calculations have also been performed by Ruttink and Holmes [66] coming to the same result that there is no activation barrier on the reaction path of reaction (13). If the rate of reaction (13) would be rather small, the reaction would be still sufficiently efficient due to the large H_2 abundancy that the theoretical abundance ratio for $[HCO^+]/[HOC^+]$ could be close to the observed value.

REFERENCES

[1] W.W. Duley and D.A. Williams, "Interstellar Chemistry", Academic Press, 1984

[2] R.C. Woods, T.A. Dixon, R.J. Saykally, and P.G. Szanto, Phys. Rev. **35**, 1269 (1975)

[3] W.P. Kraemer and G.H.F. Diercksen, Astrophys. J. **205**, L97 (1976)

[4] L.E. Snyder and D. Buhl, Nature (London) **228**, 267 (1970)

[5] L.E. Snyder and D. Buhl, Astrophys. J. **163**, L47 (1971)

[6] L.E. Snyder and D. Buhl, Bull. A.A.S. **3**, 388 (1971)

[7] H.B. Jansen and P. Ros, Chem. Phys. Lett. **3**, 140 (1969)

[8] S. Forsen and B. Roos, Chem. Phys. Lett. **6**, 128 (1970)

[9] P.J. Bruna, S.D. Peyerimhoff, and R.J. Buenker, Chem. Phys. **10**, 323 (1975)

[10] N.L. Summers and J. Tyrrell, J. Amer. Chem. Soc. **99**, 3960 (1977); Theoret. Chim. Acta **47**, 223 (1978)

[11] E. Herbst, J.M. Norbeck, P.R. Certain, and W. Klemperer, Astrophys. J. **207**, 110 (1976)

[12] R.H. Nobes and L. Radom, Chem. Phys. **60**, 1 (1981)

[13] D.J. DeFrees, G.H. Loew, and A.D. McLean, J. Chem. Phys. **73**, 4521 (1980)

[14] W.P. Kraemer, P. Hennig, and G.H.F. Diercksen, in "Les Spectres des Molecules Simples au Laboratoire et en Astrophysique", 21st International Colloquium on Astrophysics, 1977, Liege (1980)

[15] P. Hennig, W.P. Kraemer, and G.H.F. Diercksen, Technical Report MPI/PAE Astro 135, November 1977

[16] C.S. Gudeman and R.C. Woods, Phys. Rev. Lett. **48**, 1344 (1982); Phys. Rev. Lett. **48**, 1768 (1982)

[17] W.P. Kraemer and P.R. Bunker, J. Mol. Spectrosc. **101**, 379 (1983)

[18] R. Beardsworth, P.R. Bunker, P. Jensen, and W.P. Kraemer, J. Mol. Spectrosc. **118**, 40 (1986)

[19] D.J. DeFrees, P.R. Bunker, J.S. Binkley, and A.D. McLean, J. Mol. Spectrosc., in press

[20] P.R. Bunker, P. Jensen, W.P. Kraemer, and R. Beardsworth, J. Mol. Spectrosc., in press
[21] G.A. Blake, P. Helminger, E. Herbst, and F.D. DeLucia, Astrophys. J. **264**, L69 (1983)
[22] M. Bogey, C. Demuynck, and J.L. Destombes, J. Mol. Spectrosc. **115**, 229 (1986)
[23] T. Amano, private communication
[24] S.S. Prasad and W.T. Huntress, Astrophys. J. Suppl. **43**, 1 (1980)
[25] T.E. Graedel, W.D. Langer, and M.A. Frerking, Astrophys. J. Suppl. **48**, 321 (1982)
[26] H.I. Schiff and D.K. Bohme, Astrophys. J. **232**, 740 (1979)
[27] T.W. Hartquist and A. Dalgarno, in "Giant Molecular Clouds in the Galaxy" (P.M. Solomon and M.G. Edmunds, Eds.), Pergamon, 1980
[28] N.N. Haese and R.C. Woods, Astrophys. J. **246**, L51 (1981)
[29] R.G. Daniel, E.R. Keim, and J.M. Farrer, Astrophys. J. **303**, 439 (1986)
[30] S. Green, Astrophys. J. **240**, 962 (1980)
[31] M. Yoshimine and W.P. Kraemer, Chem. Phys. Lett. **90**, 145 (1982)
[32] W.P. Kraemer, P.R. Bunker, and M. Yoshimine, J. Mol. Spectrosc. **107**, 191 (1984)
[33] G. Fogarasi, J.E. Boggs, and P. Pulay, Mol. Phys. **50**, 139 (1983)
[34] W.P. Kraemer, B.O. Roos, P.R. Bunker, and P. Jensen, J. Mol. Spectrosc., in press
[35] B.O. Roos, P.R. Taylor, and P.E.M. Siegbahn, Chem. Phys. **48**, 157 (1980)
[36] P.E.M. Siegbahn, Int. J. Quantum Chem. **23**, 1869 (1983)
[37] M.E. Jacox, Chem. Phys. **7**, 424 (1975)
[38] R.F. Curl, P.G. Carrick, and A.J. Merer, J. Chem. Phys. **82**, 3479 (1985); J. Chem. Phys. **83**, 4278 (1985)
[39] C.A. Gottlieb, E.W. Gottlieb, and P. Thaddeus, Astrophys. J. **264**, 740 (1983)
[40] H. Kanamori and E. Hirota, to be published
[41] K. Kawaguchi, T. Amano, and E. Hirota, private communication
[42] E. Herbst, J.M. Norbeck, P.R. Certain, and W. Klemperer, Astrophys. J. **207**, 110 (1976)
[43] R.C. Woods, C.S. Gudeman, R.L. Dickman, P.F. Goldsmith, G.R. Huguenin, W.M. Irvine, A. Hjalmarson, L.A. Nyman, and H. Olofsson, Astrophys. J. **270**, 583 (1983)
[44] A.J. Illies, M.F. Jarrold, and M.T. Bowers, J. Chem. Phys. **17**, 5847 (1982)
[45] D.A. Dixon, A. Komornicki, and W.P. Kraemer, J. Chem. Phys. **81**, 3603 (1984)
[46] T.B. McMahon and P. Kebarle, J. Chem. Phys. **83**, 3919 (1986)
[47] R.J. Saykally, T.A. Dixon, T.G. Anderson, P.G. Szanto, and R.C. Woods, Astrophys. J. **205**, L101 (1976)

[48] K.V.L.N. Sastry, P. Helminger, E. Herbst, and F.C. DeLucia, Chem. Phys. Letters **84**, 286 (1981)
[49] C.S. Gudeman, M.H. Begemann, J. Pfaff, and R.J. Saykally, J. Chem. Phys. **78**, 5837 (1984)
[50] S.C. Foster and A.R.W. McKellar, J. Chem. Phys. **81**, 3424 (1984)
[51] T.J. Sears, J. Opt. Soc. Am. **B2**, 786 (1985); J. Chem. Phys. **82**, 5757 (1985)
[52] P. Botschwina, Chem. Phys. Lett. **107**, 535 (1984)
[53] D.J. DeFrees and A.D. McLean, J. Chem. Phys. **82**, 333 (1985)
[54] J.E. Del Bene, M.J. Frisch, K. Raghavachari, and J.A. Pople, J. Phys. Chem. **86**, 1529 (1982)
[55] S. Ikuta, Chem. Phys. Lett. **109**, 550 (1984)
[56] W.P. Kraemer, A. Komornicki, and D.A. Dixon, Chem. Phys. **105**, 87 (1986)
[57] S.G. Lias, J.F. Liebman, and R.D. Levin, J. Phys. Chem. Ref. Data **13**, 695 (1984)
[58] T.B. McMahon and P. Kebarle, J. Am. Chem. Soc. **107**, 2612 (1985)
[59] S. Green, Astrophys. J. **277**, 900 (1984)
[60] R.H. Nobes and L. Radom, Chem. Phys. **60**, 1 (1981)
[61] W. Wagner-Redeker, P.R. Kemper, M.F. Jarrold, and M.T. Bowers, J. Chem. Phys. **83**, 1121 (1985)
[62] E. Herbst and C.M. Leung, M.N.R.A.S. (1985)
[63] D.J. DeFrees, A.D. McLean, and E. Herbst, Astrophys. J. **279**, 322 (1984)
[64] M.F. Jarrold, M.T. Bowers, D.T. DeFrees, A.D. McLean, and E. Herbst, Astrophys. J. **303**, 392 (1986)
[65] R. Lindh, W.P. Kraemer, and B.O. Roos, unpublished results
[66] J.L. Holmes, private communication.

STRUCTURES AND SPECTROSCOPIC PROPERTIES OF SMALL NEGATIVE MOLECULAR IONS - THEORY AND EXPERIMENT

P. Botschwina
FB Chemie der Universität Kaiserslautern
D-6750 Kaiserslautern
West Germany

ABSTRACT. The workshop, which involved a single contribution by the organizing person, was devoted to the results of theoretical and experimental work on small molecular anions with emphasis on structures and spectroscopic properties. The discussion dealt mainly with problems of computational effort necessary for accurate ab initio calculations and prospects of interesting theoretical and experimental work for the near future.

1. INTRODUCTION

Over the last two years quite remarkable progress has been made in the high-resolution spectroscopic investigation of molecular anions in the infrared using various kinds of tunable lasers. Neumark et al. [1] observed the vibration-rotation spectrum of NH^- by autodetachment spectroscopy in a laser-ion beam spectrometer. Saykally and coworkers [2,3] detected the hydroxide ion (OH^-) by velocity modulation laser spectroscopy with a color center laser and Liu and Oka [4] found two rotational lines of the same anion with a diode laser system. The vibration-rotation spectrum of OD^- was recently recorded and analyzed by Rehfuss et al. [5]. The hydrogen-bonded species FHF^- and FDF^- were investigated by Kawaguchi and Hirota [6] by means of diode laser spectroscopy with the magnetic field modulation technique and the two stretching vibrational bands of the amide anion (NH_2^-) were studied by Tack et al. [7,8]. Following earlier work on C_2^- in the optical region [9,10] Rehfuss et al. [11] were able to investigate the A-X electronic transition in the infrared and to derive accurate spectroscopic constants for the first excited state of this ion.

In almost all of these cases, the experimental work benefitted from predictions of ab initio calculations. The investigation of NH^- [1] was facilitated by the results of PNO-CEPA calculations by Rosmus and Meyer [12]. Quite recently, more extended calculations for NH^- and CH^- were published by Mänz et al. [13]. Excellent agreement was found between the results of ab initio calculations on OH^- by Werner et al. [14] and the later obtained precise experimental data [2-5]. The adiabatic electronic excitation energy for the A-X transition of C_2^- was

P. Ausloos and S. G. Lias (eds.), Structure/Reactivity and Thermochemistry of Ions, 261–270.

predicted to be 3470 cm^{-1} [15] which is considered to be in good agreement with the experimental value of 3930.3 cm^{-1} [11] regarding the difficulties in calculating accurate values for electronic transition energies. The band origins ν_1 (symmetric stretch) and ν_3(asymmetric stretch) of NH_2^- were predicted with errors of 14 and 26 cm^{-1}, respectively, from variational calculations with an anharmonic CEPA potential surface [15] while somewhat less accurate semiempirical predictions (scaled harmonic SCF vibrational frequencies) were previously published by Lee and Schaefer [16].

In the following, a more detailed discussion of theoretical and experimental results will be given for NH_2^- and FHF^-. In addition, several predictions will be reported for further small anions such as CN^-, PH_2^-, N_3^- , and HCO_2^-.

2. THE CYANIDE ANION (CN^-)

With an adiabatic electron affinity of as much as 3.82 eV [17,18], the cyanide anion is one of the most stable diatomic anions. Following earlier theoretical investigations of its spectroscopic properties [19-22] it was recently observed in the gas-phase by means of diode laser spectroscopy, but no detailed analysis is yet available [23]. The complications in the analysis of the experimental data arise from the fact that CN^- is produced vibrationally very hot in the discharge so that several hot bands contribute to the vibration-rotation spectrum.

A summary of theoretical predictions made for the cyanide anion by the present author [22] is given in Table I. The fundamental vibrational frequency of $^{12}C^{14}N^-$ is predicted to be 2052 cm^{-1} with an uncertainty of about 0.3 %. This value is in good agreement with the "free ion" frequency of 2038 cm^{-1} which was extrapolated by Sherman and Wilkinson [24] from infrared spectra of CN^- isolated in alkali halide lattices. The calculated equilibrium bond length is R_e = 1.177 Å which is only slightly longer than the corresponding experimental value of 1.172 Å for the cyano radical [25]. The electric dipole moment of $^{12}C^{14}N^-$ in the vibrational ground state (referring to the molecular center-of-mass coordinate system) was obtained to be 0.6333 D. This agrees perfectly with the previous ab initio value of 0.64 D by Ha and Zumofen [21] who used a much smaller basis set in their CI calculations. The transition moments for $\Delta v = 1$ transitions are relatively small, about half as large as those for isoelectronic carbon monoxide. Likewise, the radiative lifetimes for excited vibrational states are quite long, ranging from 124 msec for v = 1 to 33 msec for v = 4.

The equilibrium proton affinity of the cyanide anion was obtained to be 1489 kJ mol^{-1}. At 298 K, the calculated proton affinity is 1466 kJ mol^{-1} which is in excellent agreement with the experimental value of 1464 + 4 kJ mol^{-1} as reported by Betowski et al. [26]. Quite recently, DeFrees and McLean [27] arrived at practically the same theoretical value as ours by means of fourth-order Møller-Plesset perturbation theory.

TABLE I. Vibrational term energies, rotation and centrifugal distortion constants, rotationless dipole matrix elements, and radiative lifetimes for $^{12}C^{14}N^{-}$ a)

v	G(v)	B_v	D_v	R_v^v	R_v^{v+1}	τ_v
0	1035.1	56.2248	185.94	0.6333	-0.0546	
1	3086.9	55.7319	185.91	0.6214	-0.0771	124
2	5114.3	55.2386	185.90	0.6095	-0.0941	63
3	7117.5	54.7447	185.92	0.5976	-0.1083	43
4	9096.4	54.2502	185.97	0.5857	-0.1206	33

a) G(v) in cm^{-1}, B_v in GHz, D_v in kHz, R_v^v and R_v^{v+1} in D, and τ_v in msec.

For details of the calculations see Ref. 22.

3. THE AMIDE ANION (NH_2^-)

Theoretical [15-16, 28] and experimental [7,8] spectroscopic properties for NH_2^- are listed in Table II. An approximate experimental equilibrium structure [8] was obtained from the measured ground state rotational constants, measured vibration-rotation coupling constants α_1 and α_3 and assumed α_2 values (taken to be equal to those for the amino radical). This structure is in very good agreement with the author's CEPA equilibrium geometry [15], differences amounting to 0.002 Å for r_e and 0.1° for the equilibrium bond angle. The bending vibrational frequency ν_2 of $^{14}NH_2^-$ is predicted to be 1462 cm^{-1} which is smaller than the corresponding experimental value of the amino radical [29] by only 35 cm^{-1}. The calculated equilibrium quartic centrifugal distortion constants, given in Watson's S-reduced representation [30], are all smaller than the experimental ground state values [8]. In particular, there is a noticably large difference of 79 MHz or 15 % between D_K^e and D_K^o. A difference of this magnitude appears to be quite reasonable since, as is well known for isoelectronic water, the bending contribution to the zero-point vibrational motion makes a significant effect on this centrifugal distortion constant.

Our calculations on $^{14}NH_2^-$ yield a moderately strong Fermi resonance between the first overtone of the bending vibration $2\nu_2$, calculated at 2894 cm^{-1}, and the symmetric stretching vibration ν_1, calculated at 3108 cm^{-1} ,which is 14 cm^{-1} smaller than the experimental value. In the latter case, the harmonic oscillator product function |020> gets a coefficient of 0.24 in the normalized total vibrational wavefunction.

TABLE II. Theoretical and experimental spectroscopic properties for $^{14}NH_2^-$.[a)]

	theoret. [16,28]	theoret. [15]	exp. [8]
ν_1 (cm^{-1})	3179	3108	3121.9
ν_2 (cm^{-1})		1462	
ν_3 (cm^{-1})	3237	3164	3190.3
A_e (cm^{-1})	23.456	22.770	22.7578
B_e (cm^{-1})	13.211	13.054	13.1112
C_e (cm^{-1})	8.449	8.297	8.3401
r_e(Å)	1.022	1.030	1.028
α_e(°)	102.4	102.0	101.9
D_J (MHz)		30.3	32.4
D_{JK}(MHz)		-108.7	-114.2
D_K (MHz)		540.2	619.1
d_1 (MHz)		-12.4	-14.7
d_2 (MHz)		-0.4	-1.4

a) Theoretical centrifugal distortion constants are equilibrium values, calculated from the equilibrium structure and the harmonic force field; experimental ones are ground state values.

4. PH_2^-

The SCF method and Meyer's Coupled Electron Pair Approximation (CEPA) [31] were used for the calculation of anharmonic potential energy surfaces for PH_2^- and a large basis set of 82 contracted Gaussian-type orbitals was employed. With these surfaces and Watson's form of the vibrational Hamiltonian for non-linear molecules [32], vibrational term energies and wavefunctions were calculated variationally [33-35]. The calculated CEPA-1 equilibrium geometry is r_e = 1.429 Å and α_e = 91.88°. Calculated harmonic and anharmonic vibrational frequencies for PH_2^- and PD_2^- are listed in Table III. The bending vibrational frequency of PH_2^- is predicted to be 1069 cm^{-1}, 31 cm^{-1} less than the experimental value for PH_2 [36]. Due to the close coincidence of the stretching vibrational frequencies ν_1 and ν_3, there occurs strong Darling-Dennison resonance between the first overtones $2\nu_1$ and $2\nu_3$. The difference between ν_1 and ν_3 is only 5 cm^{-1} for PH_2^-, but the first overtones are shifted apart by as much as 103 cm^{-1}.

TABLE III. Calculated harmonic and anharmonic vibrational frequencies for PH_2^- and PD_2^- (in cm^{-1}).

Band	PH_2^- SCF	PH_2^- CEPA-1	PD_2^- SCF	PD_2^- CEPA-1
ω_1 (s. str.)	2396	2296	1718	1647
ω_2 (bend)	1183	1092	850	785
ω_3 (as. str.)	2384	2293	1714	1648
ν_2	1162	1069	840	773
$2\nu_2$	2330	2127	1659	1541
ν_3	2287	2182	1664	1590
ν_1	2287	2187	1686	1591
$3\nu_2$	3490	3176	2527	2304
$\nu_2+\nu_3$	3435	3234	2497	2355
$\nu_1+\nu_2$	3421	3242	2483	2357
$4\nu_2$	4647	4213	3367	3061
$\nu_1+\nu_3$	4495	4260	3284	3132
$2\nu_3$	4494	4264	3282	3131
$2\nu_2+\nu_3$	4580	4284	3329	3113
$\nu_1+2\nu_2$	4543	4292	3302	3115
$2\nu_1$	4592	4367	3334	3180
ZPE[b)]	2945	2798	2122	2018

[a)] Strong Darling-Dennison resonance occurs between the first overtones $2\nu_1$ and $2\nu_3$. In addition, there is fairly strong Fermi resonance between vibrational states of type (v_1, v_2+2, v_3) and (v_1+1, v_2, v_3).

[b)] Zero-point energy.

5. THE AZIDE ANION (N_3^-)

CEPA-1 calculations with a basis set of 99 contracted Gaussian-type orbitals [37] yield a linear centrosymmetrical equilibrium structure for N_3^- with an equilibrium bond length of 1.191 Å. This value is probably too large by about 0.005 Å, mainly due to incompleteness of the basis set. X-ray and neutron diffraction studies of metal azides yielded NN bond lengths of 1.12 to 1.17 Å [38] so that the azide anion experiences bond length contraction in condensed phases. Among the two IR active fundamentals, the asymmetric stretching vibration ν_3 is probably most easily accessible to laser absorption spectroscopy. Its band origin was calculated at 1950 cm^{-1} which is in close agreement with the "free ion" estimates of Sherman and Wilkinson [24]. The ν_3 band is predicted to be extremely intense: the transition dipole moment amounts to 0.51 D and the integrated molar absorption intensity is as large as $\Gamma = 6.6 \cdot 10^4\ cm^2\ mol^{-1}$. Even the combination band $\nu_1+\nu_3$, for which $\Gamma = 1512\ cm^2 mol^{-1}$ was calculated, is almost as strong as the fundamental CH stretching vibration of hydrogen cyanide.

6. THE FORMATE ANION (HCO_2^-)

The following equilibrium geometry of HCO_2^- (C_{2v} symmetry) was obtained from CEPA-1 calculations with a basis set of 97 contracted Gaussian-type orbitals: r_e (CH) = 1.128 Å , R_e (CO) = 1.262 Å, and α_e (OCO angle) = 130.2°. As is indicated by the results of analogous calculations for carbon dioxide, the calculated R_e is probably too large by about 0.012 Å, although the errors in the calculated CO equilibrium bond lengths for CO_2 and HCO_2^- need not be exactly the same. An X-ray diffraction study of sodium formate yielded a CO bond length of 1.246 A and an OCO angle of 126.3° [39]. In the crystal, the formate anion thus experiences a slight contraction of the CO bond length and an increase of the OCO bond angle by 4°. The CH equilibrium bond length of HCO_2^- was estimated by Kidd and Mantsch [40] to be 1.11 A. This value, which was obtained from an empirical relationship between CH bond length and CH stretching vibrational frequency (the latter value taken from polycrystalline sodium formate) and the data compiled by McKean et al. [41], is certainly too small for the free anion.

The vibrational frequencies of the three totally symmetric modes of the formate anion were calculated from an anharmonic three-dimensional potential energy surface and an approximate vibrational Hamiltonian which neglects the anharmonic interaction with the modes of B_1 and B_2 symmetry. The results for HCO_2^- and DCO_2^- (the latter in parentheses) are (in cm^{-1}): ν_1(~CH, CD) = 2532 (1898), ν_2(~symm. CO stretch) = 1318 (1292), and ν_3(~symm. bend) = 730 (724). The present ν_2 and ν_3 values differ by less than 50 cm^{-1} from the experimental ones for polycrystalline sodium formate [40]. The present ν_1 value for HCO_2^- of 2532 cm^{-1} is, however, considerably smaller than the corresponding value for polycrystalline sodium formate of 2830 cm^{-1} [40]. High values of more than 2800 cm^{-1} have beeen also measured for aqueous solutions and various metal formate crystals [42]. The CH stretching potential is thus apparently strongly influenced by environmental effects in condensed phases.

According to the present calculations, the CH stretching vibration of HCO_2^- is strongly anharmonic. The ratio $(\omega_1-\nu_1)/\nu_1$ is calculated to be as large as 0.072, much larger than the value of 0.0385 assumed by Gregory et al. [43].

7. FHF^- AND FDF^-

The hydrogen bifluoride anion belongs to the strongly hydrogen-bonded systems of fundamental interest. Numerous spectroscopic investigations have been performed on it in the liquid and solid phase (see Ref. 6 for several references), but only quite recently a high-resolution infrared diode laser study of FHF^- and FDF^- in the gas-phase became feasible [6]. Selected anharmonic vibrational frequencies were calculated from SCF potentials [44-46] and a CI-SD potential [47], but only small basis sets were employed. More extended calculations on the dissociation energy, equilibrium geometry, and harmonic vibrational frequencies were published by Frisch et al. [48]. A thorough treatment of the strongly anharmonic vibrational problem and IR intensities on the basis of high-quality ab initio potential and dipole moment surfaces is, however, not yet available in the literature. Such an investigation is presently performed in our laboratory and we present here some preliminary results.

A basis set of 82 contracted Gaussian-type orbitals and the CEPA-1 method were used to calculate the potential and dipole moment surface of the hydrogen bifluoride anion. The calculated equilibrium bond length is R_e (F-F) = 2.287 Å which is close to the MP2 value of Frisch et al. [48] of 2.283 Å. These values may be compared with the experimental R_0 value, obtained from the ground state rotational constant, which is 2.304 Å [6]. Stretching vibrational frequencies and transition dipole moments were calculated within a two-dimensional model which neglects the anharmonic interaction with the bending modes [49], and some results are given in Table IV which includes also results of previous ab initio calculations.

The frequency of the symmetric stretching vibration, which is only slightly anharmonic, was calculated at 597 cm^{-1}, 26 cm^{-1} lower than the SCF value. These ab initio values and the previous ones [44-48] are in good agreement with experimental values obtained in condensed phases ranging around 600 cm^{-1} and the gas-phase estimate of Kawaguchi and Hirota [6] of 617 cm^{-1}, calculated from the observed quartic centrifugal distortion constant. In agreement with Almlöf's work [44] we find ν_1 of FDF^- to be slightly larger than ν_1 of FHF^- which must be attributed to vibrational anharmonicity (see also Ref. 34 for a similar example). The present SCF value for ν_3(asymmetr. stretch) of FHF^- of 1334 cm^{-1} is smaller than the previous values [44, 46, 47] by 148-324 cm^{-1}. Correlation effects increase it by about 200 cm^{-1}, but - as preliminary variational calculations with a three-dimensional potential (including bending motion) indicate - it will be reduced by inclusion of stretch-bend interaction and a value around 1400 cm^{-1} appears to be reasonable. A value of this magnitude is consistent with the experimental values obtained in condensed phases. Kawaguchi and Hirota [6], however, attribute a parallel band with origin at 1849 cm^{-1} to the asymmetric stretching vibration of FHF^- . Although they have fairly strong arguments for their assignment, a difference of about 400 cm^{-1} between experiment and theory appears to be unreasonably large. It is also hard to imagine that the effect of an argon matrix produces a red-shift of almost 500 cm^{-1} on ν_3(see Table IV of Ref. 6). Clearly further experimental and theoretical work is required for this interesting species.

Due to large vibrational amplitudes and a steep increase of the electric dipole moment with proton motion along the internuclear axis, the ν_3 and $\nu_1+\nu_3$ bands of FHF^- and FDF^- are very strong. For the two bands of the former isotopomer we calculated transition dipole moments of 0.76 and 0.38 D. The corresponding values for FDF^- are 0.72 and 0.24 D. On the basis of these results one cannot rule out that the observed bands at 1849 cm^{-1} (FHF^-) and 1397 cm^{-1} (FDF^-) have to be attributed to the combination tones $\nu_1+\nu_3$ which have about 20 % of the intensities of the fundamentals.

TABLE IV. Calculated stretching vibrational frequencies (in cm^{-1}) for FHF^- and FDF^- [a)]

Band	[44] SCF	[47] SCF	[47] CI-SD	this work SCF	this work CEPA-1[b)]
ν_1	660(666)	619	605	623(626)	597(604)
$2\nu_1$				1231(1227)	1181(1193)
ν_3	1497(1038)	1483	1611	1335(904)	1565(1087)
$3\nu_1$				1816(1780)	1746(1758)
$\nu_1+\nu_3$	2070(1628)	2042	2162	1877(1458)	2065(1621)
$4\nu_1$				2371(2295)	2276(2257)
$2\nu_1+\nu_3$	2632(2194)	2589	2703	2401(1988)	2520(2121)

a) Values for FDF are given in parentheses. Lohr and Sloboda [46] obtained ν_3= 1669 cm^{-1} for FHF^- from an SCF potential (4-31G basis) and an anharmonic model including asymmetric stretching and bending vibrations.

b) The CEPA-1 strétching potential is not yet complete so that the results are preliminary (~50 cm^{-1} numerical uncertainty for bands involving ν_3).

REFERENCES

1. D. M. Neumark, K. R. Lykke, T. Andersen, and W. C. Lineberger, J. Chem. Phys. 83, 4364(1985).
2. J. C. Owrutsky, N. H. Rosenbaum, L. M. Tack, and R. J. Saykally, J. Chem. Phys. 83, 5338(1985).
3. N. H. Rosenbaum, J. C. Owrutsky, L. M. Tack, and R. J. Saykally, J. Chem. Phys. 84, 5308(1986).
4. D.-J. Liu and T. Oka, J. Chem. Phys. 84, 2426(1986).
5. B. D. Rehfuss, M. W. Crofton, and T. Oka, J. Chem. Phys. (in press).
6. K. Kawaguchi and E. Hirota, J. Chem. Phys. 84, 2953(1986).
7. L. M. Tack, N. H. Rosenbaum, J. C. Owrutsky, and R. J. Saykally, J. Chem. Phys. 84, 7056(1986).
8. L. M. Tack, N. H. Rosenbaum, J. C. Owrutsky, and R. J. Saykally, J. Chem. Phys. (in press).
9. G. Herzberg and A. Lagerqvist, Can. J. Phys. 46, 2363(1968).
10. R. D. Mead, U. Hefter, P. A. Schulz, and W. C. Lineberger, J. Chem. Phys. 82, 1723(1985).
11. B. D. Rehfuss, D.-J. Liu, B. M. Dinelli, M.-F. Jacod, M. W. Crofton, and T. Oka, to be published.
12. P. Rosmus and W. Meyer, J. Chem. Phys. 69, 2745(1978)
13. U. Mänz, A. Zilch, P. Rosmus, and H.-J. Werner, J. Chem. Phys. 84, 5037(1986).
14. H.-J. Werner, P. Rosmus, and E.-A. Reinsch, J. Chem. Phys. 79, 905 (1983).
15. P. Botschwina, J. Mol. Spectrosc. 117, 173(1986).
16. T. J. Lee and H. F. Schaefer, J. Chem. Phys. 83, 1784(1985).
17. J. Berkowitz, W. A. Chupka, and T. A. Walter, J. Chem. Phys. 50, 1497(1969).
18. R. Klein, R. P. McGinnis, and S. R. Leone, Chem. Phys. Letters 100, 475(1983).
19. J. E. Gready, G. B. Bacskay, and N. S. Hush, Chem. Phys. 31, 467 (1978).
20. P. R. Taylor, G. B. Bacskay, and N. S. Hush, J. Chem. Phys. 70, 4481(1979).
21. T.-K. Ha and G. Zumofen, Mol. Phys. 40, 445(1980).
22. P. Botschwina, Chem. Phys. Letters 114, 58(1985).
23. M. Gruebele and R. J. Saykally, private communication.
24. W. F. Sherman and G. R. Wilkinson, in "Vibrational Spectroscopy of Trapped Species", edited by H. E. Hallam (Wiley, London, 1973).
25. K. P. Huber and G. Herzberg, "Molecular Spectra and Molecular Structure. IV. Constants of Diatomic Molecules" (Van Nostrand, New York, 1979).
26. D. Betowski, G. Mackay, J. Payzant, and D. Bohme, Can. J. Chem. 53, 2365(1975).
27. D. J. DeFrees and A. D. McLean, J. Comput. Chem. 7, 321(1986).
28. T. J. Lee and H. F. Schaefer, unpublished results (quoted in Ref. 8).
29. K. Kawaguchi, C. Yamada, E. Hirota, J. M. Brown, J. Buttenshaw, C. R. Parent, and T. J. Sears, J. Mol. Spectrosc. 81, 60(1980) and references therein.

30. J. K. G. Watson, in "Vibrational Spectra and Structure", Vol. 6, edited by J. R. Durig (Elsevier, Amsterdam, 1977).
31. W. Meyer, J. Chem. Phys. 58, 1017(1973).
32. J. K. G. Watson, Mol. Phys. 15, 479(1968).
33. R. J. Whitehead and N. C. Handy, M. Mol. Spectrosc. 55, 356(1975).
34. P. Botschwina, Chem. Phys. 40, 33(1979).
35. P. Botschwina, Dissertation, Kaiserslautern, 1980.
36. G. Duxbury, in "Molecular Spectroscopy", Vol. 3 (The Chemical Society, London, 1975) and references therein.
37. P. Botschwina, J. Chem. Phys. (in press).
38. K. Jones, in "Comprehensive Inorganic Chemistry", Vol. 2, edited by A. F. Trotman-Dickenson (Pergamon, Oxford, 1973).
39. P. L. Markila, S. J. Rettig, and J. Trotter, Acta Crystallogr. B31, 2927(1975).
40. K. G. Kidd, and H. H. Mantsch, J. Mol. Spectrosc. 85, 375(1981).
41. D. C. McKean, J. L. Duncan, and L. Batt, Spectrochim. Acta A29, 1037(1973).
42. Gmelin Handbuch der Anorganischen Chemie, 8th edition, Carbon, Vol. 14, Part C4 (Springer, Berlin, 1975).
43. A. R. Gregory, K. G. Kidd, and G. W. Burton, J. Mol. Struct. 104, 9(1983).
44. J. Almlöf, Chem. Phys. Letters 17, 49(1972).
45. S. A. Barton and W. R. Thorson, J. Chem. Phys. 71, 4263(1979).
46. L. L. Lohr and R. J. Sloboda, J. Phys. Chem. 85, 1332(1981).
47. A. Støgard, A. Strich, J. Almlof, and B. Roos, Chem. Phys. 8, 405 (1975).
48. M. J. Frisch, J. E. DelBene, J. S. Binkley, and H. F. Schaefer, J. Chem. Phys. 84, 2279(1986).
49. P. Botschwina, Chem. Phys. 68, 41(1982).

ab initio CALCULATIONS ON ORGANIC ION STRUCTURES

Nikolaus Heinrich
Institute of Organic Chemistry
Technical University Berlin
D-1000 Berlin 12
West Germany

1. Introduction

The role of hydrogen bridging in liquid and crystal structures is a well-established fact in chemistry. There has been a multitude of theoretical studies of hydrogen-bonded complexes such as water dimers, hydrogen halide dimers as well as higher clusters of various composition. In a number of cases, good agreements with experimental binding energies have been achieved.[1] However, only a few examples are presently known of hydrogen bridged structures in gas phase ion chemistry. Recent experimental studies, for example, were able to verify the existence of stable H-bridged $[CH_2=CHOH/CH_3OH]^{+\cdot}$ and $[CH_2=CHOH/H_2O]^{+\cdot}$ complexes in the mass spectrometer.[2] Their remarkable thermochemical stability towards decomposition leads to the tempting suggestion that these species may play a more important role in gas phase ion chemistry than anticipated in the past. Thus, the determination of stabilities, equilibrium geometries, and the chemistry of hydrogen-bridged ion-dipole complexes provide a challenging field for both experimentalists and theoreticians. In particular, there remains some debate as to what level of theory is adequate for the description of these unconventional structures and their energetics. Basis set effects on equilibrium geometries, the role of electron correlation effects, and the need of corrections for basis set superposition errors have been discussed on a workshop conducted at the NATO ASI conference in Les Arcs, 1986.

2. Metastable Keto Ion Dissociations

The first contribution presented by the author of this article focussed on the role of hydrogen-bridged ion-dipole complexes as central intermediates in metastable keto ion dissociations. Loosely bound ion-radical pairs have been postulated in alkane$^{+\cdot}$ and ketone$^{+\cdot}$ dissociations [3], but up to date there has been no theoretical proof for these species to exist.
The most abundant fragmentations of metastable ions of the simplest ketone, the acetone cation radical, correspond to the losses of CH_4 and CH_3.[4] The dissociation of the keto ion into CH_3CO^+ and a CH_3 radical has been thought to

P. Ausloos and S. G. Lias (eds.), Structure/Reactivity and Thermochemistry of Ions, 271–278.

occur continously endothermic via a direct C-C cleavage. In the literature, almost no attention has been spent for the mechanism of the CH_4 loss.

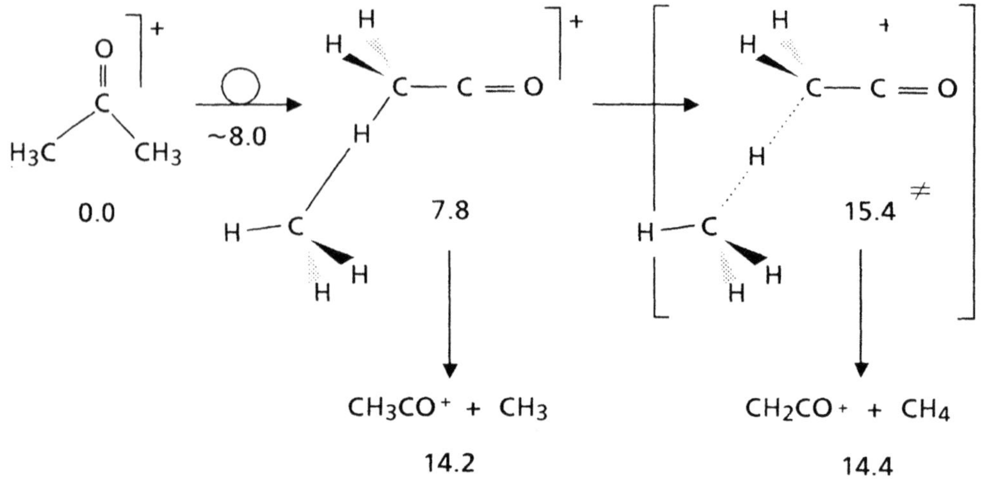

Fig.1. Proposed mechanism for the acetone ion dissociation. Relative energies (MP2 (full core) /6-31G*//6-31G* + ZPVE) are given in kcal/mol.

According to our *ab initio* calculations [5] using split valence 3-21G and 6-31G* basis sets for geometry optimizations, we found that the continous elongation of the C-C bond in ionized acetone leads to a loosely bound, H-bridged ion-dipole complex between the acylium ion CH_3CO^+ and the CH_3 radical below the threshold of decomposition. The CH_3 migration is thought to occur within the electrostatic field of the CH_3CO^+ ion. The potential energy surface in that region turns out to be extremely flat.
The subsequent dissociation of the complex into its components may compete with a shift of the bridging hydrogen towards the CH_3 moiety to form CH_4 and ionized ketene ($CH_2{=}C{=}O^{+\cdot}$). The energy required to overcome this barrier is 7.6 kcal/mol (MP2 (full core) /6-31G*//6-31G* + ZPVE) above the complex energy which is 1.0 kcal/mol and 1.2 kcal/mol above the dissociation products $CH_2{=}C{=}O^{+\cdot}$ + CH_4 and CH_3CO^+ + CH_3 , respectively. Note, that the complex is about 7.8 kcal/mol less stable than ionized acetone and the transition state connecting both is expected to be very close to 8 kcal/mol.
At this level of theory it can be stated that the energy requirements of the two reaction channels are of similar magnitude. The present results, however, cannot account for the relative abundancies of CH_4 versus CH_3 loss, the former being the more abundant fragmentation product, since there is no information on the particular rate constants.
Metastable acetamide ions give rise to the predominant formation of $[CH_5N]^{+\cdot}$ ions by CO loss. Almost no NH_2 nor NH_3 loss has been found.[6] By means of collisional activation (CA) spectroscopy, the structure of the $[CH_5N]^{+\cdot}$ ion could be identified to correspond to the distonic ion $CH_2NH_3^{+\cdot}$. The mechanistic proposal based on our *ab initio* calculations, again, involves a hydrogen-bridged ion-dipole complex between CH_3CO^+ and the NH_2 radical (see Fig. 2) which is

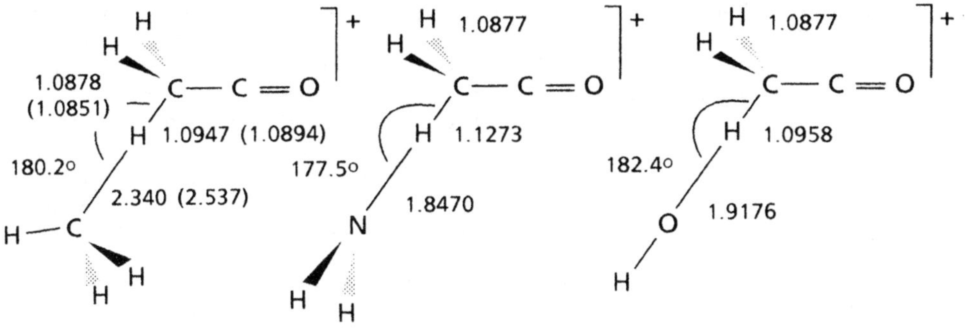

Fig.2. Selected geometry parameters of 3-21G (6-31G*) optimized [$CH_3CO^+/X^{\cdot}$] complexes ($X=CH_3$, NH_2, OH). Bond lengths are given in Å.

generated by a NH_2 shift well below any dissociation threshold. In contrast to the methyl case, the transition state corresponding to the shift of the bridging hydrogen is still lower in energy than the sum of the energies of the dissociation products CH_3CO^+ $+NH_2$ or $CH_2=C=O^{+\cdot}$ $+NH_3$, respectively. The complex between NH_3 and ionized ketene $CH_2=C=O^{+\cdot}$ may further rearrange into the covalently bound species $NH_3CH_2CO^{+\cdot}$ and from this intermediate the distonic ion $CH_2NH_3^{+\cdot}$ is subsequently formed by CO loss (see Fig. 3). Note, however,

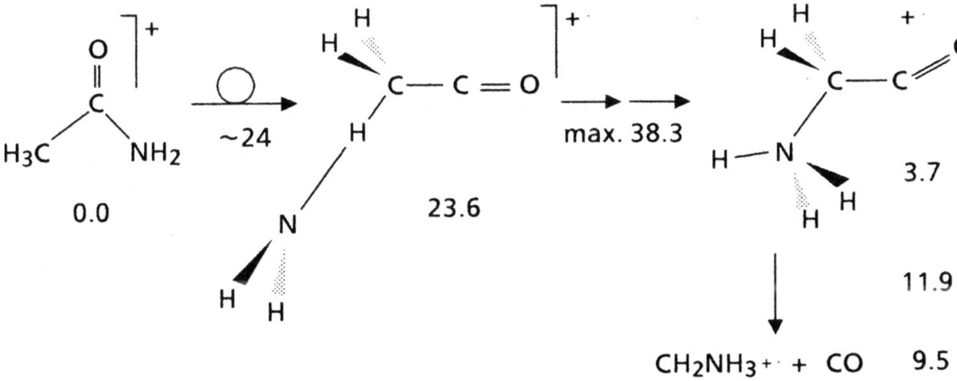

Fig.3. Mechanistic proposal for the CO loss from ionized acetamide. Relative energies (MP3/6-31G**//3-21G+ZPVE) are given in kcal/mol.

that final conclusions concerning the energetics should be drawn from the more sophisticated MP3/6-31G**//6-31G* calculations which are presently being carried out.

Our first calculations on OH versus H_2O loss from ionized acetic acid [7] indicate the same mechanism to be operating as proposed for the dissociation of ionized acetone. The hydrogen-bridged species (see Fig. 2), again, acts as the common precursor for both exit channels.

In the systems $CH_3COX^{+\cdot}$ currently being investigated, one prerequisite for the formation of stable hydrogen-bridged intermediates is that the substituent X (i.e. the migrating group) is a sufficiently polarizable dipolar component so that it can be electrostatically coordinated towards the CH_3CO^+ moiety even at relatively large distances. Note that CH_3 is slightly pyramidalized in the transition state of the migration, thus continually providing a weak dipolar interaction towards the positively charged periphery of CH_3CO^+. Hence, for systems like ionized acetaldehyde ($CH_3CHO^{+\cdot}$) and acetyl fluoride ($CH_3CFO^{+\cdot}$) we were not able to locate a stable H-bridged species on the potential surface.

The results presented here are preliminary ones and have to be improved in several respects. From the studies of smaller H-bridged complexes, we know that small basis sets like 3-21G or 4-31G tend to overestimate the electric multipole moments of the respective subsystems of the complex. This brings about overestimated SCF interaction energies and underestimated intermolecular distances. [8] It is therefore necessary to optimize these structures by means of polarized basis sets of at least 6-31G* quality, but is extremely time consuming for molecular systems of the size considered here. The potential surfaces around the hydrogen-bridged complexes mentioned here turn out to be rather flat: Large geometrical changes often bring about negligable changes in relative energies and, therefore, extreme care must be taken on the SCF convergence criteria and the optimization thresholds. The vibrational frequencies of the various stationary points should be evaluated from the analytical second derivatives of the energy with respect to all internal coordinates.
The effects of electron correlation have to be properly taken into account when H-bridged species are discussed in comparison to other, more conventional isomers. Although it is impossible to give any predictions on the magnitude of correlation effects, we have indication that in some cases loosely bound ion-dipole complexes were substantially stabilized compared to the covalently bound isomers.
The results presented here were not corrected for basis set superposition errors. There is still controversal debate as to what correction if any is sufficient depending on the basis sets used.
Despite the limitations of the theoretical model used, there is clear theoretical evidence for hydrogen-bridged species as central intermediates in keto ion dissociations. Further investigations on other basic processes in gas phase ion chemistry such as alkane$^{+\cdot}$ dissociations are under study. A more detailed survey of the problems briefly outlined in this chapter is currently in preparation and will be published elsewhere.

3. Structures and Stabilities of Hydrogen Bridged Ion Water Complexes

Ab initio calculations on structures and energetics of H-bridged [ethylene$^{+\cdot}$/ H_2O], [propene$^{+\cdot}$/H_2O] and [ketene$^{+\cdot}$/H_2O] complexes have been the topic of the second contribution of the workshop session and were presented by R. Postma and B. van Baar (Utrecht, The Netherlands).
A hydrogen-bridged complex between ionized ethylene ($CH_2{=}CH_2^{+\cdot}$) and H_2O was proposed to be generated by CH_2O loss from ionized methoxy ethanol. [9] The structures, energy differences, and barriers of interconversion of the

hydrogen-bridged ion and the distonic ion $CH_2CH_2OH_2^{+\cdot}$ have been the subject of extensive calculations on the SDCI/6-31G**//4-31G level of theory. It turns out, that the H-bridged species corresponds to a local minimum on the potential surface and is stabilized by about 10 kcal/mol towards dissociation into its components. According to Fig. 4, interconversion into the distonic ion 1 may

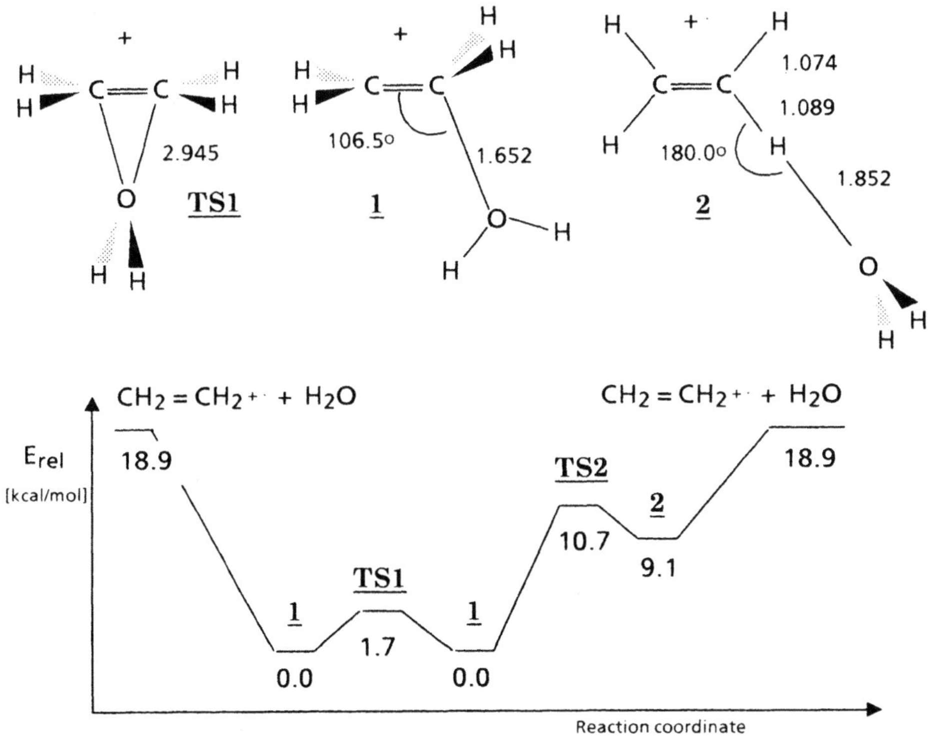

Fig. 4. Selected geometry parameters (bond lengths in Å) of [ethylene$^{+\cdot}$ /H_2O] complexes. Relative energies (SDCI/6-31G**//4-31G + ZPVE) are given in kcal/mol.

occur well below the dissociation limit. At internal energies of 11 - 12 kcal/mol above the distonic ion 1, the water molecule can move freely around and along the positively charged ethylene$^{+\cdot}$ entity without dissociation. The unimolecular and collisionally induced dissociation characteristics of ions 1 and 2 will therefore be indistinguishable. Potential scans on the distance between the bridging hydrogen and the oxygen atom (i.e. the hydrogen-transfer coordinate) indicate a steep increase in energy: The hydrogen transfer to form a complex $[CH_2{=}CH\cdots H_3O]^{+\cdot}$ cannot occur below the dissociation threshold to form $CH_2{=}CH_2^{+\cdot}$ and H_2O. Thus, positional exchange among the hydrogens on carbon and oxygen will not take place which is in agreement with experimental observations from labelling studies.

Note that the stabilization from hydrogen-bridging follows in the order of the polarization (i.e. charge alternation) of the hydrogen-bridge. The strongest stabilization has been found for the O-H-O bridged isomer of the [vinylalcohol$^{+\cdot}$

/H_2O] system [2] (36.1 kcal/mol), which is in the order of the solvation energy for the reaction $H_3O^+ + H_2O \rightarrow [H_2O \cdots H \cdots OH_2]^+$ (33 kcal/mol). The C-H-O bridged isomers [vinylalcohol$^{+\cdot}$ /H_2O] have significantly lower stabilization energies of about 23 kcal/mol. These are, however, still more pronounced than the stabilization found for the C-H-O bridged [ethylene$^{+\cdot}$ /H_2O] system (10 kcal/mol), due to the polarizing effect of the OH group in the σ-frame of the $CH_2{=}CHOH^{+\cdot}$ moiety. This may be compared at least qualitatively with the stabilization energy of 6.4 kcal/mol found in the C-H-C bridged $[CH_3CO/CH_3]^{+\cdot}$ complex mentioned above.

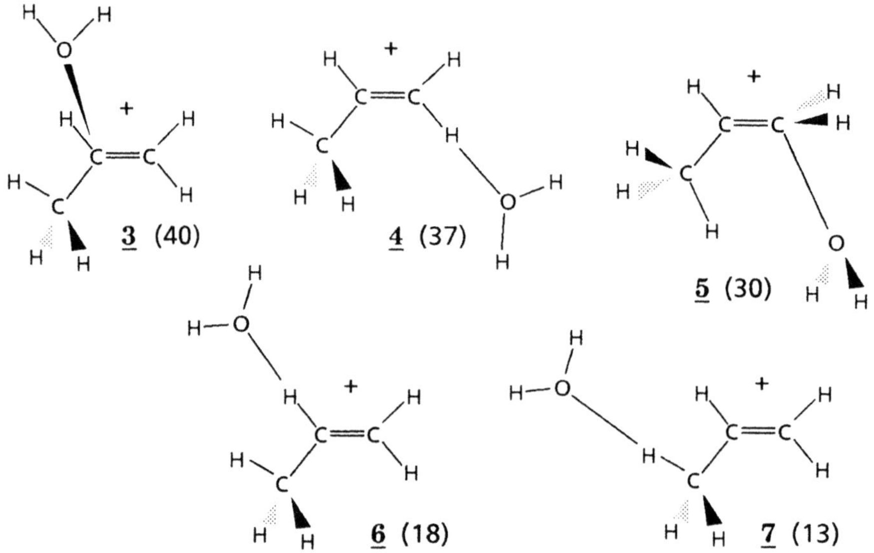

Fig. 5. Propene$^{+\cdot}$ /H_2O structures 3 - 7. Relative stabilization energies (HF/4-31G) towards decomposition are given in kcal/mol,

Preliminary calculations (HF/4-31G) on the [propene$^{+\cdot}$ /H_2O] system [10] indicate, again, that interconversion of the distonic ions 3 and 5 and the H-bridged structures 4, 6 and 7 can occur below the threshold of decomposition. The ion structures 3 - 7, presently under further studies, are shown in Fig. 5 together with their stabilization energies towards decomposition into propene$^{+\cdot}$ and water. A more elaborated discussion on this system will be the subject of a future publication.

An ion m/e 60 $[C_2H_4O_2]^{+\cdot}$ having the structure of an ion-dipole complex between ionized ketene ($CH_2{=}C{=}O^{+\cdot}$) and H_2O was proposed to be generated by the loss of formaldehyde from the molecular ion of 1.3-dihydroxy acetone.[11] Labelling experiments indicate that there is no positional exchange among the hydrogens on carbon and oxygen below the threshold of decomposition into ketene$^{+\cdot}$ and H_2O, the only major fragmentation process in the metastable time frame. Three possible minimum structures optimized with a 4-31G basis set are shown in Fig. 6. All of them are bound by purely

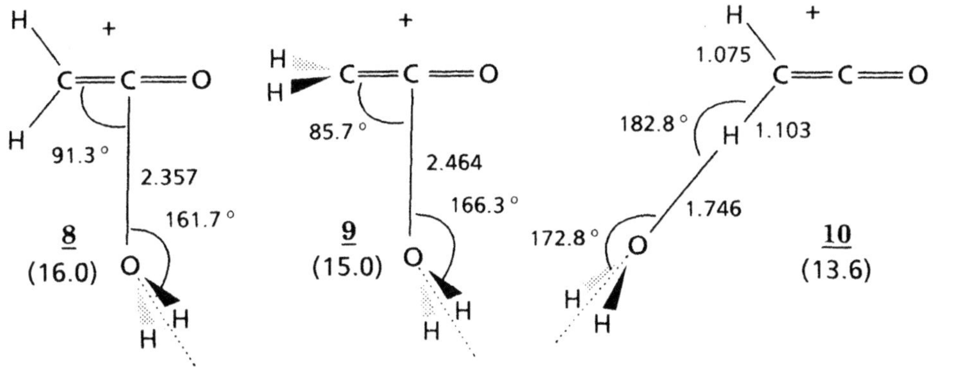

Fig. 6. Ketene+·/H_2O structures. Stabilization energies towards decomposition into ketene+· + H_2O (SDCI/6-31G**//4-31G+ZPVE) are given in parenthesis [kcal/mol].

electrostatic interactions between the positively charged centres of the ketene+· unit and the oxygen atom of the neutral dipolar component H_2O. The potential energy surface turns out to be very flat: At internal energies in excess of 2.4 kcal/mol above 8, well below the threshold of decomposition of 16 kcal/mol (SDCI/6-31G**//4-31G+ZPVE), the three species may freely interconvert. The problems concerning the choice of the theoretical models for a proper description of loosely bound complexes have already been mentioned. Note that for species 8, the geometry optimization using 6-31G* results in a structure very similar to the 4-31G optimized structure shown in Fig. 6. In contrast, 3-21G optimizations on 8 result in a significantly different species, the H_2O now being

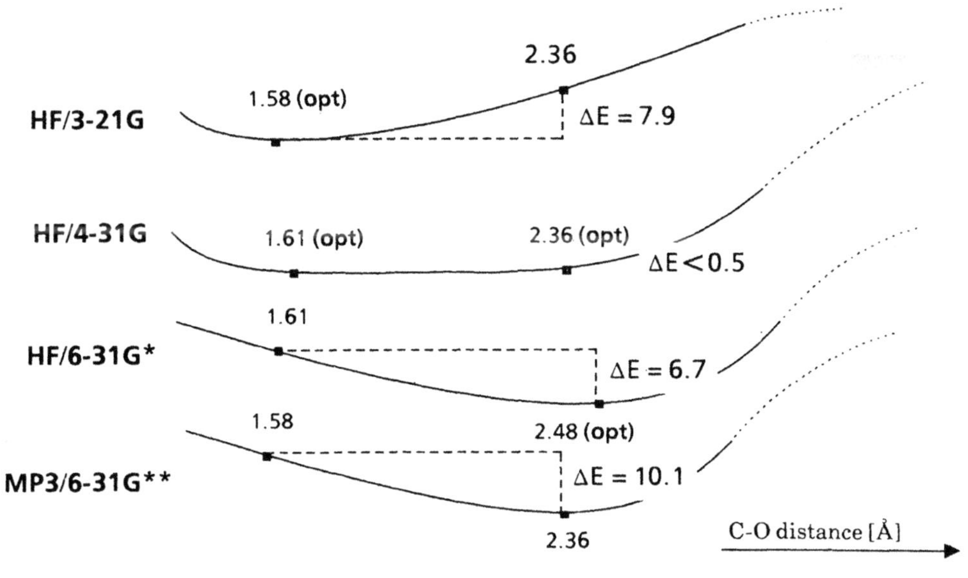

Fig. 7. Comparison of different levels of theory for the [ketene+·/H_2O] complex 8. The C-(OH_2) distance has been taken as the reaction coordinate. Energy differences ΔE given in kcal/mol.

covalently bound to the carbonyl carbon with a relatively short C-O distance of 1.58 Å. On the 3-21G surface there is no loosely bound minimum structure. However, MP3/6-31G**-single point calculations on the loose (4-31G) and tight (3-21G) structures indicate the former being 10.1 kcal/mol more stable. A qualitative comparison of the different levels of theory is given in Fig. 7.
This figure may convincingly illustrate that much must be taken for the choice of the theoretical model suited to give a reliable description of these rather unconventional species. A thorough comparative study on basis set effects on structures and stabilization energies of $[ketene/HX]^{+\cdot}$ complexes is currently in progress.[12]

4. References

[1] For a review, see: A. Beyer, A. Karpfen, P. Schuster *Top. Curr. Chem.* **120**, 1 (1984)

[2] a) J. K. Terlouw, W. Heerma, P. C. Burgers, J. L. Holmes *Can. J. Chem.* **62**, 289 (1984). b) R. Postma, P. J. A. Ruttink, F. B. van Duijneveldt, J. K. Terlouw, J. L. Holmes *ibid.* **63**, 2798 (1985)

[3] See: a) R. D. Bowen, D. H. Williams *J. Am. Chem. Soc.* **102**, 2752 (1980) and cited literature therein. b) C. E. Hudson, D. J. McAdoo *Int. J. Mass Spectrom. Ion Processes* **59**, 325 (1984)

[4] C.Lifshitz, E.Tzidony *Int. J. Mass Spectrom. Ion Phys.* **39**, 181 (1981)

[5] N. Heinrich, H. Schwarz, *manuscript in preparation.*

[6] N. Heinrich, T. Drewello, T. Weiske, H. Schwarz, W. P. M. Maas, N. M. M. Nibbering, *to be published.*

[7] For experiments, see: J. L. Holmes, F. P. Lossing *J. Am. Chem. Soc.* **102**, 3732 (1980)

[8] See: a) P. Hozba, B. Schneider, P. Carsky, R. Zahradnik *J. Mol. Struct. (THEOCHEM)* **138**, 377 (1986). b) M. J. Frisch, J. E. Del Bene, J. S. Binkley, H. F. Schaefer *J. Chem. Phys.* **84**, 2279 (1986)

[9] R. Postma, P. J. A. Ruttink, B. van Baar, J. K. Terlouw, J. L. Holmes, P. C. Burgers *Chem. Phys. Lett.* **123**, 409 (1986)

[10] R Postma, P. J. A. Ruttink, B. van Baar, J. K. Terlouw, J. L. Holmes, P. C. Burgers, *to be published.*

[11] R. Postma, P. J. A. Ruttink, J. K. Terlouw, J. L. Holmes *J. Chem. Soc., Chem. Commun.* 683 (1986)

[12] N. Heinrich, W. Koch, R. Postma, *to be published.*

FORMATION OF ANIONS IN THE GAS PHASE

Charles H. DePuy and Veronica M. Bierbaum
Department of Chemistry and Biochemistry
University of Colorado
Boulder, CO 80309-0215
U.S.A.

ABSTRACT. Only a few anions can be produced efficiently and exclusively in the gas phase by electron impact, so indirect methods, usually involving gas phase ion-molecule chemistry must be employed. In this paper we summarize some of the methods which have been used in the flowing afterglow (FA) and ion cyclotron resonance (ICR) spectrometer for producing anions, particularly carbanions, for subsequent investigation of their chemical reactions.

INTRODUCTION

In recent years there has been an increase in the study of the chemistry of anions in the gas phase, with an especial interest by organic chemists in the gas phase chemistry of carbanions. Carbanions are much more commonly used by organic chemists in synthesis because they ordinarily do not rearrange (in marked contrast to carbocations) and because many of their reactions are stereospecific. In solution, however, carbanions are usually produced in highly non-polar solvents in which they are tightly bound to their inorganic counter ion and are frequently aggregated. Obviously there is potentially a great deal of information about the intrinsic properties of carbanions (and other anions) to be learned by studying their gas phase chemistry. The challenge then is to devise ways of producing specific anions in the gas phase, and in sufficient numbers, so that their reactions and their reaction rates can be determined.

ELECTRON IMPACT

In positive ion chemistry direct electron impact is the usual method for producing the ion or ions of interest. Direct electron attachment is seldom useful for the generation of negative ions except in the case of molecular anions of species with high electron affinities (e.g., NO_2^-). One usually wants to study M-1 ions (A^-) of acids HA. The most obvious, and usually the best, way to produce such an ion is by

P. Ausloos and S. G. Lias (eds.), Structure/Reactivity and Thermochemistry of Ions, 279–291.

proton abstraction from HA by a strong base B^-. In Table 1 a list of the gas phase acidities of some common organic and inorganic acids is given (1). Methane is among the least acidic of all molecules, and the methyl anion is only very weakly bound and extremely difficult to form.

TABLE 1

GAS PHASE ACIDITIES OF SOME COMMON MOLECULES

Molecule	ΔH_{acid}	Molecule	ΔH_{acid}
CH_4	416.6	CH_3CN	372.1
NH_3	403.6	SiH_4	371.5
C_6H_6	399.8	HF	371.3
H_2O	390.8	$(CH_3)_2CO$	368.8
C_3H_6	390.7	H_2S	353.5
CH_3OH	381.4	HCN	353.1
C_2H_5OH	376.1	CH_3COOH	348.5
$(CH_3)_3COH$	373.3	HCl	333.4
$(CH_3)_2SO$	372.7	HBr	323.6

However, electron impact on ammonia will form the amide ion by dissociative electron attachment (eq. 1), and amide ion is quite stable as a gas phase base. The most useful of all gas phase bases is the hydroxide ion; it can be produced by electron impact on water, but the presence of water in the apparatus can be a nuisance, since water often forms hydrates with anions. A better way is to form O^- by dissociative electron attachment to N_2O followed by rapid hydrogen atom abstraction from methane or other hydrocarbons (eq. 2, 3). For certain exchange studies DO^- is a useful ion and C_6D_{12} (a relatively inexpensive compound because it is used as an nmr solvent) may replace the methane. In the ICR, methoxide ion (CH_3O^-) is a common initial ion because it can be formed efficiently (eq. 4) by electron impact on methyl nitrite (CH_3ONO), but this method is not as useful in the FA.

All halide ions are easily obtained by electron impact on halogen-containing molecules, organic or inorganic. The most versatile halide ion by far is F^-; it is a relatively strong base and so can be used to abstract a proton from many other molecules, but it is most useful because of its high affinity for silicon (vide infra). In our experience, NF_3 and CF_4 make the best precursors for its formation by electron impact (eq. 5).

$$NH_3 + e^- \rightarrow NH_2^- + H \quad (1)$$

$$N_2O + e^- \rightarrow O^- + N_2 \quad (2)$$

$$O^- + CH_4 \rightarrow HO^- + CH_3 \quad (3)$$

$$CH_3ONO + e^- \rightarrow CH_3O^- + NO \quad (4)$$

$$NF_3 + e^- \rightarrow F^- + NF_2 \quad (5)$$

Of course there is a wide variety of other anions which can be produced in good yield by electron impact. McDonald (2) has shown, for example, that in a FA $Fe(CO)_5$ can be ionized to form $Fe(CO)_4^-$, $Fe(CO)_3^-$ or other iron carbonyl anions depending upon the energy of the electrons used for the ionization. But for most organic anion studies ionization begins with HO^-, NH_2^-, CH_3O^- or F^-.

PROTON ABSTRACTION

Exothermic proton transfers are almost always extremely fast in the gas phase, generally occurring at every collision. Therefore initial generation of a strong base and production of the anion of interest by proton transfer is by far the most common strategy. To produce most organic anions in the FA, proton abstraction by hydroxide ion is usually best (eq. 6, 7). For more basic anions, NH_2^- is used (eq. 8). For more acidic compounds, methoxide or fluoride ion may suffice.

$$HO^- + CH_2{=}CH{-}CH_3 \rightarrow CH_2{=}CH{-}CH_2^- + H_2O \quad (6)$$

$$HO^- + CH_3CN \rightarrow {}^-CH_2{-}CN + H_2O \quad (7)$$

$$NH_2^- + C_6H_6 \rightarrow C_6H_5^- + NH_3 \quad (8)$$

Proton abstraction, particularly by hydroxide ion, is not without its hazards, primarily because of the presence along the reaction pathway of long-lived complexes. Even in highly exothermic proton transfers the resulting anion-water complex may live long enough for reversal of the proton transfer. This can be demonstrated, for example, in the reaction of DO^- with substituted propylenes (3) where the allyl anion product contains some deuterium (eq. 9). This result demonstrates that proton transfer is reversible within the complex.

$$CH_2{=}C(X){-}CH_3 + DO^- \rightarrow CH_2{=}C(X){-}CHD^- + H_2O \quad (9)$$

Although at first sight this result might not appear serious, one must be aware that in fact it can lead to prototropic shifts in the process of forming the anion, which might not have the structure one would anticipate. For example (4) we found that proton abstraction from 1,5-hexadiene by HO^- leads, in large part, to the conjugated (and hence more stable) anion rather than to the expected allylic anion (eq. 10).

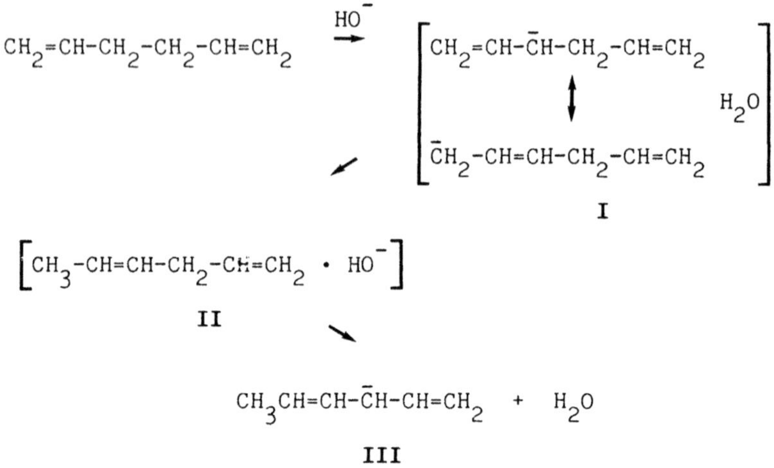

(10)

The initially formed allyl anion **I** is still complexed to water; before separation a proton can add to the carbon 1, forming the isomeric 1,4-hexadiene molecule and hydroxide ion **II**. Proton abstraction from this diene will occur most readily from the 3-carbon to form the highly conjugated ion **III**. Such rearrangement is not possible with a base without an exchangeable proton (for example methoxide or fluoride ions) and it appears to be less common with amide ion, probably because the complex is shorter lived, and proton transfers are more exothermic.

FORMATION OF ANIONS FROM SILANES

While proton abstraction is a convenient way to produce anions in the gas phase, there are several reasons why it may not be useful in specific cases. The possibility of rearrangement during the process of proton abstraction has already been mentioned. In many cases, especially in organic molecules, there may be more than one acidic proton, and proton abstraction may lead to a mixture of ions, or exclusively to an ion which is not of interest. Often the desired acid, HA, is unknown, not readily available, or simply too dangerous to want to use in the laboratory. To overcome some of these problems we developed the procedure of generating anions from trimethylsilyl derivatives by reaction with fluoride ion (5). For example reaction of acetaldehyde with base produces exclusively the enolate ion (eq. 11) while reaction of acetyltrimethylsilane with fluoride ion (6) generates the much more basic (and interesting) acetyl anion (eq. 12).

$$HO^- + CH_3\text{-}\overset{O}{\overset{\|}{C}}\text{-}H \longrightarrow \bar{C}H_2\text{-}\overset{O}{\overset{\|}{C}}\text{-}H + H_2O \quad (11)$$

$$F^- + CH_3\text{-}\overset{O}{\overset{\|}{C}}\text{-}Si(CH_3)_3 \longrightarrow CH_3\text{-}\overset{O}{\overset{\|}{C}}{}^- + (CH_3)_3SiF \quad (12)$$

Experience shows that this is a good method for the formation of carbanions which are no more basic than water. The allyl anion, for instance, can readily be generated in this way. Trimethylphenylsilane gives small amounts of phenide ion, but mostly an adduct is formed. Sometimes better yields of the desired anion result when amide ion is used instead of fluoride ion.

There is another, often unappreciated, advantage of the trimethylsilyl route to anions. In studying the rate of reaction of an anion, it sometimes happens that the product of the reaction can regenerate more of the reactant anion by proton transfer. For example in studying the hydrogen-deuterium exchange reaction between the allyl anion (produced by reaction of propylene with amide ion) and D_2O, we found that some deuterated allyl anion reacted with the excess propylene to regenerate undeuterated allyl anion, thus complicating the kinetics. If the allyl anion is generated from trimethylallylsilane, however, this does not occur and the interpretation of the reaction is much more straightforward.

CHEMICAL SYNTHESIS OF ANIONS

By far the most versatile method for generating anions is by chemical synthesis in the flow tube or ICR cell. Here the FA has an especial advantage because of the spatial and temporal separation along the flow tube of ion generation and reaction regions. Thus multistep syntheses can be undertaken, ion A^- being converted to B^- and then to C^-. Since ion A^- can be completely consumed before the precursor of C^- is added to the flow tube, there can be great diversity in the types of reactions which can be undertaken. Similar chemical syntheses are currently being carried out in the Fourier-transform ICR (FT/MS) by taking advantage of the long ion trapping time of these instruments and making use of pulsed valves (7). A neutral reagent can be added to accomplish one reaction (for example A^- forming B^-), then it can be removed and a second neutral reagent added to convert B^- to C^-. Generally speaking, most chemical reactions with which we are familiar from investigations in solution can be used equally well in the gas phase. We will illustrate with some syntheses of inorganic and organic anions.

For many simple inorganic anions the corresponding conjugate acid is either unavailable or hazardous. Consider such anions as N_3^-, HS^-, NCO^- or NCS^-; all of the corresponding acids HN_3, H_2S, HNCO and HNCS present experimental and/or toxicity problems. Yet each can be generated rapidly and easily (8) within the FA flow tube by simple chemical reactions shown below (eq. 13-16).

$$NH_2^- + N_2O \rightarrow N_3^- + H_2O \quad (13)$$

$$HO^- + CS_2 \rightarrow HS^- + COS \quad (14)$$

$$NH_2^- + CO_2 \rightarrow NCO^- + H_2O \quad (15)$$

$$NH_2^- + COS \rightarrow NCS^- + H_2O \quad (16)$$

All of these reactions are relatively rapid, so that it is easy to prepare the ion of interest at the very beginning of the flow tube and still have adequate reaction distance to study its chemistry. These processes all seem to occur by the same general mechanism, with small variations among the different neutral substrates. For example we picture NCO^- to be formed by the following sequence of reactions (eq. 17).

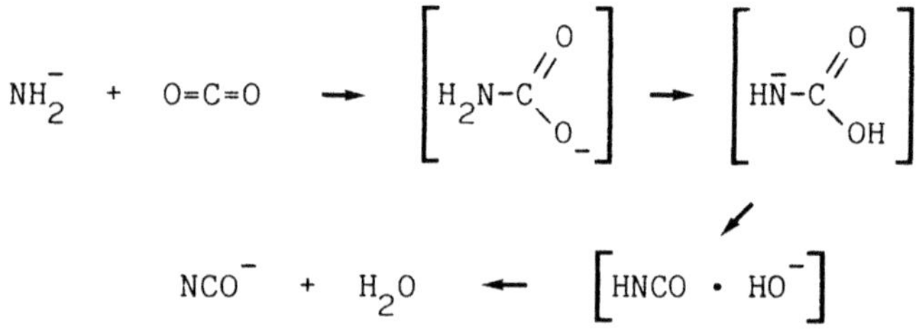

(17)

Organic chemistry provides even more scope for ion synthesis. A particular favorite is the synthesis of diazoanions (9), illustrated in eq. 18 for the case of the vinyldiazomethane anion.

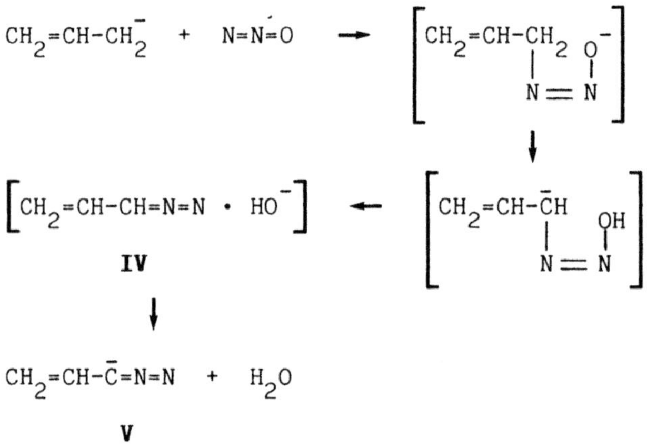

(18)

Neutral vinyldiazomethane, from which the anion could presumably be formed by proton abstraction, is not a substrate which would be easy to synthesize in the laboratory, and in any event it would be hazardous and toxic.

These reactions can be used to illustrate several points about gas phase ion synthesis. In the first place there are four overall steps in the process, if we begin with the two needed to prepare the hydroxide ion. However the overall complex transformation from the allyl anion to the diazoanion is especially interesting. A completely analogous reaction is well known in synthetic organic chemistry, and the mechanism has been tested in the gas phase by appropriate deuterium labeling. It is the last step (**IV** to **V**) which is the most important one from the standpoint of gas phase ion synthesis, for if the hydroxide ion were to depart from **IV** before removing a proton, all we would have is a synthesis of HO^-! As mentioned before, however, the product complex is relatively long-lived, and since hydroxide ion is such a strong base it almost always can find a proton on the organic molecule to abstract.

Even the best organic synthesis can give rise to side products, and this one is no exception. Along with the vinyldiazomethane anion (m/z = 67) a second product ion (m/z = 39) is observed in about 15% yield. We showed this to be the allenyl anion. A reasonable mechanism for its formation is shown in eq. 19, in which the proton on carbon 2 rather than on carbon 1 is removed by the hydroxide ion, followed by loss of nitrogen.

$$\underset{\mathbf{IV}}{\left[CH_2{=}CH{-}CH{=}N{=}N \cdot HO^-\right]} \rightarrow \left[\begin{array}{c} CH_2{=}\bar{C}{-}CH{=}N{=}N \\ \updownarrow \\ CH_2{=}C{=}CH{-}N{=}N^- \end{array}\right] + H_2O \qquad (19)$$

$$\downarrow$$

$$\underset{m/z\ 39}{CH_2{=}C{=}CH^-} + N_2$$

We took advantage of the clue provided by the formation of this minor side product to devise two new gas phase ion syntheses. In the first one (10) we simply replaced the hydrogen on the diazo carbon by another carbon, thus blocking the formation of the diazoanion and forcing the reaction completely into the channel analogous to eq. 19. A gas phase synthesis of vinyl anions resulted (eq. 20).

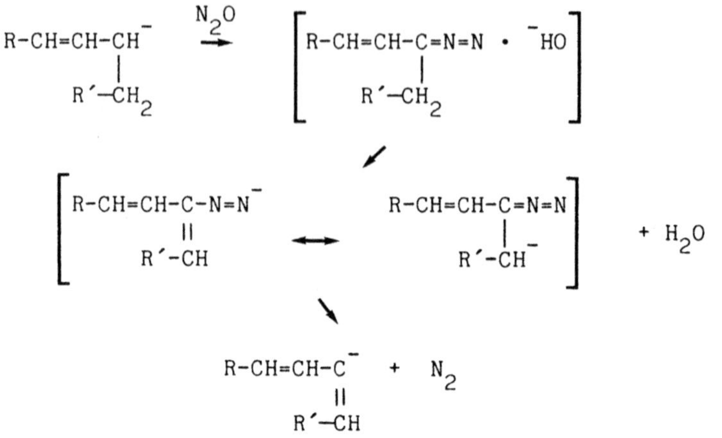

(20)

In the second synthesis (11) we generated a carbanion in which the carbon alpha to the diazo group is replaced by a nitrogen atom. This ion fragments similarly with N_2 loss to generate $CH_2{=}N^-$ (eq. 21)

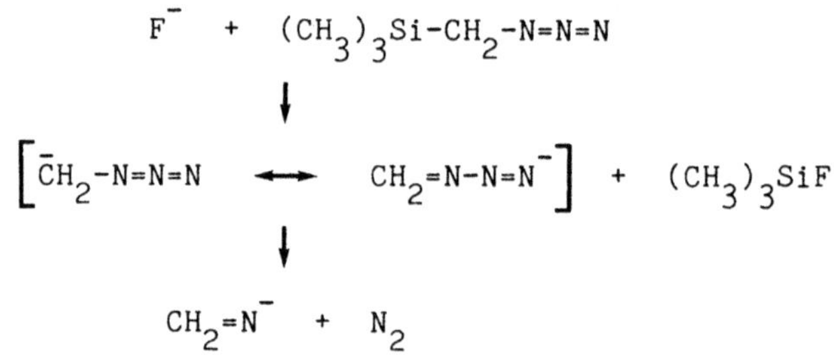

(21)

Currently there is a great deal of interest in the chemistry of solvated ions, as for example $HO^-(H_2O)$ and $HO^-(H_2O)_2$, and analogous hydrates of alkoxide ions or of alkoxides which are solvated by other alcohols. The Riveros reaction (12) is an excellent way of preparing certain solvates in an ICR (eq. 22).

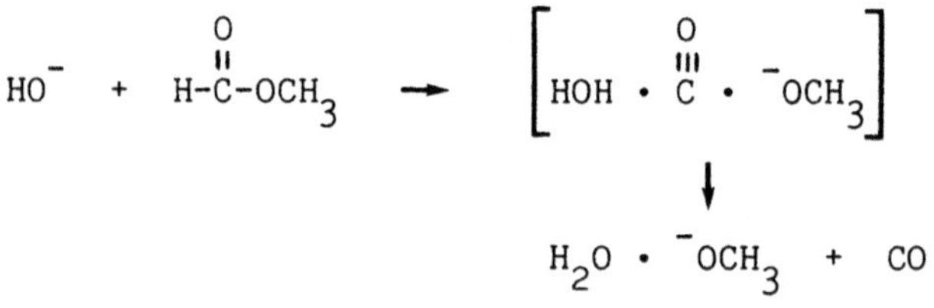

(22)

The direct hydration of an anion in the gas phase is a three-body process and hence slow, even in the FA. For example, the effective bimolecular rate constant for the formation of $HO^-(H_2O)$ at 0.5 torr He is about 1.2×10^{-12} cm^3 s^{-1}. One approach to obtaining clusters in the FA is to install a diaphragm across the flow tube to constrict the flow and create a region of high pressure (say 5 torr) in which three body reactions can occur relatively rapidly (13). We have used a chemical approach modeled in a sense after the Riveros reaction. We noted, for example, that elimination reactions of ethers with HO^- to form an alkene and an alkoxide ion are nearly thermoneutral, but the reaction becomes exothermic when the alkoxide-water cluster is the product. So, for example, reaction of diethylether with hydroxide ion (14) gives large amounts of hydrated ethoxide ion (eq. 23).

$$HO^- + CH_3CH_2OCH_2CH_3 \rightarrow [H_2O \cdot CH_2{=}CH_2 \cdot {}^-OCH_2CH_3] \downarrow H_2O \cdot {}^-OC_2H_5 + CH_2{=}CH_2 \quad (23)$$

A related, but still rather mysterious, reaction may be used to generate large amounts of hydroxide-water complexes in the FA. If a cyclic ether (tetrahydrofuran or dioxane, for example) is added to the flow tube and allowed to react with hydroxide ion, hydroxide hydrates are formed. The degree of clustering depends upon the water content of the ether. Very dry THF gives very large signals of the monohydrate; as more water is added higher clusters are formed (eq. 24).

HO^- +  $\xrightarrow{H_2O}$ $HO^- \cdot H_2O$, $HO^- \cdot (H_2O)_2$, $HO^- \cdot (H_2O)_3$ (24)

These are meant only to be examples of the way complex anions can be synthesized in the gas phase for further study of their chemistry. Many other examples could be given, both from our own laboratory and from those of other groups active in gas phase anion chemistry.

OTHER METHODS

There are some promising new methods for the generation of ions, particularly highly reactive ions, that deserve mention. With the examples we have described so far one can at least imagine methods to produce anions of almost any structure as long as they are less basic than NH_2^-. However methods for preparing more basic anions are almost non-existent, since there is no gas phase base available for carrying out proton abstraction, and the silane method of anion formation is also unsuccessful. Another complication is that even if such a strongly basic anion were to be formed, it would be sufficiently reactive to abstract a proton from its precursor, and so it would be necessary to isolate it immediately after formation.

Squires and coworkers (15) have used collision-induced dissociation in the ICR to generate the highly basic cyclopropyl anion. They

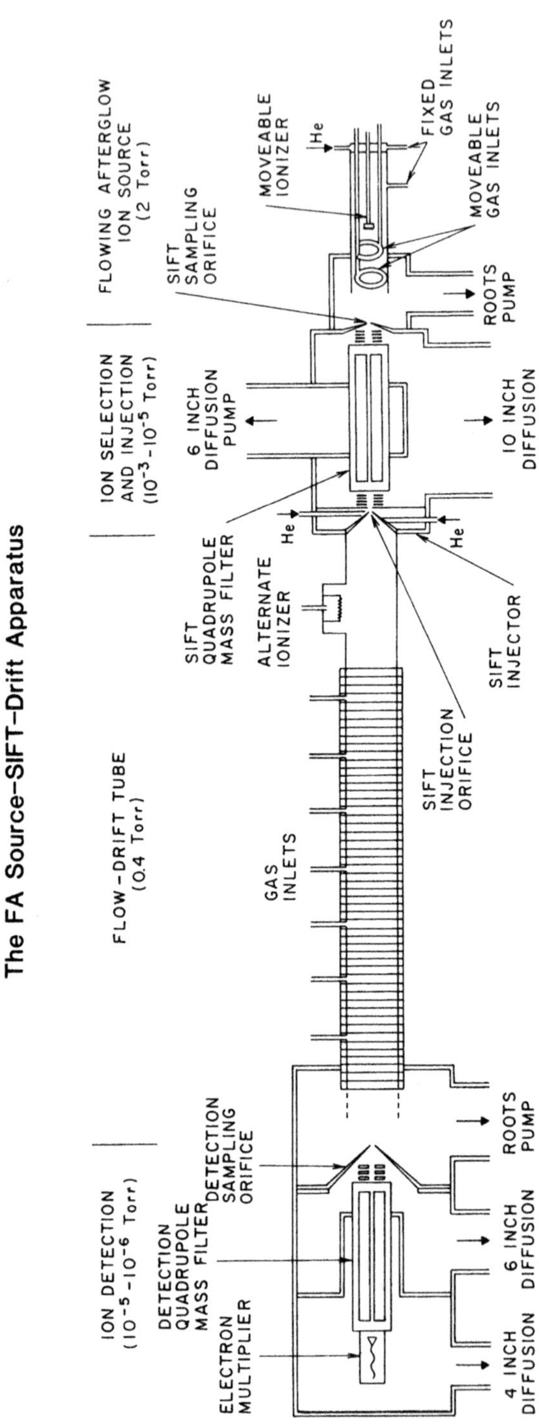

Figure 1. The selected ion flow drift tube with flowing afterglow ion source.

began with the cyclopropyl carboxylate ion and found that upon collision with argon atoms it decarboxylates, as shown in eq. 25.

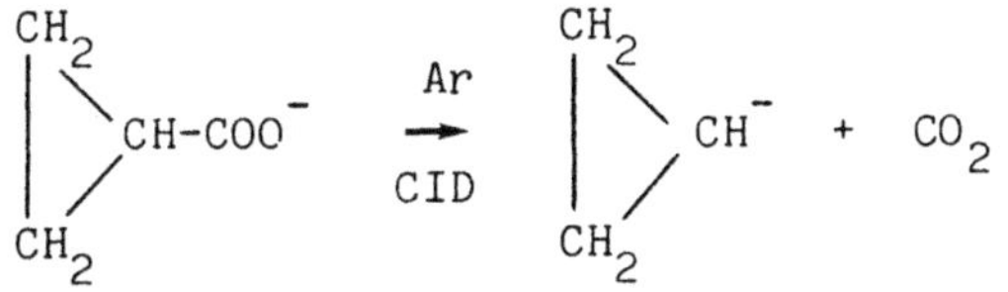

(25)

In our own laboratory we have recently constructed a flowing afterglow-selected ion flow tube (FA-SIFT) shown in Fig. 1. In this instrument ions are produced in the flowing afterglow ion source and extracted into a high vacuum region where the neutrals are removed by pumping; ions of a given mass/charge are selected with a quadrupole mass filter and injected into the second flow tube where their chemical reactions can be studied. The first flow tube is equipped with a moveable ionizer so that ionization can be achieved, if desired, very close to the nose cone. Thus highly reactive ions can be extracted before they are destroyed by their precursors. We have found that direct electron impact on methane, for example, produces small amounts of CH_2^- which can be injected into the second flow tube. In Figure 2a is shown the spectrum of all the ions produced by electron impact on methane, and in Figure 2b the spectrum of the CH_2^- injected after mass selection. HO^- arises from reaction of CH_2^- with impurity H_2O in the second flow tube.

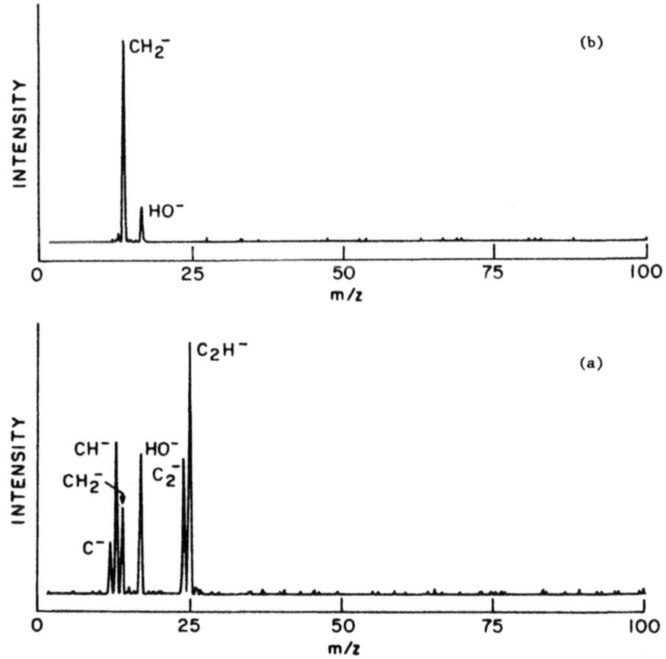

Figure 2. (a) Total ion spectrum resulting from electron impact on methane; (b) Spectrum of CH_2^- after mass selection and injection.

In a sense, then, this represents a complete circle in the methods of gas phase anion production; it may mean that the brute force method of direct electron impact can be applied to the formation of anions as well as cations.

ACKNOWLEDGMENTS

We gratefully acknowledge support of this work by the National Science Foundation, the U.S. Army Research Office and the Petroleum Research Fund, administered by the American Chemical Society.

REFERENCES

1. J.E.Bartmess and R.T.McIver, Jr. in Gas Phase Ion Chemistry; M.T.Bowers, ed.; Academic Press, New York, 1979; Vol. 2, Chapter 11.

2. R.N.McDonald, A.K.Chowdhury and P.L.Schell, J.Am.Chem.Soc. **106** 6095(1984).

3. R.R.Squires, V.M.Bierbaum, J.J.Grabowski and C.H.DePuy, J.Am.Chem.Soc., **105**, 5185(1983).

4. C.H.DePuy, V.M.Bierbaum, G.K.King and R.H.Shapiro, J.Am.Chem.Soc. **100**, 2921(1978).

5. C.H.DePuy, V.M.Bierbaum, L.A.Flippin, J.J.Grabowski, G.K.King, R.J.Schmitt and S.A. Sullivan, J.Am.Chem.Soc. **102**, 5012(1980).

6. C.H.DePuy, V.M.Bierbaum, R.Damrauer and J.A.Soderquist, J.Am.Chem.Soc. **107** 3385(1985).

7. D.B.Jacobson and B.S.Freiser, J.Am.Chem.Soc. **107** 7399(1985).

8. V.M.Bierbaum, J.J.Grabowski and C.H.DePuy, J.Phys.Chem. **88** 1389(1984).

9. V.M.Bierbaum, C.H.DePuy and R.H.Shapiro, J.Am.Chem.Soc. **99** 5800(1977).

10. S.R.Kass, J.Filley, J.M.Van Doren and C.H.DePuy, J.Am.Chem.Soc. **108** 2849(1986).

11. S.R.Kass and C.H.DePuy, J.Org.Chem. **50** 2874(1985).

12. K.Takashima and J.M.Riveros, J.Am.Chem.Soc. **100** 6128(1978).

13. D.W.Fahey, H.Bohringer, F.C.Fehsenfeld and E.E.Ferguson, J.Chem.Phys. **76** 1799(1982).

14. C.H.DePuy and V.M.Bierbaum, J.Am.Chem.Soc. **103** 5034(1981).

15. S.W.Froelicher, B.S.Freiser and R.R.Squires, J.Am.Chem.Soc. **108** 2853(1986).

PROTON TRANSFER REACTIONS OF ANIONS

Charles H. DePuy and Veronica M. Bierbaum
Department of Chemistry and Biochemistry
University of Colorado
Boulder, CO 80309-0215
U.S.A.

ABSTRACT. Proton transfer reactions are, obviously, among the most important in all of chemistry, and they play a key role in the chemistry of gas phase anions. Their primary importance is in their use in the formation of anions when a strong base, e.g. HO^-, reacts with an acid. Endothermic proton transfers are also extremely important in gas phase anion chemistry, especially that of carbanions, for these can lead to hydrogen-deuterium exchange reactions and these reactions can, in turn, give vital information about ion structure and reaction mechanism. Simple and complex H/D exchange is discussed in terms of mechanism and energetics, and examples are given of its use in deducing ion structure.

INTRODUCTION

One of the central problems in all branches of ion-molecule chemistry is that of ion structure, since all that is ordinarily measured in a mass spectrometric experiment is the mass-to-charge ratio of the ionic species. In organic chemistry, especially, an ion of a given mass may have, in principle, many isomeric structures. This problem has given rise to many mass spectrometric techniques including, for instance, collision-induced dissociation. Our main approach to the study of ion structure has been a chemical one, i.e., the devising of chemical reactions which can be carried out in the flowing afterglow (FA) which will allow us to deduce the structure of the anion under investigation (1). Of the many methods we have developed, those involving hydrogen-deuterium exchange have proven to be the most generally useful.

SIMPLE HYDROGEN-DEUTERIUM EXCHANGE

A simple example of hydrogen-deuterium exchange in gas phase ion-molecule chemistry (2) is shown in eq. 1.

P. Ausloos and S. G. Lias (eds.), Structure/Reactivity and Thermochemistry of Ions, 293–303.

$$CH_2{=}CH{-}CH_2^- \xrightarrow{D_2O} CD_2{=}CH{-}CD_2^- \qquad (1)$$

In this experiment the allyl anion is formed by proton abstraction from propylene by HO^- (formed by electron impact on N_2O/CH_4) and then D_2O is added downstream. Hydrogen-deuterium exchange occurs rapidly, with up to four deuterium atoms incorporated into the anion. One hydrogen is not exchanged, and obviously this must be the unique hydrogen on carbon 2. This result may be contrasted with those obtained for carbanions derived from longer-chain alkenes. For example, as is shown in eq. 2, when the anion derived from 1-hexene is allowed to react with D_2O all eleven hydrogens can be exchanged for deuterium!

$$CH_3CH_2CH_2CH{=}CHCH_2^- \xrightarrow{D_2O} C_6D_{11}^- \;\cdots \qquad (2)$$

We envisage these exchanges to occur in the following way (Scheme I). The anion and the exchange reagent are attracted to one another in the usual way by ion-dipole and ion-induced dipole forces until they enter a long-lived reaction complex **I** which will contain 10-20 kcal/mol excess energy. This same amount of energy must be lost on dissociation of the complex, so except for small effects due to isotopes the overall reaction is thermoneutral.

SCHEME I

$$CH_2{=}CH{-}CH_2^- + D_2O \rightarrow [CH_2{=}CH{-}CH_2^- \cdot D_2O] \ (\textbf{I})$$

$$\downarrow$$

$$[CH_2{=}CH{-}CHD^- \cdot HOD] \ (\textbf{III}) \leftarrow [CH_2{=}CH{-}CH_2D \cdot DO^-] \ (\textbf{II})$$

$$\downarrow$$

$$CH_2{=}CH{-}CHD^- + HOD \ (\textbf{IV})$$

Within this ion-dipole complex part of the energy may be "borrowed" temporarily to carry out an endothermic deuteron transfer from the D_2O to the carbanion forming a new ion-dipole complex **II**. Since propylene and water have very similar acidities (3), the reaction is only slightly endothermic in this example. However, in other cases (e.g., $^-CH_2CN$ with D_2O) exchange occurs when the reaction is as much as 19 kcal/mol endothermic. The exchange is completed by proton abstraction by DO^- to form a new complex **III** followed by separation to form the monodeuterated anion and HOD **IV.** Note that the allyl anion is a resonance hybrid with the negative charge shared by the 1- and 3-carbons (eq. 3).

$$CH_2{=}CH{-}CHD^- \longleftrightarrow {}^-CH_2{-}CH{=}CHD \qquad (3)$$

When **IV** reacts with a second molecule of D_2O it can do so at either end of the molecule. Ultimately all four hydrogens on the 1- and 3-carbons can exchange for deuterium, but the single hydrogen on carbon 2 cannot exchange because it is never on a carbon which shares the negative charge.

If the delocalized anion has more than three carbons, then all the hydrogens may be able to exchange because the double bond can migrate along the chain, allowing each carbon in turn to become allylic. This is shown in Scheme II for the anion derived by proton abstraction from 1-hexene.

SCHEME II

$$\left[CH_3CH_2CH_2CH{=}CHCH_2^- \cdot D_2O\right]$$

$$\downarrow$$

$$\left[CH_3CH_2CH_2CH{=}CHCH_2D \cdot DO^-\right] \longrightarrow CH_3CH_2CH_2CH{=}CHCHD^- + HOD$$

V **VI**

$$\downarrow$$

$$CH_3CH_2\bar{C}HCH{=}CHCH_2D + HOD$$

VII

A deuteron may be transfered from D_2O to the 1-carbon to form **V.** The resulting DO^- may then abstract a proton from either end of the double bond. It it does so from carbon 1 an unrearranged anion is formed (**VI**). But if it does so from carbon 4 a new carbanion is formed in which the carbon-carbon double bond has migrated to a new position (**VII**). Continued deuteron transfer and proton abstraction will move the double bond along the chain until each hydrogen finds itself eventually on an allylic carbon and hence exchangeable.

If an anion is too weakly basic it will not exchange with D_2O, but will do so with more acidic exchange reagents (4), for example with CH_3OD. An anion which is stabilized by conjugation with two double bonds is unaffected by D_2O (eq. 4a) but exchanges four protons readily when CH_3OD is added to the flow tube (eq. 4b).

$$CH_2{=}CH{-}\bar{C}H{-}CH{=}CH_2 \xrightarrow{D_2O} \text{No reaction} \tag{4a}$$

$$\xrightarrow{CH_3OD} CD_2{=}CH{-}\bar{C}H{-}CH{=}CD_2 \tag{4b}$$

Note in this example that exchange takes place at the ends of the conjugated system; protonation at carbon 3 would lead to a less stable unconjugated diene and is less favorable (eq. 5)

$$\left[CH_2{=}CH{-}\bar{C}H{-}CH{=}CH_2 \cdot D_2O\right] \not\rightarrow \left[CH_2{=}CH{-}CHD{-}CH{=}CH_2 \cdot DO^-\right] \tag{5}$$

For still less basic anions $(CH_3)_3COD$, CF_3CH_2OD or other alcohols may be used to bring about exchange (4).

One can see from these results that a great deal of information may be obtained about the structure of an organic anion simply by noting which reagents bring about exchange, and how many hydrogens are exchanged. For example, consider the three isomeric ions in which a phenyl group is attached to $C_3H_4^-$, namely the 1-phenylallyl anion (**VIII**), the 2-phenylallyl anion (**IX**) and the phenylcyclopropyl anion (**X**), all of m/z = 117.

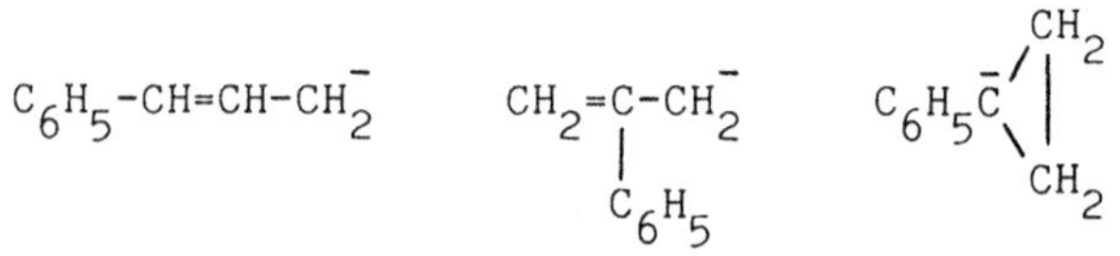

VIII **IX** **X**

These ions can readily be distinguished by hydrogen-deuterium exchange (5,6). First the response of the three ions to adding a small amount of D_2O to the flow tube was investigated; **VIII** and **X** do not exchange, while **IX** rapidly exchanges four hydrogens for deuterium. Next, CH_3OD was added; **X** was neutralized by this reagent, and a peak at m/z = 31 corresponding to CH_3O^- appeared in the spectrum. In contrast, **VIII** neither exchanged nor was neutralized. However, when the more acidic exchange reagent t-butyl alcohol O-d was used, **VIII** readily exchanged two hydrogens.

These results are those expected considering the structures of the three ions. **IX** is analogous to the allyl anion, and the allylic protons would be expected to exchange. In **VIII** the anion is conjugated both with a double bond and with a benzene ring; it should be less basic than the other two ions and exchange should not occur with D_2O but should occur with a stronger deuterated acid. In **X** there are no exchangeable hydrogens an the anion site, so no exchange should be observed; it would be expected to be neutralized by CH_3OD.

These results also shed some light on the factors which influence the rates of hydrogen-deuterium exchange. Clearly one important factor is the relative acidity of the two reagents, H_2A, whose anion HA^- is to be exchanged, and DB, the exchange reagent. The first step in the exchange process is the endothermic deuteron transfer from DB to HA^-;

if this is too endothermic there is insufficient energy in the complex to allow it to occur, and no exchange can be observed.

In considering how endothermic this initial deuteron transfer is one must be careful not to evaluate only the gas phase acidities (3) of the species involved, for these represent the acidities of the separated ions and we must instead consider ions within a complex (7). The anion from acetonitrile exchanges with D_2O; the gas phase acidity of water is 391 kcal/mol, that of acetonitrile 372 kcal/mol, for a difference of 19 kcal/mol. Yet the more weakly acidic acetylene (375 kcal/mol for a difference of 16 kcal/mol) does not exchange. However in neither case are we producing the bare DO^- ion. As eq. 6 and 7 show, in the first case we are forming deuteroxide ion complexed to acetonitrile, in the second case deuteroxide ion complexed to acetylene.

$$\left[N{\equiv}CCH_2^- \cdot D_2O\right] \rightarrow \left[N{\equiv}CCH_2D \cdot DO^-\right] \quad (6)$$

$$\left[HC{\equiv}C^- \cdot D_2O\right] \rightarrow \left[HC{\equiv}CD \cdot DO^-\right] \quad (7)$$

In contrast to acetylene, acetonitrile has a large dipole moment and a high polarizability. Therefore, a hydroxide ion complexed to acetonitrile will be easier to form than the same anion complexed to acetylene and eq. 6 is expected to be less endothermic than eq. 7.

We can see then how even this simplest type of H/D exchange can be of great use to an organic ion chemist. From a combination of the structure of the reagent which induces the exchange and the number of hydrogens which are ultimately exchanged one can learn a great deal about the ion under investigation.

MULTIPLE EXCHANGE WITHIN A SINGLE COMPLEX

When D_2O is used as the exchange reagent, and exchange is studied carefully, one can see further complications in the reaction. Specifically, ions can undergo more than one exchange during a single encounter with a D_2O molecule (8). The process is shown in Scheme III. The reaction begins as usual with the formation of a complex, then deuteron transfer to form DO^- and then proton abstraction to form the deuterated anion and HOD (Complex **XI**). This complex can dissociate, but if it is sufficiently long-lived there can be a second deuteron transfer followed by a second proton abstraction, resulting in the formation of the dideutereo anion (**XII**) in a single encounter with D_2O. We have found this to be a general phenomenon. When a series of 2-substituted allyl anions related to **IX** were studied we found that approximately 25% of the ions produced after a single encounter with a D_2O molecule contained two deuterium atoms. Since the multiple exchanges like those in Scheme III can wash deuteriums out as well as incorporate them (as well as exchange hydrogen for hydrogen and deuterium for deuterium), the amount of multiple exchange is actually much larger than it appears from this figure. Indeed for many molecules two or more transfers per encounter is the rule.

SCHEME III

$$\left[CH_2{=}CX{-}CH_2^- \cdot D_2O\right] \rightarrow \left[CH_2{=}CX{-}CH_2D \cdot DO^-\right]$$

$$\downarrow$$

$$\left[CH_2{=}CX{-}CHD_2 \cdot HO^-\right] \leftarrow \left[CH_2{=}CX{-}CHD^- \cdot HOD\right]$$

XI

$$\downarrow$$

$$\left[CH_2{=}CX{-}CD_2^- \cdot H_2O\right] \rightarrow CH_2{=}CX{-}CD_2^- + H_2O$$

XII

It is not only in these thermoneutral reactions that multiple exchanges of this type are seen; they also occur (8) in exothermic proton transfers of the type shown in eq. 8.

$$CH_2{=}CX{-}CH_3 + DO^- \rightarrow CH_2{=}CX{-}CHD^- + H_2O \qquad (8)$$

Even in the highly (19 kcal/mol) exothermic proton abstraction from 2-cyanopropene (X = CN) to DO^-, 16% of a deuterated species is seen, and in one case (X = OCH_3) more deuterated than undeuterated anion is obtained.

These results with organic anions led us to investigate simpler systems (7). Some of the exchanges we studied first are shown in eq. 9-11.

$$DO^- + NH_3 \rightarrow HO^- + NH_2D \qquad (9)$$

$$DO^- + H_2 \rightarrow HO^- + HD \qquad (10)$$

$$DO^- + C_6H_6 \rightarrow HO^- + C_6H_5D \qquad (11)$$

By analyzing the kinetics of these processes we were able to deduce the correct structure of the lowest energy form of the hydroxide ion - hydrogen cluster ($H_2O \cdot H^-$) simultaneously with and independently of the discovery of the H_3O^- ion in the ICR. More recently (9) we have examined what might be considered the most important of all proton transfer reactions, that between hydroxide ion and water in the gas phase (eq. 12).

$$HO^- + H_2O \rightleftarrows \left[HO^- \cdot H_2O\right] \qquad (12)$$

When H_2O and HO^- enter an ion-dipole complex, we wanted to know how long the complex lives before it dissociates to reactants, and how many proton transfers occur during that lifetime. To measure the first of these we determined the three-body association rate for the formation of $HO^- \cdot H_2O$. In Fig. I the bimolecular association rate is shown as a function of helium pressure.

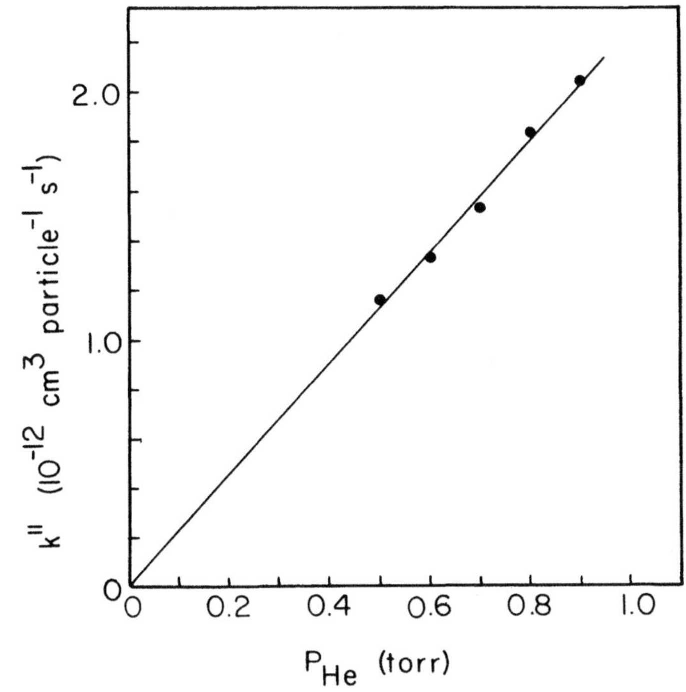

Figure 1. The apparent bimolecular rate coefficient vs. helium pressure for the three-body reaction of hydroxide ion with water in helium.

From these data one can calculate that

$$k^{III} = 7.0 \times 10^{-29} \ cm^6 particle^{-2} s^{-1}$$

and, making the standard assumptions, that the complex has a lifetime of about 400 ps.

An investigation of the extent of H/D exchange during this time involved a double labeling experiment (9). In our selected ion flow tube we prepared $D^{18}O^-$, injected it into the flowing afterglow, and allowed it to react with H_2O. As can be seen in Scheme IV, extensive but not complete exchange among the hydrogens and the deuterium occurred. We have carried out a similar study of the NH_2^-/NH_3 reaction and, in unpublished work, of HS^-/H_2S. We are currently studying hydrogen and oxygen exchange in water cluster ions, as shown in eq. 13.

$$DO^- \cdot D_2O \ + \ H_2{}^{18}O \ \longrightarrow \ products \qquad (13)$$

SCHEME IV

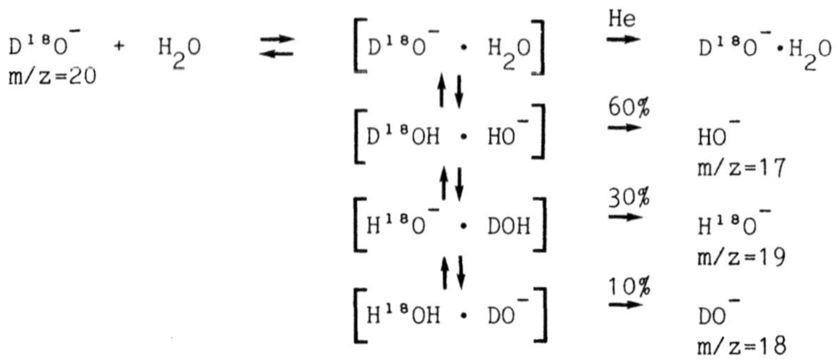

"EXTRAORDINARY EXCHANGE"

There is a second type of multiple H/D exchange within a single complex which we initially called "extraordinary exchange" to distinguish it from the type discussed above. As we mentioned, when the 2-phenylallyl anion **IX** is allowed to react with a small amount of D_2O, four hydrogens exchange rapidly. If, however, the D_2O flow is increased, additional hydrogens exchange (5). These must be hydrogens on the benzene ring, and eventually at least four more exchanges can clearly be seen (eq. 14).

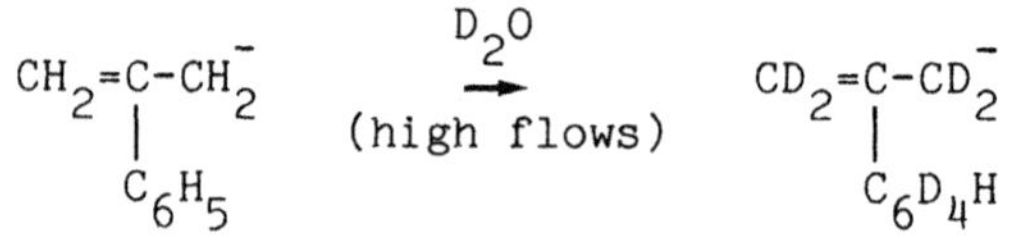

(14)

This is by no means an isolated phenomenon; with increased D_2O flows all the phenyl hydrogens in the phenylcyclopropyl anion **X** exchange and exchange of cyclopropyl hydrogens (5) also occurs. In the 2-vinylallyl anion **XIII** one can clearly see sequential exchange of the various types of hydrogens (5). First the four allyl hydrogens exchange, then, more slowly, a fifth hydrogen exchanges and finally, more slowly still, two more hydrogens exchange, so that eventually all seven hydrogens are replaced by deuterium (eq. 15).

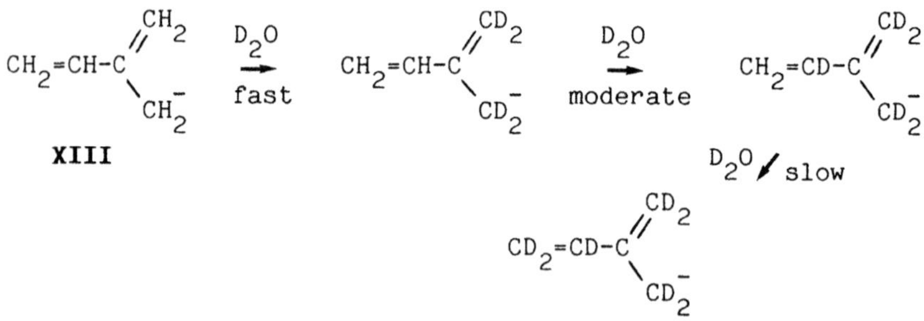

XIII

(15)

"Extraordinary exchange" can be understood by the series of reactions shown in Scheme V.

SCHEME V

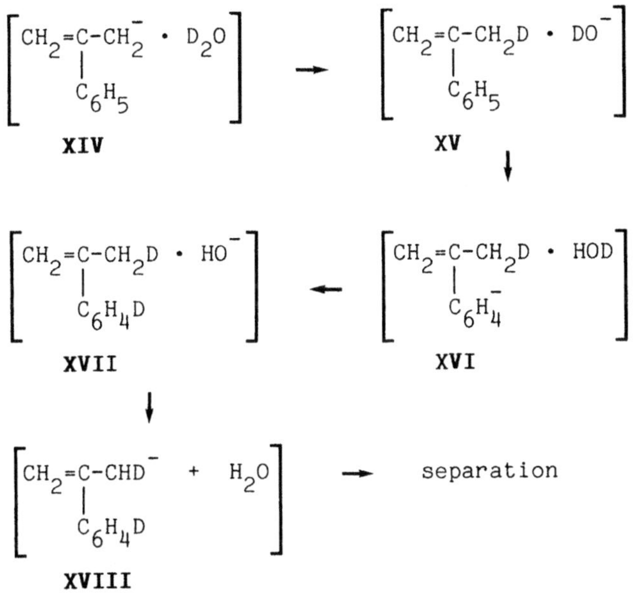

This scheme can best be understood if we consider some approximate but reasonable energetics associated with the various steps. Suppose the initial complex **XIV** between the 2-phenylallyl anion and D_2O contains 15 kcal/mol ion-dipole and ion induced-dipole energy; this same amount of energy will be required to separate the complex after exchange (ignoring small isotope effects). Suppose the initial deuteron transfer from D_2O to the anion to form **XV** is 5 kcal/mol endothermic. Then **XV** will still contain 10 kcal/mol excess energy. Phenyl hydrogens are approximately 8 kcal/mol less acidic than water; therefore there will still be sufficient energy within the complex to allow DO^- to abstract a proton from the benzene ring to form **XVI.** Neither **XV** nor **XVI** contains the necessary 15 kcal/mol energy for separation. This is regained by deuteron transfer to form **XVII** and proton transfer to form **XVIII.** By this mechanism two protons of quite different acidity are exchanged for the two deuterons in D_2O.

The phenylcyclopropyl anion **X** exchanges its phenyl hydrogens by the same mechanism, but it does so quite a bit more rapidly than does the 2-phenylallyl anion **IX** (6). We can understand this rate increase if we recognize from other evidence that **X** is a somewhat stronger base than **IX.** The initially formed complex should have the same energy, but only 2, rather than 5 kcal/mol of energy will be consumed in the initial deuteron transfer reaction. The resulting complex

(DO^-·phenylcyclopropane) will now have 13 rather than 10 kcal/mol excess energy, and proton abstraction from the benzene ring should be faster. Indeed there is sufficient energy left in this complex to abstract, occasionally, one of the cyclopropyl hydrogens and exchange it for deuterium. A similar scheme is used to understand exchange in the 2-vinylallyl anion. It is known from other evidence that the single hydrogen attached to the 1-position of the vinyl group is more acidic than the two hydrogens at the 2-position, so the former would be expected to exchange more rapidly.

In order for this type of exchange to be observed the exchange reagent must have two or more exchangeable deuterons; thus it cannot occur with deuterated alcohols. We have observed it with ND_3 (10) as shown in eq. 16.

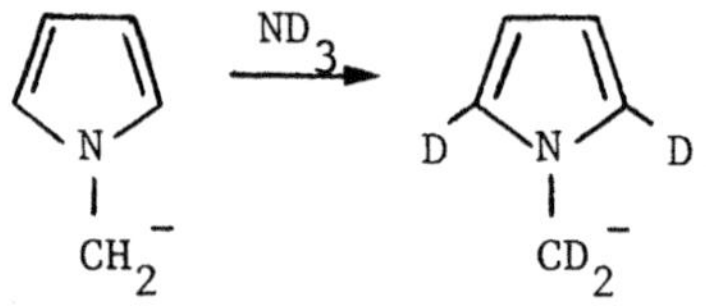

(16)

The two different types of exchange can be seen clearly when applied to the same anion. An example is the anion obtained by proton abstraction from thiophene (10). This anion exchanges a single deuterium when allowed to react with CH_3OD (17a) but all three protons are exchanged with D_2O (17b). In the first case a deuteron is added and an equally acidic proton is abstracted from the other carbon alpha to the sulfur, resulting in a single exchange. If a proton were extracted from one of the beta carbons by the CH_3O^-, this same proton would have to be returned and no exchange would occur.

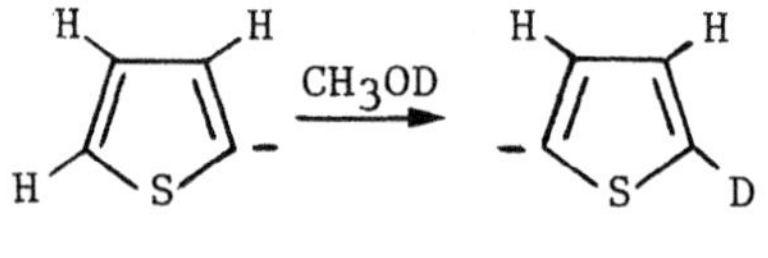

(17a)

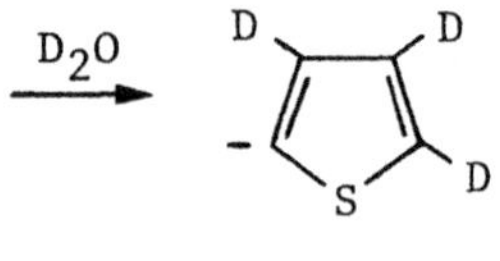

(17b)

CONCLUSIONS

It can be seen from the discussion of H/D exchange revealed by these organic examples that an ion-dipole complex is a relatively long-lived species in which a large number of reactions can be occurring before breakup. By taking advantage of the complexity of structure possible with organic compounds and the subtlety of isotopic labeling, a great deal can be revealed about these hitherto unrecognized reactions.

ACKNOWLEDGMENTS

We gratefully acknowledge support of this work by the National Science Foundation, the U. S. Army Research Office and the Petroleum Research Fund, administered by the American Chemical Society.

REFERENCES

1. C.H.DePuy and V.M.Bierbaum, Accts.Chem.Res. **14** 146(1981).

2. J.H.Stewart, R.H.Shapiro, C.H.DePuy and V.M.Bierbaum, J.Am.Chem.Soc. **99** 7650(1977).

3. J.E.Bartmess and R.T.McIver, Jr., in Gas Phase Ion Chemistry, M.T. Bowers, ed.; Academic Press: New York, 1979; Vol. 2, Chapter 11.

4. C.H.DePuy, V.M.Bierbaum, G.K.King and R.H.Shapiro, J.Am.Chem.Soc. **100** 2921(1978).

5. R.R.Squires, C.H.DePuy and V.M.Bierbaum, J.Am.Chem.Soc. **103** 4256(1981).

6. A.H.Andrist, C.H.DePuy, and R.R.Squires, J.Am.Chem.Soc., **106** 845(1984).

7. J.J.Grabowski, C.H.DePuy, and V.M.Bierbaum, J.Am.Chem.Soc., **105**, 2565(1983).

8. R.R.Squires, V.M.Bierbaum, J.J.Grabowski and C.H.DePuy, J.Am.Chem.Soc. **105** 5185(1983).

9. J.J.Grabowski, C.H.DePuy, J.M.Van Doren and V.M.Bierbaum, J.Am.Chem.Soc., **107** 7384(1985).

10. S.R.Kass, G.P.Bean and C.H.DePuy, in preparation.

ASSIGNMENT OF ABSOLUTE GAS PHASE BASICITIES OF SMALL MOLECULES

T.B. McMahon
Department of Chemistry and
Guelph-Waterloo Centre for Graduate Work in Chemistry
University of Waterloo
Waterloo, Ontario
CANADA N2L 3G1

INTRODUCTION

One of the most compelling areas of endeavour in gas phase ion chemistry over the period of the last 15 years has been the study of proton transfer reactions and gas phase basicities.[1] Proton transfer reactions, when exothermic, are usually fast, occurring at or near the collision rate. Further, mildly endothermic proton transfer reactions are also often readily observable for rate constants as low as 10^{-12} cm^3 $molecule^{-1}$ sec^{-1}. The ease of study and the importance of proton transfer as the fundamental basis of acid-base chemistry have led many investigators over the years to examine these reactions using the techniques of flowing afterflow (FA), high pressure mass spectrometry (HPMS), and ion cyclotron resonance spectroscopy (ICR). For bases B_1 and B_2 of comparable gas phase basicity a study of equilibrium proton transfer, eqn(1), may yield an equilibrium constant, eqn(2), from whch the free energy change, $\Delta G°_1$, for proton transfer in the forward direction may be obtained, eqn(3).

$$B_1H^+ + B_2 \underset{k_r}{\overset{k_f}{\rightleftharpoons}} B_2H^+ + B_1 \qquad (1)$$

$$K_1 = \frac{I_{B_2H^+}}{I_{B_1H^+}} \cdot \frac{P_{B_1}}{P_{B_2}} = \frac{k_f}{k_r} \qquad (2)$$

$$\Delta G°_1 = -RT \ln K_1 \qquad (3)$$

In a very few cases proton transfer equilibria have been examined as a function of temperature allowing a direct experimental determination of the enthalpy and entropy changes associated with the reaction, eqn(4).

$$\Delta G°_1 = \Delta H°_1 - T\Delta S°_1 \qquad (4)$$

P. Ausloos and S. G. Lias (eds.), Structure/Reactivity and Thermochemistry of Ions, 305–320.

However far more commonly the entropy changes for proton transfer have been estimated based primarily on simple statistical thermodynamic arguments and a value of ΔH°_1 thus derived. The value of ΔH°_1 is the proton affinity (PA) difference between B_1 and B_2,

$$\Delta H^\circ_1 = PA(B_1) - PA(B_2) \tag{5}$$

eqn(5) where proton affinity is defined as the enthalpy change for the deprotonation reaction, eqn(6), and is expressed in terms of enthalpies of formation, ΔH_f° of reactants and products by eqn(7). From the study of a large number of equilibria such as eqn(1)

$$BH^+ \rightarrow B + H^+ \tag{6}$$

$$PA(B) = \Delta H^\circ_6 = \Delta H^\circ_f(B) + \Delta H^\circ_f(H+) - \Delta H^\circ_f\ (BH^+) \tag{7}$$

relative basicity scales have been constructed for a wide variety of molecules from simple diatomics to complex organic, inorganic and organometallic species.

One of the overriding problems in the field of gas phase basicities however has been the assignment of absolute values to the proton affinities of molecules. In principle if ΔH°_f is known for each of the species in eqn(6) the absolute proton affinity is automatically given by eqn(7). While data for neutrals, $\Delta H^\circ_f(B)$, is usually readily available and $\Delta H^\circ_f(H^+)$ is well established, there is a general dearth of highly accurate experimental data for the protonated species, $\Delta H^\circ_f(BH^+)$. The principle source of this latter data is mass spectrometric ionization energy (IE) or appearance energy (AE) measurements. However it has become increasingly apparent in recent years that such measurements have many problems associated with them with the result that very few species BH^+ have been sufficienty rigorously examined to yield good $\Delta H^\circ_f(BH^+)$ data. It is the purpose of the present manuscript to evaulate such data as a basis for assignment of absolute proton affinities to some simple molecules.

METHODS FOR DETERMINATION OF $\Delta H^\circ_f(BH^+)$

A. APPEARANCE ENERGY MEASUREMENTS OF STABLE MOLECULES

In principle $\Delta H^\circ_f(BH^+)$ data is available from either monoenergetic electron impact or photoionization threshold measurements for production of BH^+ and a neutral, radical fragment, F, from a stable molecule, M, eqn(8). The type of data generated

$$M \xrightarrow[e^-]{h\nu \text{ or}} BH^+ + F + e^- \tag{8}$$

generally takes a form such as that in Figure 1 and one of the immediate identifiable problems is that of determining exactly where the threshold energy, Eth, is reached, corresponding to the

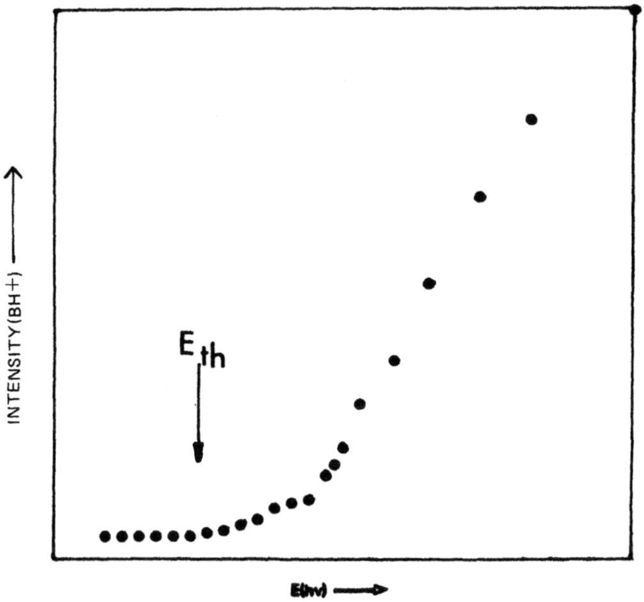

Figure 1 Typical photoionization efficiency curve.

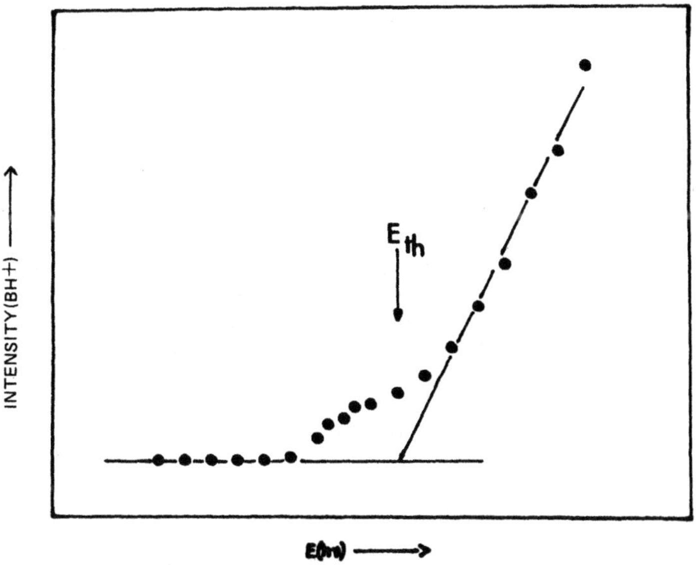

Figure 2 Blowup of the threshold region of a photoionization efficiency curve.

lowest energy possible production of BH^+. Since the amount of BH^+ produced will fall off sharply as the threshold is approached the sensitivity of the apparatus must be sufficient to detect extremely weak BH^+ signals. The final state of the products, BH^+ and F, of the ionization process must also be considered since the minimum energy at which fragmentation of M^+ can occur should leave BH^+ and F with an effective temperature of OK. As a result very early photoionization measurements frequently made the assumption that the observed threshold corresponds to eqn(8) occurring at OK. For example in an examination of fragmentation of hydrogen and methyl halides Dibeler et al[2] derived H-X and CH_3-X bond energies at OK assuming the appearance energy data to yield OK thresholds. It was recognized very early however that, despite the difficulty in doing so, the effect of the initial thermal energy of the molecule M had to be accounted for since this energy may effectively participate in the decomposition of the initially formed molecular ion, M^+.[3] The correction of the threshold experimental data for the internal energy of M is most rigorously accomplished by an examination of the appearance energy as a function of sample gas temperature;[4,5] however this procedure is almost never carried out. Instead Chupka et al[6,7] have recommended a method involving examination of the shape of the metastable peak due to decomposition of the molecular ion, M^+, in its lowest energy decomposition pathway. This curve is usually a step function with thermal energy of M producing a low energy "tail" which becomes a constant plateau at the appearance energy of the fragment ion. For example in an examination of the photoionization appearance energy of CH_2OH^+ from CH_3OH, eqn(9), Refaey and Chupka[6] applied a thermal energy correction in this way to give a value of $\Delta H^\circ_f(CH_2OH^+)$ of 170.2 kcal mol^{-1} leading to a proton affinity of CH_2O of 170.9 kcal mol^{-1}. Comparison of relative basicity data

$$CH_3OH \xrightarrow[11.67\ eV]{h\nu} CH_2OH^+ + H \qquad (9)$$

referenced to other absolute basicities suggest that this is likely an accurate value.

Chupka later suggested[3] that the effect of thermal energy participation in the fragmentation process could be deconvoluted by an extrapolation of the linear post threshold portion of the ionization curve. A blow-up of the post threshold region frequently reveals a shape such as that shown in Fig(2). The extrapolation of the linear portion is then suggested to correspond to the appearance energy threshold at the temperature of the experiment. Such data can then be used to calculate $\Delta H^\circ_{f\,298}(BH^+)$.

Traeger[8] has addressed this suggestion by carrying out a rigorous thermodynamic analysis of the fragmentation process based on the thermochemical cycle shown in Scheme I. The quantity ΔH°_R corresponds to the experimentally measured threshold energy, E_{th}, which results in fragments at a "quasi temperature", T^*, corresponding to OK internal energy, OK translational energy with

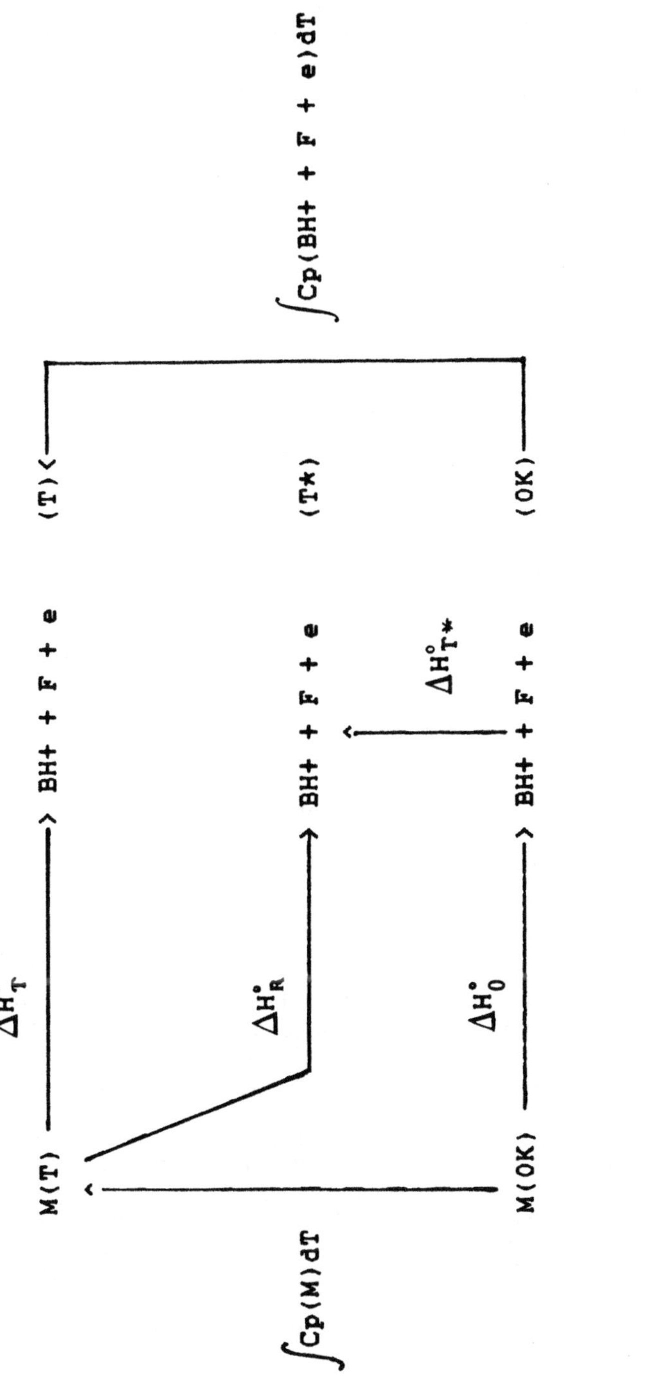

SCHEME I

respect to the centre of mass but with a translational energy of the centre of mass corresponding to a temperature T. Thus the difference in enthalpy between $BH^+ + F + e^-$ at OK and at T* is given by eqn(10), where E_i is the average internal thermal energy of M effective in dissociation. With the assumption that vibrational and rotational

$$\Delta H°_{T*} = \int C_p(M)\, dT - E_i \tag{10}$$

energy are effective in the dissociation whereas translational energy is not, E_i is given by eqn (11) and $\Delta H°_{T*}$ reduces to eqn (12). In this case the threshold

$$E_i = \int C_p^{rot}(M)dT + \int C_p^{vib}(M)dT \tag{11}$$

$$\Delta H°_{T*} = \int C_p^{trans}(M)dT \approx \frac{5}{2} RT \tag{12}$$

energy is given by eqn(13). Adopting the stationary electron

$$\Delta H°_R = \Delta H°_{fT}(BH^+) + \Delta H°_{fT}(F) + \Delta H°_{fT}(e-) - \Delta H°_{fT}(M)$$
$$- \int Cp(BH^+ + F + e^-)\, dT + \frac{5}{2} RT \tag{13}$$

convention ($\int^T Cp(e^-)dT = 0$) it is apparent that $\Delta H°\ T$ is given by eqn(14) and $\Delta H°\ fT\ (BH^+)$ can then be derived from eqn(15).

$$\Delta H°_T = \Delta H°_R + \int Cp(BH^+ + F)dT - \frac{5}{2} RT \tag{14}$$

$$\Delta H°_{fT}(BH+) = \Delta H°_R - \Delta H°_{fT}(F) + \Delta H°_{fT}(M)$$
$$+ \int Cp(BH^+ + F)dT - \frac{5}{2} RT \tag{15}$$

With the assumption that the appropriate heat capacity data are available accurate thermochemical data may be obtained for BH^+ and the absolute proton affinity of B thus derived. Traeger[8] has shown also the general validity of Chupka's suggestions for use of the linear extrapolation to determine OK thresholds and for the suggestion that internal energy of M is fully available for the dissociation process.

Despite the fact that these thermochemical considerations can successfully be applied to "simple" threshold fragmentation there are still several factors which can obscure the true thermochemical threshold leading to erroneous $\Delta H°_f(BH^+)$ values. These factors are: (i) ion-pair processes; (ii) kinetic shifts; (iii) competetive shifts and (iv) reverse activation energy. Ion pair processes are relatively rare except in the case that F is an elecronegative element or group when they may lead to an anomalously low appearance energy. Kinetic shifts involve the fact that unimolecular decomposition of M^+ may be extremely slow near threshold and thus go unobserved on the time scale of the ion residence time in the apparatus. Competetive shifts involve a similar rationale but based on the fact that a fragmentation other than the desired one occurs much faster at the threshold since its own threshold energy is lower. In each of these type of shifts an amount of energy in excess of threshold is required to make the desired unimolecular decomposition observable. Reverse activation energy is illustrated in Figure 3 and involves a barrier to the recombination of BH^+ and F to form M which dictates that an amount of energy in excess of the thermochemical threshold is required for M^+ to form BH^+ and F.

In principal each of these problems is overcome through use of the photoélectron-photoion coincidence technique (PEPICO).[9] In this experiment fragment ions BH^+ produced by photoionization are detected only in coincidence with zero energy electrons. At energies in excess of threshold those fragments formed in coincidence with zero energy electrons must then have all of the excess energy present as either internal or translational energy. Ion pair processes are unobserved since no electron is produced. Kinetic or competetive shifts may be detected by varying the ion residence times in the ion source prior to time of flight mass analysis to determine unimolecular dissociation rates of parent ions M^+. Finally reverse activation energies may be detected since some of the excess energy will appear as kinetic energy of the fragment ion which is detected in the time of flight analysis.

Thus the PEPICO technique represents a nearly ideal method for unambiguous determination of appearance energies. Unfortunately the number of systems studied of direct relevance to construction of an absoluted proton affinity scale are few.

B. IONIZATION ENERGY MEASUREMENTS OF FREE RADICALS

In many cases it is possible to produce a free radical BH by pyrolysis of some suitable precursor X, eqn(16) and to subsequently determine the ionization energy by monoenergetic electron impact,[10] photoionization[11] or by photoelectron spectrosocpy,[12] eqn(17).

$$X \xrightarrow{\Delta} BH + y\cdot \tag{16}$$

$$BH \xrightarrow[e^-]{h\nu \text{ or}} BH^+ + e^- \tag{17}$$

Most of the difficulties in relating experimental thresholds to thermochemical data found in appearance energy measurements are absent for ionization energy measurements. The most serious drawbacks of this technique are however the frequently poorly known $\Delta H^\circ_f(BH)$ values and the difficulty associated with knowing the

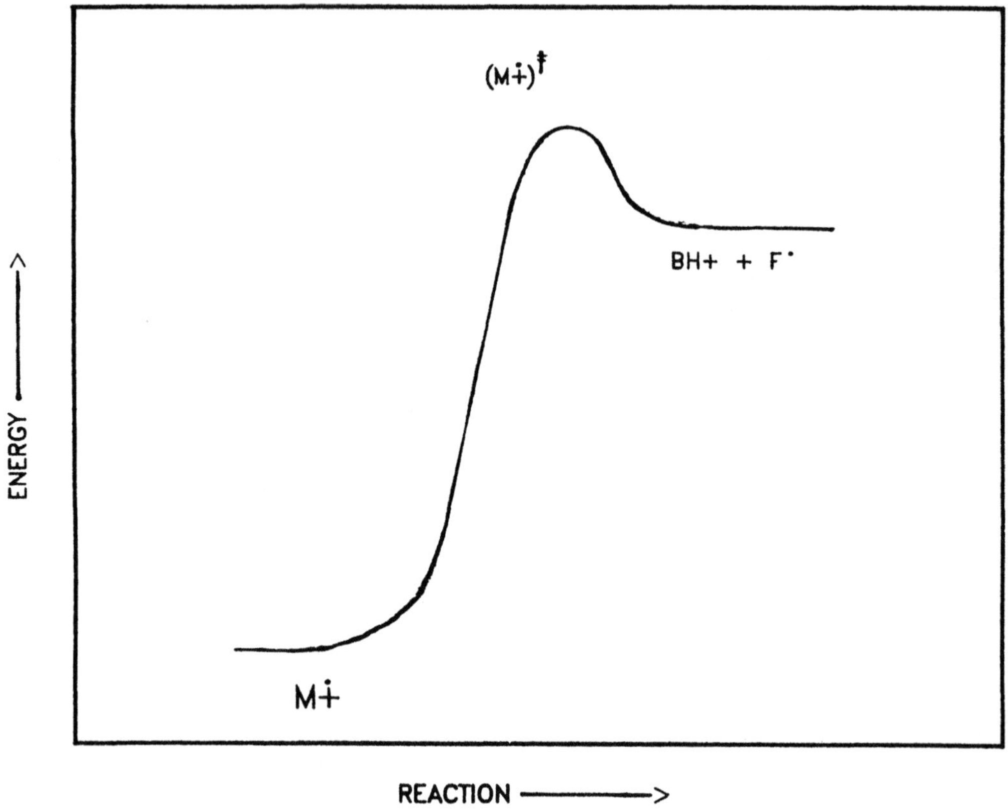

Figure 3 Potential energy surface illustrating a reverse activation energy for the process $M^+ \rightarrow BH^+ + F$

temperature of BH radicals produced either by pyrolysis, photolysis or chemical means.

C. APPEARANCE ENERGY MEASUREMENTS OF VAN DER WAALS CLUSTERS

One problem frequently confronting any determination of $\Delta H°_f(BH^+)$ from even the best technique is that BH^+ may not be readily obtainable from any stable molecular precursor, M. Thus while species such as HCO^+, CH_2OH^+ and $C_2H^+_5$ can be readily derived from CH_2O, CH_3OH and C_2H_5Cl respectively no suitable stable precusor exists for such important ions as H_3O^+, H_3S^+, NH^+_4 and H_2Cl^+. An elegant experiment devised by Lee and coworkers[13] circumvents this problem through use of a supersonic expansion to generate van der Waals dimers and higher order aggregates, eqn(18), which subsequently can be subjected to photoionization appearance energy measurements involving BH^+, eqn(19).

$$2\,B \xrightarrow[\text{expansion}]{\text{supersonic}} B_2 \tag{18}$$

$$B_2 \xrightarrow{h\nu} BH^+ + D \tag{19}$$

Since these measurements are appearance energy determinations all of the caveats described in A above apply. In addition further uncertainties involve an imprecise knowledge of the temperature of B_2 in the supersonic beam and the dimer binding energy which is necessary to determine $\Delta H°_f(B_2)$ and hence $\Delta H°_f(BH^+)$. Despite these difficulties very impressive data for H_2F^+, H_2Cl^+, H_2Br^+, H_3O^+, H_3S^+ and $NH_4{}^+$[17] have been obtained.

D. <u>AB INITIO</u> CALCULATIONS

It is rapidly becoming apparent, even to non theoreticians, that the level of sophistication of <u>ab initio</u> calculations being carried out by several groups in such that proton affinity values obtained in this manner for small molecules are comparable in accuracy, if not better than, the best experimental determinations. It must be emphasized however that only those calculations employing large basis sets, and including effects due to configuration interaction, electron correlation and basis set superposition error as well as making corrections from OK (the "temperature" of the calculation) to 298K will give reliable proton affinities.

E. DATA

Values of absolute proton affinity assigned based on both experimental and theoretical means are summarized in Table I.

TABLE 1

Absolute values of proton affinity assigned by experiment and ab initio calculation.

Species (M)	PA (expt)[a] kcal mol^{-1}	PA (ab initio)[b] kcal mol^{-1}
NH_3	202.9	204.0
iC_4H_8	195.2	---
C_3H_6	179.5	---
CH_2O	171.7	174.5
H_2S	168.7[c]	171.7[d]
H_2O	167.8	165.1
C_2H_4	162.6	---
CO	141.9	143.2;140.0[e]
HBr	141	140.3[f]
HCL	134.8	---
CO_2	130.9	130.7[g]
HF	(95)[h]	116.9
O	116.3	118.1
H_2	101.3	101.6
O_2	100.9	100.1

a) unless otherwise noted all values are taken from citations in reference 17.

b) unless otherwise noted all ab initio values are taken from ref. 32.

c) H.F. Prest, W.B. Tzeng, J.M. Brom, Jr., and C.Y. Ng J. Am. Chem. Soc. 105, 7531 (1983).

d) D.S. Marynick, K. Scanlon, R.A. Eades and D.A. Dixon, S. Phys. Chem., 85, 3364 (1981).

e) ref. 31.

f) P. Botschwina, A. Zilch, P. Rosmus, H.J. Werner and E.A. Reinsch, J. Chem. Phys. 84, 1683 (1986).

g) M.F. Frisch, H.F. Schaefer and S.J. Binkley, J. Phys. Chem. 89, 2192 (1985).

h) The authors have acknowledged the inaccuracy of the value.

Because of the apparent superiority of the PEPICO technique for unambiguous appearance energy determination and because of the concordance of values obtained for both PEPICO[14] and other photoionizations experiments[8,15,16] it is herein proposed that the values of $\Delta H°_f(C_2H_5^+)$ and $\Delta H°_f(iC_3H_7^+)$ used to determine $PA(C_2H_4)$ and $PA(C_3H_6)$ be taken as the best primary standards for anchoring relative proton affinity scales.

PROTON AFFINITY SCALES

As noted above the vast majority of proton affinity assignments have been accomplished by observation of proton transfer equilibria and referencing the resulting scales to one or more "absolute standards." Self consistency is the fundamental requirement of any final set of proton affinity assignments and hence the best way to be confident that all experimental work is free from error is for the relative proton affinities of two standards determined from proton transfer equilibria to agree with the difference in proton affinity determined by ionization energy or appearance energy measurements or by sophisticated ab initio calculations. This has been the rationale followed in the recent evaluated compilation of proton affinities[17] particularly for the basicity range above water for which several possible standards and several relative scales existed. However even with this abundance of data while the order of relative and absolute measurements agree qualitatively there is still not good quantitative agreement. For example the appearance energy determinations of $\Delta H°_f(iC_3H_7^+)$ and $\Delta H°_fC_4H_9^+)$[15,16] yield proton affinities of 179.5 kcalmol^{-1} and 195.2 kcalmol^{-1} for propene and isobutene respectively, giving a difference in basicity of 15.7 kcal mol^{-1}. By contrast the difference obtained in the basicities from proton transfer equilbrium measurements is 13.8 kcal mol^{-1} [18], 15.0 kcal mol^{-1} [19] or 12.9 kcal mol^{-1} [20] depending upon which scale of equilibrium measurements is used. Similarly based on ladders of proton transfer equilibria the difference in proton affinities of H_2O and NH_3 has been variously assigned as 31.5 kcal mol^{-1} [20], 32.0 kcal mol^{-1} [19] and 34.7 kcal mol^{-1} [17] while the photoionization measurements of van der Waal dimers place the basicity difference as 36.4 kcal mol^{-1} [13,21]. Thus it appears that there may be a general tendency for equilibrium measurements to underestimate the total difference in proton affinities over a large gap composed of many equilibria.

Another region of the gas phase basicity scale which has received less attention is that for compounds with proton affinities less than H_2O. In this region the possible absolute proton affinities derived from appearance energy measurements are C_2H_4, CO, HCl and CO_2. As noted above $\Delta H°_f(C_2H_5^+)$ appears extremely well established from PEPICO experiments yielding a proton affinity of C_2H_4 of 162.6 ± 1 kcal mol^{-1}. The proton affinity of CO has been assigned as 141.9 kcal mol^{-1} in the recent NBS compilation based on appearance energy measurements by Guyon et. al.[22] Thus the difference in "absolute" values of the proton affinities of C_2H_4 and CO is 20.7 kcal mol^{-1}. The recent proton affinity scale of McMahon and Kebarle[23] for proton transfer equilibria involving weak bases assigns a difference between C_2H_4 and CO of only 17.0 kcal mol^{-1} however. This discrepancy is much larger than either of those above for the stronger basicity scales and moreover the equilibrium experiments were done using high pressure mass spectrometry the technique which had most successfully reproduced the absolute H_2O-NH_3 difference. In order to assess the accuracy of proton affinities in this basicity range the uncertainties in both types of experimental data must be addressed. The proton transfer equilibria measurements are more difficult than those involving stronger bases because of the persistent presence of H_2O, as an impurity. Thus proton transfer to H_2O, present even at the part per thousand level, leads to a loss in signal intensity in the protonated species of interest. Despite this, measures to remove H_2O from the gas circumvented serious difficulties with the net result that individual free energy measurements may be assigned an uncertainty of ±0.2 kcal mol^{-1}. In construction of a ladder of proton affinity measurements if the ladder is ascended a single step at a time the uncertainties in relative free energies would accumlate. Considering the maximum number of steps in the ladder between CO and C_2H_4 to be 7 the cumulative error becomes ± 1.4 kcal mol^{-1}. Considering the minimum number of steps between C_2H_4 and CO as 4 the uncertainty would then be only ± 0.8 kcal mol^{-1}. However the difference between any two compounds in the ladder is obtained by many series of overlap equilibria; i.e. the difference between compounds A and B may be done in a single step, or many steps. In cases such as this the uncertainty in the difference between A and B no mattter which path is taken is rarely greater than ± 0.2 kcal mol^{-1}. The ΔG_{400} difference between C_2H_4 and CO thus remains as 17.4 ± 0.2 kcal mol^{-1} no matter which combination of steps is chosen. Considering even a maximum uncertainty in $\Delta G°_{400}$ of ± 1.0 kcal mol^{-1} and adding a further ± 1.0 kcal mol^{-1} as the largest possible error in correcting $\Delta G°_{400}$ to $\Delta H°$ through estimation of $\Delta S°$ the proton affinity of CO relative to C_2H_4 becomes 145.6 ± 3.0 kcal mol^{-1}. However these error limits are very likely excessively large.

The possibility of some error in the appearance energy determinations of $\Delta H°_f(HCO^+)$ must then also be considered. In an electron impact study of appearance energies of HCO^+ from a variety of precursors, RCHO, Haney and Franklin[23] noted varying degrees of kinetic energy release near threshold. In the 1976 NBS compilation of Gaseous Ion Energetics Rosenstock et al[24] concluded that this was due to a reverse activation energy and recommended correction of the photoionization appearance energy measurements of Warneck[25] by the amounts derived by Haney and Franklin. (Traeger has since pointed out however that translational energy measurements significantly above threshold cannot be applied as a direct experimental correction to experimental appearance energies.) If such corrections are made the resulting $\Delta H°_f$ (HCO^+) data lead to a proton affinity of CO of 145 kcal mol^{-1} in excellent agreement with the equilibrium value.

The case for a reverse activation energy in dissociation of $RCHO^+$ to HCO^+ + R is supported by recent <u>ab initio</u> calculations by

Radom[26] for the potential energy surface for CH_2O^+. These calculations predict a reverse activation barrier of 5.5 kcal mol^{-1}. If this value is applied to the appearance energy data by Guyon et al[22] a proton affinity of CO of 147 kcal mol^{-1} is obtained, again in support of the equilibrium value.

The most recent mass spectrometric study of HCO^+ system by Traeger[27] is also revealing. In a remeasurement of the appearance energies of HCO^+ from CH_2O, CH_3CHO, C_2H_5CHO, HCOCHO and HCO_2H Traeger concludes from the shapes of photoionization efficiency curves near threshold and from the wide range of values of $\Delta H°_f$ (HCO^+) that result that "there is either an excess energy associated with the threshold fragmentation or that the appearance energy assignment is affected by the threshold structure in the photoionization efficiency curve." Of the compounds examined the appearance energy of HCO^+ from HCO_2H gave the lowest $\Delta H°_f$ (HCO^+), which is also the system anticipated to give the smallest competetive shift. From these data Traeger concludes that $\Delta H°_f$ (HCO^+) derived from the HCO_2H measurement represents an upper limit to the heat of formation. The value of $\Delta H°_{f298}$ (HCO^+) obtained of 197 kcal mol^{-1} corresponds to a proton affinity of CO of 142.0 kcal mol^{-1} which then must be regarded as a lower bound. If there is an excess energy correction for this value due either to a competetive shift or reverse activation energy the proton affinity will shift towards the proton transfer equilibrium value. In addition to the <u>ab initio</u> evidence cited above support for a reverse activation barrier for $RCHO^+$ decomposition may also be found in a photoelectron spectroscopic study by Dyke.[28] In an examination of the He I photoelectron spectrum of HCO an extended vibrational series was observed corresponding to excitation of the deformation mode in the ion. It was also concluded that the adiabatic ionization energy of HCO was not observed (a rare occurrence in photoelectron spectra of free radicals) due to the very large geometry change on proceeding from bent HCO to linear HCO^+ which resulted in a negligible Franck-Condon factor for adiabatic ionization. The very large geometry change in proceeding from bent $RCHO^{+\cdot}$ to linear HCO^+ is the most likely source of a reverse activation barrier in the dissociation of $RCHO^{+\cdot}$. Unlike the apparent majority of bond breakages in molecular ions this system may then have an important barrier on the potential surface.

Note must also be taken of the various <u>ab initio</u> calculations of the proton affinity of CO. Nobes and Radom[29] carried out a MP3/6-31G** calculation of the proton affinity of CO yielding a value 143.1 kcal mol^{-1} at 298K. An isodesmic calculation for the proton affinity relative to that of H_2 of 101.3 kcal mol^{-1} yielded a value of 145.3 kcal mol^{-1}. DeFrees, McLean and Herbst[30] also carried out MP3/6-31G** calculations for the proton affinity of CO relative to H_2 to also give a value of 145.3 kcal mol^{-1} relative to H_2 as 101.3 kcal mol^{-1}. Dixon, Komornicki and Kraemer[31] employed a triple zeta polarized basis set including effects due to electron correlation to give a proton affinity of 140.0 kcal mol^{-1}. Most recently De Frees and McLean[32] have carried out calculations at the MP4SDTQ/6-311 ++ G (3df, 3pd) level including corrections for basis set superposition error to give a value of 143.2 kcal mol^{-1}.

IV CONCLUSION

It has been the intended purpose of this manuscript to initially evaluate the data at hand for assignment of absolute values of proton affinities of molecules in the gas phase. It is apparent that some disagreement remains between assignments based on appearance energies, proton transfer equilibria and ab initio calculation. With regard to the region of the basicity scale below H_2O it would be highly desirable to have PEPICO data on HCO^+ from an appropriate precursor to eliminate some of the possible sources of error in appearance energy measurements. At present in this laboratory a remeasurement of proton transfer equilibria for the region between CO and C_2H_4 is in progress including temperature variation to eliminate errors due to estimation of ΔS. It is to be hoped that within time there will be a concordance of experimental and theoretical data to allow assignment of gas phase proton affinities with confidence.

REFERENCES

1. C.R. Moylan and J.I. Brauman. Ann. Rev. Phys. Chem. 34, 187 (1983).

2. M. Krauss, J.A. Walker and V.H. Dibeler, J. Res. N.B.S. 72 A, 281 (1968).

3. W.A. Chupka, J. Chem. Phys. 54, 1936 (1971).

4. K.E. McCulloh, Int. J. Mass Spectrom. Inorg. Phys. 21, 333 (1976).

5. K.E. McCulloh and V.H. Dibeler, J. Chem. Phys. 4445 (1976).

6. K.M.A. Refaey and W.A. Chupka, J. Am. Chem. Soc. 103, 3647 (1981).

7. W.A. Chupka. J. Chem. Phys. 30, 191 (1959).

8. J.C. Taeger and R.G. McLoughlin J. Am. Chem. Soc. 103, 3647 (1981).

9. T. Baer in "Gas Phase Ion Chemistry", vol. 1, M.T. Bowers (ed) Academic Press, New York (1979).

10. F.P. Lossing and G.P. Semeluk, Can. J. Chem. 48, 955 (1970).

11. C. Lifshitz and W.A. Chupka J. Chem. Phys. 47, 3439 (1967).

12. F.A. Houle and J.L. Beauchamp J. Am. Chem. Soc. 101, 4067 (1979).

13. P.W. Tiedemann, S.L. Anderson, S.T. Ceyer, T. Hirooka, C.Y. Ng, B.H. Mahan and Y.T. Lee J. Chem. Phys. 71, 1605 (1979).

14. T. Baer J. Am. Chem. Soc. 102, 2482 (1980).

15. J.C. Traeger Int. J. Mass. Spectrom Ion Phys. 32, 309 (1980).

16. H.M. Rosenstock, R. Buff, M.A.A. Ferreira S.G. Lias, A.C. Parr, R.L. Stockbauer and J.L. Holmes J. Am. Chem. Soc. 104, 2337 (1982).

17. S.G. Lias, J.F. Liebman and R.D. Levin J. Phys. Chem. Ref. Data 13, 695 (1984).

18. S.G. Lias, D.M. Shold and P. Ausloos. J. Am. Chem. Soc. 102, 2540 (1980).

19. R. Yamdagni and P. Kebarle J. Am. Chem. Soc. 98, 1320 (1976).

20. J.F. Wolf, R.H. Staley, I. Koppel, M. Taagepera, R.T. McIver, Jr., J.L. Beauchamp and R.W. Taft. J. Am. Chem. Soc. 99, 5417 (1977).

21. S.T. Ceyer, P.W. Tiedemann, B.H. Mahan and Y.L. Lee J. Chem. Phys. 70, 14 (1979).

22. P.M. Guyon, W.A. Chupka and J. Berkowitz, J. Chem. Phys. 64, 1419 (1976).

23. T.B. McMahon and P. Kebarle J. Am. Chem. Soc. 107, 2612 (1985).

24. H.M. Rosenstock, K. Draxl, B.W. Steiner and J.T. Herron J. Phys. Chem. Ref. Data 6, (Supp 1) (1977).

25. C.S. Matthews and P. Warneck J. Chem. Phys. 51, 854 (1969).

26. W.J. Bouma, P.C. Burgers, J.L. Holmes and L. Radom J. Am. Chem. Soc. 108, 1767 (1986).

27. J.C. Traeger, Int. J. Mass Spectrom Ion Proc. 66, 271 (1985).

28. J.M. Dyke, N.B.H. Jonathan, A. Morris and M.J. Winter Mol. Phys. 39, 629 (1980).

29. R.H. Nobes and L. Radom Chem. Phys. 60, 1 (1981).

30. D.J. DeFrees, A.D. McLean and E. Herbst. Ap. J. 279, 322 (1984).

31. D.A. Dixon, A. Komornicki and W.P. Kraemer, J. Chem. Phys. 81, 3603 (1984).

32. D.J. DeFrees and A.D. McLean J. Comput. Chem. 7, 321 (1986).

KINETICS AND EQUILIBRIA OF ELECTRON TRANSFER REACTIONS: $A^- + B = A + B^-$. DETERMINATIONS OF ELECTRON AFFINITIES OF A AND B AND STABILITIES OF ADDUCTS A_2^- AND $(A{\bullet}B)^-$

Swapan Chowdhury and Paul Kebarle
Chemistry Department,
University of Alberta,
Edmonton, Canada T6G 2G2

ABSTRACT. The kinetics and equilibria of electron transfer reactions: (1) $A^- + B = A + B^-$ can be determined with a pulsed electron beam high pressure mass spectrometer. Determination of the equilibrium constant K_1 and its temperature dependence lead to ΔG_1^0, ΔH_1^0 and ΔS_1^0 values. A series of interconnected equilibria involving different A and B lead to ΔG_1^0, ΔH_1^0 and ΔS_1^0 ladders (scales) which can be converted to absolute ΔG_7^0, ΔH_7^0 and ΔS_7^0 values for the electron capture (7): $A + e = A^-$, using known values for a selected standard compound (SO_2). Measurements for over 100 compounds were performed. These are mostly benzenes, naphthalenes, anthracenes substituted with electron withdrawing substituents, benzo, naphtho and anthroquinones with various substituents, and other conjugated molecules with electron withdrawing groups. All these have low lying vacant π^* orbitals which accommodate the extra electron. A number of perfluoro compounds with low lying π^* or σ^* orbitals were studied also. The kinetics of most of the reactions (1) were also studied. The vast majority of the exothermic reactions proceeded at collision rates. Reactions involving SF_6 and perfluorinated cycloalkanes were slow with rates increasing as the exothermicity of the reaction is increased. The slow rates are believed to be due to a large geometry change between the perfluoro neutral A and negative ion A^-. The barriers due to the geometry change can be evaluated with the Marcus equations. To obtain agreement with experiment one must also consider the geometry and energy changes of the complexes $A^-{\bullet}B$ and $A{\bullet}B^-$ caused by the different bonding of A^- to B and B^- to A.

I. INTRODUCTION

The material and organization of this article follow the two lectures presented at the NATO meeting in Les Arcs, France 1986. The topic are arranged in the following order: Section II, "Electron affinities from measurement of the electron transfer equilibria eqn. (1)". This section discusses briefly the actual determinations,

P. Ausloos and S. G. Lias (eds.), Structure/Reactivity and Thermochemistry of Ions, 321–366.

compares the EA's deduced from the equilibria (1) with EA's from the

$$A + B = A + B^- \qquad (1)$$

literature and then examines the effect of molecular properties on the magnitude of the electron affinity. In section III, "Binding energies in $A^-\bullet B$ and A_2^-, first the determinations of the binding energies from the equilibria (2) and (3) are described. Then relationships between the magnitude of the binding energies and the molecular properties of

$$A^- + B = A^-\bullet B \qquad (2)$$

$$A^- + A = A_2^- \qquad (3)$$

A and B are examined. The results are found to be of significance to several areas including the kinetics of slow exothermic electron transfer reactions (1). Finally, in Section IV, "Kinetics of electron transfer reactions: $A^- + B = A + B^-$", the results and interpretation for fast and slow exothermic electron transfer reactions are presented.

II. ELECTRON AFFINITIES FROM ELECTRON TRANSFER EQUILIBRIA (1)

The methods by which electron affinities can be determined have been recently reviewed by Christophorou et al.[1] Some 31 different methods are described and available electron affinities are given in an extensive tabulation in this 200 page article. In spite of the availability of a large number of methods, the tabulation[1] demonstrates that for some molecules there is a bewildering variety of EA values, spreading at times over 1 or even 1.5 eV. Furthermore, the electron affinities of many molecules of interest had not been determined and are therefore not included in the article.[1]

Two other very useful recent reviews of electron affinities, "Photodetachment in negative ion beams" by Lineberger et al [2] and "Electron photodetachment from gas phase molecular anions" by Brauman et al [3] provide data compilations and also deal specifically with the photodetachment techniques. Electron photodetachment threshold studies represent the most important technique for electron affinity determinations. For diatomic [2] and small polyatomic molecules,[3] where the photodetachment spectrum can often be completely resolved, one obtains at high resolution, rotational and vibrational spacings from which one can deduce not only a highly accurate adiabatic electron affinitities but also rotational and vibrational constants, and thus bond distances and bond energies in the negative ion. An example of the power of the method is given in the discussion of photodetachment from OH^- and OD^- by Lineberger.[2] However, for polyatomic molecules and particularly when the negative ion contains internal excitation induced by the method used for its production (hot bands) the observed threshold curve often can not be interpreted. This is further aggravated for cases where the geometries of the neutral and the negative ion are quite different so that the vertical transitions lead to excitation of more than one normal vibration. The photodetachment spectrum in such cases is so complicated that an electron affinity determination can not be obtained.

Gas phase ion-molecule equilibria measurements, first developed and demonstrated in this laboratory [4-6] with a high ion source pressure mass spectrometer, see Fig. 1 and 2, and later also executed in ion cyclotron [7] and flowing afterglow apparatus [8] have proven to be an extraordinary prolific source of ion thermochemical data such as ion-ligand binding energies [9], proton affinities of neutrals (gas phase basicities) and negative ions (gas phase acidities) [10] and hydride and chloride affinities of carbocations.[11]

The electron affinities were measured with the pulsed electron beam, high ion source pressure mass spectrometer apparatus used in the earlier work[4-10] and discussed else where [6(ii)]. Shown in Fig. 1 is a simplified diagram of the ion-chamber, and in Fig. 2 a detailed drawing of the essential components. A typical example of experiments leading to determinations of the equilibrium constant K_1 given by eq. (4) are shown in Fig. 3. The pulsed electron ion source, see Fig. 1 in

$$A^- + B = A + B^- \qquad (1)$$

$$K_1 = \frac{[B^-][A]}{[A^-][B]} \qquad -\Delta G_1^o = RT\ln K_1 \qquad (4)$$

$$\Delta G_1^o = \Delta H_1^o - T\Delta S_1^o \qquad (5)$$

such experiments contains typically a ~5 torr of a bath gas, most often methane, and millitorr pressures of the compounds A and B which have positive electron affinities. A short (~10-20 μsec) pulse of electrons accelerated by 2000 V enters the ion source through a slit and leads to production of positive ions and secondary electrons by electron impact with the bath gas. The secondary electrons are brought to near thermal energies by collision with the bath gas. Electron capture by A and B leads to excited $(A^-)^*$ and $(B^-)^*$ which are subsequently thermalized by collisions with the bath gas. As A^- and B^- diffuse towards the walls they engage in electron transfer with neutral B and A molecules and the electron transfer equilibrium (1) is achieved under suitably selected conditions. In the ion-time dependence shown in Fig. 3a, the compound A (4-NO_2nitrobenzene) which is of lower electron affinity than B (2,3-dichloronapthoquinone) is present in much larger concentration than B. Therefore electron capture occurs predominantly by A and the A^- concentration observed at short times (0.1 msec) after the electron pulse is dominant. Rapid electron transfer to B leads to decrease of A^- and increase of B^-. Some 0.7 msec after the pulse the ratio A^- to B^- has become constant which in the logarithmic plot shown in Fig. 3 corresponds to a constant vertical distance between A^- and B^-. In Fig. 3b, the concentrations of both A and B are chosen to be significantly higher and equilibrium is established very rapidly. In general, when

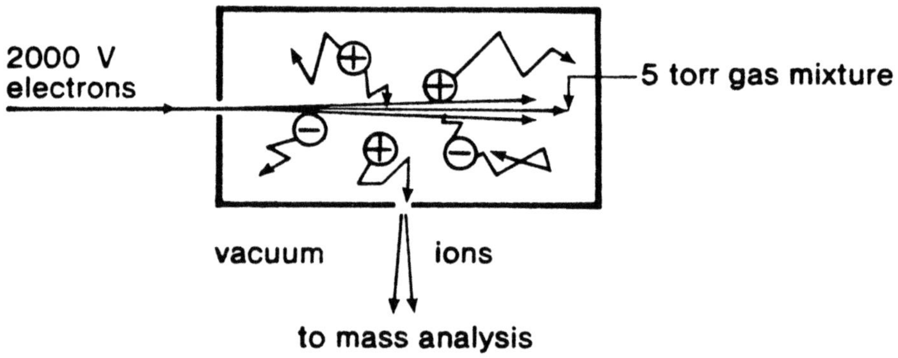

Figure 1. Ion trapping at high pressure. Ions generated by electron impact at high pressure (~5 torr) experience many thousands of collisions before reaching the wall or the small exit slit. This results in collisional quenching to thermal levels and achievement of reactive ion molecule equilibria.

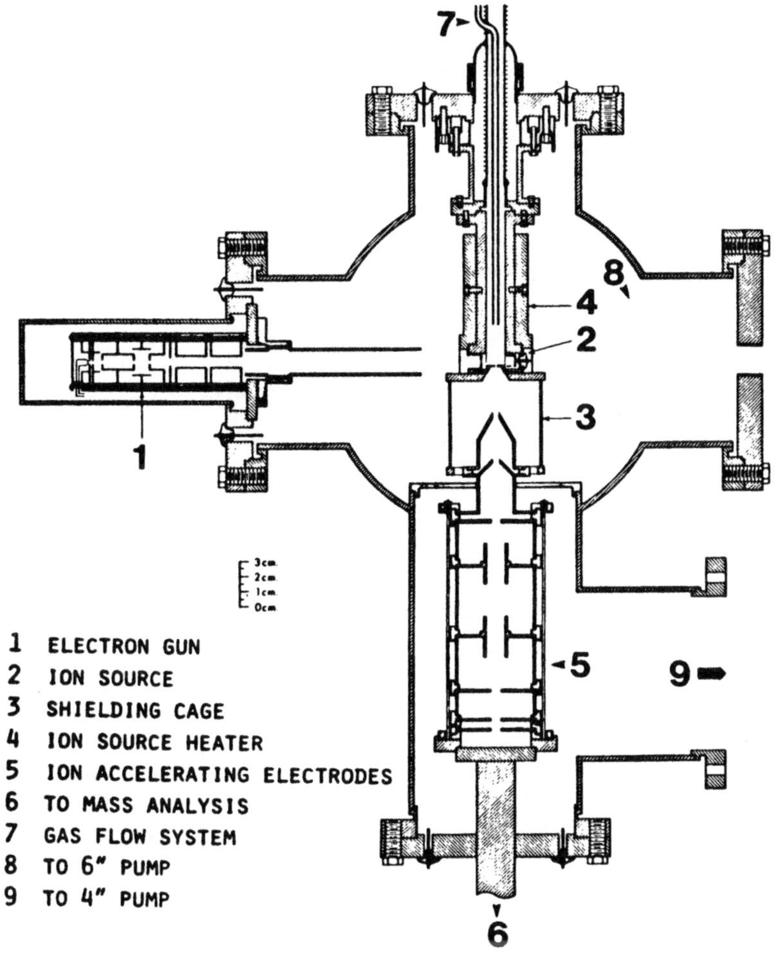

Figure 2. The apparatus

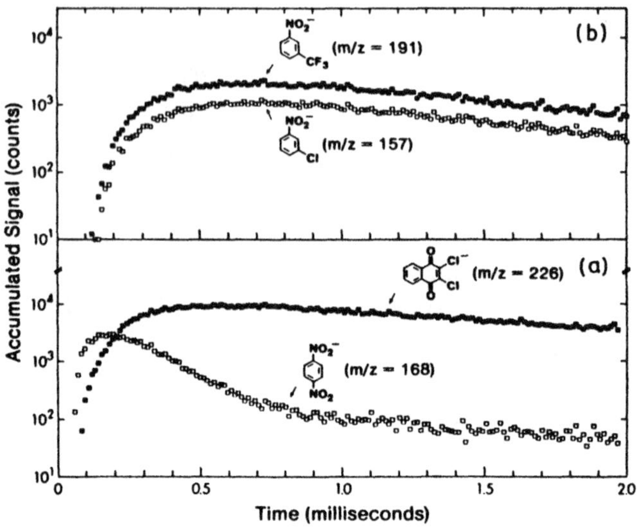

Figure 3. Ion intensities as a function of time after a 10 μsec electron pulse. Ion source temperature 150°C. Figure (a) equilibrium (1) is reached after about 0.9 msec. 2.8 torr CH_4, 0.54 mtorr p-NO_2nitro-benzene, 0.49 mtorr 2,3-dichloronaphthoquinone. Figure (b), ions have reached equilibrium (1) already at onset of ion signal. Constant $[A^-]/[B^-]$ ratio corresponds to constant vertical distance between A^- and B^-. 3.5 torr CH_4, 0.53 mtorr m-CH_3nitrobenzene, 7.6 mtorr m-Clnitrobenzene.

equilibrium measurements were performed, the concentrations of A and B were chosen to be such that equilibrium establishes rapidly, while for kinetics determinations conditions were chosen to make the logarithmic decay of A^- due to unidirectional electron transfer quite visible as in Fig. 3a.

Equilibrium constants K_1 determined from runs like those shown in Fig. 3 lead to ΔG_1^o values via eqn. 4. Separate measurements of such ΔG_1^o values using compounds A and B which interconnect lead to a free energy ladder (or scale) which is shown in Fig. 4.

The equilibrium constants K_1 for a given A, B pair can be measured at different temperatures and the enthalpy change ΔH_1^o and entropy change ΔS_1^o can be obtained from van't Hoff plots and

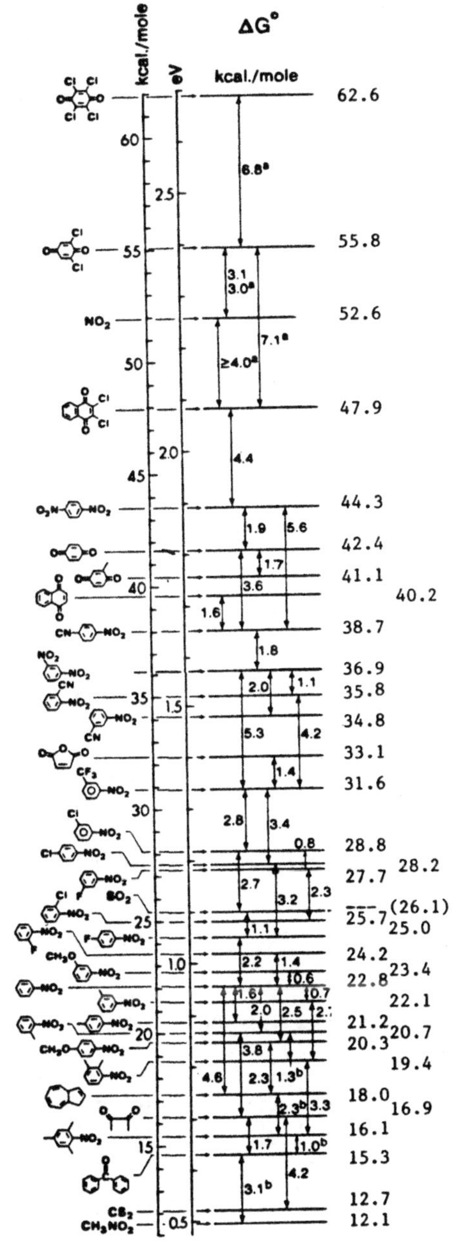

Figure 4. ΔG^{o} from equilibria (1): $A^{-} + B = A + B^{-}$ measurements. ΔG_{1}^{o} given between double arrows connecting A and B for each equilibria measured. All equilibria unless otherwise indicated measured at 150°C; (a) 215°C, (b) 80°C. Free energy changes for electron attachment, ΔG_{7}^{o} for $A + e = A^{-}$ were obtained by anchoring the scale to the electron attachment free energy,[21] $-\Delta G_{7}^{o} = 26.1$ kcal/mol for $SO_2 + e = SO_2^{-}$ obtained from electron photodetachment study due to Hall.[12]

equation (5). Representative van't Hoff plots are shown in Fig. 5. The ΔH_1^o and ΔS_1^o data deduced from such plots and connected in ΔH_1^o and ΔS_1^o ladders are shown in Fig. 6 and 7. In all, ΔG_1^o determinations involving more than 100 (hundred) compounds into a ΔG^o ladder have been performed. Since the van't Hoff plot determinations are more time consuming, van't Hoff plots for only some fifty pairs of A and B were obtained so far.

Table I

Entropy (ΔS^o) and Enthalpy (ΔH^o) change for reaction $e + SO_2 = SO_2^-$ at 423°K[a]

	SO_2	SO_2^-
O-S-O angle	120°	
	(119°)[b]	(113.8°)[b]
r_{S-O} (Å)	1.432	
	1.444[b]	1.497[b]
ν_1 (cm^{-1})	1150.5, (1361)[b]	984, (1131)[b],
ν_2	525, (597)[c]	496, (513)[b],
ν_3	1336, (1561)[c]	1040, (1232)[b],

$\Delta H^o = \Delta H_f^o(SO_2^-) - \Delta H_f^o(SO_2) = -25.3$ kcal/mol 0°K (Hall).[12]

$= -25.2$ kcal/mol 423°K

$\Delta S^o = S^o(SO_2^-) - S^o(SO_2) = \Delta S^o{}_{rot} + \Delta S^o{}_{vib} + \Delta S^o{}_{spin}$

$= 0.4 + 0.25 + 1.38 = 2$ e.u.

$\Delta G^o(423) = \Delta H^o - T\Delta S^o = -26.1$ kcal/mol.

[a] Data for SO_2 except where otherwise noted, from Herzberg, G. 1945, "Infrared and Raman Spectra" p. 337, Van Nostrand Reinhold.

[b] Hirao, K. 1985, J. Chem. Phys. **83**, 1433; Hirao, K. (Private Communication).

[c] Milligan, D.E. and Jacox, M.E. 1971, J. Chem. Phys. **55**, 1003.

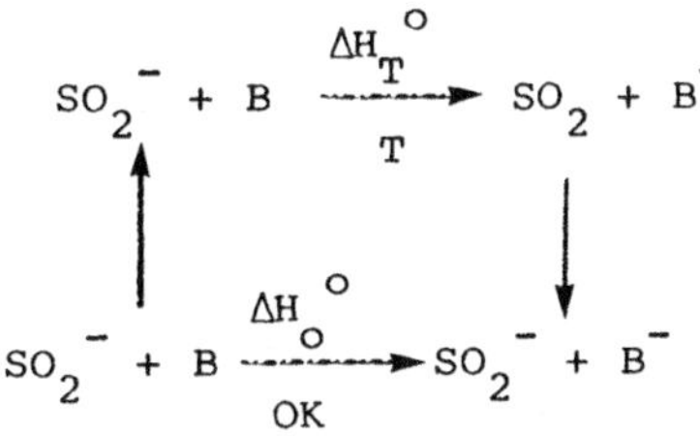

(6a)

Absolute electron affinities can be obtained by anchoring the enthalpy and free energy scales to the electron affinity of one compound whose absolute electron affinity is known from other measurements. The molecule SO_2 was selected as such a standard. The electron affinity of SO_2 has been determined by Cellota, Bennett and Hall [12] by the photodetachment method. Data for the vibrational frequencies and geometries of SO_2 and SO_2^- available in the literature, are summarized in Table I. The thermodynamic cycle (6a) shown below illustrates the relationships that can be used, to connect the enthalpy changes ΔH at 0° K which correspond to the exact adiabatic electron affinity difference to the measured ΔH_1^o values at a given temperature T (298K). To complete the cycle one requires the heat capacities of SO_2, SO_2^- and B and B^-. In general one expects that the heat capacities of the neutral and the negative ion will be quite similar such that a cancellation occurs and ΔH_o^o is within less than 1 kcal/mol of ΔH_{298}^o. The heat capacities for most of the negative ions are not available, however for SO_2, using the data from Table I one evaluates for the formal cycle (6b) only a 0.1 kcal/mol difference between 0°K and 298°K.

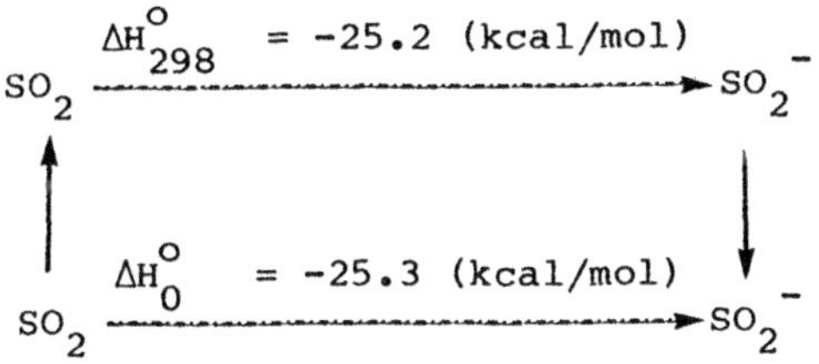

(6b)

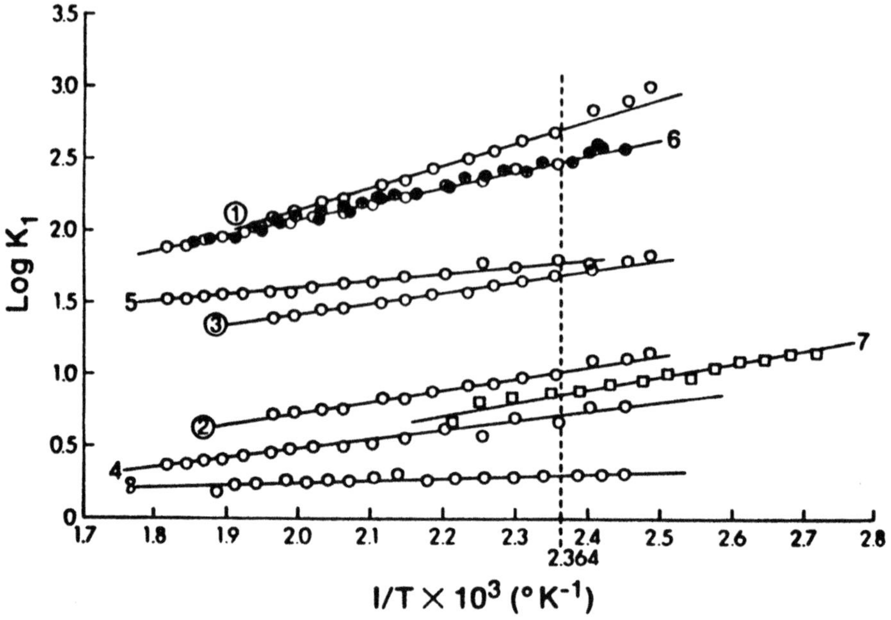

Figure 5. van't Hoff plots of equilibrium constants for reaction (1) $A^- + B = A + B^-$. Numbers beside plots identify reactions as follows: 1 (A = SO_2, B = 3-CF_3NB); 2 (A = SO_2, B = 4-ClNB); 3 (A = 4-ClNB, B = 3-CF_3NB); 4 (A = SO_2, B = 3-FNB); 5 (A = NB, B = SO_2); 6 (A = NB, B = 3-FNB); 7 (A = azulene, B = 2-CH_3NB); 8 (A = 3-FNB, B = 4-ClNB). Three sets of experimental points for reaction 6 show reproducibility obtained in three measurements separated by two months on the same instrument, performed by same operator.

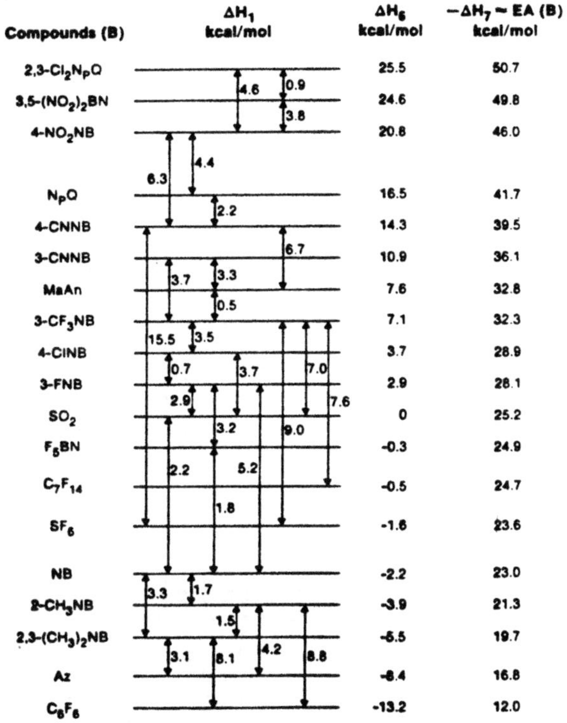

Figure 6. Enthalpy ladder for ΔH_1^o changes from van't Hoff plots. ΔH_6^o corresponds to change for reaction (6): $SO_2^- + B = SO_2 + B^-$. ΔH_7^o is the enthalpy change for reaction $e + B = B^-$ at 150°C (423K). Similar values are expected also for 298K. Stationary electron convention was used in evaluation of ΔH_7^o.

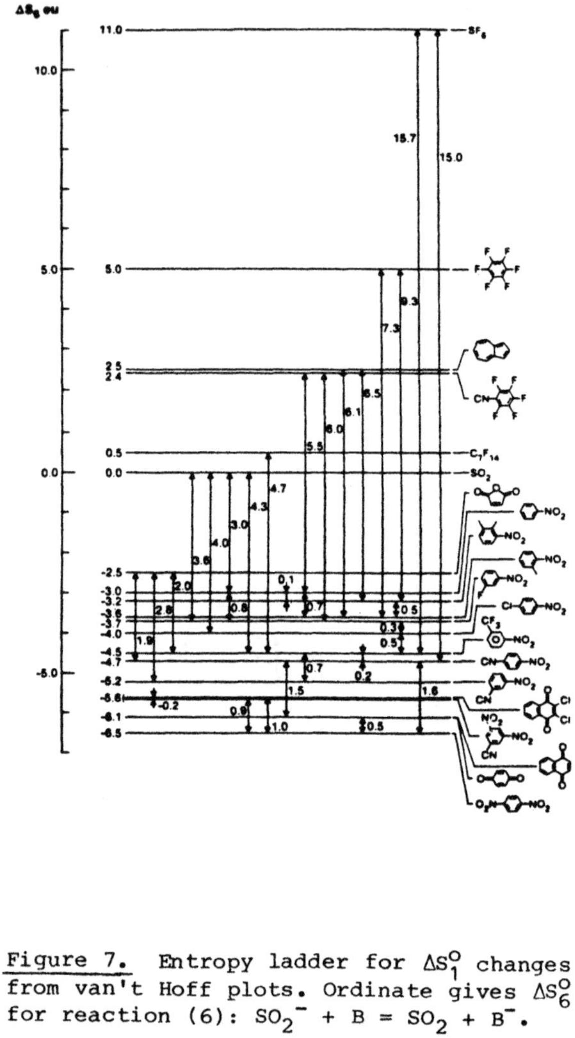

Figure 7. Entropy ladder for ΔS_1^o changes from van't Hoff plots. Ordinate gives ΔS_6^o for reaction (6): $SO_2^- + B = SO_2 + B^-$.

Absolute electron affinities obtained from the equilibria measurements (1) and anchoring to the electron affinity of SO_2 b Hall [12] are shown in Table II. Gas phase electron affinity determinations from the literature are also included in Table II. On the whole, there is a good agreement between the present determination and the literature values. Cases for which there is significant disagreement will be discussed in the next part of this section.

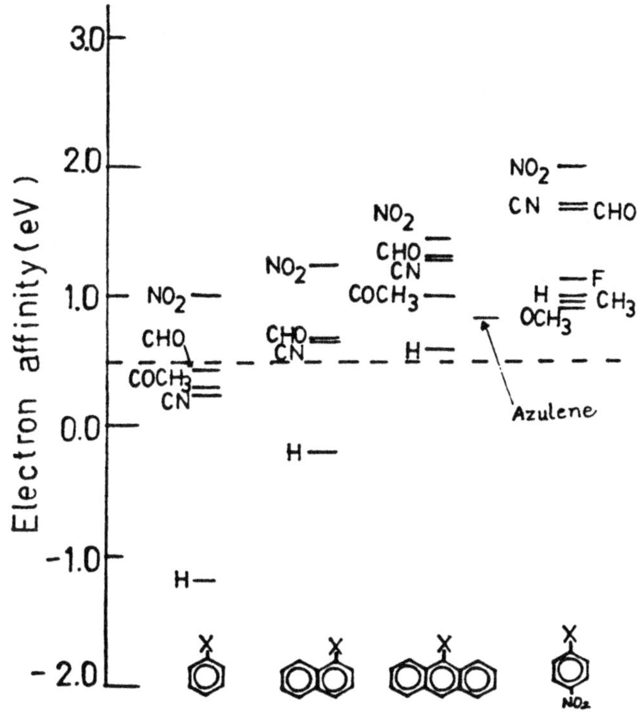

Figure 8. Substituent effects on electron affinities of 4 different classes of compounds are shown. The electron affinity values above the dotted line were obtained experimentally through electron transfer equilibria, while those under the line were obtained from: C_6H_6 and naphthalene, Jordan , benzonitrile, benzaldehyde and acetophenone, Chen.[32] The dotted line indicates that electron affinities less than 0.5 eV could not be measured by electron transfer equilibria measurements, probably the ions are lost rapidly due to thermal electron detachment under the experimental conditions.

The first determinations of electron transfer equilibria (1) were reported by Fukuda and McIver [13], who used the pulsed electron, trapped ion ion cyclotron resonance (ICR) apparatus developed by McIver.[14] However Fukuda and McIver [15] chose not to anchor their results to the absolute literature electron affinity value of a stable molecule, but selected to connect to the known EA value of the t-butoxy radical via the bracketing technique. Unfortunately, the bracketing experiments [15] led to an erroneous result, and thus to an erroneous anchoring of the McIver data.[15] The relative determinations due to McIver et al [16] anchored by us to SO_2, shown in Table II, are generally in very good agreement with the present results.

Molecular properties of singlet ground state molecules that form stable negative ions on electron capture

Examining the compounds in Table II we find that the vast majority involves molecules with conjugated electronic structures i.e. systems with relatively low lying LUMO's provided by the lowest π^* type orbital. In addition to this feature the compounds contain often also electron withdrawing groups which lead to a further lowering of the energy of the singly occupied molecular orbital, SOMO in the negative ion.

Typical examples are the aromatic compounds with electronegative substituents. The electron affinities of a representative group are shown in Fig. 8. In the unsubstituted benzene, naphthalene, anthracene as expected and predicted by Huckel molecular orbital calculations (HMO-Heilbronner [17]) expansion of the aromatic system leads to increase of the EA. Electron withdrawing substituents CN, CHO, CH_3CO, NO_2 lead to increase of the electron affinity. The substituent effects are examined in greater detail in Fig. 9 and 10. Shown in Fig. 9 is the effect of substituents X on the electron affinity of nitrobenzenes $X\text{-}C_6H_4NO_2$ versus the effect of the same substituents on the gas phase acidity of the phenols XC_6H_4OH. The increasing gas phase acidity of the phenols being represented by the change of (free) energy for the reaction (8)

$$C_6H_5O^- + X\text{-}C_6H_4OH = C_6H_5OH + X\text{-}C_6H_4O^- \tag{8}$$

The relationship observed in Fig. 9 is quite remarkable. Not only in the observed substituent order practically exactly the same but also the slope of the correlation is very close to being equal to unity i.e. the substituent effect is also nearly of exactly the same magnitude for the two systems. A plot like that shown in Fig. 9 was first published by Fukuda and McIver[16]. Yet the stabilization of the phenoxide anions occurring on deprotonation of the phenols involves delocalization of the π type orbital system while the electron affinity changes correspond to delocalization of the π^* SOMO electron. Evidently, the substituent effects closely parallel each other in spite of this difference.

A more comprehensive survey of the substituent effects is given in Figure 10, where the electron affinities of substituted benzenes, naphthalenes, anthracenes and nitrobenzenes are plotted versus the Hammet-Taft $\sigma_p^-(g)$ parameter (Taft [18] based on the gas phase phenol acidities). The electron affinities of the first member of a given series i.e. where X = H are seen to increase in the order benzene, naphthalene, anthracene, cyanobenzene, nitrobenzene, a trend already discussed on basis of Fig. 8. For each series the increase of the electron affinity due to the electron withdrawing substituent i.e. the slope ρ in Fig. 10 decreases as the electron affinity of the first member increases. This effect is not surprising. The increasing electron affinities of the first members X = H corresponds to progressively lower electron density in SOMO. Thus, in the higher

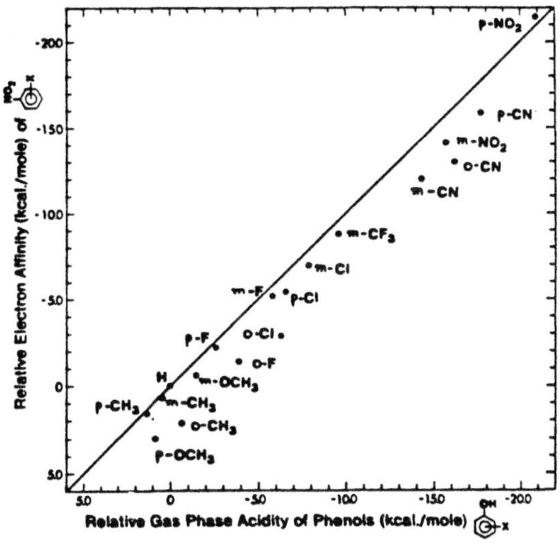

Figure 9. Comparison of relative electron affinities of substituted nitrobenzenes X-NB, EA(X-NB) - EA(NB) with relative gas phase acidities of substituted phenols X-PhOH, $D(X\text{-}PhO^-\text{-}H^+)$ - $D(PhO^-H^+)$. Near unit slope demonstrates similarity of electronic effects of substituents on the nitrobenzene radical anion and the phenoxy anion. Ortho substituents are noticeably shifted further downwards than other substituents i.e. produce a lesser stabilization of nitrobenzene anion relative to the corresponding phenoxide anion. This probably reflects decreased conjugation in nitrobenzene anions due to twisting and out of plane bending due to steric repulsion between bulky NO_2 group and ortho substituent.

electron affinity compounds the electron withdrawing X, attached to a ring carbon has less SOMO electron density to operate on and thus has a smaller substituent effect.

In addition to the above effect, there is an interesting reversal of the substituent effect between the CHO and CN group. For the low EA i.e. high SOMO density compounds, CHO leads to a significantly higher EA than CN, while for the high electron affinity compounds, the "normal" order [18], CN stronger than CHO, electron withdrawing effect is observed, see Figure 10.

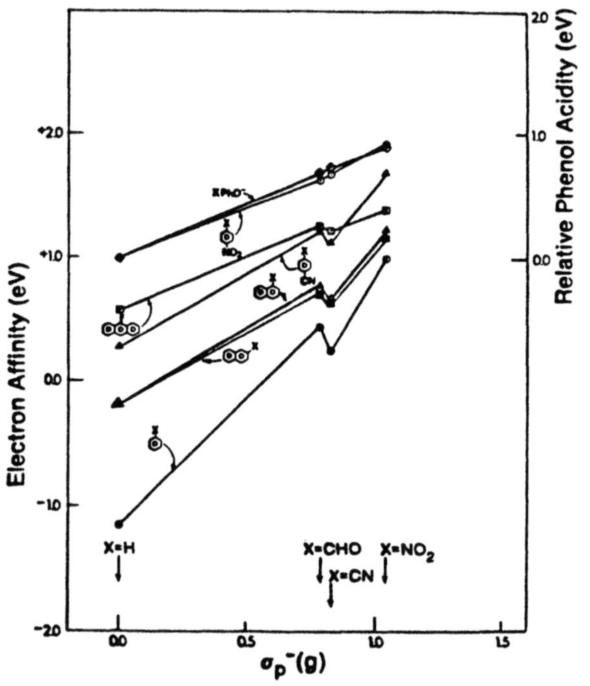

Figure 10. Electron affinities of benzenes, naphthalenes, anthracenes, cyanobenzenes and nitrobenzenes with substituents X (H, CHO, CN, NO_2), versus Hammet type substituent constants $\sigma_p^-(g)$ based on gas phase acidities phenols PhOH, Taft.[18] The free energy change for the proton transfer PhO^- + X-PhOH = PhOH + X-PhO^- is shown on the right side ordinate. Electron affinities with cross inside symbol i.e. , ⊕ due to Chen and Wentworth,[32] benzene and napthalene are due to Jordan et al 1977, Accts. Chem. Res. **11**, 341. other values are from present laboratory.

The Hückel type SOMO orbitals for the benzene negative ion are shown in scheme I. Radom[19] has performed STO-3G calculations for several singly substituted benzene radical anions. He points out that electron withdrawing substituents like CHO, CN, NO_2 will lead to preference for the $2b_{1u}$ π^* orbital as the SOMO orbital since this orbital can provide high electron density to ipso position of the substituent i.e., the $2b_{1u}$ orbital can much better provide π type electrons to feed the π electron withdrawing X and also provide nearby electron charge to interact with the dipole of X (field effect). The calculations of Radom [19] predict for the benzoanions $C_6H_5X^-$, a higher SOMO electron density on X = CHO (0.528e) relative to X = CN (0.160e). The separation of Taft and coworkers [18] of the substituent effect (Table V) [18] into resonance R and inductive-field I effects, based on phenol gas phase acidities and calculations assigns a large electron withdrawing R effect (-9.2) and a small electron withdrawing I(-6.6) effect to CHO group while for CN the opposite assignment of a small R (-4.5) and a large I (-12.1) is made. We note that the larger R effect for CHO is parallel to the Radom calculated larger SOMO π density for the CHO group of the benzaldehyde anion relative to CN in the benzonitrile anion.

The observation that the CHO substituent has a stronger stabilizing effect relative to CN in the radical anions with large SOMO π^* electron densities in the ring, and the reversal of the effect as the density is decreased, see Figure 1, indicates that the π withdrawal (R effect) is much more strongly attenuated than the

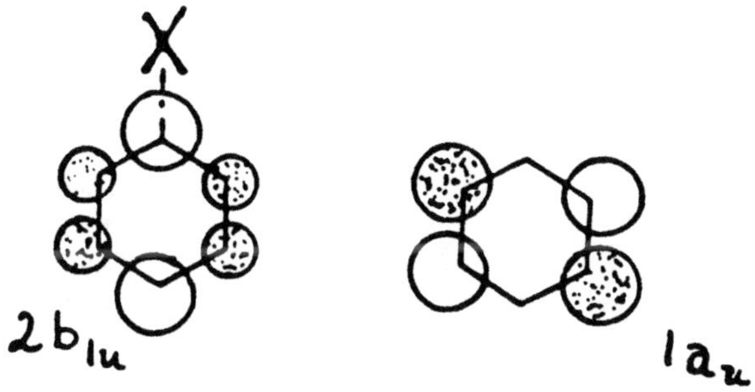

Scheme I

inductive i.e. field effect, with decrease of the electron density. Most of the data in Table 2 are from Kebarle et al [20-28] other gas phase results [29-32].

The electron affinity values given in Table II are from Kebarle et al[20-28]. A comparison is made in Table II with available gas phase experimental electron affinity.[29-39] The stationary electron convention[40-41] rather than the thermal electron convention[43] was used[21] to obtain the data in Table II.

Also given in Table II are the electron affinities of a number of quinones. The ability of these compounds to take up electrons has been long known and has been recognized to be of great importance in biological redox processes. The substituent effects evident from the data in Fig. 4 and 6 are very similar to those observed for the aromatic systems (see Fig. 8-10). It is interesting to note that the electron affinity of naphthoquinone is lower than that of benzoquinone [20-22] (see also table II). Thus while for the substituted aromatic compounds expansion of the ring system from benzene to naphthalene to anthracene leads to increase of the electron affinity, for the quinones the opposite effect is observed. Unfortunately no modern theoretical calculations on the quinone and napthoquinone radical ions seem to have been made. An HMO calculation by Heilbronner [17b] predicted decreasing electron affinities in the order: quinone, napthoquinone, anthroquinone i.e. an order in agreement to that observed in the present measurements.[22] Electron affinities deduced [32] from half wave reduction potential measurements and charge transfer spectroscopic measurements in solution also predict EA(benzoquinone) EA(naphthoquinone).

SF_6, C_6F_{12} and $C_6F_{11}CF_3$ represent an interesting group of compounds. Only the electron affinity of SF_6 has been studied extensively in the past by various methods.[35,39,45-47] These compounds have very large electron capture coefficient [39] and the negative ions are very unreactive.[39,45-46] These properties make SF_6 and perfluorocoycloalkanes important gaseous dielectrics and electron scavengers.[39,45-46] The unreactivity of SF_6^- and other perfluorocycloalkane negative ions towards electron transfer and autodetachment, makes the electron affinity determination of these compounds difficult. A number of different values ranging from 0.4 to 1.5 eV (see Table IV, Streit [35]) is available for EA of SF_6. Measurements through electron transfer equilibria from this laboratory [27] provide a value of 1.05 ± 0.1 eV = EA(SF_6), in good agreement with the result of Streit [35] (1.0 ± 0.2 eV). The entropy change for electron attachment to SF_6 and the exothermic electron transfer kinetics involving SF_6 and perfluorocycloalkanes will be discussed in separate sections later in this article. The electron affinity of $C_6F_{11}CF_3$, EA = 1.06 ± 0.1 eV was determined also.[27]

The electron affinity of hexafluorobenzene is of special interest in connection with the perfluoro effect.[49] Fluorine substitution can be expected to be strongly stabilizing to σ-type orbitals, while the combined field effect stabilization and opposing π-donation can have only a moderately net stabilizing effect. C_6F_6 forms a stable $C_6F_6^-$ ion and if the electron is in π*-type of orbital one would expect a Jahn-Teller type distortion.[19,49] This would lead to a loss of symmetry which could be detected experimentally. Two electron affinity determinations for C_6F_6 were available in literature.[42,50] The more recent one due to Lifshitz et al [42] gives $EA(C_6F_6) \geq 1.8$ eV.

Electron transfer equilibria measurements from this laboratory [28] lead to the following electron affinities for C_6F_6, and substituted perfluorobenzenes. C_6F_6 (0.52 eV), C_6F_5CN (1.1 eV), $(C_6F_5)_2$ (0.91 eV), $C_6F_5CH_3$ (0.94 eV), $C_6F_5COCF_3$ (0.94 eV), $(C_6F_5)_2CO$ (1.61 eV), 1,4-$(CN)_2C_6F_4$ (1.89 eV). The determinations for C_6F_6 and C_6F_5CN are based on the determination of the temperature dependence of K_1 and ΔH_1^o and ΔS_1^o, values are obtained from van't Hoff plots. For other compounds assumption $\Delta H_1^o \approx \Delta G_1^o$ was made. The value for $EA(C_6F_6) = 0.52$ eV measured in this laboratory, see Table II, is much lower than the $EA(C_6F_6) \geq 1.8$ eV obtained by Lifshitz et al [42] with the endothermic negative ion electron transfer method. A similar large difference is observed for perfluorotoluene, $EA(C_6F_6CF_3) = 0.94$ eV [28] (present investigation, table II) versus $EA(C_6F_5CF_3) \geq 1.7$ eV by Lifshitz et al.[42].

The electron affinities of other C_6F_5X compounds given in Table II are also very much lower than 1.8 eV [37] and are of magnitudes consistent with qualitative predictions of substituent effects (for details see Tables IV and V in reference 28). The consistency of the present result for C_6F_6 and C_6F_5X (as shown in the previous paragraph) in terms of substituent (X) effects is a very strong evidence for the reliability of the present results which makes the high values [37] almost certainly wrong. Further, $C_6F_6^-$ was observed to undergo thermal electron detachment above 150°C. Thermal electron detachment from negative ions A^- can be expected at 150°C,[27] only if the electron affinities of molecules A are much lower than 1 eV. The occurrence of electron detachment from $C_6F_6^-$ at 150°C [28] independently supports a low electron affinity value for C_6F_6.

Entropies of electron attachment reaction : $A + e^- = A^-$

By measuring the temperature dependence of K_1 for equilibria 1, ΔH_1^o and ΔS_1^o values were obtained via van't Hoff plots. These were interconnecting such that a continuous scale of ΔS_1^o and ΔH_1^o values were obtained. A ΔS^o scale is given in Fig.7. It is anchored to the ΔS^o for the reaction: $(SO_2 + e^- \rightarrow SO_2^-)$.[21] Available geometries and vibrational frequencies for SO_2 and SO_2^- (Table I) permit the evaluation of $S^o(SO_2^-) - S^o(SO_2) = \Delta S_7^o(SO_2)$. Combining this value with the scale for ΔS_1^o data, one obtains the $\Delta S^o(B^-) - S^o(B) = \Delta S_7^o(B)$ differences for the compounds B which are part of the ladder see Fig. 7 and Table II.

The substituted nitrobenzenes and quinones are observed to have negative ΔS_7^o i.e. experience a decrease of entropy on becoming negative ions. Since all the resulting negative ions are doublets and the ΔS_7^o due to the multiplicity change is +1.4 e.u, the observed overall decrease of entropy for these compounds must be due to a stiffening of vibrations in the negative ions. The $-\Delta S_7^o$ is seen to increase with increasing electron affinities of the nitrobenzenes and quinones. Increase of electron affinity is achieved by increased substitution with electron withdrawing groups. The extra electron in the nitrobenzene radical anion is in a π^* type orbital which extends over the nitro group and the aromatic ring. The presence of this electron in NB^- may lead to a stiffening of the internal rotation of the NO_2 group relative to the phenyl group. The rotational barrier in NB has been quoted as equal to 6 kcal/mol.[50] From the reduced moment of inertia one can evaluate [51] the entropy for the restricted rotation S^o(restr.) = 4.2 e.u.

The presence of an additional electron in NB^- may double or triple the internal barrier and this would lead [51] to a loss of 1 to 2 e.u in this rotation. Additional stiffening may occur in the C-N, N-O and ring bending vibrations of NB^-. These combined effects may be responsible for the observed negative ΔS^o_7 of the nitrobenzenes.

The quinones show also negative entropy changes on electron attachment. Of the thirty vibrations of benzoquinone, there are some five CO bending modes [52,53] ranging from 240 to 500 cm^{-1}. A stiffening of these modes in the presence of the electron in the benzoquinone negative ion can be expected. This could lead to a loss of a few e.u.

The entropy change for electron attachment to SF_6 is 13.0 cals/deg (see Table II). The vibrational frequencies and the symmetry of SF_6^- are unknown. The theoretical calculation of Hay [47] predicts 844 cm^{-1} for the symmetric stretch in SF_6 and 652 cm^{-1} for that in SF_6^-. This corresponds to a decrease of the frequency in the negative ion to 77%. To estimate the other frequencies in SF_6^- one can take the available [54] experimental six stretching and nine bending vibrations for SF_6 and reduce them to 77%. The vibrational entropy change ΔS^o_7(vib) ≈ +6.5 e.u. (at 150°C, the mid-temerature of the van't Hoff plot) can then be evaluated with these frequency changes. The increased moments of inertia in SF_6^- due to the larger S-F distance lead to the value ΔS^o_7(rot) = 1.1 e.u. The sum of these changes 7.6 e.u., falls short of 13 e.u., but is essentially within the experimental error limits. We conclude that the regular octahedron symmetry SF_6^- is compatible with the measured ΔS^o_1 = 13 e.u. if one assumes that the experimental value is too high by at least 5 e.u.

It is also possible that the structure of SF_6^- is different. Hay[47] assumed that the regular octahedron symmetry of SF_6 is retained in SF_6^- and that there is no Jahn-Teller distortion, but did not investigate structures of lower symmetries. A low symmetry structure resembling $SF_5..F^-$ has been proposed by Brauman.[45] This or another structure of low symmetry still remain as distinct possibilities for SF_6^-.

The entropy change for electron attachment to C_6F_6, was found to be 7.4 e.u. (Table II). Subtracting 1.4 e.u. caused by the change of multiplicity (singlet to doublet) [21] there remain ~6 e.u. This change could be due to loosening of the vibrational frequencies in $C_6F_6^-$ and/or to a decrease of symmetry from C_6F_6 to $C_6F_6^-$. It was demonstrated that [28] electron transfer to C_6F_6 occurs with near 100% collision efficiency even when the reaction is near thermoneutral. This result indicates that the energy difference between the $C_6F_6^-$ obtained by a vertical transition from C_6F_6 and the ground state geometry of the $C_6F_6^-$ is not large.[27] Since a significant softening of the vibrational frequencies in $C_6F_6^-$ would be expected to go together with a large energy difference, it is more likely, particularly considering the high symmetry of C_6F_6, symmetry number $\sigma = 12$, that the ~6 e.u. are largeley due to a loss of symmetry in $C_6F_6^-$. A change from $\sigma = 12$ to $\sigma = 1$ corresponds to a $\Delta S^o_{rot\ sym} = 5$ e.u. Thus, probably a reduction of symmetry to $\sigma = 1$ or 2 occurs in $C_6F_6^-$. This result is consistent with the lower symmetry of the theoretically predicted geometries for $C_6F_6^-$, see for example Shchegoleva.[54]

III. BINDING ENERGIES IN $A^-\bullet Sl$, $A^-\bullet B$ AND A_2^- ADDUCTS

a. Bonding of charge delocalized ions to protic and aprotic solvent molecules and comparison with that in the bulk solution.

The bond energy of $A^-\bullet Sl$ where Sl are protic or dipolar aprotic solvent molecules was determined by measuring the equilibria (8).

$$A^- + Sl = A^-\bullet Sl \qquad (8)$$

The equilibrium constants K_8 are evaluated from the ion intensities after the ion ratio $A^-\bullet Sl/A^-$ has been constant. ΔG^o values are then obtained from equation (9).

$$K_8 = \frac{[A^-\bullet Sl]}{[A^-][Sl]} \qquad \Delta G^o_8 = -RT\ln K_8 \qquad \Delta G^o_8 = \Delta H^o_8 - T\Delta S^o_8 \qquad (9)$$

ΔG^o_8 values were determined for some 30 different A^- ions, mostly substituted nitrobenzenes, XNB^-, some substituted benzoquinones and napthoquinones, azulene and meleic anhydride, bonding to the solvent molecules: methanol, acetonitrile, dimethyl sulfoxide, dimethylformamide and tetrahydrofuran.[55] The temperature dependence of the equilibrium constants K_8 for some of the equilibria were determined also and ΔH^o_8 and ΔS^o_8 values are obtained from the van't Hoff plots.[55].

A plot of the binding energies in A^--Sl versus the electron affinities of A is shown in Fig. 11. It was found [56] that (see Table I [55]) ΔG^o_8 for Sl = MeCN is generally a litte (~0.5 kcal/mol) more exothermic than that for DMF and that for Me_2SO is more exothermic by ~3 kcal/mol than that for MeCN. On the other hand, the $-\Delta G^o_8$ values for the protic MeOH do not maintain a constant difference with the aprotic Sl, i.e. for methanol the exothermicity decreases faster with increasing electron affinity of A than is the case for the aprotic Sl. The relationships are illustrated in Figure 11, from which the data for DMF are omitted to reduce crowding in the figure. Examining the figure one finds also the interesting result that the binding

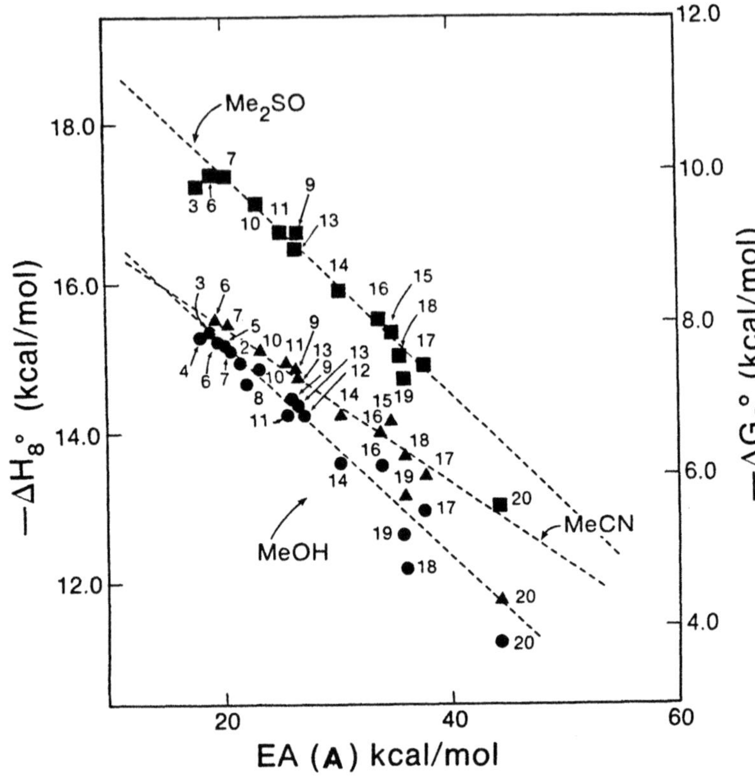

Figure 11. Plot of ΔG_8^o at 70°C and ΔH_8^o for reaction $A^- + Sl = A^-{\bullet}Sl$ versus electron affinity of A. The three plots are for Sl = dimethylsulfoxide ; acetonitrile ; and methanol . The actual A^- are identified by the number. All A^- used in the plot were substituted nitrobenzenes XNB; substituent X is; 2,3 diCH_3 (#1); 3-OCH_3 (#2); 4-OCH_3 (#3); 2-CH_3 (#4); 3-CH_3 (#5); 4-CH_3 (#6); H (#7); 2-F (#8); 3-F (#9); 4-F (#10); 2-Cl (#11); 3-Cl (#12); 4-Cl (#13); 3-CF_3 (#14); 2-CN (#15); 3-CN (#16); 4-CN (#17); 2-NO_2 (#18); 3-NO_2 (#19); 4-NO_2 (#20). The ΔH_8^o values are approximate and were obtained by assuming that ΔS_2^o = −25 cal/mol degree, which is an average of the experimental values for ΔS_8^o, (see ref. 55).

energies, − ΔG_8^o, decrease approximately linearly with increasing electron affinity of A.

The solvation energies of the ions A^-, $\Delta G_s^o(A^-)$ in a given liquid solvent $\Delta G_s^o(A^-)$ can be obtained by a comparison of the gas phase electron affinities of A with polarographic halfwave reduction potentials $\varepsilon_{1/2}$. The energy released in the reduction increases with the electron affinity of A. However due to the presence of the solvent also the solvation energies of A and A^- influence the energy change [32(i)], as shown in equations (10) and (11).

$$\varepsilon_{1/2} + C_1 = EA - \delta\Delta G_s^o \tag{10}$$

$$\delta\Delta G_s^o = \Delta G_s^o(A^-) - \Delta G_s^o(A) \tag{11}$$

$$\varepsilon_{1/2} + C_2 \approx EA \tag{12}$$

The constant C_1 contains the voltage of the standard cell coupled to the dropping mercury electrode. Chen and Wentworth,[32] assuming that $\delta\Delta G_s^o$ does not change as A is changed, simplified (10) to (12). Calibrating (12) to some known gas phase electron affinities, they determined the value of the constant C_2 and then used (12) to evaluate "gas phase" EA's from experimental $\varepsilon_{1/2}$. A plot of these solution derived "gas phase" EA's (solvent = MeCN) versus gas phase EA's obtained in this laboratory,[20,21] is shown in Figure 12 (see also Figure 6 Grimsrud [20]). Had the $\delta\Delta G_s^o$ been constant, the plot would have led to a slope equal to unity. The observed slope is 0.72. Since an approximately straight line is observed, the $-\delta\Delta G_s^o$ must be decreasing approximately linearly with increasing electron affinity of A, as shown in equation (13a)

$$-\delta\Delta G_s^o \approx C_3 - C_4\, EA(A) \tag{13a}$$

$$-\Delta G_s^o(A^-) \approx C_5 - C_4\, EA(A) \tag{13b}$$

$$(C_4 = 0.28 \text{ for Sl = MeCN})$$

It is known that $-\Delta G_s^o(A^-)$ is much bigger than $-\Delta G_s^o(A)$, therefore the changes of $\Delta G_s^o(A)$ as A is changed will not have a large effect on $\delta\Delta G_s^o$. This assumption leads to equation (13b) which predicts a linear decrease of $-\Delta G_s^o(A^-)$ with increase of EA(A). Recalling the result in Figure 11 we note that both the binding free energy for A^--Sl and the solvation energy $-\Delta G_s^o(A^-)$ in Sl decrease approximately linearly with increasing EA(A).

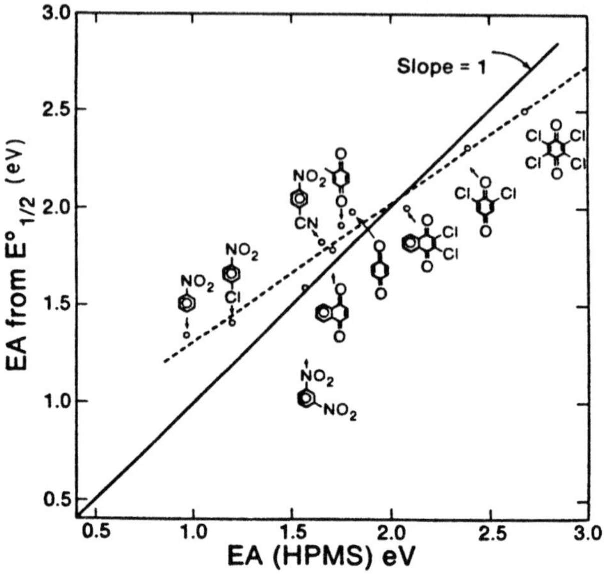

Figure 12. Plot of "gas phase" electron affinities deduced from polarographic half wave potentials $\varepsilon^o_{1/2}$ in acetonitrile versus gas phase electron affinities from Grimsrud.[20,21] Slope observed is less than unity. This shows that solvation energy of anions A^-, $-\Delta G^o(A^-)$, decreases with increasing electron affinity of A.

Shown in Fig. 13 is the plot of $\varepsilon_{1/2}$ measured in different solvents against electron affinities of a group of aldehydes. As can be seen clearly, the slopes are not only less than one but also they are different for measurements in different solvents. The slope for measurements in protic solvents are lower than that for dipolar aprotic solvent, DMF. The above observation is in line with the binding energy pattern shown in Fig. 11 for one molecule solvation and also the solvation energies predicted from Fig. 12 in bulk solution.

The above results can be explained qualitatively for the A^--Sl binding energies, i.e. for the one molecule solvation of the ion. An electron affinity increase in compounds M is generally obtained by adding electron withdrawing groups to the aromatic or conjugated system. The effect of these groups is to delocalize the negative charge. A particularly simple case is nitrobenzene and 4-dinitrobenzene. The negative ions are shown in Scheme II. The introduction of the second nitro group increases the electron affinity, delocalizes the charge and thus increases the effective radius between the solvent molecule and the effective centre of negative charge on the ion. An increase of the radius necessarily leads to a decrease of the bond energy, since the bond is largely due to electrostatic type interactions.

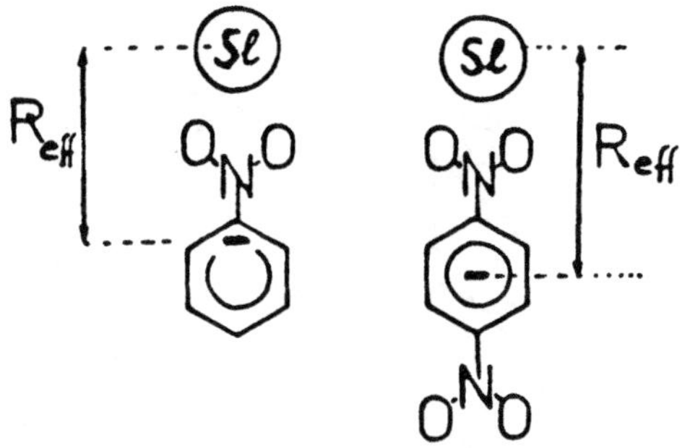

Scheme II

The greater decrease of the A^{-}-Sl bond energy with increase of EA(A), i.e. increase of effective radius R_{eff}, for the protic solvent MeOH relative to the dipolar aprotic MeCN, See Figure 11, is also easily rationalized in qualitative terms. While the molecular dipole in MeOH (1.7 Debye) [56] is largely located on the OH group and the smallness of the hydrogen allows a close approach to the negative centre (hydrogen bonding), the molecular dipole in MeCN (4 Debye) [56] is due to a positive charge distribution which is diffuse, with most of the charge on the cyanocarbon and some on the methyl hydrogens. In effect this locates most of the dipole on the cyano group, and this group cannot be approached closely by the negative ion as shown in Scheme III. It is obvious that an increase of effective radius occurring for the change from NB^{-} to $4\text{-}NO_2NB^{-}$ will cause a relatively smaller decrease of the electrostatic bonding interaction with MeCN where R_{eff} is already large relative to MeOH whose R_{eff} is smaller.[55]

b. Binding energies in $A^{-}\cdot B$ and A_2^{-}

Binding energies of charge delocalized radical anions, A^{-} (where $A = NO_2$, substituted nitrobenzenes, quinones etc) with same neutral molecules, A or different molecules, B were obtained by measuring the equilibrium constant K_{14} as outlined in section IIIa.

$$A^{-} + A = A_2^{-} \qquad (14)$$

$$A^- + B = A^- \cdot B$$

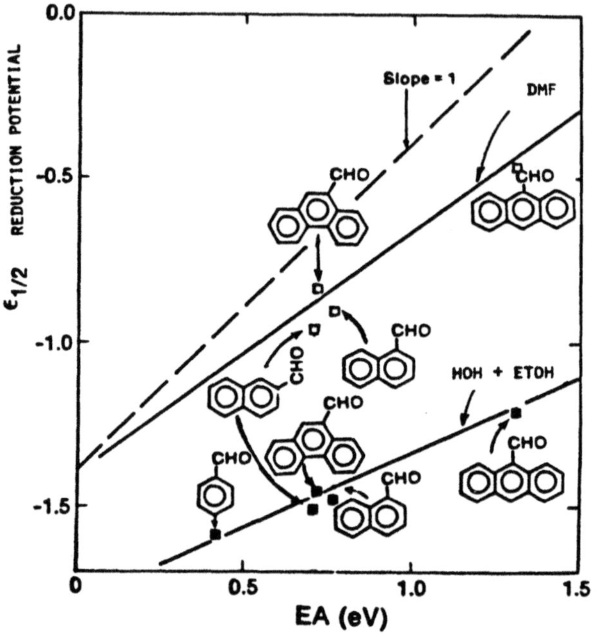

Figure 13. Plot of $\varepsilon_{1/2}$, half wave reduction potential against the electron affinities measured in this laboratory for some selected aldehydes. The slopes are lower than one. Slopes obtained for measurements in protic solvents is much lower (than unity) than that in dipolar aprotic solvent.

Shown in Figure 14 is the plot of binding energies of NO_2^- to A versus the electron affinities of A. One observes a good correlation i.e. an approximate proportionality in the increase of binding energies $NO_2^- \cdot A$ with increase of electron affinities A. . The stabilities of the $X\text{-}NB \cdot NO_2^-$ adducts (Fig. 14) are in the order X: NO_2, CN, CF_3, F ≈ Cl, H, CH_3. This is the same order as is observed for the electron affinities of X-NB. The relative NO_2^- binding abilities of the o-, m-, p- isomers of X-NB also show some interesting regularities. For the strongly stabilizing substituents X = NO_2, CN the observed stability order is ortho > meta >> para.

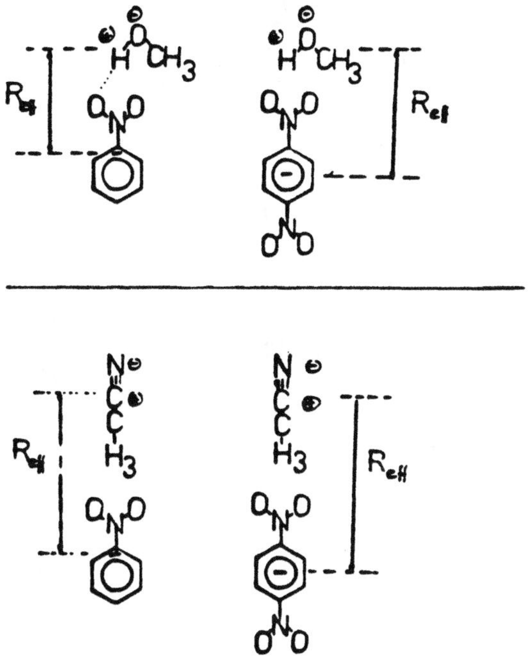

Scheme III

The same stability order is found for weakly stabilizing substituents, F and Cl. However, for electron donating substitiuents, X = CH_3, OCH_3 etc. an opposite order para > meta > ortho is obtained.

Earlier measurements [58] and theoretical calculations [58] pertaining to adducts Cl^-•A between a variety of molecules A and Cl^- showed that the negative ion interacts with the most protic hydrogen in the molecule A. This was true not only for oxygen acid adduct molecules, but also for compounds such as benzene and substituted

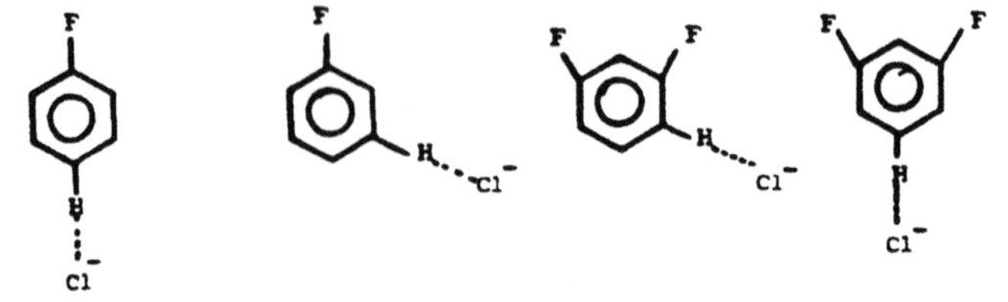

(10.3) II (10.8) III (10.9) IV (12.9)

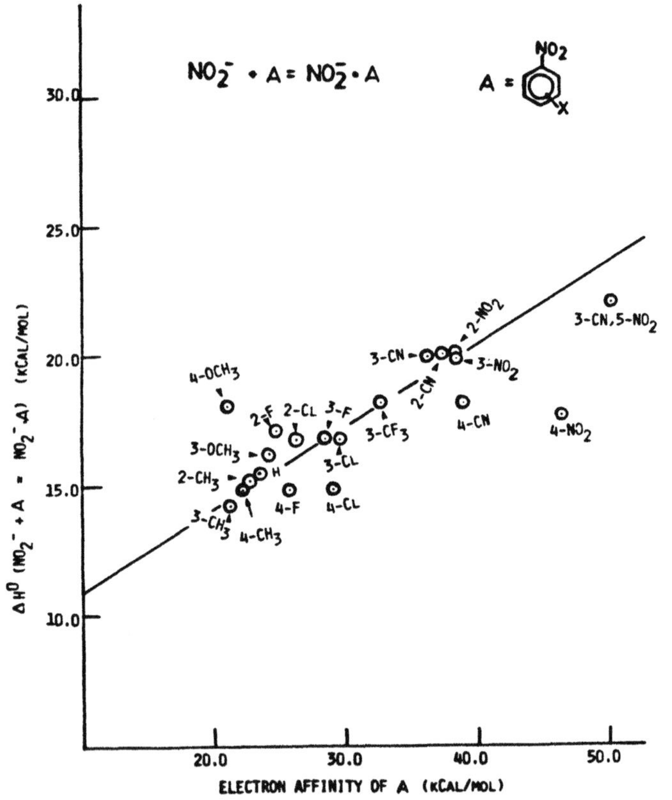

Figure 14. Plot of $\Delta H^o(NO_2^- + A = NO_2^- \cdot A)$ against electron affinities of A shows a correlation between binding energies in $NO_2^- \cdot A$ and electron affinities of A.

benzenes. The most stable adducts predicted by STO-3G calculation [58] are shown in structures I to IV involving fluorobenzene and 1,3-

difluorobenzene. The numbers in brackets are the calculated bond energies in kcal/mol.

The only molecule for which measurements were made with Cl^- and NO_2^- is NB. The binding free energy for the Cl^-•NB was 10.5 kcal/mol 59f while that for NO_2^-•NB (Fig. 14) is 8.9 kcal/mol.[57] Thus, the bonding to the two ions is of quite similar magnitude, with NO_2^- giving a slightly weaker interaction. The more delocalized negative charge in NO_2^- compared to Cl^- is probably responsible for the weaker interaction in NO_2^-•A.

Considering the information presented above, which indicates that the A•NO_2^- adducts are held together at least partly by hydrogen bonding forces, it becomes possible to interpret the observed relationship between NO_2^-•A binding energies and the electron affinities of A, Fig. 14. On formation of A^- the electron enters the lowest unoccupied (π*) orbital in A whose energy is decreased and thus EA(A) increased, by electron withdrawing substituents. The same type of substituent also increases the protic character of the hydrogens in the conjugated system of A and thus increases the (hydrogen) bonding ability of A toward negative ions like Cl^- and NO_2^-.

However, since the molecules A examined in Fig. 14, often have large dipole moments, additional stabilization could occur from classical ion dipole interaction between the negative ion NO_2^- and neutral molecules A. If ion-dipole stabilization is present in NO_2^-•A adducts, it will be reflected in the stability of NO_2^-•A involving isomers of A, where there is a large difference in dipole moments between isomers.

The dipole moments of the dinitrobenzene are: μ_D(O-NO_2NB) = 6.0 D, μ_D(m-NO_2NB) = 3.8 D and μ_D(P-NO_2NB) = 0.0 D.[56] Thus one would expect the increase of stability of NO_2^-•A in the same order of increase in dipole moments of A. The expected stability order from ion-dipole stabilization energy is in agreement with the observed binding energies of NO_2^- to three dinitrobenzenes. One can explain the observed stability order of other NO_2^-•X-NB isomeric adducts (shown in Fig. 14), in a similar manner.

Shown in Fig. 15 is the plot of binding energies of NO_2^- to A against the binding energies of o-NO_2NB^- to A. The points in Fig. 15 fall roughly on a straight line. The results shown in Fig. 15, therefore, predicts that the binding energies in NO_2^-•A and o-NO_2NB^-•A are similar. However, the latter are weaker. The relatively lower binding energies of o-NO_2NB^- to A compared to that of NO_2^- to A is not surprising. The negative charge of NO_2^- is rather localized in three atoms while that in o-NO_2NB^- the charge is more delocalized with the ring and this will lead to weaker interaction.

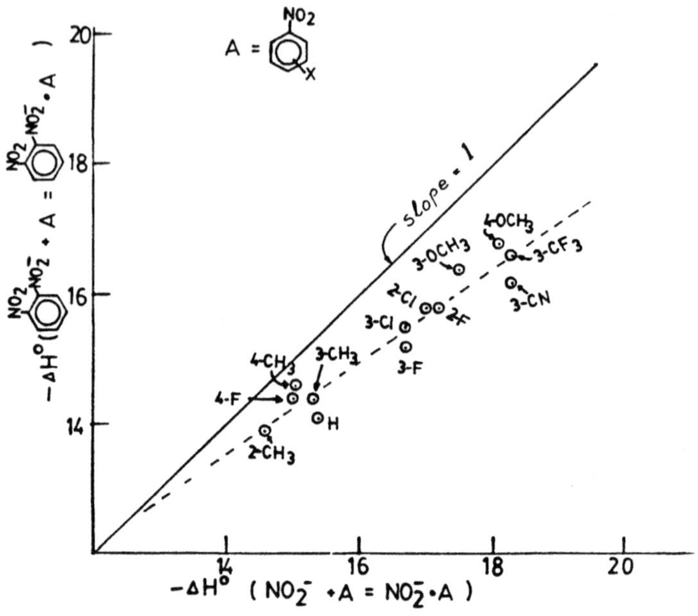

Figure 15. Plot of ΔH^o(O-NO_2NB^- + A = O-NO_2NB^-•A) against electron affinities of A. NB stands for nitrobenzene.

The results shown in Figs. 14 and 15 are for complexes where the negative ion is kept the same (NO_2^- or o-NO_2NB^-) while the neutral molecules, A were varied. An increase of binding energies in NO_2^-•A or o-NO_2NB^-•A was observed with the increase of electron affinities of A. Shown earlier in Figure 11 are results where the neutral molecule Sl, is kept the same while the negative ions A^- (where A = substituted nitrobenzenes) were varied and an opposite effect, the decrease in binding energies in A^-•Sl is observed with the increase of electron affinities of A. The, binding energies of A^- to A do not show any correlation with the electron affinities of A as shown in Figure 16 (for details see ref. 59ii). The lack of correlation in Fig. 16 is probably due to the cancellation of the two opposing effects demonstrated earlier in Figs. 11, 14 and 15. However, one can understand the change in binding energies in A_2^- for the cases

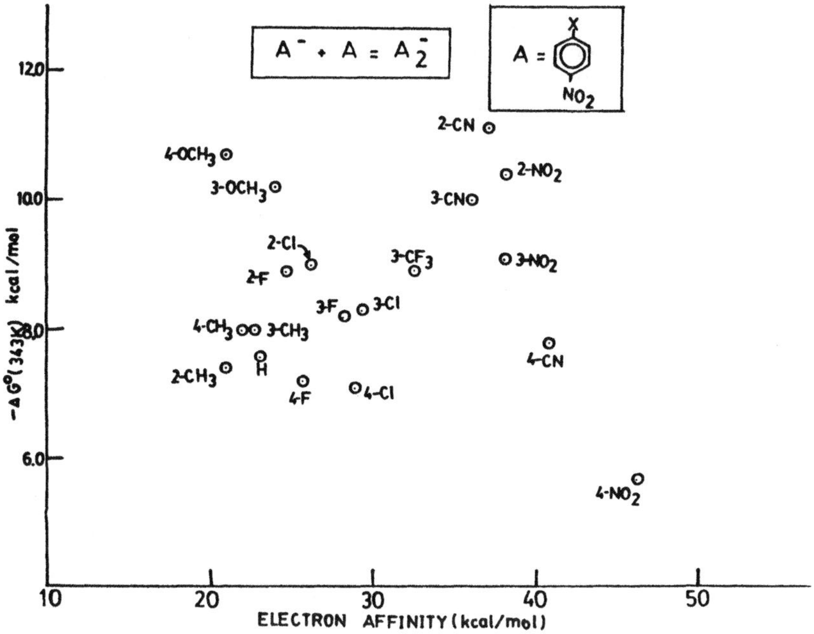

Figure 16. Plot of $-\Delta G^{o}(A^{-} + A = A_2^{-})$ at 343K versus electron affinities of A.

where the different A are isomers. For all electron withdrawing substituents NO_2, CN, F, Cl etc ortho isomers from the strongest clusters and the order of the strength is for ortho ≥ meta >> para while for electron donating substituents an opposite order para ≥ meta > ortho is observed (Fig. 16). These orders in stability of A_2^{-} are in line with the dipole moments of isomers of A [56], and this means that the strength of the molecular dipole of A in the $A^{-}\cdot A$ adduct is the factor that dominates the bonding.

In examining factors influencing the bonding of the adducts $A^{-}\cdot B$ and A_2 we did not consider the possible role of charge transfer from the negative ion to the neutral. The observed increase of bonding in $A^{-}\cdot B$ with increasing electron affinity of B and decreasing electron affinity of A may be a consequence of charge transfer from A^{-} to B.

Modern ab initio calculations and more experimental results involving many different negative ions and neutral molecules will be required before the bonding and structures of the ion-neutral complexes are well understood. The interesting preliminary results provided above should encourage such efforts.

Table II: Thermochemical Data[a] for electron capture reaction:

$$e + B = B^-$$

B^b	(ΔS^o)	$(-\Delta H^o)\approx EA(B)$		EA Literature	(ΔG^o)	$-\Delta G^{o}$[16]
	(e.u.)	kcal	eV	eV	(kcal/mol)	(kcal/mol)[16]
TCNE		73.0	3.17±.2	2.3±.3[29]	73.0	
Cl_4BQ^d	-3.5	64.0	2.78	2.76±0.2[30]	62.6	62.3
F_4BQ^d	-3.5	62.3	2.70	2.92±0.2[30]	60.9	55.2
$2,6\text{-}Cl_2BQ^d$	-3.5	57.2	2.48		55.8	56.1
$NO_2{}^f$	-2.6	52.8	2.29	2.36±.1[36]	52.6	-
$2,3\text{-}Cl_2NpQ$	-3.5	50.6	2.19		49.2	-
$3,5\text{-}(NO_2)_2BN$	-3.5	49.9	2.16		48.5	-
$4\text{-}NO_2NB$	-4.5	46.2	2.00		44.3	44.4
BQ	-4.0	44.1	1.91	1.89±0.2[31]	42.4	43.0
$2\text{-}CH_3BQ^d$	-4.0	42.7	1.85	1.91[32]	41.1	41.3
NpQ	-4.0	41.8	1.81		40.2	40.1
$1,4\text{-}(CN)_2F_4C_6{}^d$	+4.4	41.8	1.81		43.5	-
$3,5\text{-}(CF_3)_2NB^b$	-2.5	41.2	1.79		40.2	-
4-CNNB	-2.7	39.5	1.72		38.7	-
$3\text{-}NO_2NB^d$	-3.2	38.2	1.65		36.9	37.0

2-NO_2NB[d]	-3.2	38.2	1.65	1.66[32]	36.9	-
2-CNNB	-3.2	37.1	1.61		35.8	-
3-CNNB[d]	-3.2	36.1	1.56		34.8	35.2
MaAn	-0.4	33.2	1.44	1.39±.2[31]	33.1	32.7
3-CF_3NB	-2.5	32.6	1.41		31.6	31.5
3-BrNB[d]	-1.7	30.2	1.32		29.5	-
4-BrNB[d]	-2.0	29.9	1.29		29.0	-
3-ClNB[d]	-1.7	29.5	1.28		28.8	28.4
4-ClNB	-2.0	29.0	1.26		28.2	27.7
trans-1,2-$(CN)_2$ ethylene	-1.0	28.8	1.24		28.4	
3-FNB	-1.7	28.4	1.23		27.7	27.2
4-CHOBN		28.1	1.22		28.1	
PhAn[d]	-0.4	27.9	1.21	1.18[32]	27.8	27.4
2,4-F_2NB[d]	-1.7	27.0	1.17		26.3	-
2-BrNB[d]	-1.6	26.8	1.16		26.2	-
4-CH_3COBN	-2.0	25.9	1.13		25.0	
SO_2[e]	+2.0	25.2	1.09	1.1±0.04[12]	26.1	(26.1)
4-CNBN	+1.0	25.4	1.10	1.1[32]	25.8	
F_5BN	+4.4	25.0	1.08		27.0	-
3,5-$(CF_3)_2$BN		26.3	1.14		26.3	
2-ClNB	-1.6	26.3	1.14		25.7	25.1
4-FNB[d]	-2.0	25.8	1.12		25.0	24.7
2-FNB[d]	-1.6	24.8	1.07		24.2	24.9
3-OCH_3NB[d]	-1.7	24.1	1.04		23.4	
NB	-1.0	23.2	1.01	>0.8±0.2[32]	22.8	23.1

SF_6[g]	+13.0	23.7	1.03	1.0±0.2[35]	29.3	-
C_7F_{14}[g]	+2.5	24.9	1.08		26.0	-
3-CHOBC		23.2	1.00		23.2	
3-CH_3NB[d]	-1.7	22.8	0.99		22.1	22.1
2-CNBN	+0.5	22.0	0.95	0.95[32]	22.3	
4-CH_3NB[d]	-2.0	22.0	0.95		21.2	21.7
2-CH_3NB	-1.6	21.3	0.92		20.7	21.4
4-OCH_3NB[d]	-2.0	21.1	0.91		20.3	
3-CNBN		21.0	0.91	0.90[32]	21.0	
2,3-$(CH_3)_2$NB	-1.1	19.8	0.86		19.4	20.7
4-CF_3BN		17.5	0.76		17.5	
Azulene	+4.5	16.0	0.69	0.66[32]	18.0	
2,6-Cl_2BN		16.1	0.60		16.1	
2,3-butane-dione[d]	+2.0	16.0	0.69	0.63±0.05[34]	16.9	
2-CF_3BN		16.1	0.70		16.1	
2,4,6-$(CH_3)_3$NB[d]	-1.1	16.3	0.70		16.1	17.5
3-CF_3BN		15.5	0.67		15.5	
4-FBPh[d]	+2.0	14.9	0.64		15.8	17.9
1-naphtho-nitrile		15.6	0.68		15.6	
BPh[d]	+2.0	14.4	0.62	0.64±0.5[34]	15.3	16.7
2-naphtho-nitrile		14.9	0.65		14.9	
CS_2[d]	+2.0	11.8	0.51	0.60±0.09[34]	12.7	
CH_3NO_2[d]	+2.0	11.2	0.48	0.45±0.09[34]	12.1	
C_6F_6	+7.4	12.0	0.52	>1.8[37]	14.8	

$(C_6F_5)_2$		20.9	0.91		20.9
$C_6F_5CF_3$		21.6	0.94	≥1.7[37]	21.6
$C_6F_5COCH_3$		21.7	0.95		21.7
$(C_6F_5)_2CO$		37.1	1.61		37.1
1-NO_2Npe	-0.9	28.3	1.23		27.9
2-NO_2Npe	-0.7	27.1	1.18		26.8
1,3-$(NO_2)_2$Npe		39.7	1.72		39.7
1,5-$(NO_2)_2$Npe		40.7	1.76		40.7
1-CHONpe	+3.7	16.2	0.70	0.68[38]	17.8
2-CHONpe	+3.7	14.8	0.64	0.62[38]	16.4
Anth		13.2	0.57		13.2
9-NO_2 Anth	-2.9	33.0	1.43		31.8
9-CN Anth	-1.6	29.3	1.27		28.2
9-CHO Anth	-2.7	30.2	1.31		29.0

(a) Compounds for which ΔS^o_7 is not given, was not obtained and $\Delta G^o_7 = \Delta H^o_7 \approx$ EA is assumed.

(b) Notations used: BN: benzonitrile; Bph: benzophenone; MaAn: maleic anhydride; PhAn: phthalic anhydride. TCNE: tetracyanoethylene, BQ: 1,4-benzoquinone, NB: nitrobenzene, Npe: naphthalene and Anth: anthracene.

(c) ΔG^o measured at 80°C.

(d) Entropy estimated assuming entropy equalities: TCNE = 4-NO_2NB; 2,3 Cl_2NpQ = Cl_4BQ; F_4BQ = 2,6-Cl_2BQ; BQ = 2-$_3$BQ; $F_4(CN)_2C_6$ = F_5CNC_6; 3,5-$(CF_3)_2$NB = 3-CF_3NB; 3-NO_2NB = 2-NO_2NB = 3-CNNB = 2-CNNB; 3-CH_3NB = 3-BrNB = 3-ClNB = 3-FNB = 3-OCH_3NB; 2-CH_3NB = 2-BrNB = 2-ClNB = 2-FNB 4-BrNB = 4-ClNB = 4-FNB = 4-OCH_3NB = 4-CH_3NB; MaAn = PhAn; 2,3-butane dione = CH_3NO_2 = CS_2 = Bph = SO_2. The ΔH^o_7 for these compounds were estimated from ΔG^o_7, which is based on the experimental ΔG^o_6, and $\Delta G^o_7 = \Delta H^o_7 - T\Delta S^o_7$.

(e) Anchor compound, see Table I on whose ΔS^o_7 and ΔH^o_7 the whole Table II is based.

(f) ΔS^o_7 for NO_2 was calculated from available[36] frequencies and geometries for NO_2 and NO_2^- using procedure outlined in Table I.

(g) From Grimsrud.[27]

IV KINETICS OR ELECTRON TRANSFER REACTION IN GAS PHASE $A^- + B = A + B^-$

a) Well behaved, fast reactions

By proper selection of the composition of the reaction mixtures one can measure the electron transfer equilibria or the kinetics of the electron transfer reactions. Shown in Figure 17a is the plot of log intensities of two ions 4-FNB^- and BQ^-. BQ^- has 18.3 kcal/mol higher EA than 4-FNB^-. The linear decay 4-FNB^- is thus due to the electron transfer to BQ. The slope when converted to the natural logarithm by multiplying with 2.303 equals ν_1(obs), the observed pseudo first order rate constant for A^- disappearance. This is given by equation (15) where ν_{d,A^-} is the first order rate constant for loss of A^- by diffusion to the walls and k_1 is the rate constant for the electron transfer reaction (1). Choosing different concentrations of B equal to $[B^-]$ in a series of experiments at constant T one obtains a series of slopes (obs) as shown in Figure

$$\nu_1(\text{obs}) = k_1[B] + \nu_{d,A^-} \qquad (15)$$

17b. The ν_1(obs) obtained from this series can then be plotted versus the known [B].[20] The straight line obtained gives a slope which equals k_1 and an intercept equalling ν_{d,A^-},[20] (see eqn (15). When large [B] are used ν_{d,A^-} is very much smaller than k_1[B]. Under such conditions, quite accurate (±20%) rate constants k_1 can be obtained by determining ν_1(obs) at a given large [B] concentration and then assuming $\nu_{d,A^-} \approx \nu_{d,B^-}$ and evaluating k_1 from (15). $\nu_d \cdot B^-$ can be easily evaluated from the constant slope of B^- observed when reaction (1) is complete, see Fig. 17.

Table III. Rate Constants for Exothermic Electron Transfer Reactions (Eq. 1): $A^- + B = A + B^-$[a]

	A[b]	B[b]	$k_{1(expt)}$[c]	$k_1(ADO)$[c]
1	NB	p-NO_2NB	1.3	1.1
2	NB	m-NO_2NB	2.1	1.8
3	m-CF_3NB	p-NO_2NB	1.3	1.0
4	m-FNB	m-NO_2NB	1.9	1.8
5	NB	m-CF_3NB	1.8	1.8
6	p-MeNB	m-CF_3NB	1.9	1.7
7	NB	NO_2	0.4	0.6
8	p-CH_3NB	NO_2	0.5	0.6
9	p-ClNB	NO_2	0.5	0.6
10	m-NO_2NB	NO_2	0.5	0.6
11	p-NO_2NB	NO_2	0.4	0.5
12	BQ	NO_2	0.3	0.5
13	azulene	m-CF_3NB	1.0	
14	p-FNB	BQ	1.4	
15	p-NO_2NB	2,5-Cl_2BQ	1.0	

[a] This table contains a few of the total 94 rate constants measured (see references 20,21,22,25-28,59)

[b] NB = nitrobenzene, BQ = benzoquinone

[c] k in cc. $molecules^{-1}$ sec^{-1}.

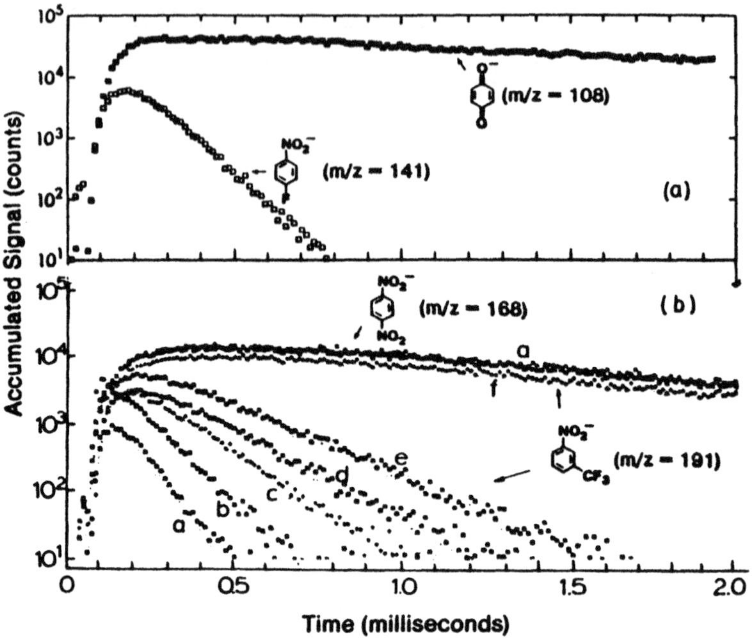

Figure 17. (a) Concentration changes due to electron transfer from p-Fnitrobenzene anion to higher electron affinity benzoquinone. Slope of linear portion of nitrobenzene anion decay can be used for determination of rate constant for electron transfer. 2 mtorr flouronitrobenzene, 0.3 mtorr benzoquinone, 4 torr CH_4. Figure (b), electron transfer from m-CF_3 notrobenzene anion to higher electron affinity p-NO_2 nitrobenzene. Slopes of linear portion of decaying CF_3-nitrobenzene anion = ν_1(obs) are used for determination of rate constant k_1 m-CF_3nitrobenzene 15 mtorr, CH_4, 4 torr, p-dinitrobenzene at different pressures: a = 0.5; b = 0.33; c = 0.25; d = 0.167; e = 0.116; f = 0 mtorr. Temp. 150°C. Accumulated signal for p-dinitrobenzene anion is almost identical for cases (a) - (f), so only case (a) is shown.

Rate constants for electron transfer reactions (1) were measured for ~90 reactions involving 50 different molecules.[20,22,25,27,28,57] The measured k_1(exp) can be compared with the ADO theory [20,25,27,28,57] predicted collision rate constants k_{ADO}. Good agreement of k_1 with k_{ADO} is observed (only some of the representative reaction rate constants are given in Table III) for reactions involving A and B which are substituted nitrobenzenes, cyanobenzenes, anthracenes, perfluoro benzenes etc. Notable is the comparison for B = m-dinitrobenzene and B = p-dinitrobenzene. Due to the presence of a dipole in the meta compound and its cancellation in the para case, ADO predicts k_{ADO} (meta) > k_{ADO} (para) and this trend is quantitatively followed[20] by the experimental rate constants (see reactions (1) - (4) Table III).

b) Slow electron transfer reactions involving SF_6 and $C_6F_{11}CF_3$.

$$A^- + B \underset{k_{-1}}{\overset{k_1}{\rightleftharpoons}} A + B^- \qquad (1)$$

As discussed in the previous section (IVa), it was found, practially in all cases, that the rate constants k_1 in the exothermic direction of reaction (1) was near the Langevin-ADO collision limit, $k_1 \approx 10^{-9}$ cc.molecule^{-1} s^{-1}.[20,22,25,28] However, when SF_6 and perfluorocycloalkanes, like C_6F_{12}, $C_6F_{11}CF_3$ were used k_f was found to be very small, often less than 10^{-13} cc. mole-cule^{-1} s^{-1}.[27,59] Shown in Table IV are the exothermic forward rate constants k_1 for electron transfer from $A^-(=SF_6)$ to B, where B = substituted nitrobenzenes and quinones, along with the exothermicities of the reactions. As can be seen in Table IV the rate constant becomes larger as the exothermicity $-\Delta H_1$ is increased and reaches the ADO limit only after the exothermicity has become as high as 38.7 kcal/mol.[27,59]

The temperature dependence of reactions involving exothermic electron transfer from SF_6^- was determined in a series of experiments.[27] When the rate constant was near ADO no or little temperature dependence was observed. At lower exothermicities, the rate constants were lower than ADO limit and a negative temperature dependence was observed. The lowest exothermic reaction studied (B = m-CF_3NB, $\Delta H = -7.9$ kcal/mol), the rate constants were slowest and a positive temperature dependence was observed.[27]

The relevant rate constants k_1 for reactions (1) are summarized in Table IV. Included in the table are the calculated heights of the internal barrier $E^{\ddagger}$ and the enthalpy changes ΔH_1^o, for the reactions (1).

$$SF_6^- + B = SF_6 + B^- \qquad (1)$$

The reaction coordinate adopted[27] in an earlier work (1) is shown in Fig. 18a. The double minimum potential with the internal barrier $E^{\ddagger}$ is the standard reaction coordinate[60,61] used for bimolecular gas phase ion-molecule reactions. At low pressures the collision complex $A^-\cdot B$ is not deactivated by collisions i.e. there is energy and angular momentum conservation as the collision pair proceeds from reactants to products. It can be shown[60,61] that the overall forward rate constant k_1 is given by equation (16) (for notations see Fig. 18). When $E^{\ddagger}$ is small such that the energy gap $\Delta E_0^{\ddagger}$ is negative and its absolute value is large

$$k_1 = \frac{k_c k_p}{k_p + k_b} \qquad (16)$$

$k_p >> k_b$, and the reaction proceeds at collision rates and has the lack of temperature dependence observed for these rates. (See B = tetrafluoroquinone Table IV). For larger $E^{\ddagger}$ where $\Delta E_0^{\ddagger}$ is still negative and its absolute value equal to a few to several kcal/mol, $k_p < k_b$ and the temperature dependence is negative i.e. k_2 decreases with increase of temperature (B = 4-NO_2NB and 4-CNNB, see table IV). Finally as $\Delta E_0^{\ddagger}$ approaches zero or becomes positive $k_p <<< k_b$ and the temperature dependence becomes positive.[27,60,61] For electron transfer reactions in the absence of solvent molecules the barrier $E^{\ddagger}$ is generally assumed to be due to geometry changes between the neutral and negative ion for A and B. The barriers $E^{\ddagger}$ given in Table IV were calculated by Richardson [62] on basis of the Marcus equations [63] and the theoretical geometries of Hay [47] for SF_6 and SF_6^-. Richardson assumed essentially minimal geometry changes for B to B^-. Similar values for $E^{\ddagger}$ were obtained in the earlier work [27] by a qualitative estimate based on similar geometry change assumptions.

Table IV. Rate constants k_f and calculated [a] energy Barriers $\Delta H^{*}_{1,2}$ for reactions:

$$SF_6^- + B = SF_6 + B^-$$

B	ΔH_1^b	$(k_f \times 10^{10})^b$	
m-CF_3NB	-7.9	7.4	0.005
m-NO_2NB	-13.9	5.4	0.6
p-CNNB	-15.84.8	1	
p-NO_2NB	-22.1	2.9	10
F_4BQ	-38.70.3	14	

(a) D. E. Richardson.[62]
(b) Energy in kcal/mol, k_f in cm^3 molecules^{-1} sec^{-1}

The parabolic potential wells shown in Fig. 18 for $SF_6^-\cdot B$ and $SF_6\cdot B^-$ are drawn to scale potential energies for SF_6^- and SF_6 based on Hay's [47] calculations. The reaction coordinate in this region

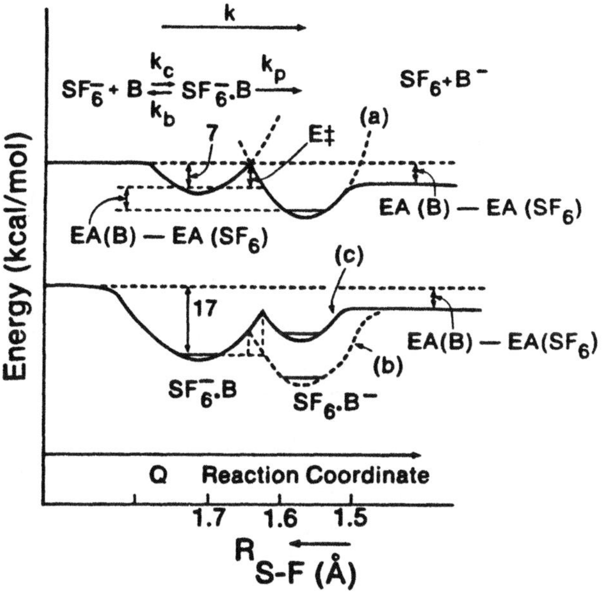

Figure 18. Reaction coordinate for reaction (1) $SF_6^- + B = SF_6 + B^-$, where B = 3-CF_3 nitrobenzene. Internal barrier $E^\ddagger$ due to geometry change between SF_6 and SF_6^-. (a) Reaction coordinate used in earlier work.[27] Values of bond energies $D(SF_6^- - B) \approx D(SF_6 - B^-) \approx 7$ kcal/mol were assumed. (b) Reaction coordinate after measurement of $D(SF_6^- - B) \approx 17$ kcal/mol and with assumption $D(SF_6^- - B) \approx D(SF_6 - B^-)$. (c) Reaction coordinate with measured $D(SF_6 - B^-)$. The barrier $E^\ddagger$ is assumed to be only due to geometry changes between SF_6^- and SF_6 which consist of a symmetric shortening of the S-F bonds in the region of $E^\ddagger$, therefore in the region of $E^\ddagger$ (only) the reaction coordinate is represented by the S-F bond length.

is assumed to be the S-F distance. This is in line with the expected minimal geometry change from B to B^-. The resulting barrier $E^{\ddagger}$ in the Figure thus represents an approximate graphical solution for the Marcus barrier, given in Table IV.

For a qualitative discussion of reactions (1), the electron transfer involving the least exothermic reaction (B = 3-CF_3NB) and 0, see Fig. 18a. The value of the adduct binding energy was not available at the time, however on the basis of a comparison of related binding energies it was concluded that $D(SF_6^- - B) \approx 7$ kcal/mol is a possible value, which was consistent with the expected reaction coordinate.

Recently we measured the binding energy of $SF_6^-{\cdot}B$ and $SF_6{\cdot}B^-$ for different B molecules,[59] by studying the equilibria

$$SF_6^- + B = SF_6^-{\cdot}B \tag{17}$$

$$B^- + SF_6 = B^-{\cdot}SF_6 \tag{18}$$

The $-\Delta H^o_{17}$ for B = 3-CF_3NB was found to be equal to 17 kcal/mol,[60] which is much larger than the earlier assumed,[27] $-\Delta H^o_{17} \approx$ 7 kcal/mol. The new result is incorporated into a new reaction coordinate shown in Fig. 18b. Evidently, the barrier $E^{\ddagger}$ of 7.4 kcal/mol is now much too small so that the reaction coordinate 18b leads to a large $-\Delta E^{\ddagger}_0$ and is thus inconsistent with the observed very slow kinetics. However, the reaction coordinates 18b, as well as 18a, were obtained by making the assumption that the binding energies $D(SF_6^- - B) \approx D(SF_6 - B^-)$, i.e. that the difference between the zero point energies of the two adducts corresponds to the overall exothermicity ΔH_1 of the reaction (see Fig. 18). The complex $SF_6^-{\cdot}B$ was found [59] to be weaky bonded, i.e. it could not be observed even at room temperature. An upper limit for the binding energy of m-$CF_3NB^-{\cdot}SF_6$ could be estimated, $D(B^- - SF_6) < 8$ kcal/mol.[60] This means that $D(SF_6^- - B) - D(SF_6 - B^-) > 10$ kcal/mol.[60]

Taking this difference into account one can draw a reaction coordinate energy diagram which uses the individual adduct binding energies. This diagram is shown in Fig. 18c, where the assumption has been made that $D(SF_6 - B^-)$ is the same for B = 3-CF_3NB and 3-NO_2NB. Actually, because of increased charge delocalization in 3-NO_2NB^- one expects $D(SF_6$ - 3-$NO_2NB^-)$ to be slightly smaller.

The accurate accounting of the adduct binding energies in the reaction coordinate Fig. 18c, changes the picture again. The electron transfer reaction based on the zero point levels of the two complexes has now become endothermic rather than exothermic and the internal barrier $E^{\ddagger}$ due to SF_6 geometry change is thereby raised, such that $-\Delta E^{\ddagger}_0$ is quite small again. Furthermore since there is a change in the adduct binding geometries, this change must also become part of the reaction coordinate, an effect not taken into account in the evaluation of $E^{\ddagger}$ Table I and Fig. 1 which considers only S-F bond length changes. Consideration of the adduct bonding changes will lead to a further increase of the internal barrier $E^{\ddagger}$ so that $E^{\ddagger} \geq D(SF_6^- - B)$ becomes possible and the kinetic model becomes again compatible with the observed kinetics.

The difference in the bonding of $SF_6^- \cdot B$ and $B^- \cdot SF_6$ is easily rationalized.[59i] The bonding of negative ions to molecules B was discussed earlier see section IIIb. The nature of bonding of SF_6^- to B would also be similar. However, the very weak binding energies in $B^- \cdot SF_6$ are due to the specific charge distribution in SF_6. Thus, the $S^+ - F^-$ bond dipoles in SF_6 result in a dodecapole with negative charge on the outside of the molecule which shields the SF_6 from interaction with negative centres in B^-.

The rate constants involving perfluorinated cycloalkanes like C_6F_{12} and $C_6F_{11}CF_3$ were found to be very slow.[20,27] The slow rates observed most likely are due to a large change of geometry on going from the neutral molecule to the negative ion, however, no geometric information from theoretical calculations are available at this time for the negative ions of those two compounds. The rate constants for these systems can also be predicted by a similar model as used above for SF_6.

REFERENCES

1. Christodoulides, A. A., McCorkie, D. L., and Christophorou, L. G., 1985, Elec. Mol. Inter. Appls. **2**, pp.423.
2. Mead, R. D., Steavens, A., Lineberger, W. C., 1984, 'Photodetachment in negative ion beams' in "Gas Phase Ion Chemistry" M. T. Bowers ed. vol. 3, pp.214.
3. Drzaic, P. S., Marks, J., and Brauman, J. I., 1984, 'Electron Photodetachment from Gas Phase Molecular Ions' in "Gas Phase Ion Chemistry", M. T. Bowers ed. vol. 3, pp.168.
4. (i) Kebarle, P. and Hogg, A. M., 1965, J. Chem. Phys. **42**, 798. (ii) Kebarle, P. and Goodbole, E. W., 1963, J. Chem. Phys. **39**, 1131.
5. Hogg, A. M., Haynes, R. N., Kebarle, P. 1966, J. Am. Chem. Soc. **89**, 6393.
6. (i) Kebarle, P., Arshadi, M., Scarborough, J. 1969, J. Chem. Phys. **50**, 1049. (ii) Kebarle, P., 1975, in "Interaction Between Ions and Molecules", P. Ausloos ed. pp. 459.
7. Bowers, M. T., Aue, D. H., Webb, H. M., McIver, R. T., 1971, J. Am. Chem. Soc. **93**, 4314.
8. (i) Fehsenfeld, F. C., Ferguson, E. E., 1973, J. Chem. Phys. **59**, 6272. (ii) Bohme, D. K., Hemsworth, R. S., Rundle, H. W., Schiff, H. I., 1973, J. Chem. Phys. **58**, 3504.
9. (i) Caldwell, G., Chowdhury, S. and Kebarle, P., 1987, J. Phys. Chem. (submitted). (ii) Larson, J. W., McMahon, T. B., 1983, J. Am. Chem. Soc. **105**, 2944. (iii) Caldwell, G. and Kebarle, P., 1984, J. Am. Chem. Soc. **106**, 967. (iv) Kebarle, P., Caldwell, G., Magnera, T., Sunner, J., 1985, Pure & Appl. Chem. **57**(2), 339.

10. (i) Taft, R. W., 1975, in 'Proton Transfer Reactions' ed. E. F. Caldin, and V. Gold. London: Chapman and Hall, p.31. (ii) D. K. Bhome, 1975, in "Interaction Between Ions and Molecules" P. Ausloos Ed. New York, p.489. (iii) Kebarle, P. 1977, Ann. Rev. Phys. Chem. **28**, 445. (iv) Bartmess J. E. and McIver, R. T., 1979, In 'Gas Phase Ion Chemistry" M. T. Bowers ed. vol. 2, pp. 88.
11. Sharma, R. B., Sen Sharma, D. K., Hiraoka, K. and Kebarle, P., 1985, J. Am. Chem. Soc. **107**, 3747.
12. Cellota, R. J., Bennett, R. A. and Hall, J. L., 1974, J. Chem. Phys. **60**, 1740.
13. Fukuda, E. K., and McIver, R. T., 1982, "Lecture Notes In Chemistry", Hartman, H. Eds. Springer, Berlin p. 164.
14. McIver, R. T., 1978, Rev. Sci. Instrum. **49**, 111.
15. Fukuda, E. K., and McIver, R. T., 1982, J. Chem. Phys. **77**, 4942.
16. (i) Fukuda, E. K. and McIver, R. T., 1983, J. Phys. Chem. **87**, 2993. (ii) Fukuda, E. K. and McIver, R. T., 1985, J. Am. Chem. Soc. **107**, 2291.
17. (i) Heilbronner, E. and Bock, H. 1968, "Das HMO-Modell und Siene Anwendung" p.333, Verlat Chemie, Bergstr. (ii) Heilbronner, E. (Private communication).
18. Fujio, M., McIver, R. T. and Taft, R. W. 1981, J. Am. Chem. Soc. **103**, 4017.
19. Birch, A. J., Hinde, A. L. and Radom, L. 1980, J. Am. Chem. Soc. **102**, 3310.
20. Grimsrud, E. P., Caldwell, G., Chowdhury, S. and Kebarle, P. 1985, J. Am. Chem. Soc. **107**, 4267.
21. Chowdhury, S., Heinis, T., Grimsrud, E. P., and Kebarle, P. 1986, J. Phys. Chem., **90**, 2747.
22. Heinis, T., Chowdhury, S., Scott, S. and Kebarle, P. J. Am. Chem. Soc. (submitted).
23. Chowdhury S. and Kebarle (in preparation).
24. Kebarle, P. (unpublished data).
25. Chowdhury, S. and Kebarle, P. 1986, J. Am. Chem. Soc. **108**.
26. Chowdhury, S., Heinis, T. and Kebarle, P. 1986, J. Am. Chem. Soc. **108**.
27. (i) Grimsrud, E., Chowdhury, S. and Kebarle, P. 1985, J. Chem. Phys., **83**, 1059; (ii) ibid, **83**, 3983.
28. (i) Chowdhury, S., Grimsrud, E. P., Heinis, T. and Kebarle, P., 1986, J. Am. Chem. Soc. **108**, 3630. (ii) Chowdhury, S., Nicol, G. and Kebarle, P. 1986, Chem. Phys. Letts. **127**, 130.
29. Lyons, L. E. and Palmer, L. D. 1976, Aust. J. Chem. **29**, 1919.
30. Cooper, C. D., Frey, W. F., Compton, R. N. 1978, J. Chem. Phys. **69**, 2367.
31. Cooper, C. D., Naff, W. T., Compton, R. N., 1975, J. Chem. Phys. **63**, 2752.
32. Chen, E. C. M., Wentworth, W. E., 1975, J. Chem. Phys., **63**, 3183.
33. Becker, R., and Chen, E., 1966, J. Chem. Phys. **45**, 240.
34. Chen, E. C. M., Wentworth, W. E., 1983, J. Phys. Chem. **87**, 45.

35. Streit, G.E., 1982, J. Chem. Phys. **77**, 826.
36. Herbst, E., Patterson, I. A., and Lineberger, W. C., 1974, J. Chem. Phys. **61**, 1300.
37. Lifshitz, C., Tiernan, T. O., and Hughes, B. M., 1973, J. Chem. Phys. **59**, 3182.
38. Wentworth, W. E., Kao, L. W., Becker, R. S., 1975, J. Phys. Chem., **79**, 1161.
39. (i) Chen, C. L. and Chantry, P. J., 1979, J. Chem. Phys. **71**, 3897. (ii) Christophorou, L. G. 1978, Adv. Elec. Elec. Phys. **46**, 55.
40. Rosenstock, H.M.; Draxl, K.; Steiner, B.W.; Herron, J.T. 1977 J. Phys. Chem. Ref. Data Supl. 1, **6**.
41. Lias, S.; Liebman, J. F.; Levin, R. D. 1985 ibid, **13**, 695.
42. Traeger, J. C. and McLaughlin, R. G. 1981, J. Am. Chem. Soc. **103**, 3647.
43. Stull, D. R. and Prophet,H. 1971 JANAF Thermochemical Tables National Bureau of Standards (U.S.).
44. Fehsenfeld, 1971, J. Chem. Phys. **54**. 438; ibid 1970, **53**, 2000.
45. Drzaic, P. S. and Brauman, J. L. 1982, J. Am. Chem. Soc. **104**, 13.
46. Lifshitz, C. 1983, J. Phys. Chem., **87**, 3474.
47. Hay, P. J. 1982, J. Chem. Phys. **76**, 502.
48. Heilbronner, E. 1976, "Molecular Spectroscopy", Inst. of Petroleum, London.
49. Page, F. M. and Goode, G. C. 1969 'Negative ions and the Magnetron' (Wiley, New York).
50. Stull, D. R., Westrum Jr., E. F. Sinke, G. C. 1969, "The Chemical Thermodynamics of Organic Compounds p. 458 (John Wiley and Sons).
51. Pitzer, K. S. "Quantum Chemistry" 1954 Table A18-4, Prentice Hall.
52. Becker, E. D., Charney, E., Anno, T. 1965, J. Chem. Phys. **42**, 942.
53. Herzberg, G. 1945, "Infrared and Raman Spectra", p.337 van Nostrand Reinhold 1945.
54. Shchegoleva, L. N., Bilkis, I. I. and Schastner, P. V. 1983, Chem. Phys. **82**, 343.
55. Chowdhury, S., Grimsrud, E. P. and Kebarle, P. 1986, J. Phys. Chem. (submitted).
56. McClellan, 1963 "Tables of Experimental Dipole Moments" W. H. Freeman and Co., San Francisco.
57. Grimsrud, E., Chowdhury, S. and Kebarle, P., 1986, Int. J. Mass Spec. Ion Proc. **68**, 57.
58. French, M. W., Ikuta, S. and Kebarle, P. 1982, Can. J. Chem. **60**, 1907.
59. (i) Chowdhury, S. and Kebarle, P. 1986, J. Chem. Phys (in press). (ii) Chowdhury, S., Grimsrud, E., and Kebarle, P. (in preparation).
60. Olmsted, W. N. and Brauman, J. I., 1977, J. Am. Chem. Soc. **99**, 4219; Farneth, W. E., Brauman, J. I., 1976, ibid, **98**, 7891.

61. Hiraoka, K. and Kebarle, P. 1975, J. Chem. Phys. **63**, 394. Sen Sharma, D. K., and Kebarle, P., 1982, J. Am. Chem. Soc. **104**, 19. Caldwell, G., Magnera, T. F., and Kebarle, P. 1984, J. Am. Chem. Soc. **106**, 959.
62. Richardson, D. E., 1986, J. Phys. Chem. **90**, 3697.
63. Sutin, N. 1983, Progress in Inorganic Chemistry, **30**, 441.

ION THERMOCHEMISTRY: SUMMARY OF THE PANEL DISCUSSION

John E. Bartmess
Department of Chemistry
University of Tennessee
Knoxville, TN 37996, U. S. A.

There has been some question regarding the original anchoring of the gas phase acidity scale based on ICR measurements[1] in the region of the aliphatic alcohols. The data for methanol from this scale differ by 2.2 kcal/mol from the value from the thermochemical cycle based on bond strength and electron affinity.[1,2] Moylan and Brauman have questioned the values for the relative acidities of HF, $tBuCH_2OH$, and $PhCH_2OH$.[2] The acidity scale presented in 1979 is based on two assumptions. First, the ion temperature is assumed to be equal to the cell temperature of 320K, so that the slope of a plot of the relative ICR acidities vs. the D-EA absolute values (converted to $\Delta G°_{acid}$ as in equation 1) should be unity. Secondly, hydrogen fluoride, with the most accurately known acidity from the D-EA thermochemical cycle, is properly related to the rest of the acidity ladder, so that it may serve as a single point anchor for the whole scale.

$$\Delta G°_{acid}(AH) = DH°(A\text{-}H) + IP(H^{\cdot}) - EA(A^{\cdot}) + T\Delta S°_{acid} \quad (1)$$

In order to find the source of the discrepancies, we have re-examined the primary data for the ICR acidity scale. The original ladder of relative acidities was constructed from the individual equilibria "by eye": the single equilibrium measurement for two acids was assumed to be the most accurate, and other multiple equilibria pathways between those two acids were used to bias that value slightly if the numbers did not add up exactly right. This method may result in an expansion or contraction of the scale, depending on what choices are made for given values. For example, the free energy corresponding to the D-EA difference for methanol and phenol is 36 kcal/mol,[1] while the relative ICR scale gives 28.1 kcal/mol. We have written a computer program that constructs the equilibria ladder on a rational basis, averaging the difference between two acids over all pathways out to n successive overlaps, where n is up to 10. A number of different weighting methods for the less direct paths have been tried, such as by length in kcal/mol, number of paths, or the square of these. The results of all these schemes consistently give 24.5±0.5 kcal/mol for the $\delta\Delta G°_{acid}$ difference between methanol and phenol, with negligible

P. Ausloos and S. G. Lias (eds.), Structure/Reactivity and Thermochemistry of Ions, 367–380.

variation in the results for n greater than 4. This indicates that there was bias in the original ladder construction, but makes the fit of the ICR data to the D-EA values even worse.

A least-squares correlation of these "new" ICR $\delta\Delta G_{acid}$ values (at an assumed 320K) vs. the $\delta\Delta G^{o}_{acid}$ values calculated from equation (1) results in a non-unity slope, implying an effective temperature of 409K for the species in the ICR cell. Hydrogen fluoride is not included in the correlation, since its value is clearly outside the error limits. There is justification given below for this. While there is some question of the exact temperature of the cell walls in the original experiments, due to heating by the electron beam filament in the original McIver cell design,[1] the 409K value derived here is well outside the range of macroscopic temperatures ever measured by thermocouple on the cell itself. Using the 409 effective temperature and the least-squares anchoring as described, we thus have a re-anchored scale. As further upport for this realignment of the acidity scale, we find that:

(1) Acidities obtained at 500 to 600K in a pulsed high pressure mass spectrometer[3] now agree with the "new" ICR scale to within 0.4 kcal/mol for acids that are common to both scales.

(2) If the acidity of HF is set not by the ICR equilibrium experiments, but only by the D-EA number, it now becomes 0.9 kcal/mol more acidic than neopentyl alcohol. It was 0.5 kcal/mol less acidic in free energy than $tBuCH_2OH$ in the old scale. This is consistent with the observations of Moylan and Brauman concerning the kinetics of fluoride reacting with $tBuCH_2OH$ and $PhCH_2OH$,[2] and with the acidity overlaps of HF, HCF_3, and H_2O_2 measured in the flowing afterglow[4]. We believe that the pressure of the HF in the ICR spectrometer was not accurately measured during the original experiments, due to vacuum system conditioning.

What is the source of this temperature discrepancy? There has long been debate on whether the ions in the ICR spectrometer are thermalized, with respect to those in "high pressure" techniques such as the flowing afterglow and pulsed CI mass spectrometry. The present results seem to be evidence for that contention, but the source of the excess energy is not clear. It is possible that the "equilibria" measured are not true equilibria, but steady state populations.[5] However, the cationic equilibria scales in the ICR spectrometer agree quite well with those from high pressure techniques over large energy differences[6]. This may be due to the presence of barriers on the reaction surface being prevalent in the π-π proton transfers that predominate in the anion reactions. For the cationic basicity scale, it is expected that any barriers will be much smaller, since most proton transfers are between non-bonding lone pairs that are not delocalized. A second source of excess energy may be from the axial electrostatic trapping well.[7]

In order to quantitate such internal temperature effects, an ad hoc group has been formed at this meeting to determine reactions and equilibria which can function as "thermometer" reactions. These are reactions with rate or equilibrium constants that are sensitive and well established functions of temperature. The reactions suggested are merely first round possibilities; based on the experimental outcome in the collaborators' laboratories, other reactions may be put forward as better "thermometers".

Two rate constants are suggested as non-temperature dependent standards that should be run in all participating labs, to verify the numerical accuracy of the instrumentation:

$$CH_4^{+\cdot} + CH_4 \rightarrow CH_5^{+} + CH_3^{\cdot} \qquad k_{coll} = 13.35x10^{-10}$$

$$Ar^{+\cdot} + H_2 \rightarrow ArH^{+} + H^{\cdot} \qquad k_{rxn} = 2/3 \text{ of } 15.10x10^{-10}$$

For the "thermometer" reactions, two rate constants and two equilibria have been suggested as first round candidates:

$$toluene{\cdot}H^{+} + m\text{-}xylene = m\text{-}xylene{\cdot}H^{+} + toluene$$

$$HC{\equiv}C^{-} + MeC{\equiv}CH = HC{\equiv}CH + MeC{\equiv}C^{-}$$

$$Ar^{+\cdot} + N_2 \rightarrow N_2^{+\cdot} + Ar$$

$$Me_2CH^{+} + Me_3CH \rightarrow Me_3C^{+} + Me_2CH_2$$

REFERENCES

1. Bartmess, J.E.; Scott, J.A.; McIver, R.T.,Jr. J. Am. Chem. Soc. 1979 101, 6046.
2. Moylan, C.R.; Brauman, J.I. J. Phys. Chem. 1984 88, 3175.
3. Cumming, J.B.; Kebarle, P. Can. J. Chem. 1978 78, 1.
4. Bierbaum, V.M.; Schmitt, R.J.; DePuy, C.H.; Mead, R.H.; Schulz, P.A.; Lineberger, W.C. J. Am. Chem. Soc. 1981 103, 6262.
5. Su, T., in "Kinetics of Ion-Molecule Reactions," P. Ausloos, Ed., Plenum Press: New York, 1978, pp.250-251.
6. Aue, D.H.; Bowers, M.T., in "Gas Phase Ion Chemistry", vol. 2, M.T.Bowers, Ed., Academic Press, 1979, Ch.9
7. Rempel, D.L.; Huang, S.K.; Gross, M.L. Int. J. Mass Spectrom. Ion Proc. 1986 70, 163.

The "New" Acidity Scale - Methanol to Phenol

Acid	$\Delta G°_{acid}$	$\Delta H°_{acid}$		Acid	$\Delta G°_{acid}$	$\Delta H°_{acid}$	
p-xylene	375.4	382.5	b	PH_3	363.6	371.1	a
$PhSCH_3$	374.9	382.4	c	tBuCH(Et)OH	363.6	370.2	a
CH_3OH	374.7	381.3	a	$PhCH_2OH$	363.6	370.2	a
$PhCH_3$	374.3	381.4	a	PhC≡CH	363.3	371.1	a
MeC≡CH	374.0	381.8	a	m-CF_3-toluene	362.9	370.1	b
$PhCH_2Me$	373.9	380.4	a	tBuCH(iPr)OH	362.3	368.9	a
m-xylene	373.4	380.5	a	m-cyanotoluene	362.2	369.4	b
$PhCHMe_2$	373.4	379.5	a	CH_3COCH_3	362.0	369.2	a
p-fluorotoluene	372.9	380.0	b	p-methylaniline	361.3	368.5	a
CH_2=$CFCH_3$	372.9	379.6	f	p-CF_3-toluene	360.9	368.1	b
nPrC≡CH	372.5	380.3	a	p-methoxyaniline	360.9	368.2	a
EtOH	371.1	377.7	a	$(tBu)_2CHOH$	360.9	367.5	a
tBuC≡CH	370.5	378.3	a	m-methylaniline	360.5	367.7	a

o-difluorobenzene	370.0	378.2	h
nPrOH	369.5	376.1	a
cycloheptatriene	369.2	375.2	a
HCF_3	369.2	377.0	a
iPrOH	368.8	375.4	a
nBuOH	368.7	375.3	a
HOOH	368.5	375.8	j
HC≡CH	368.4	376.6	a
m-fluorotoluene	368.2	375.4	b
iBuOH	368.0	374.6	a
CH_3CONMe_2	368.0	374.8	a
tBuOH	367.9	374.5	a
Me_2CHCN	367.7	375.0	a
c-$(CH_2)_2$CHCN	367.6	375.2	a
sBuOH	367.4	374.0	a
$MeCH_2CN$	367.2	374.8	a
CH_2=C(CF_3)CH_3	367.2	373.8	f
m-chlorotoluene	366.8	374.0	b
HC_2F_5	366.7	372.5	l
$Me_2CHCH_2CH_2CH_2OH$	366.6	373.2	i
n-C_6H_{13}OH	366.6	373.2	i
$MeOCH_2CH_2OH$	366.6	373.6	a
1,3-dithiane	366.6	374.0	m
$MeSOCH_3$	366.6	373.7	a
Et_2CHOH	366.2	372.8	i
$tBuCH_2OH$	366.1	372.7	a
n-C_7H_{15}OH	366.0	372.6	i
$tBuCH_2CH_2OH$	366.0	372.6	i
$nC_8H_{17}OH$	365.4	372.0	i
nBuC(Me)$_2$OH	365.2	371.8	i
CH_3CN	365.2	372.9	a
CH_3CO_2Me	365.1	371.9	a
nPrC(Me)$_2$OH	365.0	371.6	i
iPrCH(Et)OH	365.0	371.6	i
tBuCH(Me)OH	364.9	371.5	a
n-C_9H_{19}OH	364.8	371.4	i
CH_2=CHCN	364.7	370.5	n
CH_2=C(CN)CH_3	364.6	371.2	f
Et_3COH	364.5	371.1	i
$MeOCH_2CN$	364.5	372.1	a
m-formyltoluene	364.4	371.6	b
(iPr)$_2$CHOH	364.3	370.9	i
Et_2NOH	364.1	370.7	o
FCH_2CH_2OH	363.8	370.4	p
SiH_4	363.7	372.2	a
CH_3COCH_2F	360.0	366.8	d
F_2CHCH_2OH	359.9	367.1	a
aniline	359.9	367.2	a
Me_2C=NOH	359.9	366.9	a
CH_3CH=O	359.7	366.5	a
MeCH=NOH	359.3	366.3	a
$MeCH_2$CH=O	359.2	365.8	a
Ph_2CH_2	359.1	364.5	a
m-nitrotoluene	359.0	366.2	b
$CH_3SO_2CH_3$	359.0	366.6	a
p-fluoroaniline	358.1	365.4	a
$(FCH_2)_2$CHOH	357.9	364.5	q
B$(CH_3)_3$	357.7	365.1	k
tBuCH=NOH	357.5	364.5	a
F_3CCH_2OH	356.5	364.1	a
m-CH_3S-aniline	356.3	363.6	a
CF_3CH(Me)OH	356.1	363.7	e
$PhCOCH_3$	356.0	362.8	a
CF_3C(Me)$_2$OH	355.9	363.5	e
m-CH_3O-acetophenone	355.5	362.3	a
m-fluoroaniline	355.3	362.6	a
p-formyltoluene	353.9	361.1	b
p-chloroaniline	353.5	360.8	a
m-chloroaniline	353.0	360.3	a
H_2NSH	352.7	359.0	g
pyrrole	352.2	360.0	a
MeSH	351.7	358.0	a
CH_3COCH_2F	351.3	358.8	d
Me_2CHNO_2	351.2	357.3	a
CH_3NO_2	350.9	357.6	a
m-CF_3-aniline	350.7	358.0	a
$MeCH_2NO_2$	350.6	357.1	a
CH_3CF=O	350.0	356.8	d
EtSH	349.9	356.2	a
$tBuCH_2NO_2$	349.3	356.1	a
nPrSH	348.7	355.0	a
cyclopentadiene	348.5	354.7	a
iPrSH	346.9	354.1	a
tBuSH	346.8	353.1	a
p-CF_3-aniline	346.7	354.1	a
PhCH=NOH	346.5	353.5	a
H_2S	345.4	351.8	a
$PhCH_2CN$	344.8	351.4	a
HCN	343.8	351.2	a
phenol	343.2	350.1	a

a. J.E. Bartmess; J.A. Scott; R.T. Jr. McIver; J. Am. Chem. Soc. 1979 101, 6047.
b. G. Caldwell; J.E. Bartmess; unpublished results.
c. S. Ingemann; N.M.M. Nibbering; J. Chem. Soc. Perkin II 1985, 837.
d. R. Farid; T.B. McMahon; Can. J. Chem. 1980 58, 2307.

e. G. Caldwell; T.B. McMahon; P. Kebarle; J.E. Bartmess; J.P. Kiplinger; J. Am. Chem. Soc. 1985 107, 80.
f. J.E. Bartmess; R.D.Burnham; J. Org. Chem. 1984 49, 1382.
g. V.M. Bierbaum; J.J. Grabowski; C.H. DePuy; J. Phys. Chem. 1984 88, 1389.
h. J.M. Riveros; S.M.J. Briscese; J. Am. Chem. Soc. 1975 97, 230.
i. G. Boand; R. Houriet; T. Baumann; J. Am. Chem. Soc. 1983 105, 2203.
j. V.M. Bierbaum; R.J. Schmidt; C.H. DePuy; R.H. Mead; P.A. Schulz; W.C. Lineberger J. Am. Chem. Soc. 1981 103, 6262.
k. M.K. Murphy; J.L. Beauchamp J. Am. Chem. Soc. 1976 98; 1433.
l. S.A. Sullivan; J.L. Beauchamp J. Am. Chem. Soc. 1976 98; 1160.
m. J.E. Bartmess J. Am. Chem. Soc. 1980 102; 2483.
n. J.E. Bartmess; R.L. Hays; H.N. Khatri; R.N. Misra; S.W. Wilson J. Am. Chem. Soc. 1981 103, 4746.
o. J.E. Bartmess; T. Basso; R.M. Georgiadis J. Phys. Chem. 1983 87, 912.
p. J.H.J. Dawson; K.R. Jennings Int. J. Mass Spectrom. Ion Phys. 1977 25, 47.
q. R.L. Clair; T.B. McMahon Int. J. Mass Spectrom. Ion Phys. 1980 33, 21.

THE LIMITING VALUES OF IONIZATION POTENTIALS OF 1-SUBSTITUTED ALKANES - Joel F. Liebman (Dept. of Chemistry, University of Maryland, Baltimore County Campus, Baltimore MD USA)

In this note we present a simple method of obtaining the limiting value of the ionization potentials of 1-substituted alkanes with long carbon chains. These quantities are important input numbers in the method of Bachiri et al.[1] for estimating the ionization potentials of generally substituted alkanes. From the literature equations, we can derive the simple relationship

$$[IP(RX) - IP(R_\infty X)]/[IP(R_0 X) - IP(R_\infty X)] = \Phi(R) \qquad (1)$$

wherein X is the substituent, $IP(R_0X)$ and $IP(R_\infty X)$ represent the IP's for the initial and limiting members of the suitable set of 1-substituted alkanes and Φ is a universal function of the alkyl group R. While Bachiri, et al. considered these two IP's to be fitting parameters, we alternatively assert they have physical meaning and so should correspond to the experimentally measured values. Quite clearly $IP(R_\infty X)$ is unmeasurable because $R_\infty X$ is but a mathematical construct. Even taken as a limiting quantity it may prove difficult to measure because of the low volatility of the derivatives of the higher alkanes. We now seek to derive this quantity. We note that in eqn. 1 the right hand side is independent of the substituent X and therefore the left hand side must also be independent of X. Equivalently, the following expression is true to the extent that Bachiri et al. analysis is true for all substituents X and Y.

$$[IP(RX) - IP(R_\infty X)]/[IP(R_0 X) - IP(R_\infty X)] = [IP(RY) - IP(R_\infty Y)]/[IP(R_0 Y) - IP(R_\infty Y)] \qquad (2)$$

Accepting expression (2) as true for all alkyl groups, say R and R', dividing both sides of eqn. (2) with alkyl group by eqn. (2) with alkyl group R' gives the new identity

$$[IP(RX)-IP(R_\infty X)]/[IP(R'X)] = [IP(RY)-IP(R_\infty Y)]/[IP(R'Y)-IP(R_\infty Y)] \quad (3)$$

We take R = Me and R' = Et, the two smallest alkyl groups. While X is in principle arbitrary, we take it as $-CH=CH_2$. This choice was made because alkenes are generally simply synthesized and functionalized, there is already considerable energetics data on neutrals and ions alike, and because there is a unique π orbital from which the electron is ionized as opposed to the quasi-cylindrical pair of π orbitals for alkynes and the σ ionization from ketones and aldehydes.

Data on adiabatic ionization potentials is known[2] through R = n-$C_{11}H_{23}$ and so we take this alkyl group to be R_∞. In the table that follows, we will give the limiting ionization potentials for a collection of RY derivatives, the values of IP for the 1-Y-alkane with the largest number of carbons given by Bachiri et al., ny, from experiment,[3] from the estimate by Bachiri et al.,[5] and from using the new eqn. 4 (where R = Me, R' = Et and X = $-CH=CH_2$ as before). Admittedly, eqn. 4 only "looks" plausible and no proof is offered here.

$$[IP(RX) - IP(R_{ny}X)]/[IP(R'X)] = [IP(R_{ny}Y)]/[IP(R'Y) - IP(R_{ny}Y)] \quad (4)$$

Y	$IP(R_\infty Y)$	ny	$IP(R_{ny}Y)$ expt.	$IP(R_{ny}Y)$ Bachiri	$IP(R_{ny}Y)$ pred.(4)
$-C{\equiv}CH$	9.94	6	9.95	10.04	10.00
$-C{\equiv}CCH_3$	9.28	6	9.30	9.29	9.32
$-C{\equiv}CC_2H_5$	9.16	6	9.18	9.19	9.20
$-CH=CHCH_3$(E)	8.96	5	8.85	8.91	8.97
-OH	9.96	6	9.89	10.10	10.09
$-OCH_3$	9.31	4	9.54	9.52	9.39
$-OC_2H_5$	9.23	4	9.36	9.36	9.28
-CH=O	9.58	5	9.69	9.67	9.65
$-COCH_3$	9.24	4	9.26	9.35	9.30
$-COC_2H_5$	9.04	4	9.02	9.05	9.07

It is seen that in general the limiting values of the ionization potential, $IP(R_\infty Y)$, are numerically reasonable and that the results of eqn. (4) are of comparable quality to those predicted by Bachiri et al. using their parameterized expression.

References and comments:

1. M. Bachiri, G. Mouvier, P. Carlier and J. E. DuBois J. Chim. Phys. 1980 77, 899.
2. This value, and all other experimental values used in this study, is taken from the evaluated data compendium on the energetics of gaseous ions and related neutrals by J. E. Bartmess, J. L. Holmes, R. D. Levin, S. G. Lias and J. F. Liebman, J. Phys. Chem. Ref. Data, to be submitted.
3. While "experiment" is cited here, it includes the evaluated numbers

appearing in reference 2 as well as directly measured and reported quantities.

4. These aren't always the original values given in ref. 1 but those presented in reference 2 wherein the predictions of Bachiri et al. were updated by using evaluated input numbers for their equation.

THE HYDRIDE AFFINITY SCALE - Robert R. Squires (Dept. of Chemistry, Purdue University, West Lafayette IN USA)

Hydride affinities are useful and instructive thermochemical parameters for gas phase negative ion studies. The hydride affinity of a neutral compound, HA(X), is defined in eqn. 1 as the enthalpy of dissociation of the corresponding anion HX^- to free hydride and the neutral X.[1]

$$HX^- \xrightarrow{\Delta H^\circ} H^- + X \qquad (1)$$

$$HA(X) = \Delta H^\circ = \Delta H^\circ_f(H^-) + \Delta H^\circ_f(X) - \Delta H^\circ_f(HX^-)$$

These data may be readily calculated from known heats of formation for neutrals and the corresponding anions, the latter being derived from gas phase acidity measurements. From known hydride affinities, values for other neutrals may be determined experimentally by bracketing techniques, much the same way as proton affinities are bracketed. In certain cases, relative hydride affinities can also be measured via equilibrium hydride transfer reactions or other thermochemical cycles. A listing of selected hydride affinities for a variety of neutral compounds is provided in Table 1. All data were taken from the Anion Thermochemistry Data-Base[2] and the Sussex-NPL compilation[3] unless otherwise noted.

1. Bartmess, J. E.; McIver, R. T. Jr. in "Gas Phase Ion Chemistry," Bowers, M. T., Ed., Academic Press, New York, 1979.
2. Bartmess, J. E. J. Phys. Chem. Ref. Data, in press.
3. Pedley, J. B.; Rylance, J. Sussex-NPL Computer Analyzed Thermochemical Data, U. Sussex, 1977.

Table of Hydride Affinities

Neutral(X)	$\Delta H^\circ_f(X)$,kcal/mol	Anion(XH^-)	$\Delta H^\circ_f(XH^-)$,kcal/mol	HA(X)
C_2	200.0	C_2H	64.2	170.1
$Fe(CO)_4$	-178.3	$(CO)_4FeH-$	-105.6a	107.4
CH_2	101.8	CH_3-	33.0	103.5
C_6H_4	118.0	C_6H_5-	52.9	99.8
CH	142.1	CH_2-	77.8b	99.0
$(CF_3)_2CO$	-334.0	$(CF_3)_2CHO-$	-384.4	85.1
$CH_2=CHNO_2$	13.4	CH_3CHNO_2-	-32.0	80.1
$PhSO_2CH=CH_2$	-30.8	$PhSO_2CHCH_3-$	-72.2	76.1
BH_3	22.0	BH_4-	-18.3c	75.0
BEt_3	-36.5	$HBEt_3-$	-74.8d	73.0
CH_3NCO	-15.0	CH_3NCHO-	-53.0e	72.7

HC=CCN	84.0	CH_2=CCN-	47.3	71.4
CS_2	28.0	HCS_2-	-4.6f	67.3
COS	-33.1	HCOS-	-65.7f	67.3
PhNCO	-5.3	PhNCHO-	-35.0[e]	64.4
CH_2=C=O	-11.4	CH_2CHO-	-38.9	62.2
CH_2=CHCHO	-18.0	CH_3CHCHO-	-44.8	61.5
CH_2=CHCOCH_3	-31.4	CH_3CHCOCH_3-	-56.0	59.3
CF_3CHO	-189.0	$CF_3$$CH_2$O-	-213.5	59.2
CH_3CH=CHCHO	-24.8	$CH_3$$CH_2$CHCHO-	-49.2	59.1
CF_3COCH_3	-194.0	CF_3(CH_3)CHO-	-218.2	59.0
O_2	0.0	HO_2-	-23.3g	58.0
FCH_2CHO	-78.0	F$CH_2$$CH_2$O-	-101.0	57.7
CH_2=CHCN	43.2	CH_3CHCN-	20.3	57.6
CH_2=C(CH_3)CN	36.0	$(CH_3)_2$CCN-	14.0	56.7
$Fe(CO)_5$	-173.0	$(CO)_4$FeCHO-	-194.4h	56.1
F_2CHCHO	-130.0	F_2CHCH_2O-	-147.0	51.7
CO_2	-94.1	HCO_2-	-111.0	51.6
CH_2=C=CH_2	45.6	CH_2CHCH_2-	29.9	50.4
PhCH=CH_2	35.3	PhCHCH_3-	19.6	50.4
$(tBu)_2$CO	-82.6	$(tBu)_2$CHO-	-97.5	49.5
PhC(CH_3)CH_2	26.9	Ph$C(CH_3)_2$-	12.7	48.9
n-C_6H_{13}CHO	-63.1	n-C_7H_{15}O-	-74.3	45.9
PhCHO	-8.8	PhCH_2O-	-20.1	45.9
$Cr(CO)_6$	-217.1	$(CO)_5$CrCHO-	-227.1i	44.7
$Mo(CO)_6$	-219.0	$(CO)_5$MoCHO-	-229.0i	44.7
$W(CO)_6$	-211.4	$(CO)_5$WCHO-	221.4i	44.7
$(iPr)_2$CO	-74.4	$(iPr)_2$CHO-	-83.7	44.0
n-C_4H_9CHO	-55.0	n-C_5H_{11}O-	-63.8	43.5
$(CH_3)_2$CHCHO	-51.5	$(CH_3)_2$CHCH_2O-	-60.2	43.4
CH_2=O	-26.0	CH_3O-	-32.5j	41.2
(iPr)(Et)CO	-68.4	(iPr)(Et)CHO-	-76.6	42.9
Et_2CO	-61.8	Et_2CHO-	-69.3	42.2
cyclopropene	66.0	cyclopropyl-	59.0k	41.7
$CH_3$$CH_2$CHO	-44.8	$CH_3$$CH_2$$CH_2$O-	-51.9	41.8
(iPr)(Me)CO	-63.0	(iPr)(Me)CHO-	-69.2	40.9
tBuCH_2CHO	-64.4	tBu$CH_2$$CH_2$O-	-73.5	43.8
n-PrCHO	-49.6	n-PrCH_2O-	-57.4	42.5
CH_3CHO	-39.6	$CH_3$$CH_2$O-	-45.6	40.7
$CH_3$$CH_2COCH_3$	-57.6	$CH_3$$CH_2$CH($CH_3$)O-	-63.3	40.4
NO	21.5	HNO-	16.01	40.2
$(CH_3)_2$CO	-51.9	$(CH_3)_2$CHO-	-56.8	39.6
CH_2=CH-CH=CH_2	26.3	CH_3CHCHCH_2-	22.0m	30.0
C_2H_2	54.2	C_2H_3-	54.8k	34.1
CH_3N=CH_2	17.5	$(CH_3)_2$N-	26.0	26.2
CH_2=S	-21.5	CH_3S-	-12.2f	25.4
CH_2=NH	22.6	CH_3NH-	32.0	25.3
HCN	32.3	H_2CN-	43.4f	23.6
c-C_6H_6	19.8	c-C_6H_7-	31.7n	22.8
SiH_4	7.3	SiH_5-	20.0o	22.0
H_2O	-57.8	H_3O-	-39.0p	16.0
CH_2=CH_2	12.5	C_2H_5-	35.2q	12.0

Et_3SiH	-72.7	$Et_3SiH_2^-$	-48.0[o]	10.0
CO	-26.4	HCO^-	0.3[r]	8.0
NH_3	40.1	NH_4^-	66.8[s]	8.0
Atoms				
O	59.6	OH^-	-32.7	127.0
S	66.2	SH^-	-17.1	118.0
Se	48.4	SeH^-	-19.6	102.7
C	171.3	CH^-	113.6	92.4
Si	107.7	SiH^-	60.5	81.9
P	79.8	PH^-	36.9	77.6
Be	76.6	BeH^-	59.5	51.8
Ca	46.0	CaH^-	33.2	47.5
Mn	67.1	MnH^-	67.7[t]	34.1
Fe	99.3	FeH^-	100.3[t]	33.7

a. Stevens, A.E.; Beauchamp, J.L. unpublished results.
b. Leopold, D.G.; Murray, K.K.; Stevens-Miller, A.E.; Lineberger, W.C. J. Chem. Phys. 1985 83, 4849.
c. Altshuller, A.P. J. Am. Chem. Soc. 1955 77, 5455.
d. Workman, D.B.; Squires, R.R., unpublished results.
e. Estimate based on measured data for analogous systems.
f. Kass, S.R.; DePuy, C.H. J. Org. Chem. 1985 50, 2874.
g. Oakes, J.M.; Harding, L.B.; Ellison, G.B. J. Chem. Phys. 1985 83, 5400.
h. Lane, K.R.; Sallans, L.; Squires, R.R. Organometallics 1985 3, 408.
i. Lane, K.R. Ph.D. Thesis, Purdue University 1985.
j. Moylan, C.R.; Brauman, J.I. J. Phys. Chem. 1984 88, 3175.
k. Froelicher, S.W.; Freiser, B.S.; Squires, R.R. J. Am. Chem. Soc. 1986 108, 2853.
l. Ellis, H.B.; Ellison, G.B. J. Chem. Phys. 1983 78, 6541.
m. Lee, R.E.; Squires, R.R., unpublished results.
n. Lee, R.E.; Squires, R.R. J. Am. Chem. Soc. 1986 108, 0000.
o. Hajdasz, D.J.; Squires, R.R. J. Am. Chem. Soc. 1986 108, 3139.
p. Kleingeld, J.C.; Nibbering, N.M.M. Int. J. Mass Spec. Ion Proc. 1982 27, 108.
q. DePuy, C.H.; Bierbaum, V.M.; Damrauer, R. J. Am. Chem. Soc. 1984 106, 4051.
r. Murry, K.K.; Miller, T.M.; Leopold, D.G.; Lineberger, W.C. J. Chem. Phys. 1986 84, 2520.
s. Kleingeld, J.C.; Ingemann, S.; Jalonen, J.E.; Nibbering, N.M.M. J. Am. Chem. Soc. 1983 105, 2474.
t. Miller, A.E.S.; Feigerle, C.; Lineberger, W.C. J. Chem Phys. 1986 84, 4127.

ON THE NEED TO DETERMINE "EFFECTIVE" TEMPERATURES IN MEASURING THE RATES OF ION/MOLECULE REACTIONS - Michael Henchman (Dept. of Chemistry, Brandeis Univ., Waltham MA USA), David Smith, and Nigel Adams (Dept. of Space Research, Univ. of Birmingham, Birmingham UK)

During the past fifteen years, there has been increasing awareness of the need to determine a temperature as an essential part of measuring rate constants of ion/molecule reactions.[1-10] In the first two sections, we survey that need and the design of suitable thermometers[9] to satisfy the need. In the third section, we propose some possible themometers. The fourth section lists some reactions where different values for the rate constant have been obtained with different techniques - and we suggest how the apparent differences may be reconciled.

1. The Need.
For ion/molecule reactions, there is a particular need to characterize rate constant measurements with an "effective" temperature. Many reactions are studied using more than one technique and an astonishing variety of techniques is available.[1] (Binary rate constants are even measured by sampling from a satellite in the ionosphere![11]) In general, the various experimental rate constants show reassuring agreement. For some reactions, however, they do not. How should this apparent disagreement be interpreted? Is one value right and the other wrong? Or are they both wrong? Or are they both "right", even though the actual values differ? The third possibility - rate constants that differ but which are actually compatible - becomes feasible whenever the rate constant is temperature dependent and whenever the measurements are made at different temperatures. In such cases it is essential to characterize each rate constant measurement with an "effective" temperature.

Beyond the need to establish "best" values for rate constants, there are fundamental reasons why it is desirable to measure temperature. Over a wide enough pressure range, binary rate constants, measured at the same temperature, may be expected to show a pressure dependence. This can occur when the time between collisions becomes comparable to the lifetime of the collision. (Bohringer's studes[12] are a recent example of the extensive literature on this topic.) Ion/molecule reactions provide an excellent opportunity to investigate this effect because the large variety of techniques available span an enormous range of pressures (10^{-9} to 10 torr). Clearly pressure dependence can only be investigated if the measurements are made at a fixed temperature.

Can a temperature always be measured - particularly for techniques, such as the drift tube, where strictly the system is not in a thermodynamic state? The rigorous answer to this question is negative; nevertheless, an "effective" temperature can be measured and this parameter can characterize the system in a way that is imprecise but still useful. Although the question is a thermodynamic one, the answer may be formulated in terms of kinetic theory. The system exhibits a characteristic distribution (velocities, vibrational states, etc.) and the "effective" temperature is just a first moment of that distribution: to specifiy the distribution precisely, one needs an infinite set of moments, which is of course an equivalent distribution.

2. DESIGNING THERMOMETERS

Measuring the temperature or an "effective" temperature requires a thermometer. For the present purpose, a thermocouple on the wall of the reaction chamber is an inadequate thermometer because it cannot probe the electric fields within the chamber, and these have a significant influence on the distribution and hence on the "effective" temperature. "Chemical" thermometers[12] determine temperature directly by measuring some property of a chemical reaction that is temperature dependent. These properties include:

(I) The equiibrium constant of the reaction. The reaction cannot be thermoneutral but there are conflicting design criteria. Sensitivity requires a reasonable slope for the van't Hoff plot, i.e. a sizable heat of reaction, but the equilibrium constant must be small enough to lie within the dynamic range of the technique. In practice an enthalpy chnage of $\sim$2 kcal/mole is convenient. The equilbrium constant is obtained either directly from a system at equilibrium (ICR) or from measured forward and backward rates (ICR or flow tube).

(II) Either the rate constant or the product distribution of the reaction.[13] For the former, an accurate pressure measurement is needed; for the latter, it is not. Thus reactions with temperature dependent product distributions are potentially attractive thermometers.[6] However, product distributions cannot always be measured directly. In a SIFT, an extrapolation to zero reactant pressure is necessary if the product ions themselves undergo secondary reaction.

(III) The rate constant for endothermic reaction of known threshold energy, E_T. Measurement of the rate constant determines an "effective" temperature., if a functional form for the temperature dependence of the rate constant is assumed (e.g. $k(T) = k_{capture}$ * $\exp[-E_T/RT]$).[7]

Several restrictions apply to the selection of a suitable chemical thermometer. The reagent gas or gases should be sufficiently volatile at 300K to be usable for all techniques (vapor pressure >10^{-4} torr for a flow tube).; the gases should not be "sticky"; they should not require purification because unnecessary uncertainty is introduced; and it is simpler if only one reagent gas is needed. While many isotope exchange reactions show strong temperature dependences,[14] it can be difficult to measure some of them with sufficient accuracy, even on a SIFT.

3. POSSIBLE THERMOMETERS

The rate constant: $Ar^+ + N_2 \rightarrow Ar + N_2^+$

This appears to be an attractive candidate and it has been used as such in an FTICR.[15] The rate constant rises steeply with increasing temperature/energy - from $1x10^{-11}$ cm^3/molecule-s at 300K to $1.2x10^{-10}$ cm^3/molecule-s at a center-of-mass energy of 1 eV in a drift-tube experiment.[15a] The reagents are readily available and are particularly easy to handle. The simplicity of reactants allows a "translational" temperature to be measured. There are some limitations: (1) At an effective temperature of ∿1000K, the rate constant should only be $3x10^{-11}$ cm^3/molecule-s. Thus the magnitude of the rate constant lies at the lower limit of an ICR measurement throughout the whole temperature range 300-1000K. (2) While the rate constant shows an energy dependence above thermal energies, it appears to show no temperature dependence in the range 80-300K.[16b] (3) Beacuse the reverse reaction occurs at the collision rate, flow tube mesurements must use a SIFT, not a flowing afterglow; in an ICR, pulsed ejection of the nitrogen ion is necessary. (4) In a flow tube, the Ar^+ reactant appears to be in its lower energy state, $^2P_{3/2}$,[16b] but in an ICR the $^2P_{3/2}/^2P_{1/2}$ ratio should be statistical (67%:33%). The rate constant is a factor of 3 smaller for the excited state. This uncertainty could introduce confusion.

Certainly this thermometer can reveal the presence of ions with suprathermal energies.

The Equilibrium: $HS^- + HCN = CN^- + H_2S$

This thermometer has already been used to measure temperature.[4] Forward and backward rate constants were measured at 296±2K. An estimate of $\Delta S°_{298}$= +2.1 e.u. gives $\Delta H°_{298}$ = -1.0±0.2 kcal/mole. The disagreement between this measurement and an ICR measurement[17] can be explained if the effective temperature of the ICR cell is assigned to be 427K.[4] It is important to confirm by measurement the value of the entropy change.

The Equlibrium: $HCO^+ + HCl = CO + H_2Cl^+$

This could be an alternative equilibrium with a slightly larger enthalpy change, which has been measured to be $\Delta H°$ = 2.7 kcal/mole.[18]

4. CASE STUDIES

We list below several reactions for which more than one technique has been used to measure the rate constant at thermal energies. In each case there is apparent conflict. The conflict can be resolved from the known remperature dependence of the rate constant, by infering that one of the measurments was at suprathermal energies. In principle each reaction could be used as a thermometer; in several cases there are practical limitations.

a). Product distributions from the reaction

$$N^+ + O_2 \rightarrow NO^+ + O$$
$$\rightarrow O^+ + NO$$
$$\rightarrow O_2^+ + N$$

measured in a SIFDT as a function of drift field[6] are reconciled with the ICR measurement[19] if the effective energy of the latter experiment is assigned as ∿0.5 eV. The product distribution shows little energy dependence at low energies.

b). A SIFT measurement of the reaction $^{13}C^+ + {}^{12}CO \rightarrow {}^{13}CO + {}^{12}C^+$ in the temperature range 80-500K[20] can be reconciled with an ICR measurement[21] if the latter is assigned an effective temperature of ~500K.

c). A SIFT measurement of the reaction $H_3^+ + HD \rightarrow H_2D^+ + H_2$ in the temperature range 80-300K[22] can be reconciled with an ICR measurement[23] if the latter is assigned an effective temperature of ~1000K.

d). From a SIFT measurement of the endothermic reactions

$$S^+ + H_2S \rightarrow H_2S^+ + S \quad \text{and} \quad HS^+ + H_2S \rightarrow H_2S^+ + SH$$

at 295K,[7] the effective temperature of an ICR measurement[24] can be estimated to lie in the range 700-900K.

e). A SIFDT measurement[16a] of the reaction $Ar^+ + N_2 \rightarrow Ar + N_2^+$ can be reconciled with an ICR measurement[25] if the latter is assigned an effective temperature of >600K.

f). A SIFT measurement of the reaction

$$\begin{aligned} CH_4^+ + CD_4 &\rightarrow CH_4D^+ + CD_3 \\ &\rightarrow CH_3D_2{+} + CHD_3 \\ &\rightarrow CH_2D_3{+} + CH_2D_2 \\ &\rightarrow CHD_3{+} + CH_3D \\ &\rightarrow CHD_4{+} + CH_4 \end{aligned}$$

in the temperature range 80-300K[14] is compatible with an ICR measurement if the latter is assigned an effective temperature of ~1000K.

5. PRESENT NEEDS

Evidence now suggests that in certain experiments, nominally thermal measurements may have been made at "effective" temperatures as high as 500-1000K. The first task is to develop simple, crude chemical thermometers to answer - yes or no - the following question: is the "effective" temperature of this measurement >500K? In the majority of cases, we know that this cannot be correct, and we need the experimental ability always to be able to answer that question definitely. The second task is to develop more refined chemical thermometers to measure "effective" temperatures, say to ±20K. The third task is to develop chemical thermometers to measure "effective" temperatures, for example vibrational as compared to translational. Substantial progress has been made in this last case in the development of monitoring techniques for studying vibrational relaxation.[27] The fact that vibrational relaxation can be studied in flow tubes emphasizes that even after millions of collisons with the helium bath gas, thermal ions can still be vibrationally hot. The ion H_3^+ is particularly troublesome in this regard[22] and care must always be taken to monitor the vibrational energy and, if necessary, remove it.

REFERENCES

1. Henchman, M., in "Ion-Molecule Reactions", Vol. 1, J. Franklin, Ed., Plenum Press: New York, 1972; pp. 154, 162, 238-241.
2. Henchman, M., in "Interactions between Ions and Molecule," P. Ausloos, Ed., Plenum Press: New York, 1975; pp. 21-23.
3. Meisels, G.G., in "Interactions between Ions and Molecule," P. Ausloos, Ed., Plenum Press: New York, 1975; p 595.
4. Betowski, D.; MacKay, G.; Payzant, J.; Bohme, D. Can. J. Chem. 1975 53, 2365. See p. 2369.
5. Chesnavich, W.J.; Su, T.; Bowers, M.T.; J. Chem. Phys. 1976 65, 990.
6. Howorka, F.; Dotan, I.; Fehsenfeld, F.C.; Albritton, D.L. J. Chem. Phys. 1980 73, 759.
7. Smith, D.; Adams, N.G.; Lindinger, W. J. Chem. Phys. 1981 75, 3365.
8. Adams, N.G.; Smith, D., in "Reactions of Small Transient Species," A. Fontijn and M.A.A. Cyyne, Eds., Academic Press: London, 1983, p. 333.
9. Henchman, M.; Smith, D.; Adams, N.G., presented at the 31st Ann. Conference on Mass Spectrom. and Allied Topics, Boston MA, May 8-13 1983.
10. Riveros, J.M.; Jose, S.M.; Takashima, K. Adv. Phys. Org. Chem. 1985 21, 197. See pp. 208-209.
11. Torr, M.R.; Torr, D.G. Rev. Geophys. Space Phys. 1982 20, 91.
12. Bohringer, H. Chem. Phys. Lett. 1985 122, 185; Bohringer, H.; Arnold, F. J. Chem. Phys. 1986 84, 1459.
13. Steel, C.; Starov, V.; Leo, R.; John, P.; Harrison, R.G. Chem. Phys. Lett. 1979 62, 121.
14. Smith, D.; Adams, N.G., in "Ionic Processes in the Gas Phase", M.A. Almoster Ferreira, Ed., Reidel: Dordrecht, 1984, p. 41.
15. Nibbering, N.M.M., presented at the NATO Advanced Study Institute on Chemistry of Ions in the Gas Phase, Vimero, Portugal, Sept. 6-17, 1982.
16. (a) Lindinger, W.; Howorka, F.; Lukac, P.; Kuhn, S.; Villinger, H.; Alge, E.; Ramier, P. Phys. Rev. A 1981 23, 2319; (b). Smith, D.; Adams, N.G. Phys. Rev. A 1981 23, 2327.
17. McIver, R.T. Jr., Eyler, J.R. J. Am. Chem. Soc. 1971 93, 6335.
18. Smith, D.; Adams, N.G. Ap. J. 1985 298, 827.
19. Anincich, V.; Huntress, W.T.; Futrell, J.H. Chem. Phys. Lett 1977 47, 488.
20. Smith, D.; Adams, N.G. Ap. J. 1980 242, 424.
21. Huntress, W.T. Jr. Ap. J. Suppl. 33 1977, 495.
22. Smith, D.; Adams, N.G. Ap. J. 1981 248, 373.
23. Huntress, W.T. Jr.; Anincich, V. Ap. J. 1976 208, 237.
24. Huntress, W.T. Jr.; Pinizzotto, R.F. J. Chem. Phys. 1973 59, 4742.
25. Laudenslager, J.B.; Huntress, W.T.; Bowers, M.T. J. Chem. Phys. 1974 61, 4600.
26. Huntress, W.T. J. Chem. Phys. 1972 56, 5111.
27. Ferguson, E.E. J. Phys. Chem. 1986 90, 731.

ENTROPY-DRIVEN REACTIONS:

SUMMARY OF THE PANEL DISCUSSION

Michael Henchman[1]

Air Force Geophysics Laboratory
Hanscom Air Force Base
MA 01731, U. S. A.

Many exothermic ion-molecule reactions occur at the collision limit. Can *endothermic* ion-molecule reactions still be fast, if "driven" by a compensating entropy increase ? This question continues to excite controversy. To some it is anathema. For others it is as self-evident as the sweating of a feverish brow or the charring of a cake —"entropy-driven" processes that are familiar to all schoolchildren, although not perhaps in those terms.

The controversy raises interesting and fundamental questions. Under what conditions can entropy-driven reactions occur ? Consider the two extremes of pressure. At one extreme, at high pressures, we describe the experiments as being "collision-dominated", providing efficient coupling between system and surroundings. Under these conditions, at constant temperature and pressure, endothermic reactions can proceed spontaneously provided that the free energy decreases. (It is useful to follow biochemical usage and call such processes *exergonic.*[2]) The reaction will be exergonic and spontaneous if an entropy increase offsets the endothermicity. At the other extreme, consider the reaction occurring at zero pressure — for example, in a "single-collision" experiment using beams. Under these conditions, endoergic[3] reactions *cannot* proceed, because that would violate the conservation of energy. By considering these two extremes, we therefore see how endoergic reactions must be *forbidden* at low pressures but may be *allowed* at high pressures. If they are allowed, there must be an intermediate pressure at which a "switching" occurs. What is switched *off* at low pressure can be switched *on* at high pressure. The reaction is switched on at high pressure, when the entropy increase can drive the endothermic reaction. In contrast, at the low-pressure limit, the system is isolated ; it is not in a state of thermodynamic equilibrium ; and its state cannot be described by a true thermodynamic temperature.

As contributions to this chapter show, entropy-driven ion-molecule reactions have been observed. The analysis, presented above, suggests that it would be interesting to study these as a function of pressure. Where an entropy-driven reaction has been observed in a high-pressure mass spectrometer, we know that we should not expect to see it in a beam experiment ; but should we expect to see it, for example, in a FTICR ? (Viewed another way, the analysis presented here suggests a diagnostic test for identifying the thermodynamic state of a system.) Ion-molecule reactions are particularly suitable for this kind of investigation because (a) exothermic reactions generally occur without any kind of energy barrier, and (b) the experimental techniques available span an enormous range of pressure, from 10^{-9} to 10 torr.

P. Ausloos and S. G. Lias (eds.), Structure/Reactivity and Thermochemistry of Ions, 381–399.

The issues raised in this discussion can be presented in another way. Under what conditions should enthalpy or free energy provide the criterion legislating whether a reaction is forbidden or allowed ? This is an ancient question which is still being discussed today.

ΔG or ΔH ? : A Little History

"Proton-transfer reactions proceed on every collision if exothermic but not if endothermic." For the past thirty years the literature of ion-molecule reactions has been peppered with such statements of the enthalpic condition for facile reaction. About five years ago, enthalpy began to be replaced by free energy, as the legislative property.[4] This poses two questions. First, what caused this change ? Second, why was enthalpy chosen in the first place, given that the thermodynamic condition for spontaneous change for a system at constant T and P, is $\Delta G < 0$?

The first question must have an answer but I do not know it. The second question asks if enthalpy (the original choice) is indeed the correct one. Or has thirty years of repeated use, like the recital of any litany, simply induced belief ? There seems to be two plausible reasons why enthalpy was chosen in the first place. First, the experiments of that time (at source pressures of 10^{-3} to 1 torr) were identified as "single-collision" experiments. Reactions could only "go" if the reactants had enough energy ; and thermal reactants could not drive endothermic reactions. (This is of course the "zero-pressure" criterion, discussed above.) Second, reaction enthalpies were the only thermodynamic data available in the early days — and these were estimated from appearance potentials and threshold energies, often measured under non-equilibrium conditions in the presence of electric fields. Accordingly enthalpy was the *pragmatic* choice for the criterion — both intuitively plausible and available, in the absence of other thermodynamic information.

How valid are these reasons today ? First, we would identify the pressure range of the early measurements to be in the upper range of today's measurements — certainly not in the "zero-pressure" limit. Put another way, we might expect today to see some "entropy-driven" reactions under these conditions. Second, we have available today a wide range of thermodynamic information ; and we are not restricted to enthalpy for want of other choice.

In summary, the arguments, which led to the use of enthalpy in the past, provide no definitive answer in the present. Before proceeding with the question, it is important to establish that the question is worth asking.

ΔG or ΔH : When Is There a Recognizable Distinction ?

Is there a perceptible difference between using enthalpy and free energy as a criterion for reaction ? To answer this, let us categorize various types of reaction according to Δn, the change in the number of molecules in the reaction. Most ion-molecule reactions that have been studied, are bimolecular reactions yielding two products, i.e. $\Delta n = 0$. For such reactions entropy changes are typically ≤ 5 cal/K mol, making $T.\Delta S \leq 1.5$ kcal/mol at 300 K. Endothermic reactions can only be "driven" by this entropy increase for endothermicities of $\leq$ 1.5 kcal/mol. In some cases this is comparable to the uncertainty in the reaction enthalpy. It is therefore not surprising that no distinction can be clearly made for this category of reaction. Indeed, Bohme's correlations of rate constants with ΔG^o and with ΔH^o are practically indistinguishable for reactions of this type.[5] Clearly we should not look to standard bimolecular reactions of this type ($\Delta n = 0$) for identifiable examples of entropy-driven reactions.

Where two reactants yield three products ($\Delta n = +1$), typical entropy increases of ~ 20 cal/K mol yield values of T.ΔS of ~ 6 kcal/mol at 300 K. Indeed one of the first observations of an entropy-driven reaction was for just such a system, namely reaction (1) :

$$CH_3ONO + F^- \rightarrow NO^- + HF + H_2CO \qquad \Delta H^o = +6\ ;\ \Delta G^o < 0\ \text{kcal/mol} \qquad (1)$$

(In the experiment no actual ionic product was detected because NO^- undergoes collisional detachment.)[6] For reactions of this type, the endothermicity is large enough to demonstrate clearly the occurrence of entropy-driven reactions .

The above discussion provides an introduction to the six contributions which follow and which are grouped in three pairs. Mautner reports on a series of reactions of the type $\Delta n = 0$ but where the entropy change is extremely large, due to internal hydrogen bonding in the ionic reactant or product. Lias provides commentary on these results. Fernandez, Jennings and Mason, and Stone discuss a series of proton-transfer reactions involving substituted benzene rings. Finally Squires and Hierl *et al.* report on a series of solvated-ion reactions of the type $\Delta n = +1$, where sizeable entropy increases are to be expected.

ENTROPY-DRIVEN ION-MOLECULE REACTIONS (M. Meot-Ner [Mautner], Center for Chemical Physics, National Bureau of Standards, Washington, D. C., 20234, U. S. A.)

Most exothermic proton transfer and charge transfer reactions proceed with near unit efficiency ($r \approx 1$), while endothermic reactions are slow. Also, in most reactions $\Delta S°$ is small, and $\Delta H°$ and $\Delta G°$ are of the same sign. However, in some cases $\Delta S°$ is significant, and in some of these cases $\Delta H°$ and $\Delta G°$ are opposite in sign, i.e. endothermic processes may be exergonic. These are good test cases to answer the question: Is the thermochemical factor controlling the kinetics $\Delta H°$ or $\Delta G°$? Table 1 shows some such reactions.

The results show clearly that in these reactions the controlling factor is $\Delta G°$. Reactions (2), (3), (5), (6) and (7) are endothermic, but exergonic since $\Delta S°$ is positive. These are entropy-driven reactions. In contrast, in the reverse direction, reactions (4), (8) and (9) are exothermic but endergonic, since $\Delta S°$ is negative. These are entropy-inhibited reactions.

The present reactions are a special case of fast ion-molecule reactions which are characterized by reaction efficiencies $r \approx 1$, or, more generally (see below) $r_{forward} + r_{reverse} \approx 1$. This is in contrast to slow reactions, where $r_{forward} + r_{reverse} < 1$. Slow reactions are usually treated by the double-well model, and the slow rate is ascribed to reflection from a free-energy barrier (low density of states at the barrier separating the two minima in the well). Evidently, in fast reactions such reflection is absent, and they can be treated by a single-well model [equation (10)] involving a single complex which is formed by the forward or reverse direction collision :

$$A^+ + B \underset{k_b}{\overset{k_c}{\rightleftarrows}} [AB]^{+*} \underset{k'_c}{\overset{k_p}{\rightleftarrows}} B^+ + A \qquad (10)$$

TABLE 1. KINETICS AND THERMOCHEMISTRY OF SOME ION-MOLECULE REACTIONS WITH LARGE ENTROPY CHANGES[a]

A^+	B	r[b]	K/(1+K)	ΔH°	ΔS°	ΔG°	T (K)
Charge Transfer $A^+ + B \rightarrow B^+ + A$							
(2) n-$C_8D_{18}^+$	c-C_8H_{16}	0.78[c]	0.87	+4.6	+29.3	-5.0	335
(3) n-$C_9H_{20}^+$	c-$C_6H_{11}CH_3$	0.58[c]	0.93	+6.3	+39.7	-7.1	335
(4) c-$C_8H_{16}^+$	n-C_8D_{18}	0.12[c]	0.13	-4.6	-29.3	+5.0	335
Proton Transfer $AH^+ + B \rightarrow BH^+ + A$							
(5) $HO(CH_2)_3NH_3^+$	$(CH_3)_3N$	0.96[d]	0.91	+14.6	+49.7	-9.2	483
(6) $CH_3CO(CH_2)_3CO(H^+)CH_3$	2-F pyridine	0.67[e]	0.89	+9.2	+34.2	-10.0	567
(7) $H_2N(CH_2)_3NH_3^+$	2,6-di-Me pyridine	0.90[e]	0.90	+36.0	+88.2	-11.8	580
(8) $(CH_3)_3NH^+$	$HO(CH_2)_3NH_2$	0.067[e]	0.098	-14.6	-49.7	+9.2	483
(9) 2-F pyridine-H^+	$(CH_3COCH_2)_2CH_2$	0.10[e]	0.11	-9.2	-34.2	+10.0	567

a. ΔH° and ΔG° in kJ/mol ; ΔS° in J/mol K. Note that reaction (4) is the reverse of reaction (2), that reaction (8) is the reverse of reaction (5), and that reaction (9) is the reverse of reaction (6).

b. $r = k / k_{collision\ (ADO)}$

c. Reference 7.

d. Reference 8.

e. Reference 9.

The overall forward and reverse rate constants are given by (11) and (12):

$$k_f = k_c k_p / (k_p + k_b) \quad (11)$$
$$k_r = k'_c k_p / (k_p + k_b) \quad (12)$$

By definition, the reaction efficiency is given by $r = k / k_{collision}$. Therefore:

$$r_f = k_p / (k_p + k_b) \quad (13)$$
$$r_r = k_r / (k_p + k_b) \quad (14)$$

Therefore:

$$r_f + r_r = 1 \quad (15)$$

Equation (15) characterizes reactions that are "intrinsically fast", i.e., where there is no reflection in either direction by a barrier. This is the generalized nature of the conventional "fast reactions".

Applying the approximation $k_c = k'_c$ (which is usually good to within a factor of 2), and since $K = k_f / k_r$, equations (11) - (15) lead to:

$$K \approx k_p / k_b = r_f / r_r \quad (16)$$

which, together with (15), leads to (17) :

$$r = K / (1 + K) \quad (17)$$

In these fast reactions, therefore, the reaction efficiency is determined completely and uniquely by the overall reaction thermochemistry (K or $\Delta G°$). These relations result from a complex which decomposes to reactants or products according to the equilibrium ratio of reactants / products. This can be justified by reactant-like and product-like transition states, respectively, for these decompositions.

Of course, fast endothermic reactions are consistent with the conservation of energy. The reactions are allowed because the internal energy of the reactants is greater than the endothermicity. Therefore internal energy is converted to potential energy, and the products of such reactions should be vibrationally cooled. Also, at low temperatures all reactions should become slow in the endothermic direction. This is consistent with thermochemistry, since at low temperatures $T.\Delta S°$ is small and $\Delta G°$ assumes the same sign as $\Delta H°$.

In the present reactions, $\Delta S°$ was due to internal solvation of the charge or the proton in the flexible products, or the release of this solvation. In other cases, we observed similar kinetic effects due to rotational entropy changes, such as in the ionization of benzene,[9] and proton transfer between CO and N_2O.[5a] Other interesting candidates for entropy-driven reactions would be endothermic reactions involving cluster dissociation with a large entropy release, such as reaction (18), or endothermic dissociative proton transfer such as (19):

$$(ROH)_2H^+ + B \rightarrow BH^+ + 2\,ROH \quad (18)$$
$$BH^+ + ROH \rightarrow R^+ + H_2O + B \quad (19)$$

With respect to the reactions in Table 1, we note that the entropy change that drives the reaction is due to a change in conformation. Since this change affects the kinetics, the conformation change must occur in the reaction complex, synchronously with the charge or proton transfer.

COMMENTS ON *ENTROPY-DRIVEN ION-MOLECULE REACTIONS* BY M. MAUTNER (Sharon G. Lias, Center for Chemical Physics, Center for Chemical Physics, National Bureau of Standards, Washington, D. C., 20234, U. S. A.)

Mautner has presented a model in which the ion-molecule adduct formed by collision with either reactants ($A^+ + B$) or products ($C^+ + D$) :

$$A^+ + B \Leftrightarrow [X^+] \Leftrightarrow C^+ + D$$

undergoes dissociation to ($A^+ + B$) or ($C^+ + D$) to give the equilibrium ratio $[C^+][D]/[A^+][B]$. He further assumes that $Z_f \sim Z_r$ (where Z is the appropriate ion-molecule collision rate constant), so that the sum of the reaction efficiencies is constrained to be unity (i.e. the reaction efficiencies can be directly related to the probabilities of dissociation of $[X^+]$ to the left or right).

Under these assumptions, Mautner postulates that a highly endothermic reaction having a negative Gibbs free-energy change could occur with a rate constant *nearly as high as the collision rate constant.* He gives examples of a number of such "fast endothermic reactions " as evidence in support of such a mechanism ; in this list some reactions indicated to be endothermic by 8 kcal/mol or more (i.e. with $e^{-\Delta H/RT} \leq 1.5 \times 10^{-6}$) are shown to occur with high efficiencies. In effect the usual assumption that $e^{-E/RT}$ describes the fraction of a reactive population able to surmount an energy barrier of height E, is being suspended with the explanation that the "internal energy contents" of the reactants are sufficient to overcome the barrier.

Accepting some of the more extreme examples of "fast endothermic reactions", one is confronted by some other logical problems. For example, since :

$$K_{eq} = k_f / k_r = e^{-\Delta G/RT} = e^{-\Delta H/RT} . e^{\Delta S/R}$$

if $k_r \sim Z_r$ (k_r cannot have a temperature dependence, having reached the collision limit), k_r cannot be described by some expression accounting for the energy barrier through inclusion of a term, $e^{-\Delta H/RT}$, and the rate constant for the exothermic reaction would have to be described by an expression including the term $e^{-\Delta H/RT}$, i.e. would have to display a negative temperature dependence which depended on the exothermicity. The difficulty of contriving a potential surface to explain such a phenomenon is obvious.

A word of caution is in order. In every example cited of a "fast endothermic reaction" in which the reaction efficiency significantly exceeds $e^{-\Delta H/RT}$, the value for the $\Delta H_{reaction}$ has been established *only* by ion-molecule equilibrium constant determinations. If there is some problem with this experiment (i.e. observation of a steady state rather than a thermodynamic equilibrium, dissociation of product ions, etc), the endothermicity of the subject reactions cannot be considered to be well established, and one is in danger of "proving" the occurrence of "fast endothermic reactions" using a cyclic argument.

With this problem in mind I attempted several years ago to examine the temperature dependences of rate constants of corresponding exothermic and endothermic charge transfer reactions for which the enthalpy change of reaction was well known from experiments *not* involving equilibrium constant determinations. The results indicated that for the reactions studied, an exothermic reaction with a negative or zero entropy change could be well predicted by the expression $k_f \sim Z_f\, e^{\Delta S/R}$, and that the corresponding endothermic reaction rate constant (where the reaction shows a positive entropy change) was approximately equal to $Z_r\, e^{-\Delta H/RT}$. (This was equally true for reactions where $Z_f \sim Z_r$ and for reactions where the two collision rate constants were very different.)

A closer examination of the rate constants for one of Mautner's "fast endothermic reactions" seemed to be called for. A rate constant of 15.7 x 10^{-10} cm^3/molecule-s (corresponding to a reaction efficiency of about 1) had been measured for :

$$n\text{-}C_9H_{20}^+ + C_6H_5CF_3 \rightarrow C_6H_5CF_3^+ + n\text{-}C_9H_{20}$$

Accepting values for the ionization energies of alkanes based on equilibrium constant determinations, this reaction was cited as being endothermic by 1.8 kcal/mol. When I examined this reaction in an ICR, I found that the rate constant for the reaction (15.6 ± 1.0 and 14.9 ± 1.5 x 10^{-10} cm^3/molecule-s at 310 K and 390 K respectively) displayed no temperature dependence, and that the rate constant for the reverse reaction ($C_6H_5CF_3^+$ + n-C_9H_{20}) *increased* with temperature — a behavior consistent with an endothermic reaction. An Arrhenius treatment of the temperature dependence of the rate constant led to an estimate that the charge transfer from n-$C_9H_{20}^+$ to $C_6H_5CF_3$ is 0.6 - 0.9 kcal/mol *exothermic* . Interestingly, an estimate of the ΔH of reaction based on a van't Hoff plot of the equilibrium constant observed in the ICR gave results in reasonable agreement with the results from the high-pressure mass spectrometer (i.e. with Mautner's results), but the observed equilibrium constants were not equal to the ratio of the rate constants ; it was found that a large fraction (70% at 400 K) of the n-$C_9H_{20}^+$ ions formed by reaction, were undergoing dissociation, and that a reasonable approximation to the "correct" (i.e. $K_{eq} = k_f / k_r$) value could be obtained if one added the abundances of the fragment ions to that of n-$C_9H_{20}^+$, in computing K_{eq}. This is one specific example (similar results were obtained for several other systems but cannot be included here because of space limitations) of a case in which the thermochemical data used to infer the occurrence of a "fast endothermic reaction" is in error.

The concept of an "entropy-driven reaction" is not unusual. Observations of reactions having a positive enthalpy change but a negative Gibbs free energy change are commonplace in kinetics. However, the extreme examples presented by Mautner of highly endothermic reactions proceeding at rates near to the collision rate must, in my opinion, be treated with some suspicion — particularly of the thermochemistry. It should also be pointed out that the model suggested by my results[10] which apparently describes rate constants for some simple charge transfer reactions, in which k_f is described by $Z_f \cdot e^{\Delta S/R}$ (for ΔS negative or equal to zero), and in which k_r can be adequately predicted by $Z_r \cdot e^{-\Delta H/RT}$, poses certain questions having to do with the problems of relating rate constants and equilibrium constants to thermochemical parameters in systems for which the kinetics can be treated using collision theory.

THE THERMODYNAMICS OF SOME PROTON-TRANSFER REACTIONS (M.T. Fernandez, K. R. Jennings, and R.S. Mason, Department of Chemistry, University of Warwick, Coventry, CV4 7AL U. K.)

Proton transfer equilibria of the type

$$AH^+ + B = A + BH^+ \qquad (20)$$

involving halotoluenes were studied[11] at a single temperature and measurements of K led to their relative gas-phase basicities (GB), being derived from the relationship

$$\Delta G° = -RT \ln K = GB[A] - GB[B].$$

It was found that the gas-phase basicities of the isomers *ortho, meta* and *para* were different. This led to a study of the equilibria as a function of temperature and hence to the determination of the proton affinities of the halotoluenes and to the entropy changes for these reactions. It has often been assumed that, for proton transfer reactions, $\Delta S°$ is essentially zero or that it can be equated to the change in rotational entropy, which can be estimated using the equation

$$\Delta S° = R \ln (\sigma_{AH^+} \sigma_B / \sigma_A \sigma_{BH^+}) .$$

The latter requires one to assume a site of protonation and for simple systems, this presents no problem and good agreement between calculated and experimental values is obtained. For the C_6H_6 / C_6H_5F system, theory and experiment agree on a value of $\Delta S° = 15$ J K^{-1} mol^{-1} if *para* - protonation of the fluorobenzene is assumed.[12] In the halotoluene systems under study, $\Delta S°_{rot}$ is calculated to be between 0 - 6 J K^{-1} mol^{-1}.

On the other hand, for the system C_6H_5Cl / C_6H_5F , the calculated value of $\Delta S°_{rot}$ is zero if one assumes *para* - protonation of each reactant , whereas the experimental value of $\Delta S°$ is 2.5 J K^{-1} mol^{-1} , suggesting that some protonation may occur in the *ortho* - position.

When more than one site for protonation is available, all with similar proton affinities, another contribution which must be considered is the entropy of mixing, given by

$$\Delta S°_{mix} \leq R \ln (m / n)$$

where m and n are the number of isomers of the species BH^+ and AH^+ respectively. In the halotoluene systems studied, the maximum value for this contribution is R ln (5 / 2) = 7.6 JK^{-1} mol^{-1}.[12]

The measurement of equilibrium constants was carried out in a pulsed electron beam, high-pressure source,[12] having a pulse duration of up to 50 μs, a pressure range of 1 - 5 torr and a temperature range of 350 - 750 K. The reagent mixture consisted of CH_4 (99.8 %) , A and B (0.1 %) together with a trace of SF_6 in some cases to act as an electron scavenger. The intensities of AH^+ and BH^+ were followed for up to 5 - 10 ms and equilibrium was considered to be established when the ratio $I[AH^+] / I[BH^+]$ became constant as a function of time (usually after 1-2 ms). The establishing of equilibrium was ensured by the constancy of K when the total and partial pressures of reactants were varied. Values of $\Delta H°$

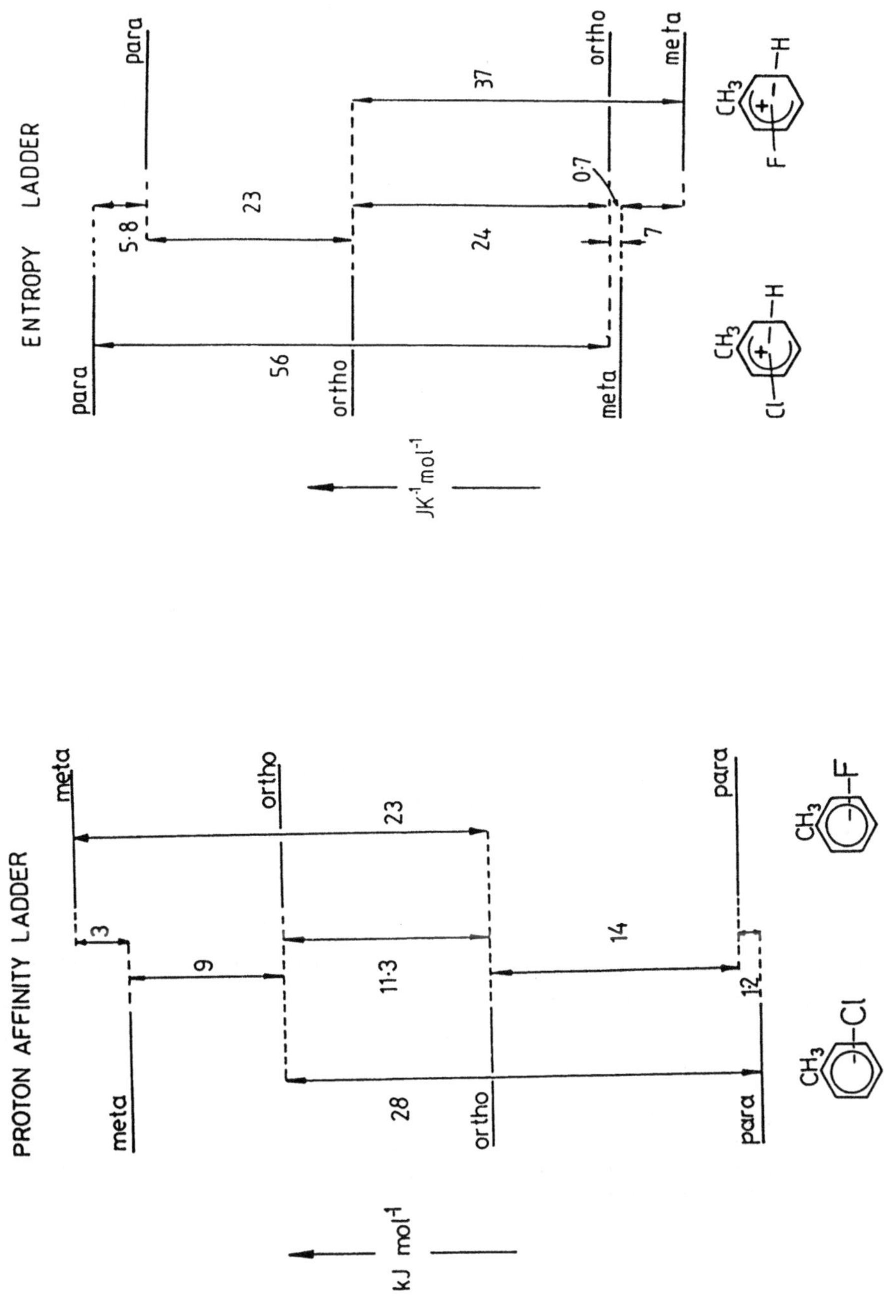

Figure 1 : Proton affinity and entropy ladders for the *ortho-*, *meta-* and *para-* halotoluenes.

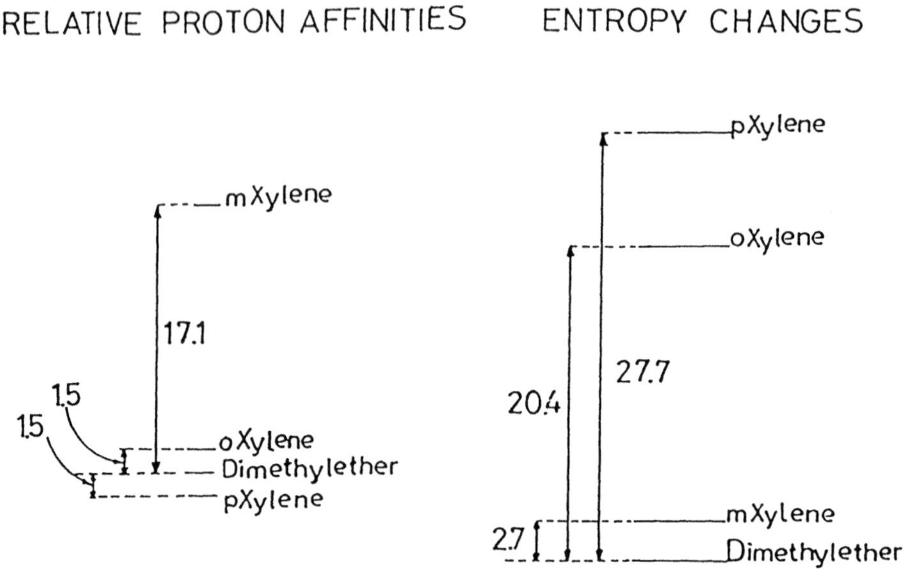

Figure 2 : Relative proton affinities and entropy changes for the isomeric xylenes.

and ΔS° were derived from van't Hoff plots and the data are presented in the form of ladders in Figures 1 and 2.

The data show very good internal consistency but some values of ΔS° are much higher than those expected from the considerations outlined above. In a similar study of systems containing the isomeric xylenes, similar trends in the relative proton affinities and entropy changes were observed :

Proton affinities : *meta* > *ortho* > *para*
Relative ΔS° : *para* > *ortho* > *meta*

Protonation was assumed to occur on the ring[13] and experiments with C_6H_5CHO / $(C_2H_5)_2O$ showed that when protonation takes place on the substituent as in this case, ΔS° = 6 J K^{-1} mol^{-1}, as expected from the normal contributions from entropies of rotation and mixing.

In summary, for isodesmic processes, such as reaction (20), no translational, vibrational or electronic contributions to ΔS° are expected and the contributions from rotation and mixing are not expected to exceed 6 and 8 J K^{-1} mol^{-1} respectively, leading to an overall ΔS° of not more than 14 J K^{-1} mol^{-1}.

The values of $\Delta S°$ observed may be rationalised in terms of the following model. In the case of *meta* - isomers, two non-adjacent protonation sites are preferred, *para* to the halogen substituent and *ortho* to the methyl group, and *vice-versa*. This leads to a high proton affinity and a low value of $\Delta S°$ with the proton firmly attached at a specific site. For the other isomers, there is no single preferred site and this leads to a lower proton affinity and a higher value of $\Delta S°$ which is assumed to arise from the motion of the proton within the plane of the aromatic ring, leading to an internal translational contribution to $\Delta S°$. Support for this comes from studies of protonated benzene in the liquid and gas phases.[14,15]

ENTROPY-DRIVEN REACTIONS (J.A. Stone, Department of Chemistry, Queen's University, Kingston, Ontario, Canada K7L 3N6)

Any non-isothermal reaction must, by definition, be exothermic in one direction and endothermic in the reverse direction. For a proton transfer reaction of the form: $B_1H^+ + B_2 \rightleftharpoons B_1 + B_2H^+$, $\Delta H = \Delta U$ and the endothermic direction is also the direction in which there is an increase in internal energy. When the reaction proceeds spontaneously in this direction, there must be an increase in entropy. In this sense all endothermic reactions may be described as entropy-driven. Most ion-neutral reactions studied are quite exothermic so that the reverse, endothermic, entropy-driven direction is not amenable to study. In order to study these entropy-driven reactions, the enthalpy change for reaction must be small. We have studied the kinetics and thermodynamics of proton transfer between tri-, tetra-, penta- and hexamethylbenzenes. The reactions provide a range of standard reaction enthalpies ($\pm$ 1.2 to $\pm$ 3.5 kcal/mol) when forward and reverse reactions are considered. The proton affinities obtained (relative to PA [ethyl acetate] = 200.7 kcal/mol as standard) are (in units of kcal/mol) : 1,3,5-trimethylbenzene (MES) = 201.0 ; 1,2,3,5-tetramethylbenzene (TMB) = 203.2 ; pentamethylbenzene (PMB) = 204.4 ; and hexamethylbenzene (HMB) = 206.6 . These proton affinity values, relative to that of benzene (PA = 181.3 kcal/mol) can be estimated with good precision (better than 1.6 kcal/mol) by assuming that the proton adds to a carbon of the ring at the most stable site available, which is determined by the number and position of the methyl substituents. Each methyl group can be assigned a stabilization energy. A methyl para to the site of protonation stablizes by 7.6 kcal/mol while for ortho and meta methyls the stabilizations are 6.4 and 2.6 kcal/mol respectively. This additivity of substituent stabilization energies supports the idea that protonation of each of these compounds leads to a single ionic structure and that a proton does not rapidly migrate between low energy sites, as Fernandez, Jennings and Mason suggest occurs in certain halogenated toluenes.[13] There are no anomalously large entropy changes in the proton transfer equilibria which again suggests that rapid intramolecular proton transfer does not occur. This lack of mobility is inconsistent with information from super-acid solution studies in which rapid intramolecular proton migration occurs at all but the lowest temperature.[16] It is possible that solvent-assisted migration may be occurring in these liquid systems.

The effect of temperature on reaction rate constants is basic to the theory of ion-molecule reactions. The rate constant k_f for the exergonic, usually exothermic, forward reaction is most easily studied. For very exergonic reactions, k_f is usually close to the collision rate constant k_c and shows only a slight temperature dependence. Exergonic reactions which have k_f significantly lower than k_c, often show a negative temperature dependence of k_f, i.e. its value decreases with increasing temperature. Such behaviour can be explained in terms of a double-well potential-energy surface.[17] The temperature coefficient of the reverse rate constant, k_b, can be obtained from the equilibrium constant for the reaction, K, and k_f, i.e. $k_b = K / k_f$. The use of this equation is limited to systems that can be studied over a range of temperatures and for which k_f is measureably less than k_c. Proton transfer among aromatics is one of the small number of such reactions that can be readily studied in a high-pressure mass spectrometer. Figure 3 shows that both k_f and k_b can have negative temperature coefficients in a given reaction.

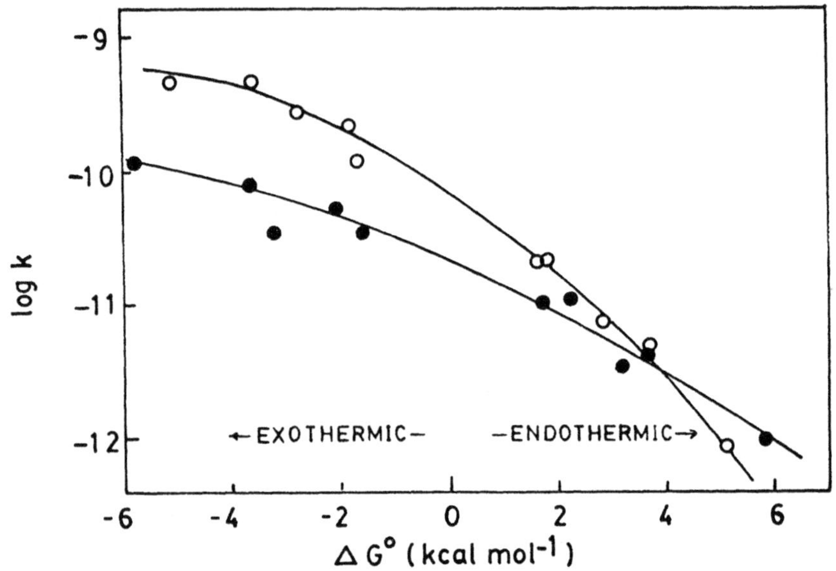

Figure 3 Rate constants for proton transfer between methyl aromatics at 400 K (o) and 600 K (•) as functions of the change in standard free energy for reaction. The data shown are for the following pairs of compounds : MES/TMB (1.77, 1.56) ; MES/PMB (3.65, 3.70) ; TMB/PMB (1.75, 2.01) ; TMB/HMB (5.03, 5.80) ; PMB/HMB (2.84, 3.20). The numbers in brackets following each pair are, in order, - $\Delta G°_{400}$ and - $\Delta G°_{600}$ for the reactions written in the exothermic directions ; the units are kcal/mol.

The changeover from positive to negative temperature dependence for k_b in the examples shown occurs when $\Delta G° = \Delta H° > + 3.5$ kcal/mol. Such behaviour is consistent with the double-well potential model in which the height of the central barrier in the reaction path relative to the reactant channel is the determining factor. When the exothermicity of the proton transfer channel is small, the behaviour of the system in both directions will be influenced by the barrier in a similar manner. The conclusion is that the sign of ΔH° for a reaction does not specify the temperature dependence of the rate constant. If a double-well potential is applicable then both exothermic and endothermic directions may show a negative temperature coefficient for the reaction rate constant. On the other hand a reaction path with a single potential well cannot show a negative temperature coefficient for the endothermic direction.

ON THE ROLE OF ENTROPY IN SOLVATED-ANION REACTIONS WITH $Fe(CO)_5$ (Robert R. Squires, Department of Chemistry, Purdue University, West Lafayette, Indiana 47907)

In the course of our studies of the gas-phase reactions of bare and partially solvated negative ions with transition metal carbonyls, we encountered several examples of fast binary reactions which were believed to be substantially endothermic, but exergonic. Our aim here is to present some of the more important details of these reactions and to point out some of the potential difficulties in identifying an "entropy-driven reaction " when solvated-ion reactants are involved.

Our experiments were carried out at 298 ± 2 K with a flowing afterglow apparatus which has been fully described.[18] A flow tube pressure of 0.4 torr (He) was employed for these experiments. We recently described the reactions of an extensive series of negative ions with $Fe(CO)_5$.[19] With strongly basic reactants such as OH^- or CH_3O^- , the primary reaction involves addition of the anion and loss of two CO ligands [equation (21)].

$$X^- + Fe(CO)_5 \rightarrow XFe(CO)_3^- + 2\,CO \qquad (21)$$
$$X^- = OH^-,\ CH_3O^-$$

In contrast, when either of these anions are associated with one or more solvent molecules, no CO loss is observed and only solvent expulsion from the metal ion product occurs [equation (22)].[20]

$$X^-(HX)_n + Fe(CO)_5 \rightarrow (CO)_4\overset{-}{Fe}\text{-}\underset{\underset{O}{\parallel}}{C}\text{-}X + [\,n\,HX \text{ or } (HX)_n\,] \qquad (22)$$
$$X^- = OH^-,\ n = 1 - 4$$
$$X^- = CH_3O^-,\ n = 1 - 3$$

The product ion structure as an iron tetracarbonyl-acyl complex, and the binding energies of negative ions to $Fe(CO)_5$ have been discussed by us.[21] Many other homoconjugate cluster ions $X^-(HX)_n$ react in this way, and in no case is any solvent retention by the product observed. The measured bimolecular rate coefficients for the reactions illustrated in equation (22) are all near the collision limit. We are confident that the observed decays of the large cluster ions are not due to precursor ion depletion effects, as discussed in the original papers.[18-21] The most interesting aspect of these reactions is the thermochemistry. By combining the measured and/or estimated $Fe(CO)_5$ binding energies of negative ions, $D[(CO)_4FeCO - X^-]$, with the known solvation energies of the clusters and other thermochemical data,[21] we could evaluate the enthalpy, entropy and free energy of a number of solvated-ion reactions with $Fe(CO)_5$. Table 2 is excerpted from our recent paper,[20] and shows that the reactions involving $OH^-(H_2O)_{3,4}$ and $CH_3O^-(CH_3OH)_3$ are exergonic, but endothermic, for the production of *separated* neutral solvent molecules as products [equation (22)]. The fate of the neutral solvent molecules cannot be determined in our experiment, and herein lies the difficulty in identifying the role of entropy in these reactions. While experimental data are lacking, high-level theoretical calculations suggest that the total solvation energy in $(H_2O)_3$ and $(H_2O)_4$[22] could be sufficient to overcome the reaction endothermicities listed in Table 2 for the $OH^-(H_2O)_3$ and $OH^-(H_2O)_4$ reactions, respectively. Thus, if the neutral waters in the $OH^-(H_2O)_{3,4}$ reactions with $Fe(CO)_5$ are

TABLE 2. THERMOCHEMICAL DATA FOR HOMOCONJUGATE CLUSTER-ION REACTIONS WITH $Fe(CO)_5$ AT 298 K[a].

Cluster Ion $X^-(HX)_n$	n	$-\Delta H_{o,n}$ [b]	$-\Delta S_{0,n}$ [b]	$D[(CO)_4FeC(O)\text{-}X^-]$ [c]	ΔH	ΔS [d]	ΔG
$NH_2^-(NH_3)_n$							
	0	...	...	60.4	-60.4	-31.2	-51.1
	1	11.9[e]	20.0[f]		-48.5	-11.2	-45.1
$OH^-(H_2O)_n$							
	0	...	...	$53.1 \leq x \leq 60.3$	-53.1	-32.8	-43.3
	1	25.0	20.8		-28.3	-12.0	-24.7
	2	42.9	42.0		-10.4	9.2	-13.1
	3	58.0	66.8		4.7	34.0	-5.4
	4	72.2	96.3		18.9	63.5	$\leq$0.0
$CH_3O^-(CH_3OH)_n$							
	0	...	...	45.7	-45.7	-32.2	-36.1
	1	21.8	21.8		-23.9	-10.4	-20.8
	2	(37.4)	(44.0)		-8.3	11.8	-11.8
	3	(50.6)	(70.0)		4.9	37.8	-6.4
$CH_3CO_2^-(CH_3CO_2H)_n$							
	0	...	...	24.0	-24.0	-34.6	-13.7
	1	27.0	26.2		3.0	-8.4	5.5

a. Energies in units of kcal/mol and entropies in eu.

b. Solvation enthalpies and entropies taken from references 18 and 21. Estimated values in parentheses, see text.

c. Estimated using equation 17 of reference 21. The values for OH^- are the experimentally determined limits.

d. Entropy changes for anion attachment to $Fe(CO)_5$ estimated from $\Delta S_{trans} + \Delta S_{rot}$: NH_2^- (-31.2 eu), OH^- (-32.8 eu), CH_3O^- (-32.2 eu), $CH_3CO_2^-$ (-34.6 eu). The ΔS value listed is the difference between the value for the relevant anion and the $\Delta S_{0,n}$ value for the relevant cluster ion, given in column four.

e. HF/4-31+G//4-31G

f. Estimate based on known values for other homoconjugate clusters.

initially expelled in associated (trimeric, tetrameric) forms, then the overall production of $(CO)_4FeCOOH^-$ may not be endothermic. Note that, were such water clusters to form, they would become rapidly dissociated at 298 K and 0.4 torr in the flow reactor, so the *overall* reaction taking place would be to produce separated water molecules.

An alternative possibility, presently favored by us, [20] is for the water molecules to initially remain associated with the metal ion product, providing sufficient *total* solvation energy to overcome the endothermicity for OH^- transfer from the cluster to the metal complex. However, each *individual* water interaction with the diffuse charge in $(CO)_4FeCOOH^-$ is weak, so rapid, sequential, dissociative loss of H_2O occurs as the product metal ion thermalizes in the helium bath prior to detection. The implication here is that efficient thermal coupling between the system (the reactants, the intermediate collision complexes, and the products) and the surroundings (the room temperature helium bath gas) be possible in order for this "entropy-driven reaction" to take place. That is, $OH^-(H_2O)_{3,4}$ and $CH_3O^-(CH_3OH)_3$ either will not react with $Fe(CO)_5$ to produce free $(CO)_4FeCOOH^-$ in a low-pressure experiment (e.g. ICR, $P = 10^{-7}$ torr), or a partially solvated metal ion product will appear.

In general, this uncertainty regarding the fate of the product solvent molecules in a solvated-ion reaction will create difficulties in clearly identifying the role of entropy in ion-molecule reactions of this class. Careful studies of the pressure and temperature dependence of the product distributions for these specific reactions, and others like them, will be critical.

As a final added note, we would like to stress the importance of evaluating the role of entropy in gas-phase ion-molecule reactions under a variety of conditions in order to ascertain whether the *observation* of so-called entropy-driven reactions is pressure dependent. This has important implications with regard to the use of thermochemical "bracketing" experiments which involve reactions with differing *numbers* of products and reactants, such as dissociative electron and proton transfers.

ON THE REACTION OF $OH^-(H_2O)$ WITH CH_3CN (Peter M. Hierl,[23] John F. Paulson, Michael Henchman,[1] Anton F. Ahrens[23] and A.A. Viggiano, Air Force Geophysics Laboratory (AFGL/LID), Hanscom Air Force Base, MA 01731, U.S.A.)

About five years ago, Paulson investigated the title reaction in a Selected Ion Flow Tube (SIFT) at 300 K in a helium buffer.[24] Two channels were observed — one exothermic and one endothermic — and the total rate constant was equal to the collision limit.

$$OH^-(H_2O) + CH_3CN \xrightarrow{30\%} CH_2CN^- + 2\,H_2O \qquad \Delta H° = +6.3 \text{ kcal/mol} \qquad (23)$$

$$\xrightarrow{70\%} CH_2CN^-.H_2O + H_2O \qquad \Delta H° = -7.2 \text{ kcal/mol} \qquad (24)$$

How could reaction (23), endothermic by ~ 6 kcal/mol [25], be occurring so efficiently ? Although the thermochemistry was, and is, uncertain, the uncertainty was not so large as to make reaction (23) thermoneutral or exothermic. We were led to the explanation that reaction (23) was an "entropy-driven" reaction ; and indeed analysis of the thermochemistry of reaction (23) reveals $\Delta G°_{300}$ = - 0.6 kcal/mol [25].

We have subsequently extended this study to cover the temperature range 240 - 363 K, using a hydrogen buffer. Only essential conclusions, pertinent to this chapter, will be presented here. The results can be fitted by two kinetic schemes. The first scheme is the one presented above — that CH_2CN^- is formed by reaction (23) by an "entropy-driven" reaction. The second scheme sets $k_{23} = 0$, disallowing the "entropy-driven" reaction. Instead it forms CH_2CN^- by collision-induced dissociation

$$CH_2CN^-.\,H_2O + M \rightarrow CH_2CN^- + H_2O + M. \qquad \Delta H° = +13.5 \text{ kcal/mol} \quad (25)$$

Kinetic analysis of reaction (25) in terms of an Arrhenius plot produces an activation energy in the range 8 - 10 kcal/mol. At the present time, we are unable to reconcile this result with the reaction enthalpy of reaction (25), for which the uncertainty is $\pm$ 1 kcal/mol.

Reaction (23) is a ($\Delta n = +1$) type process (as outlined in the introduction to this chapter). We had thought that solvated-ion reactions would be a fruitful area of kinetics for the identification of many "entropy-driven" reactions, whereby de-solvation of the reactant ion contributes to the chemical potential necessary to drive the reaction. We have found that we cannot eliminate the possibility that weakly solvated ions (bound by $<$ 15 kcal/mol) are undergoing collisional dissociation in the flow tube. Because we are unable to obtain a unique kinetic solution for the data, our preliminary conclusion is that flow-tube studies of solvated-ion reactions can contribute little at this time to the study of "entropy-driven" reactions.

CODA

The contributions to this chapter demonstrate that interesting results are being obtained at the present time ; and they show further that the interpretation of those results is not always straightforward. The following comments are added to aid that interpretation.

(1). How accurately are enthalpies of reaction known ? For example, the present uncertainty in a quantity as elementary as the proton affinity of CO is $\pm$ 2 kcal/mol.[26]

(2). Entropies of reaction can only be trusted if they are *measured* from van't Hoff plots. Calculated entropies, however plausible, can sometimes be in error, as shown, for example, in electron-transfer reactions.[27]

(3). In low pressure measurements, temperatures are uncertain. They have to be *measured*, using a chemical thermometer.[28]

(4). Generalizations about the temperature dependence of the rate constants of endothermic reactions probably should be avoided. They can increase and decrease with increasing temperature ; and they can pass through a maximum.[29] (These may be special cases but, if so, we need to be able to recognize them.)

(5). Functional forms for the rate constants of endothermic reactions, of the type $k = Z e^{-\Delta H/RT}$, have no necessary theoretical basis. They can be derived from a dynamical *model* — the so-called *line-of-centers* model.[30] They are often "derived" from the Arrhenius expression, which is *empirical.* (The Arrhenius expression is adequate for data of normal kinetic accuracy over a limited temperature range. Several recent examples, with better accuracy over an extended temperature range, show *curved* Arrhenius plots.) This particular functional form could possibly provide a valid description of the data but the validity has to be tested using the really accurate data which are now available from beam measurements.[31]

(6). The experience of Squires and Hierl *et al.,* outlined above, suggests that solvated-ion reactions are not fruitful systems for the study of "entropy-driven" reactions.

REFERENCES

(1) AFSC-URRP Visiting Professor 1984-1986. Permanent address : Department of Chemistry, Brandeis University, Waltham, MA 02254, U.S.A.

(2) See for example *McGraw-Hill Dictionary of Scientific and Technical Terms,* 3rd Edition, McGraw-Hill Inc., New York, 1984.

(3) Processes in which energy is absorbed. The term applies to single collisions and derives from nuclear physics and chemistry [G. Friedlander, J.W. Kennedy, E.S. Macias and J.M. Miller: *Nuclear and Radiochemistry,* 3rd. Edition, Wiley, New York, 1981] and is used in beam experiments [Gentry, W.R.: **1979**, in *Gas Phase Ion Chemistry* (Bowers, M.T.,ed.), Academic, New York, Vol. 2, p. 221].

(4) Hartman, K.N.; Lias, S.; Ausloos, P.; Rosenstock, H.M.; Schroyer, S.S.; Schmidt, C.; Martinsen, D. and Milne, G.W.A.: **1979**, *A Compendium of Gas Phase Basicity and Proton Affinity Measurements,* U.S. Department of Commerce, National Bureau of Standards NBSIR 79-1777.

(5) a). Bohme, D.; Mackay, G.I. and Schiff, H.I.: **1980**, *J. Chem. Phys. 73,* 4976. b). Bohme, D.: **1984**, in *Ionic Processes in the Gas Phase* (Almoster Ferreira, M.A., Ed.), Reidel, Dordrecht, p.111.

(6). King, G.K.; Maricq, M.M.; Bierbaum, V.M. and DePuy, C.M.: **1981**, *J. Am. Chem. Soc. 103,* 7133.

(7). Sieck, L.W. and Meot-Ner (Mautner), M.: **1982**, *J. Phys. Chem. 86,* 3646.

(8). Meot-Ner (Mautner), M.; Hamlet, P.; Hunter, E.P. and Field, F.H.: **1980**, *J. Am. Chem. Soc. 102,* 1393.

(9). Meot-Ner (Mautner), M.: unpublished results.

(10). Lias, S.G.: **1982**, in *Lecture Notes in Chemistry Series : Ion Cyclotron Resonance Spectrometry II,* (Hartmann, H. and Wanczek, K.-P., Eds.) Springer-Verlag, Berlin, p. 409.

(11). Mason, R.S; Bohme, D.K. and Jennings, K.R.: **1982**, *J. Chem. Soc., Faraday Trans. I, 78*, 1943.

(12). Bohme, D.K.; Stone, J.A.; Mason, R.S.; Stradling, R.S. and Jennings, K.R.: **1981**, *Int. J. Mass Spectrom. Ion Phys. 37*, 283.

(13). Fernandez, M.T.; Jennings, K.R. and Mason, R.S.: *J. Chem. Soc., Faraday Trans. II.* (submitted for publication).

(14). Olah, G.A.; Schlosbag, R.H.; Porto, R.D.; Mo, Y.K.; Kelly, D.P. and Mateeseu, G.D.: **1972**, *J. Am. Chem. Soc. 94*, 2034.

(15). Bruins, A.P. and Nibbering, N.M.M.: **1976**, *Org. Mass Spectrom. 11*, 950.

(16). Farcasiu, D.: **1982**, *Acc. Chem. Res. 15,* 46.

(17). Magnera, T.F. and Kebarle, P.: **1984**, in *Ionic Processes in the Gas Phase* (Almoster Ferreira, M.A., Ed.) Reidel : Dordrecht, p.135.

(18). Lane, K.R.; Lee, R.E.; Sallans, L. and Squires, R.R.: **1984**, *J. Am. Chem. Soc. 106,* 5767.

(19). Lane, K.R.; Sallans, L. and Squires, R.R.: **1986**, *J. Am. Chem. Soc. 108,* 4368.

(20). Lane, K.R. and Squires, R.R.: *J. Am. Chem. Soc.* in press.

(21). Lane, K.R.; Sallans, L. and Squires, R.R.: **1985**, *J. Am. Chem. Soc. 107,* 5369.

(22). Kistenmacher, H.; Lie, G.C.; Popkie, H. and Clementi, E.: **1974**, *J. Chem. Phys. 61*, 546.

(23). Permanent address: Department of Chemistry, University of Kansas, Lawrence, KS 66045.

(24). Viggiano, A.A. and Paulson, J.F: **1984**, in *Swarms of Ions and Electrons in Gases* (Lindinger, W.; Mark, T.D. and Howorka, F., Eds.) Springer-Verlag, Vienna, p.218. The product distribution reported in this reference is correct but the rate constants for reactions (23) and (24) should be 1.7 and 1.4 ($x10^{-9}$ cm^3/molec s) respectively.

(25). Hierl, P.M.; Paulson, J.F.; Ahrens, A.F.; Henchman, M. and Viggiano, A.A.: to be published.

(26). McMahon, T.B.: see the chapter in this volume.

(27). Kebarle, P.: see the chapter in this volume.

(28). See the chapter on *Temperature Determination in the Measurement of Ion-Molecule Reaction Rates* in this volume.

(29). Smith, D. and Adams, N.G.: **1984**, in *Ionic Processes in the Gas Phase* (Almoster Ferreira, M.A., Ed.) Reidel : Dordrecht, p.49. Lindinger,W. and Smith, D.: **1983**, in *Reactions of Small Transient Species* (Fontijn, A. and Clyne, M.A.A., Eds.) Academic : London, p.411.

(30). Gardiner, W.C.,Jr.: **1969**, *Rates and Mechanisms of Chemical Reactions*, Benjamin : New York, pp.83-88.

(31). Armentrout, P.: see the chapter in this volume.

ORGANIC ION/MOLECULE REACTIONS:

SUMMARY OF THE PANEL DISCUSSION

Nico M.M. Nibbering
Laboratory of Organic Chemistry
University of Amsterdam
Nieuwe Achtergracht 129
1018 WS Amsterdam
The Netherlands

INTRODUCTION

There is nowadays a large interest in studying reactions between ions (both positive and negative) and molecules in the gas phase. A major reason is that much can be learnt from such studies in the absence of solvent molecules and that in this way an insight can be obtained in both the intrinsic properties of the species involved and the rôle of solvent molecules. Another reason is that the instruments necessary for such studies undergo continuously further development and become more and more sophisticated, so that finer and more quantitative details of gas phase reactions can be uncovered.

As in the preceding NATO Advanced Study Institutes on the Chemistry of Ions in the Gas Phase (1-3), various spontaneously organized workshops were held in the afternoons or evenings around the topics presented in the plenary lectures. One of the workshops was focussed to the subject as given in the title, although widely interpreted as can be concluded from the synopses of the contributions given below.

The presentations ranged from the areas of physical organic to physical chemistry and were based upon studies which made use of various methods such as radiolysis, photodissociation, flowing afterglow and Fourier transform ion cyclotron resonance.

Ten speakers (indicated as first authors of the synopses below) were on the list for the workshop which gave rise to some pressure on them to present their contributions within the set time limits. Nevertheless, the whole workshop was a success with lively discussions and well-attended notwithstanding the fantastic views on the impressive mountains of the Mont Blanc massive and the lovely weather outside the lecture building.

P. Ausloos and S. G. Lias (eds.), Structure/Reactivity and Thermochemistry of Ions, 401–412.

PHENYL PARTICIPATION IN GAS-PHASE ACID-INDUCED SUBSTITUTION REACTIONS

(Simonetta Fornarini*, C. Sparapani* and M. Speranza, Istituto di Chimica Farmaceutica, University of Rome*, 00100 Rome and Istituto di Chimica Nucleare del C.N.R., Area della Ricerca del C.N.R. di Roma, 00016 Monterotondo Stazione, Italy)

β-phenyl substituted substrates of general formula R-CHPh-CHX-R' (R,R' = Me, H; X = OH, F, Cl) have been protonated/alkylated by gaseous charged acids, formed by γ-radiolysis of suitable parent gases at atmospheric pressure, containing small concentrations of a trapping nucleophile (MeOH). The distribution of methoxy-substituted products is consistent with a stepwise substitution process, rather than with a direct S_N2 pathway. The overall process clearly involves phenyl participation, which can occur synchronously with the leaving group departure, leading to a σ-bridged "phenonium ion", followed by nucleophilic attack by methanol at the three membered ring. The parent phenonium ion (R = R' = H) is prone to isomerization to the more stable secondary benzylic cation $Ph\overset{+}{C}H-CH_3$. When substrates containing a butyl side chain are considered (R = R' = Me), the erythro and threo isomers display a high degree of retention of stereochemistry in the methoxy-substituted products. This fact provides further evidence for the intermediacy of a cyclic species, whereby free rotation about the C_2-C_3 bond is forbidden. A quantitative estimate of the rate of phenyl participation yields a value $k \geq 10^{11}$ s^{-1}, comparable to the frequencies of rotational motions.

STEREOCHEMICAL ASPECTS OF THE PINACOLIC REARRANGEMENT IN THE GAS PHASE

(Pierluigi Giacomello*, G. de Petris*, T. Picotti, A. Pizzabiocca, G. Renzi and M. Speranza, Università di Roma "La Sapienza"*, 00185 Rome and Dipartimento di Scienze Chimiche, Università di Camerino, 62032 Camerino (Macerata), Italy)

A combination of mass spectrometric and radiolytic methods was applied to study the gas-phase pinacol rearrangement of cis- and trans--1,2-dimethylcyclopentane-1,2-diol and cis- and trans-1,2-dimethylcyclohexane-1,2-diol, promoted by D_3^+, $CH_5^+/C_2H_5^+$ and t-$C_4H_9^+$ ions in the pressure range 10^{-2}-760 Torr.

The CID mass spectra of the $[MH-H_2O]^+$ ions from the protonation of the 5- and 6-membered ring diols, under either CH_4 or iso-C_4H_{10} C.I. conditions, were indistinguishable from those of protonated 2,2-dimethylcyclopentanone and 1-acetyl-1-methylcyclopentane, respectively. This indicates that all the ions have undergone pinacol-like rearrangement within $5x10^{-6}$ s from their formation. The structural assignment was confirmed by isolation of the same ketones as neutral end products from high-pressure radiolytic experiments. Furthermore, in the presence of a strong base, $N(CH_3)_3$, good yields of 2,2-dimethylcyclohexanone accompanying the ring contraction product were also obtained from the cyclo--C-6 diols.

Under the experimental conditions employed, in the absence of competitive reactions with the solvent, dehydration of the protonated diols resulted rate-limiting for the overall rearrangement, as confirmed inter *alia* by the very low product yields from the mild $\underline{t}$-$C_4H_9^+$ acid (estimated $\Delta G° \geq -5$ kcal mol^{-1} for proton transfer to all substrates).

The use of $N(CH_3)_3$ as a trapping reagent for cationic species showed the MH^+ ions to be the only reaction intermediates surviving at least 10^{-8}-10^{-9} s in the radiolytic systems (P = 1 atm). Moreover, no epimerization product was obtained by carrying out the reaction in the presence of up to 20 Torr H_2O in the system, which provides an argument against formation of free carbenium ions as intermediates.

Competitive experiments carried out between each diol and pinacol itself, with or without added $N(CH_3)_3$, showed stereochemical effects upon the rates of rearrangement, indicating participation of the migrating group in the loss of a water molecule. Accordingly, for the 5-membered ring glycols $k_{cis} \geq k_{pinacol} > k_{trans}$, while for the 6-membered ones $k_{trans} > k_{cis} > k_{pinacol}$.

Comparison of the relative yields of C-5 and C-6 cyclic ketones from the dimethylcyclohexanediol epimers allowed as well to establish the following migrating ability scale: CH_2 > OH > CH_3.

REACTION MECHANISMS IN IONIC ARYLATION OF SIMPLE HYDROCARBONS

(Maurizio Speranza, Istituto di Chimica Nucleare del C.N.R., Area della Ricerca del C.N.R. di Roma, 00016 Monterotondo Stazione, Italy)

Detailed analysis of the mechanistic features of the gas-phase attack of arylium ions on methane, ethane, and propane, has been carried out by using the nuclear-decay method to generate the ionic reactant, either phenylium or isomeric tolyl cations, at hydrocarbon pressures ranging from 10 to 100 torr. Under these conditions, extensive fragmentation of the intermediate adducts is prevented by efficient collisional quenching. Detailed information upon their nature and isomeric composition can be obtained by isolation and structural discrimination of their neutral derivatives. In this way, arylium ions confirmed their exceedingly high affinity for hydrocarbons by directly inserting into their C-H bonds to yield the corresponding arenium intermediates. Insertion of phenylium ion into the C-H bonds of propane takes place predominantly *via* a preliminary hydride-ion transfer to form an arene-alkyl cation electrostatic adduct, wherein profound structural rearrangement of the alkyl moiety may occur. A similar mechanism is operative in the attack of isomeric tolyl cation on the secondary C-H bonds of propane, whereas that involving its primary C-H bonds proceeds *via* direct insertion with formation of the σ-bonded arenium intermediate.

These mechanisms have been discussed and compared with related processes involving the intermediacy of electrostatic ion-neutral adducts.

IS PROPYLENE OXIDE CATION RADICAL BEHAVING NON-ERGODICALLY IN ITS DISSOCIATION REACTIONS?

(Chava Lifshitz, Department of Physical Chemistry, The Hebrew University of Jerusalem, Israel)

We have studied propylene oxide cation dissociations as part of our effort to find out whether highly excited polyatomic cations behave ergodically (4,5). The method involves simultaneous activation and symmetrization of a molecule. In this case propylene oxide undergoes isomerization to a highly excited acetone cation radical. Carbon-13 labeling in $CH_3\text{-}CH\text{-}^{13}CH_2^{\cdot+}$ (with O bridging) has shown for metastable ions that the rearranged methyl in the isomerized cation radical, $CH_3\text{-}C(=O)\text{-}^{13}CH_3^{\cdot+}$ is lost preferentially (> 15:1) to the pre-existing methyl. These results were corroborated by neutral re-ionization experiments on the $^{13}CH_3^{\cdot}$ and $^{12}CH_3^{\cdot}$ fragments formed in the unimolecular dissociation taking place in the field free region of a ZAB-2F mass spectrometer. Ab initio calculations by Radom and Nobes are under way to obtain more quantitative information on the minima and maxima (transition states) along the reaction coordinate and more specifically whether isomerization to acetone involves the methoxyethylidene radical cation $CH_3\text{-}C\text{-}O\text{-}CH_3^{\cdot+}$ or the $\tilde{A}$ state of acetone.

ESTIMATION OF PHASE CHANGE ENTHALPIES FOR DERIVING THERMOCHEMICAL PROPERTIES OF ORGANIC COMPOUNDS

(Joel F. Liebman, University of Maryland, Baltimore County Campus, Catonsville, Maryland 21228, USA)

We take as our starting point a major lesson from the talks by John Holmes that the energetics of gas phase ions cannot be separated from the energetics of gas phase neutrals. A major conceptual problem arises when there is thermochemical data on the compound of interest solely in the condensed phase, i.e. the heat of formation of the compound as solid and/or liquid is available but there is no directly measured heat of sublimation or heat of vaporization data from which to derive the desired gas phase energies. What does one do? In principle, one may determine the requisite heat of sublimation or vaporization but that requires a sample of the compound and the additional confidence that it will not decompose under the necessary heating. Given the sublimation and boiling points, one can also estimate the desired quantity at these temperatures (e.g., using Trouton's rule) but it is necessary to correct it to 298 K. We started our investigations of heats of vaporization with those of hydrocarbons (6) because

1) numerous thermochemical data exists for this class of compounds
2) hydrocarbons, despite being composed of only two elements, have diverse structural features
3) they are model species for the study of substituent effects, strain and aromaticity
4) they are usually nonpolar and so more volatile.

In summary, hydrocarbons are an easier class of compounds to study than most others. From casual observation we noted that the heat of vaporization (at 298 K) of hydrocarbons was dependent primarily on the number of carbons, and using more careful statistical analysis derived the simple equation

$\Delta H_v = 1.1\bar{n}_C + .3n_Q + .7$, where $\bar{n}_C$ is the number of non-quaternary carbons and n_Q is the number of quanternary carbons.

This simple, 2-parameter equation is generally accurate to ca. +/- 1 kcal/mol. Indeed, if a little more sloppiness is tolerable, the even simpler expression

$$\Delta H_v = 1.1 n_C + RT$$

where all of the carbons in the molecule are counted, fulfills that need.

Perhaps not that surprisingly, there is no such simple relation for heats of sublimation, since now molecular shapes and packing efficiencies have become important. For example, the heats of sublimation of the isomeric and valence isomeric anthracene and phenanthrene differ by 2.5 kcal/mol, while cyclotetradecane related by "merely" the same number of carbons has a heat of sublimation ca 10 kcal/mol higher. However, remembering the simple identity

$$\Delta H_{sub} = \Delta H_{fus} + \Delta H_v$$

and that the determination of the heat of fusion is a simple, relatively low temperature (and so benign) measurement, allows us to derive the heat of sublimation by combining this new experimental measurement with our heat of vaporization relation. This, too, is generally accurate to ca 1.5 kcal/mol (7).

Of course, it must be admitted that the ion energetics community (perhaps far more than the "pure" neutral community) cares about species that contains more than just hydrogen and carbon. Preliminaty investigation shows that the heat of vaporization of non-hydrogen-bonded species obeys related regularities and near constancy of heats of vaporization independent of structure is again found. For example, for the following oxygen-containing compounds with 6 carbons, we find the accompanying heats of vaporization:

furylethylene	9.1
2-hexanone	10.8
3,3-dimethylbutanone	9.3
n-propyl ether	8.7
i-propyl ether	7.8

while for the following nitrogen-containing compounds with 6 carbons, we find:

2-methylpyridine	10.2
4-methylpyridine	10.6
1-cyanobicyclo[2.1.0]pentane	10.6
triethylamine	8.3

Simply observing that the hydrogen bonds in liquid water are worth ca 5 kcal/mol suggests all such hydrogen bonds are worth ca 5 kcal/mol. The below comparison of 6-carbon ethers and alcohols documents the approximate validity of this assertion:

propyl ether	8.7
i-propyl ether	7.8
cyclohexanol	14.8
1-hexanol	15.0

REACTIONS OF $O^{-\cdot}$ WITH METHYL FORMATE

(Hans van der Wel and N.M.M. Nibbering, University of Amsterdam, Amsterdam, The Netherlands)

Reactions of $O^{-\cdot}$ with methyl formate have been studied some years ago in a drift cell ion cyclotron resonance spectrometer (11) and recently in a Fourier transform ion cyclotron resonance (FT-ICR) instrument (12). Although in general the same observations were made, the observed abundance of the $(M-H_2)^{-\cdot}$ ions in particular was quite different in these studies. This prompted us to reinvestigate the $O^{-\cdot}$/methyl formate system with our home-made FT-ICR spectrometer and to extend our study of this system with stable isotopic labelling experiments (13).

It is shown that the $(M-H_2)^{-\cdot}$ ions are generated in low abundance if $O^{-\cdot}$ is produced from O_2 as reported in the recent FT-ICR study (12), but in a significantly higher abundance if N_2O is used as the source of $O^{-\cdot}$ as reported in the drift cell ICR study (11). It is known that the former $O^{-\cdot}$ ions are translationally considerably more excited than the latter, so that this effect could account for the observed difference of abundance of the $(M-H_2)^{-\cdot}$ ions. This has been shown to be indeed the case. The abundance of the $(M-H_2)^{-\cdot}$ ions increases if the $O^{-\cdot}$ ions generated from O_2 are translationally cooled by collisions with an inert bath gas such as Ar.

Specific deuterium labelling has indicated further that the hydrogen atoms abstracted during formation of the $(M-H_2)^{-\cdot}$ ion originate exclusively from the methyl group of methyl formate.

The $(M-H_2)^{-\cdot}$ ion has been found to react with its parent molecule through three reaction channels with a rate constant roughly of the order of 10^{-10} cm^3 $(molecule.s)^{-1}$ (reaction efficiency $\sim$ 0.2). The three reactions lead to formation of the ions m/z 45 ($\sim$ 75%), m/z 63 ($\sim$ 10%) and m/z 59 ($\sim$ 15%).

The ion m/z 59 is formed by abstraction of the formyl proton as shown by deuterium labelling, while the ion m/z 63 is the result of the Riveros reaction (14) of the CH_3O^- ion generated via α-elimination with another molecule of methyl formate to generate the methanol solvated methoxide ion.

For generation of m/z 45 as the major product ion of $(M-H_2)^{-\cdot}$ the interesting observation is made based upon the deuterium labelling that it contains the formate group of $(M-H_2)^{-\cdot}$, so that the latter formally transfers a CH group to its neutral precursor. A mechanistic rationalization of this observation is given in the Scheme below:

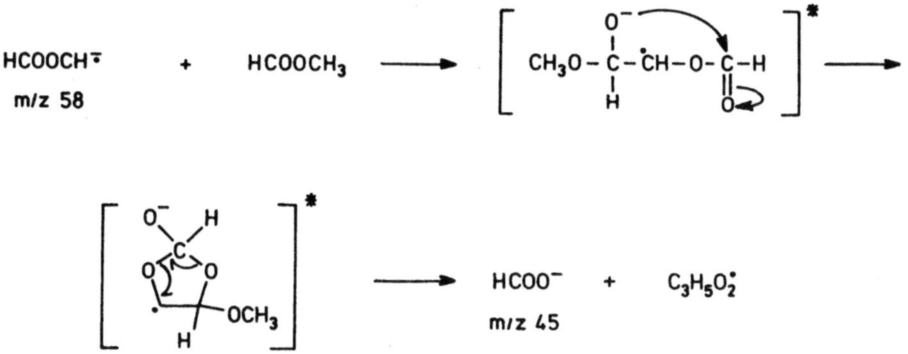

GAS PHASE REACTIONS OF THIOIC AND DITHIOIC ESTERS WITH ALKENES AND ALKYNES

(Cristina Paradisi*, M.J. Caserio and H. Keuttämaa, Dipartimento di Chimica Organica, Universita degli Studi di Padova*, 35131 Padova, Italy and Department of Chemistry, University of California, Irvine, CA 92717, USA)

Thioacylium ions, RCS^{+}, are relatively unknown and elusive species in solution. They can be generated readily in the gas phase and their ion--molecule reactions studied under typical ICR conditions (15). Of particular interest is the comparison with the better characterized acylium ions, RCO^{+}. It was shown, for example that, in contrast to the behavior of acyl derivatives, protonated thioacyl derivatives, $[CH_3C(=S)X]H^{+}$, do not thioacylate n-type nucleophiles such as water and alcohols (15). The gas phase reactivity towards π-type nucleophiles is, however, unreported.

In this context we have examined the behavior, under typical FT-ICR conditions, of two-component mixtures of methyl ethanedithioate, $CH_3C(=S)SCH_3$, 1, our model thioacylium ion precursor, and a variety of π-type nucleophiles including alkenes, alkynes, dienes and arenes. While thioacylation reactions occur to some extent with dienes and some alkynes, a more prominent process takes place with alkenes and alkynes and leads to very long-lived ionic products. An example is provided by the reaction of 1 with styrene, which on the basis of ion ejection experiments, is properly described by eq. 2.

$$\underset{m/z\ 106}{[CH_3C(=S)SCH_3]^{+\cdot}} + PhCH{=}CH_2 \longrightarrow \underset{m/z\ 195}{C_{10}H_{11}S_2^{+}} + CH_3^{\cdot} \qquad (2)$$

The reaction is quite general and takes place readily with ionized 1 and any of the alkenes and alkynes examined. These include compounds with terminal as well as internal double and triple bonds bearing alkyl and phenyl substituents.

The reaction is not unique of 1. Among compounds of general formula R-C(=X)-Y-R', those with X=S,O; Y=S; R=alkyl,CH_3O; R'=alkyl react, those with X=S,O; Y=O,N; R,R'=alkyl and also X=O; Y=S; R=alkyl; R'=H do not.

On the basis of the available evidence and also in view of the similarity with solution phase cycloaddition reactions between alkynes and S-containing 1,3-dipoles, such as 2 (16,17), it is suggested that the reaction under study proceeds via attack of an ionized 1,3-dipole, such as 3, followed by radical elimination and ring closure, as sketched in

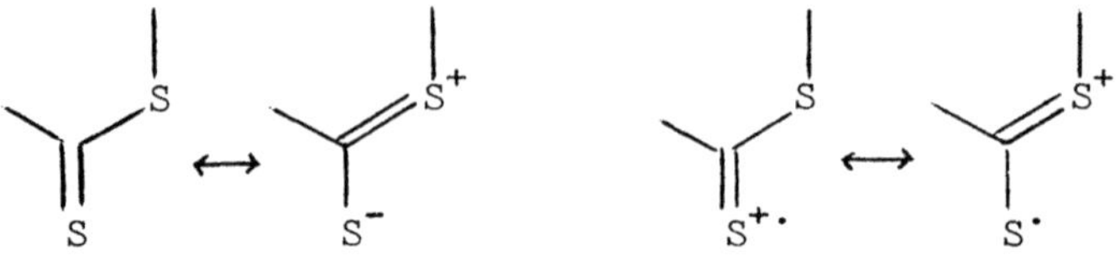

2 3

the following scheme for reaction of an alkyne:

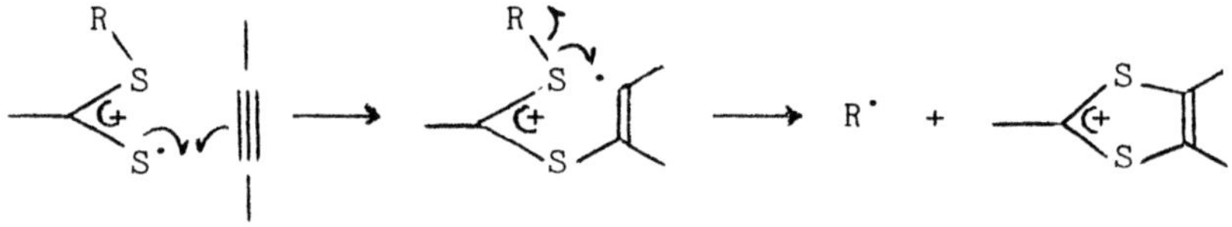

Radical elimination is the endothermic step which controls the overall reactivity. The strength of the Y-R' bond is, therefore, a crucial factor. Thus, for example, $CH_3C(=O)SCH_3$ reacts readily but its isomer $CH_3C(=S)OCH_3$, for which the overall reaction is more exothermic, does not (average bond energies of S-C and O-C in neutral compounds are (18) 280 and 360-380 kJ mol^{-1}, respectively).

PHOTODISSOCIATION, ION/MOLECULE REACTIONS AND STRUCTURES OF $C_6H_6^{+\cdot}$ AND $C_4H_4^{+\cdot}$ IONS

(Wim J. van der Hart, Department of Chemistry, University of Leiden, P.O. Box 9502, 2300 RA Leiden, The Netherlands)

1. $C_6H_6^{+\cdot}$ IONS

The structures of $C_6H_6^{+\cdot}$ ions from a number of different neutral C_6H_6 isomers were studied by photodissociation and ion/molecule reactions in rapid scan and Fourier Transform ICR spectrometers (19). Benzene and fulvene ions retain the structures of the parent neutral molecules. $C_6H_6^{+\cdot}$ ions from 1,5-hexadiyne isomerize to the benzene structure. In all other cases (1,4-hexadiyne, 1,3-hexadien-5-yne and dimethylene cyclobutene) a substantial fraction of the ions (80%, 60% and 55%, respectively) has the structure of the parent neutral molecule. For dimethylene cyclobutene 25% of the ions has ring-opened to the 1,2,4,5-hexatetraene structure. In all three cases the remaining ion fraction has the benzene structure.

These results were independent of the energy of the ionizing electron beam, indicating that the barriers for isomerization are low.

2. $C_4H_4^{+\cdot}$ IONS

Photodissociation experiments in beam instruments (20,21) on $C_4H_4^{+\cdot}$ ions from benzene and 1,5-hexadiyne seem to be in disagreement with the previous conclusion from ion/molecule reactions (22) that about 60% of the ions has the vinyl acetylene structure.

Photodissociation experiments in an ICR spectrometer with visible light from an argon-ion laser show that the fraction of photodissociating ions increases with light intensity. From further experiments it followed that on the ICR time scale photodissociation competes with the loss of internal energy of the $C_4H_4^{+\cdot}$ fragment ions by infrared light emission. Probably, the photodissociation is a complicated mixture of one- and two-photon dissociation of ions with a large amount of internal energy. The fraction of photodissociating ions at high light intensities is in good agreement with the results obtained by Ausloos (22), thus confirming the conclusion that the major fraction of the ions has the vinyl acetylene structure.

INFRARED MULTIPHOTON DISSOCIATION OF PROTONATED POLYETHERS AND HYDROXYPOLYETHERS

(John R. Eyler, Department of Chemistry, University of Florida, Gainesville, Florida 32611, USA)

We have investigated infrared multiphoton dissociation pathways of several protonated polyethers and hydroxypolyethers (23) using a Nicolet FT/MS-1000 Fourier transform ion cyclotron resonance mass spectrometer and an Apollo Model 570 cw CO_2 laser. The most general photodissociation pathways result from cleavage of a C-O bond in the ion. Examination of existing thermochemical values indicates that in general protonation lowers the dissociation energy of the C-O bond next to the "protonated oxygen". Considering this fact we have tried to explain photodissociation pathways of ethers and alkoxy alcohols by assigning them different protonation sites. The example of bis(2-(2-Methoxyethoxy)ethyl)ether demonstrates this argument. Three dissociation pathways are observed which can be explained by the following schemes:

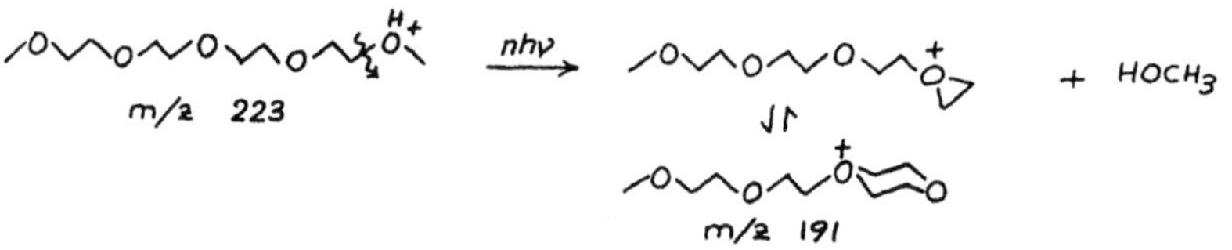

(3)

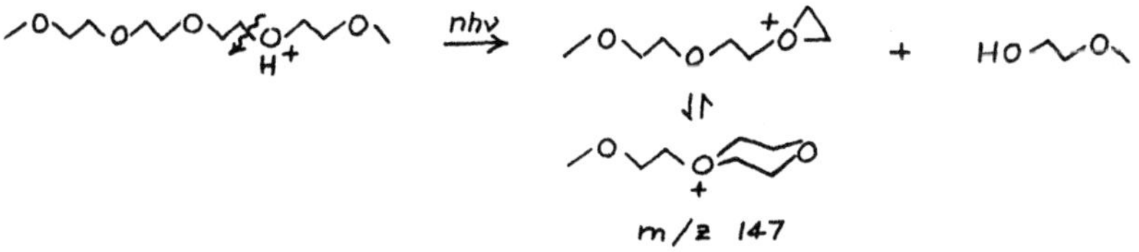

(4)

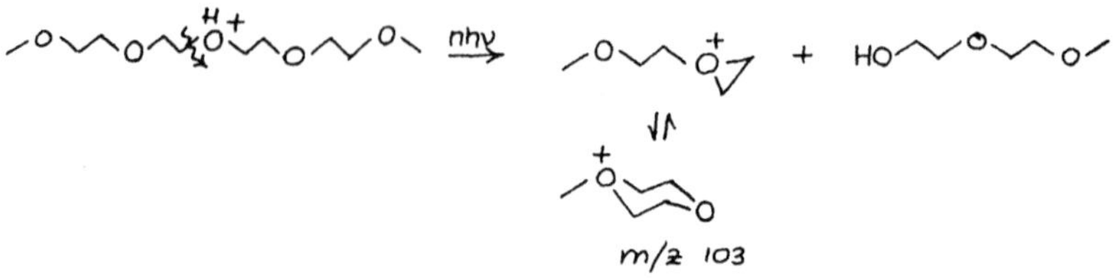

(5)

In addition to the simple cleavages shown above, other photodissociation results are best explained by rearrangements, as in 2-butoxyethanol:

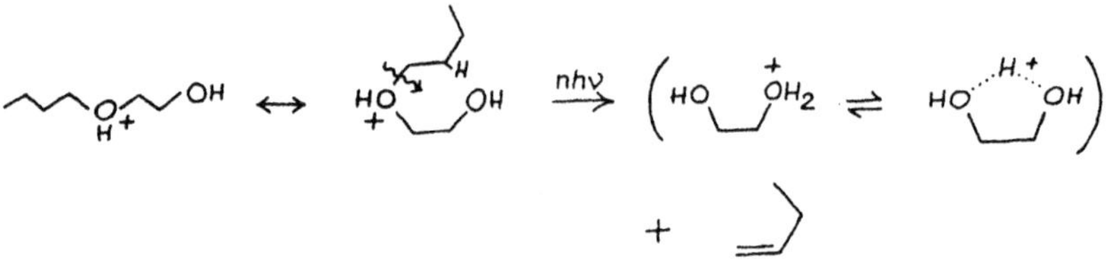

(6)

General rules derived from these dissociations are that in direct cleavage, protonated oxygen remains with the neutral product, while in a rearrangement-assisted cleavage a protonated alcohol is produced. These studies of photodissociation pathways thus yield information about the protonation sites of molecules.

The photodissociation of protonated bis-(2-methoxyethyl)ether (diglyme) produces m/z 103 and m/z 59:

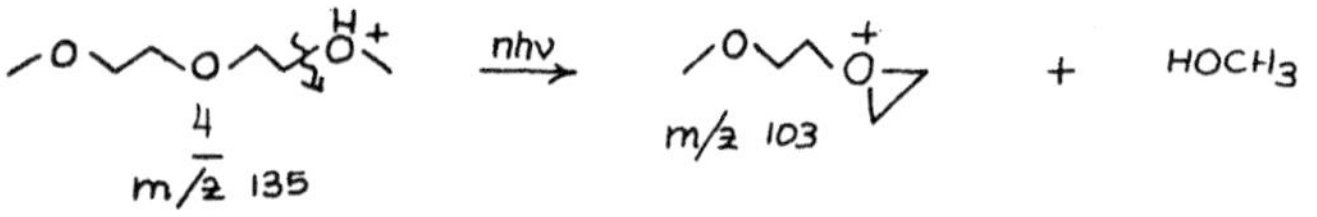

(7)

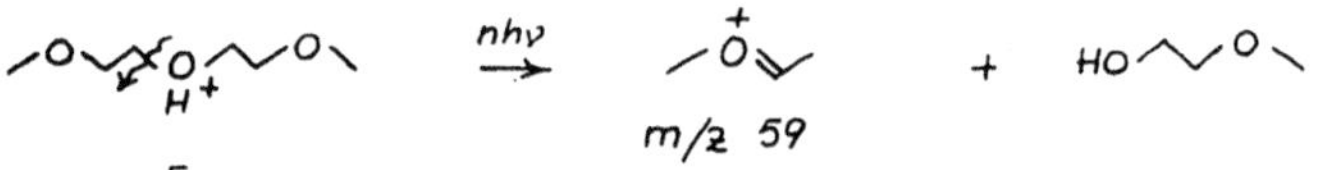

(8)

5

Since m/z 103 also photodissociates to produce m/z 59, we have ejected m/z 103 to study the direct formation of m/z 59 from m/z 135. Experiments show that the pathway to form $C_3H_7O^+$ (m/z 59) has a branching percentage of approximately 2.5%, indicating that protonation of the central oxygen is much less favored than end protonation. Protonated diglyme can form an intramolecular proton bond between oxygens (24),

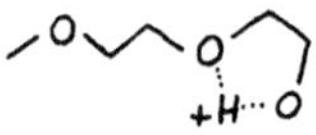

6

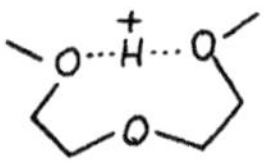

7

where the proton "bridge" in the structure 6 is more strained than in 7, and thus 7 is energetically more stable than 6. Since protonations are exothermic, the product is energetically (vibr.) excited. A relaxation of 6 by collisions with neutral diglyme molecules, especially by self-proton exchange reactions, would produce more of the structure 7. To study this possible relaxation, we inserted a variable time period after selection of the protonated molecule and before laser irradiation. With increasing "waiting time" less photodissociation leading to m/z 59 (eq. 8) was observed.

A study of self-proton exchange reactions using the $^{13}CC_5H_{15}O_3^+$ peak (m/z 136) showed that the

$$^*C_6H_{15}O_3^+ + C_6H_{14}O_3 \rightarrow C_6H_{15}O_3^+ + {}^*C_6H_{14}O_3 \qquad (9)$$

reaction has approximately the same rate constant as the decrease in production of m/z 59 (eq. 8) with "waiting time" discussed above ($k \approx 7 \times 10^{-10}$ $cm^3/(mol\ s)$). Since diglyme molecules should be good acceptors of excess vibrational energy from protonated diglyme, during the proton exchange reaction a very effective relaxation can take place leading to conversion of the protonated molecule to its more stable form (structure 7).

REFERENCES

(1) Ausloos, P. (Ed.): 1975, "Interactions between Ions and Molecules", Plenum Press, New York, N.Y.
(2) Ausloos, P. (Ed.): 1979, "Kinetics of Ion-Molecule Reactions", Plenum Press, New York, N.Y.
(3) Almoster Ferreira, M.A. (Ed.): 1984, "Ionic Processes in the Gas Phase", Reidel, Dordrecht, The Netherlands.
(4) Lifshitz, C.: 1983, J . Phys. Chem. 87, p. 2304.
(5) Lifshitz, C., Peres, T., Ohmichi, N., and Pri-Bar, I.: Int. J. Mass Spectrom. Ion Proc., in press.
(6) Chickos, J.S., Hyman, A.S., Ladon, L.H., and Liebman, J.F.: 1981, J. Org. Chem. 46, p. 4294.
(7) Chickos, J.S., Annuziata, R., Ladon, L.H., Hyman, A.S., and Liebman, J.F.: J. Org. Chem., in press.
(8) Bartmess, J.E., and McIver, R.T., Jr.: 1979, in "Gas Phase Ion Chemistry" (Bowers, M.T., Ed.), Vol. 2, Academic Press, New York, p. 87.
(9) Bartmess, J.E.: J. Phys. Chem. Ref. Data, in press.
(10) Pedley, J.B., and Rylance, J.: 1977, Sussex-NPL Computer Analyzed Thermochemical Data, U. Sussex.
(11) Dawson, J.H.J., and Nibbering, N.M.M.: 1978, Lect. Notes Chem. 7, p. 146.
(12) Johlman, C.L., and Wilkins, C.L.: 1985, J. Am. Chem. Soc. 107, p.

327.
(13) Wel, H. van der, and Nibbering, N.M.M.: Int. J. Mass Spectrom. Ion Proc., in press.
(14) (a) Isolani, P.C., and Riveros, J.M.: 1975, Chem. Phys. Lett. 33 p. 362. (b) Faigle, J.F.G., Isolani, P.C., and Riveros, J.M.: 1976, J. Am. Chem. Soc. 98, p. 2049.
(15) Caserio, M.C., and Kim, J.K.: 1983, J. Am. Chem. Soc. 105, p. 6896.
(16) Bastide, J., and Henri-Rousseau, O.: 1978, in "The Chemistry of the Carbon-Carbon Triple Bond", (Patai, S., Ed.), Part 1, John Wiley & Sons, Ch. 11.
(17) Easton, D.B.J., Leaver, D., and Rawlings. T.J., 1972, J. Chem. Soc. Perkin I, p. 41.
(18) March, J.: 1985, "Advanced Organic Chemistry. Reactions, Mechanisms and Structure", 3rd Edition, John Wiley & Sons, New York, p. 23.
(19) van der Hart, W.J., de Koning, L.J., Nibbering, N.M.M., and Gross, M.L.: Int. J. Mass Spectrom. Ion Proc., in press.
(20) Weger, E., Wagner-Redeker, W., and Levsen, K.: 1983, Int. J. Mass Spectrom. Ion Phys. 47, p. 77.
(21) Krailler, R.E., and Russell, D.H.: 1985, Org. Mass Spectrom. 20, p. 606.
(22) Ausloos, P.: 1981, J. Am. Chem. Soc. 103, p. 3931.
(23) Baykut, G., Watson, C.H., and Eyler, J.R.: manuscript in preparation.
(24) Meot-Ner (Mautner), M.: 1983, J.Am.Chem.Soc. 105, p. 4906.

EXPERIMENTAL AND THEORETICAL STUDIES OF SMALL ORGANIC DICATIONS, MOLECULES WITH HIGHLY REMARKABLE PROPERTIES

Wolfram Koch and Helmut Schwarz
Institute of Organic Chemistry
Technical University Berlin
D-1000 Berlin 12 (F.R.G.)

ABSTRACT. Using the results of both, experiments as well as high level quantum chemical calculations the exciting properties of small doubly-charged cations in the gas phase are discussed. The comparison of experimental and theoretical data shows in almost all cases good to excellent agreement. All dications studied are thermochemically extremely unstable towards charge separation reactions. These unimolecular reactions are, however, associated with high activation barriers, thus kinetically stabilizing dications. As a consequence, the generation and detection of small dications are feasible in the gas phase. However, it is very unlikely that any of those dications will ever be generated in solution as viable species; instead, they will immediately oxidize or protonate the solvent shell, which is emphasized by the extremely high negative heats of proton or charge transfer reactions calculated for all small dications.

A reversal of the stability order of structural isomers, if compared with their neutral or monocationic counterparts, is frequently observed. Moreover, dications often favour equilibrium structures which differ significantly from those of the corresponding neutral isomers, like e.g. the preference for anti-van't Hoff structures.

One of the origins of the stability reversal is the importance of donor-acceptor interactions, which is particularly pronounced for doubly-charged systems and which will be discussed in detail.

P. Ausloos and S. G. Lias (eds.), Structure/Reactivity and Thermochemistry of Ions, 413–465.

1. INTRODUCTION

"An emerging class of remarkable molecules", this attribute was given recently to small, gaseous dications by P.v.R. Schleyer /1/. Indeed, the existence of a large variety of small doubly-charged cations in the gas phase, which are stable on the μsec time scale of a mass spectrometer is now well established /2/. By means of modern experimental techniques, such as charge stripping (CS) mass spectrometry /3/ or photoion-photoion coincidence spectroscopy (PIPICO) /4/ it is now nearly a routine task to generate and detect doubly-charged cations, provided their lifetime exceeds the time necessary to pass through the mass spectrometer. However, these experimental methods are not suited for the investigation of structural aspects of such dications, while information on the energetics of their formation can be obtained. On the other hand, further characterization of these species is easily achieved by performing high level ab initio molecular orbital (MO) calculations, which can be used to elucidate the geometric and electronic structures of dications as well as their possible reactions, i.e. fragmentation and rearrangement pathways. Additionally, accurate calculations on the stability of dications, which have not yet been experimentally observed, may guide the experimentalists to focus on new, promising candidates, which are predicted by theory to exist in the gas phase /5/.

Among the experimentally known stable dications even very small ones can be found, in spite of the enormous Coulomb repulsion exerted by the two positive charges. Furthermore, the removal of electrons from a di- or polyatomic molecule does not necessarily mean weaker bonding. On the contrary, ionization may even strengthen the bonds considerably, thus leading to exotic molecules, whose neutral counterparts are often not known to exist as stable species. The most prominent example is, without any doubt, the He_2^{2+} ion. The neutral counterpart, i.e. He_2 does not form a stable molecule, however, the He_2^{2+} dication has recently been generated and detected in CS experiments by Beynon and collaborators /6/. The electrostatic repulsion in He_2^{2+} is enormous.

According to Coulomb's low, the energy released, when two point charges of 0.704 Å apart (which corresponds to the computed equilibrium distance of He_2^{2+}) separate to infinity is 472 kcal/mol! Despite this huge driving force, He_2^{2+} is predicted theoretically to have a bound (metastable) ground state ($X\,^1\Sigma^+$), and the dissociation is associated with an energy barrier, calculated to be 34.6 kcal/mol /7/. What is the nature of this barrier, to which principles do He_2^{2+} as well as many other dications owe their existence?

In early semi-empirical investigations, Hurley viewed the potential curves of diatomic dications as simple super-positions of coulombic repulsion curves with curves derived from virial-theorem based scaling of spectroscopically known states of isoelectronic neutral diatomics (e.g. H_2 for He_2^{2+} or C_2 for N_2^{2+}) /8/. That is, the bonding in XY^{2+} is regarded as being qualitatively the same as that in the neutral isoelectronic species. However, a more adequate picture of the bonding in small dications has been provided by Dorman and Morrison /9/. These workers suggested that the diabatic states describing the ground states of dications may properly be described by charge polarization type states i.e. $X\text{-}Y^{2+}$; the latter are, at small distances, strongly bound due to polarization of electrons of the formally neutral X by the Y^{2+} ion into bonding orbitals. These diabatic charge polarization states mix with the repulsive, Coulomb type diabatic states of the same symmetry, to give adiabatic potential curves (or surfaces) which exhibit minima and maxima. Basically the same concept, which was put forward in recent publications by Wetmore et al. /10/ can already be found in the early discussion of He_2^{2+} by Pauling /11/, some fifty years ago. He described the bonding interaction in He_2^{2+} in terms of resonance between valence bond structures corresponding to $He^+\text{-}He^+$ and $He\text{-}He^{2+}$, the former being repulsive, the latter being a bound charge-polarization state.

In general one may state that at small internuclear distances, the energy gained from the formation of a new bond overcomes the electrostatic repulsion, which dominates at larger distances.

In this article we present an overview of the still growing, exciting field of small, gaseous dications, which has been explored by us and several other groups. Our strategy uses a combined experimental/theoretical approach. The first section includes the discussion of various dications of hydrides, XH_n^{2+} (X=C,N,O,Si,P,S; n=1-6), whereas the following chapters are dealing with the intriguing chemistry of some larger dications, containing as heavy elements carbon, nitrogen, oxygen and fluorine atoms. Special attention is paid to the stability and structural details of isomers, the transition structures (TS) and reaction barriers connecting isomers as well as transition structures for possible fragmentation pathways. Whenever appropriate, a comparison with the corresponding neutrals and monocations will be made. The correlation of the experimentally derived ionization energies with theoretically computed values will also be discussed. Finally, in the concluding section we will describe briefly the decisive role of donor-acceptor interactions for the structures and stabilities of dications. Among the examples discussed are ylide cations (H_2CX^{2+}; X=NH_3,OH_2,FH) and dications of the type H_4CX^{2+} made up by combining H_4C^{2+}, acting as an acceptor, and various donors X (X=CO,N_2,H_2O etc.).

2. METHODS

The most widely used experimental method to study small dications in the gas phase is charge stripping mass spectrometry /3/: Monocations,m^+ formed in the ion source of a double focussing mass spectrometer (preferably of BE configuration, B magnetic and E electric sector), using any of the many existing ionization methods, are mass selected by adjusting the magnetic field B to the required value, are subsequently collided in a collision cell with a collision gas G (typically N_2 or O_2); the collision cell is located in the second field free region prior to the electric sector.

Charge stripping peaks due to reaction (1) are obtained by scanning

$$m^+ + G \longrightarrow m^{2+} + G + e^- \qquad (1)$$

the electric sector (E) voltage around E/2, where E represents the voltage required to transmit the primary ion beam of m^+ ions. The onset of the displacement of the peak from 0.5 E reflects the translational energy loss (termed Q_{min}) in process (1) and corresponds to the ionization energy (IE_{m+}) of the monocation m^+. For further details, including a discussion of the assumptions made by using this approach and leading references, the reader is referred to the recent review of Ast /3f/.

Quantum chemical calculations of the properties of molecules can be performed by either semi-empirical or ab initio MO methods. However, the former procedures are not very well suited for studies of doubly-charged cations, since these methods have been parametrized using experimental data of neutral reference molecules. In fact, it turned out, that even the calculation of mono-cationic species using MINDO/3 or MNDO may lead to erroneous results, and the virtue and strength of these methods are when using them as a qualitative guide /12/. Accordingly, almost all theoretical studies of dications employed ab initio MO methods at both Hartree-Fock and post-Hartree-Fock levels. For details of the theoretical treatments the reader is referred to the original papers, whose references are given where appropriate.

3. DICATIONS OF XH_n^{2+} HYDRIDES (X=C,N,O,Si,P,S; n=1-6)

The chemistry and physics of the CH_n^{2+} dications have formed the subject of many experimental /4a,13/ and theoretical /10b,13f,14/ investigations. In an elaborate charge stripping study, Beynon and co-workers found signals corresponding to stable dications for all CH_n^{2+} ions, up to n=5 /13c/. The measured Q_{min} values were in fair agreement with the theoretical data, computed by Pople et al. /14c/ and Siegbahn /14b/. The only qualitative discrepancy occured for the CH^{2+} ion; the experimental observation of a stable CH^{2+} ion is in distinct contrast to the theoretical finding of Pople et al. who calculated for this ion a purely repulsive potential curve. In a recent, very accurate multi-reference CI investigation, Wetmore et al. re-investigated the CH^{2+} potential curve /10b/. In fact, they located a potential dip for the CH^{2+}

ground state species but which is still much to shallow to accomodate any vibrational levels. Thus, the discrepancy between experimental observations and theoretical predictions is still not fully resolved and deserves further studies.

Stahl et al. presented a thorough re-investigation of the CS experiments on CH_4^{2+} and $CH_5^{2+\cdot}$ /13f/. In agreement with the earlier results of Beynon et al. /13a-c/, they found stable CH_4^{2+} and $CH_5^{2+\cdot}$ dications; however, the Q_{min} values were about 1 eV higher than those given in the elder study. Their values of 18.9 and 21.6 eV for oxidizing $CH_4^{+\cdot}$ and CH_5^{+}, respectively, are now in good agreement with the best theoretical adiabatic ionization energies of 19.8 /14b/ and 21.9 eV /13f/, respectively.

The theoretically determined equilibrium structures of the CH_2^{2+} and $CH_3^{2+\cdot}$ dications show a linear, $D_{\infty h}$ symmetric geometry for the former and a Jahn-Teller distorted C_{2v} species for $CH_3^{2+\cdot}$. The most unusual geometry is found for the methane dication: While neutral CH_4 has a perfect T_d symmetry and the cation radical $CH_4^{+\cdot}$ a C_{2v} structure, due to Jahn-Teller distortions, the only minimum on the CH_4^{2+} potential energy surface corresponds to a planar, anti-van't Hoff geometry with D_{4h} symmetry (1). Like the isoelectronic methyl cation, 1 has no π electrons; its six σ-electrons are used to construct four CH bonds. The D_{4h} arrangement gives maximum overlap of the two half-empty carbon 2_p-orbitals and the four hydrogens in a localized orbital picture.

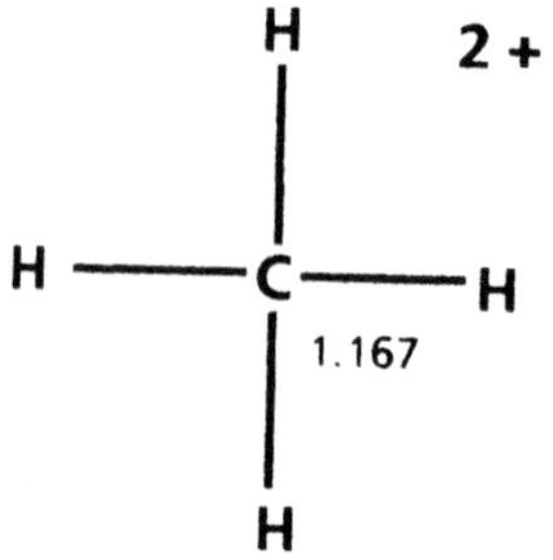

1

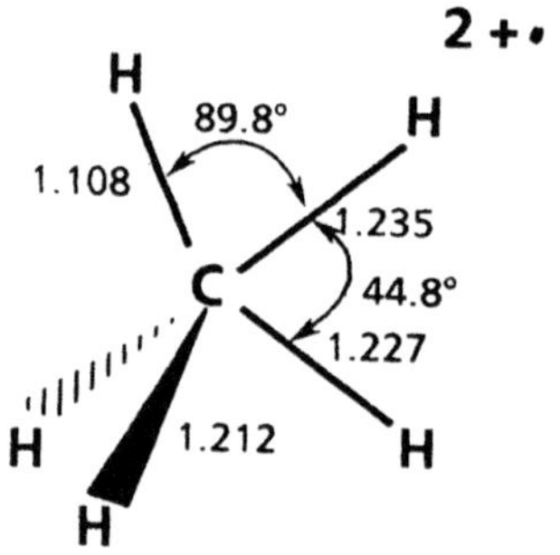

2

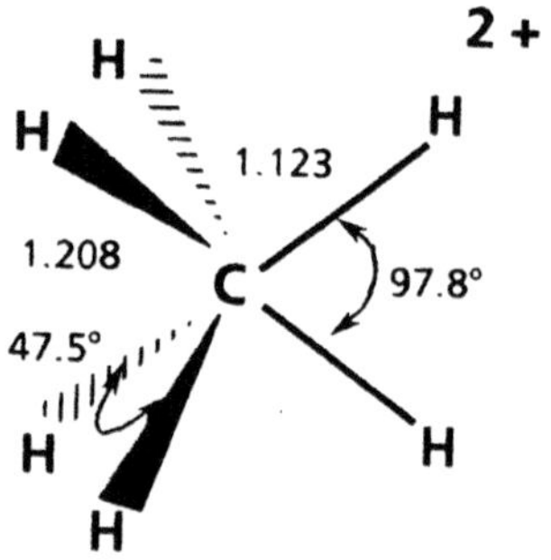

3

Although the deprotonation reaction of 1, to give the methyl cation, is exothermic by 109.2 kcal/mol, a barrier of 19.7 kcal/mol prevents 1 from dissociating spontaneously /14b/. Analogously, the methonium dication, 2, which adopts a C_s symmetric structure is prevented from spontaneous deprotonation by a barrier of 24 kcal/mol. The exothermicity of this reaction amounts to 77 kcal/mol /13f/. In a purely theoretical study, Lammertsma et al. proposed the CH_6^{2+} ion 3 as an experimentally accessible ion /14d/. They calculated a C_{2v} minimum structure, which profits from two 3 centre-2 electron (3c-2e) interactions, and is best described as a complex of CH_2^{2+} and two H_2 molecules. For the highly exothermic deprotonation reaction (ΔH_R= -63.1 kcal/mol) a barrier of 35.4 kcal/mol is predicted /14d/. However, for the time being no experimental data of this truly unusual dication are available.

In these , as well as in almost all other dications discussed in the present article, the positive charges are dispersed to the hydrogen periphery which is favoured both electrostatically and by the electropositive nature of hydrogen.

The NH_n^{2+} dications show a picture different from that of the carbon hydride analogues. In CS experiments only $NH_2^{2+\cdot}$ and NH_3^{2+} were found as stable dications. Neither NH^{2+} nor $NH_4^{2+\cdot}$ could be detected experimentally /13c/. This finding agrees well with theoretical investigations which predict an extremely shallow potential dip for the diatomic NH^{2+} ion /15/. The calculated barrier of 2.2 kcal/mol vanishes completely after the inclusion of zero point vibrational energies. Thus, this ion cannot be viewed as a stable molecule; should it ever be generated it will dissociate spontaneously into N^+ (^1S) + H^+, which is also a highly exothermic process (Figure 1).

The triatomic $NH_2^{2+\cdot}$ dication is characterized by a linear, $D_{\infty h}$ structure for the global minimum, in contrast to the C_{2v} structures predicted for the isoelectronic molecules $BH_2^{\cdot}$ and $CH_2^{+\cdot}$ /15/. The change in symmetry is most probably due to the lower electrostatic repulsion of the highly positively charged hydrogens in the linear arrangement of $NH_2^{2+\cdot}$. Although the computed activation barrier for the proton loss amounts to only 9.8 kcal/mol, obviously this barrier is sufficiently high to kinetically stabilize $NH_2^{2+\cdot}$; the vertical ionization of NH_2^+ ($^3B_1$) leads to a $NH_2^{2+\cdot}$ species whose energy lies well below the energy of the transition structure of the deprotonation reaction (Fig. 1).

The planar ammonia dication is predicted to have a high barrier of 35.2 kcal/mol for the most favourable deprotonation reaction /15/, in accord with the experimental generation and detection of this ion /13c/. In a very recent publication, Boyd et al. discussed the fragmentation behaviour of NH_3^{2+} by analyzing the peak shapes and the kinetic energy releases, T; the latter may be used as an approximation of the activation barrier for the corresponding reverse reaction /16/. For the unimolecular reaction (2) they found two distinct reaction channels,

$$NH_3^{2+} \longrightarrow NH_2^+ + H^+ \qquad (2)$$

connected with different kinetic energy releases, being 4.5 $\pm$ 0.1 and 5.8 $\pm$ 0.1 eV. The former value is in excellent agreement with the theoretical prediction of 4.26 eV for the barrier of the reaction NH_2^+ ($^1A_1$) + H^+ $\longrightarrow$ NH_3^{2+} ($^1A'_1$) /15c/. It should be noted, however, that Boyd et al. did not attribute the lower T value to the deprotonation leading to the excited singlet NH_2^+ ($^1A_1$) but to the reaction leading to the ground state triplet, which, of course, must be connected with a singlet-triplet intersystem crossing /16/.

The $NH_4^{2+\cdot}$ dication is, as expected, Jahn-Teller distorted from regular T_d symmetry, preferring a D_{2d} structure /15/, in distinct contrast to the isoelectronic methane cation, which has C_{2v} symmetry. Again, the larger H-H distances in the D_{2d} structure lower the electrostatic repulsion compared to that expected for a C_{2v} geometry. An activation barrier of only 7.8 kcal/mol has been theoretically estimated for the deprotonation reaction /15c/. This barrier is not high enough to kinetically stabilize $NH_4^{2+\cdot}$, since vertical ionization of NH_4^+ would lead to a state located energetically well above the transition structure for loss of H^+ (Fig. 1). This analysis explains the absence of stable $NH_4^{2+\cdot}$ dications in the CS experiments /13c/.

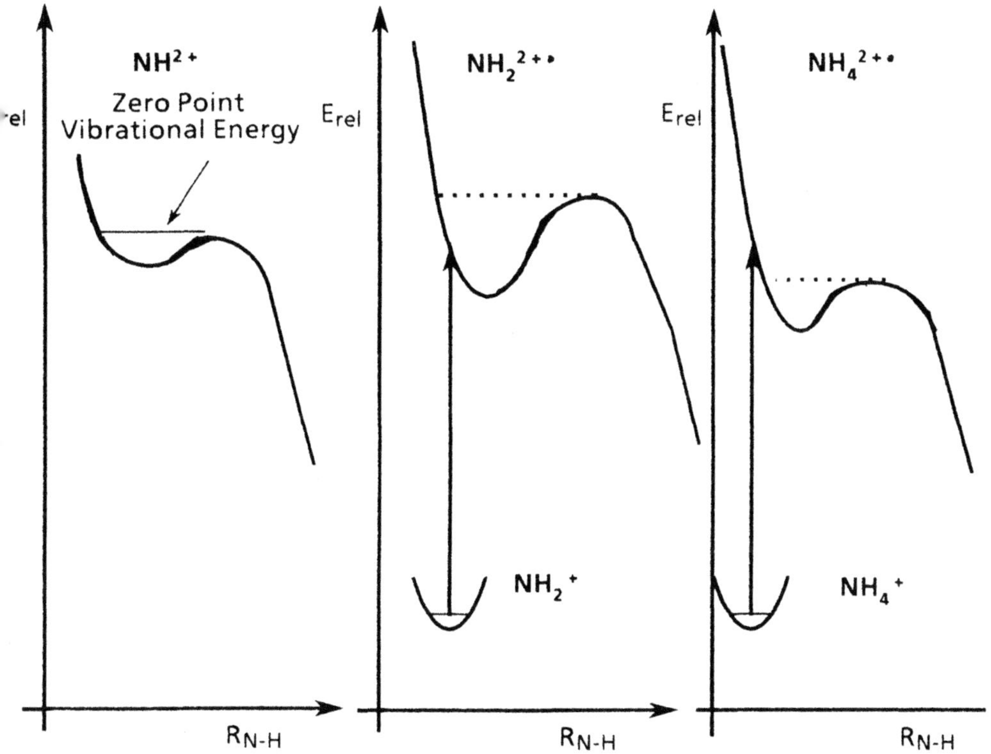

Figure 1: Potential energy curves for NH^{2+}, $NH_2^{2+\cdot}$ and $NH_4^{2+\cdot}$

Among the dications of the oxygen hydrides, OH_n^{2+}, the only stable species so far detected experimentally /2a,13c/ is that of $OH_3^{2+\cdot}$, which is easily accessible by charge stripping from OH_3^+, requiring an ionization energy of 23.5 eV /2a/. The planar $OH_3^{2+\cdot}$ dication radical, which is isoelectronic with the methyl radical, exists in a deep potential well /2a,15a,b/. Although the deprotonation reaction is exothermic by 87 kcal/mol, spontaneous dissociation is hindered by a barrier of 21 kcal/mol.

The potential curves for both, the $^4\Sigma^-$ as well as the $^2\Pi$ state of OH^{2+} dications are predicted to be purely repulsive /2a,15a,b,17/, in line with the experimental absence of stable OH^{2+} species /13c/. The water dication, OH_2^{2+}, being isoelectronic with methylene, exists as minima both for the linear triplet and the bent singlet state, the former being 68.3 kcal/mol more stable than the latter /2a/. Nevertheless, neither the triplet nor the singlet state are likely to have significant lifetimes , since the barriers for the deprotonation reactions are as small as 1.5 and 2.5 kcal/mol, respectively.

No doubt, from a chemical point of view the most spectacular dication among the OH_n^{2+} hydrides corresponds to the tetrahedral OH_4^{2+} ion (4). This unusual ion, which is isoelectronic with methane, is

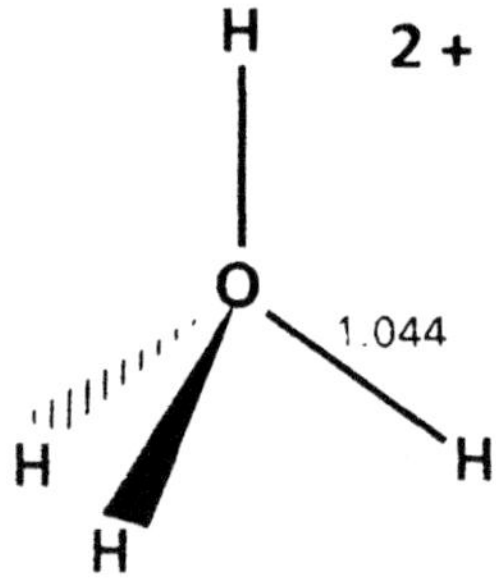

4

characterized by surprisingly short OH bonds (1.044 Å) and high harmonic frequencies (a_1: 2751; e: 1517; t_2= 1376, 2680 cm^{-1}). Deprotonation of OH_4^{2+} to give the stable hydronium cation, is thermodynamically favoured (ΔH_R= -59 kcal/mol), but a substantial barrier of 39.6 kcal/mol confers kinetic stability to this dication /2a/. However, all attempts to generate OH_4^{2+} so far failed, due to the lack of an appropriate mono-cationic precursor. In contrast to the large stability of OH_4^{2+}, no minimum could be located on the $OH_5^{2+\cdot}$ potential energy surface /18/.

While $OH_3^{2+\cdot}$ is found to exist and OH_4^{2+} predicted to be stable in the gas phase, both species are very unlikely to ever be generated as viable dications in solution, because of the extremely high thermodynamic instability. The enthalpy of formation, ΔH_f^o, is estimated to be 685 kcal/mol for $OH_3^{2+\cdot}$ and 564 kcal/mol for OH_4^{2+} /2a/. Moreover, the electron respective proton transfer reactions (3) and (4) have theoretically predicted enthalpies of reaction, ΔH_r^o, of -253 (3) and -225 kcal/mol (4), respectively /2a/. Thus, even if $OH_3^{2+\cdot}$ or OH_4^{2+} could be generated in solution, they would immediately either transfer

$$H_3O^{2+\cdot} + H_2O \longrightarrow H_3O^+ + H_2O^{+\cdot} \quad (3)$$

$$H_4O^{2+} + H_2O \longrightarrow 2\ H_3O^+ \quad (4)$$

a proton to their solvent shell or strip a negatively charged particle from it. Thus, the mechanism recently proposed by Olah et al. /19/ for the proton exchange in OH_3^+ in superacid medium invoking OH_4^{2+} species as intermediates seems to be quite unlikely.

The SiH_n^{2+} dications have been extensively studied, both theoretically /20/ and experimentally /20b,21/. In CS experiments performed by Beynon et al. /21/ and Stahl et al. /20b/ the first three members of the SiH_n^{2+} series, i.e. $SiH^{2+\cdot}$, SiH_2^{2+}, and $SiH_3^{2+\cdot}$ have been observed as stable ions. For the $SiH^{2+\cdot}$ dication two different Q_{min} values have been reported, depending on the neutral precursor molecule used and the electron impact ionization energy employed. This is an indication that CS ionization of the SiH^+ ion to generate the dication involves not only the ground state of SiH^+ ($^1\Sigma^+$) but also the first excited, i.e. the $^3\Pi$ state. This assumption is supported by calculations /20c/ which give the following energies: SiH^+ ($X\,^1\Sigma^+$) $\longrightarrow$ $SiH^{2+\cdot}$ ($^2\Sigma^+$): IE_{exp} = 18.8 eV, IE_{theor} = 18.3 eV; SiH^+ ($a\,^3\Pi$) $\longrightarrow$ $SiH^{2+\cdot}$ ($^2\Sigma^+$): IE_{exp} = 16.5 eV, IE_{theor} = 16.1 eV. There is also experimental evidence that the $SiH^{2+\cdot}$ dication exists not only in its ground state, but also in an excited state /20b/: When the deprotonation reaction $SiH^{2+\cdot} \longrightarrow Si^{+\cdot} + H^+$ is monitored, a peak shape ("composite peak") analysis of the kinetic energy release associated with the formation of H^+ and $Si^{+\cdot}$ indicates, that the intercharge distance $Si^{+\cdot}\cdots H^+$ of the two singly charged species at the time of separation are ca. 6 Å for the ground state and 1.7 Å for the excited state of $SiH^{2+\cdot}$. The corresponding energy barrier of the ground state reverse reaction has been experimentally estimated to be 2.42 eV /20b/. These experimentally derived results are in good agreement with high level CI calculations /20c/, which predict a deep potential well for the $X\,^2\Sigma^+$ ground state of $SiH^{2+\cdot}$. The barrier for proton loss amounts to 1.07 eV and occurs at an H-Si distance of 5.55 Å. The barrier for the reverse reaction is calculated to be 2.50 eV. The calculations also show a small potential well for the first excited, the $^2\Pi$ state of $SiH^{2+\cdot}$, in line with the experimental observations (Fig. 2).

The computed potential curves for $SiH^{2+\cdot}$ provide also a good example for illustrating the importance of charge polarization states for stabilizing dications. At distances below 5.5 Å, the $X\,^2\Sigma^+$ state may be described as $Si^{2+}\cdots\cdots H^{\cdot}$, i.e. a bound charge polarization state. However, at 5.5 Å an avoided crossing with the $B\,^2\Sigma^+$ state (which is purely repulsive, corresponding to Si^+-H^+) takes place, thus allowing the $X\,^2\Sigma^+$ state to dissociate to the most favourable dissociation limit, i.e. $Si^{+\cdot}$ (^2P) + H^+. The existence of a deep potential well for SiH^{2+}

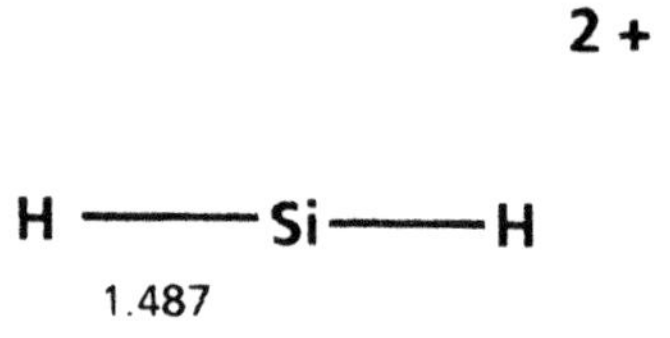

5

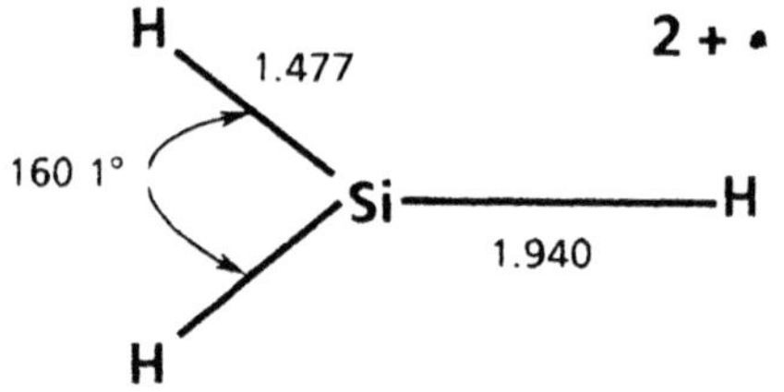

6

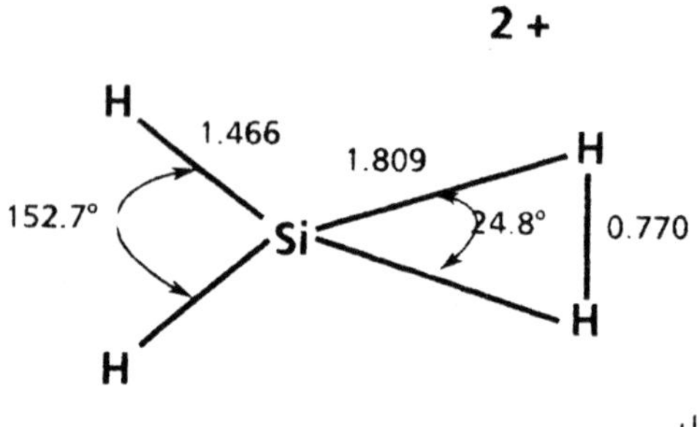

7

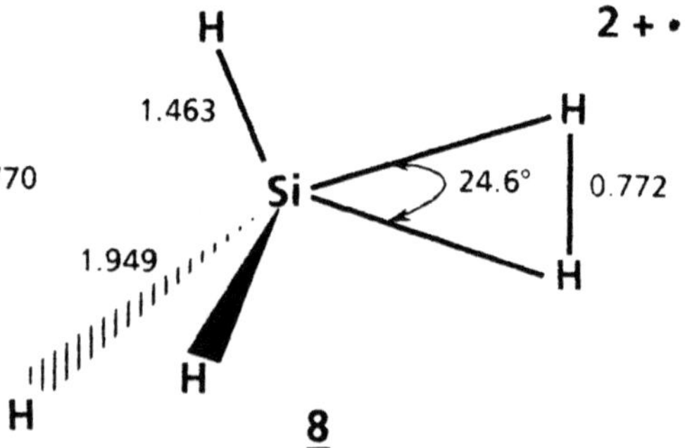

8

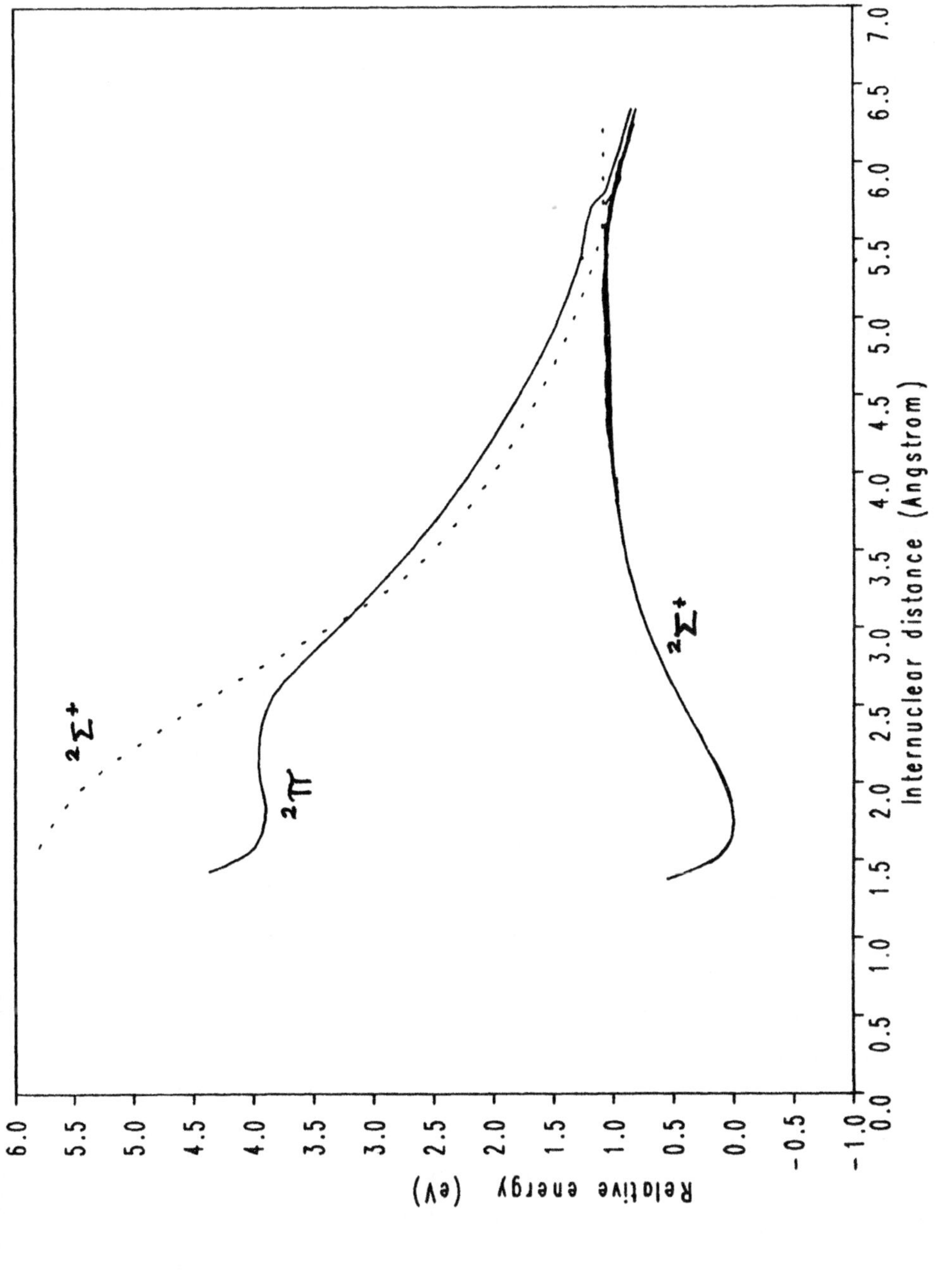

Figure 2 Potentialenergy curves of some lower states of SiH++•

is also in distinct contrast to the above mentioned results found for CH^{2+}.

For SiH_2^{2+} the most stable species corresponds to a linear singlet (5), which is also prevented by a barrier from spontaneous dissociation /20a,b/.

The $SiH_3^{2+\cdot}$ dication is, like the valence isoelectronic $CH_3^{2+\cdot}$ ion, Jahn-Teller distorted from D_{3h} symmetry. The theoretically predicted minimum geometry (6) /20a,b/ can be understood as a loosely bound complex of SiH_2^{2+} and $H^\cdot$, which is supported by the long (1.940 Å) and weak (anti-symmetrical Si-H vibration: 810 cm^{-1}) SiH bond. Furthermore, the geometry of the SiH_2^{2+} unit in 6 is quite similar to the structure of isolated SiH_2^{2+} (5), and 86 % of the positive charges of 6 are located on this moiety. The higher SiH_n^{2+} species, SiH_4^{2+} (7) and $SiH_5^{2+\cdot}$ (8) are as yet unobserved due to the lack of suitable precursors. These dications can also be viewed as complexes of SiH_2^{2+} with H_2 and $H^\cdot$, respectively. It is noteworthy, that the global minimum of SiH_4^{2+} is completely different from the planar structure of the methane dication (1). A planar D_{4h} symmetric SiH_4^{2+} structure does not even exist as a minimum on the SiH_4^{2+} potential energy surface.

For the series of the PH_n^{2+} dications no experimental information is available. Elaborate calculations of Pope et al. /15b/ show that PH_n^{2+} ions (n=1-4) are kinetically stable dications, in contrast to the NH_n^{2+} ions, for which NH^{2+} and $NH_4^{2+\cdot}$ do not exist. The largest difference in the shape of the potential curves is encountered for the NH^{2+} and PH^{2+} ions. Whereas the former does not exist in a potential well, the barrier of deprotonation of PH^{2+} is as high as 64.5 kcal/mol /15b/. While $NH_4^{2+\cdot}$ prefers a D_{2d} structure /15/, the corresponding $PH_4^{2+\cdot}$ ion (9) has C_{3v} symmetry with one long P-H bond (1.882 Å). The structure with D_{2d} symmetry is no minimum for $PH_4^{2+\cdot}$ but characterized by a doubly-degenerate imaginary frequency /15b/.

Among the SH_n^{2+} dications the first three members have been observed as stable ions in CS experiments /13c/. This is, again, in line with the theoretical data, which predict high activation barriers for the exothermic deprotonation reactions /15b/. Again, the stability of

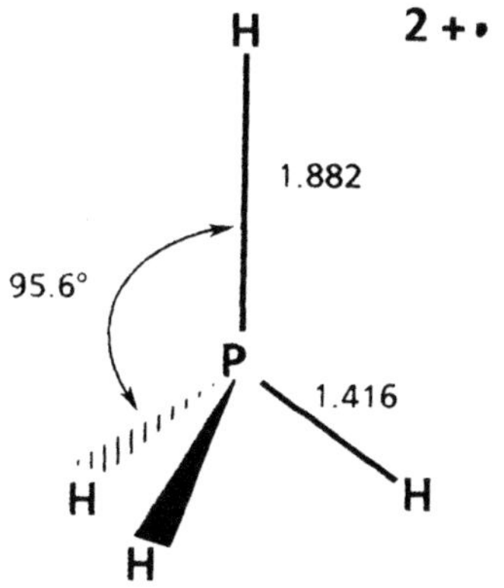

9

$SH^{2+\cdot}$ is in distinct contrast to the purely repulsive curves found for $OH^{2+\cdot}$. Furthermore, while the ground state of $OH^{2+\cdot}$ is the $^4\Sigma^-$ state, with the $^2\Pi$ state next in energy /17/, for $SH^{2+\cdot}$ the $^2\Pi$ state is found to correspond to the ground state /15b/.

A common feature of all second row dicationic hydrides is that the corresponding deprotonation reactions are much less exothermic and are connected with higher barriers than those calculated for their first row counterparts. This is summarized in Table I. Additionally, a comparison of the experimentally and theoretically derived ionization energies of the appropriate monocations is included, which shows in general good agreement. The greater stability of the second row hydride dications is due to the better capability of second row atoms to accomodate the positive charge, which itself is a consequence of their lower electronegativity and high polarizability compared with first row elements /22/.

Table I: Calculated deprotonation energies and activation barriers for XH_4^{2+} dications and theoretical and experimental ionization energies of the corresponding monocations.

Molecule	Deprotonation Energy[a)]	Ref.	Activation Barrier[a)]	Ref.	Adiabatic Ionization Energy[b)]	Ref.	Vertical Ionization Energy[b)]	Ref.	Q_{min}[b)]	Ref.
$CH^{2+\cdot}$	---		---		---		23.6	14c	22.8	13c
CH_2^{2+}	-71.4	14c	30.8	14c	21.1	14c	---		19.6	13c
$CH_3^{2+\cdot}$	-108.5	14c	2.9	14c	23.5	14c	---		18.9	13c
CH_4^{2+}	-105.7	14c	19.7	14b	19.8	14b	21.8	13f	18.9	13f
$CH_5^{2+\cdot}$	-86.6	13f	23.7	13f	21.9	13f	23.9	13f	21.6	13f
NH^{2+}	-136.3	15c	---		26.4	15c	26.9	15c	25	13c
$NH_2^{2+\cdot}$	-108.1	15c	9.8	15c	23.2[c]/24.6[d]	15c	24.7[c]/25.2[d]	15c	23.3	13c
NH_3^{2+}	-62.9	15c	35.2	15c	23.3	15c	23.7	15c	22.2	13c
$NH_4^{2+\cdot}$	-116.5	15c	7.8	15c	24.0	15c	26.8	15c	24.5	13c
$OH^{2+\cdot}$	---		---		---		29.5	2a	29.0	13c
OH_2^{2+} ($^3\Sigma_g^-$)	-128.8	2a	1.5	2a	24.5[e]/23.5[f]	2a	26.6[e]/24.4[f]	2a	23.5	13c
OH_2^{2+} ($^1A_1$)	-126.5	2a	2.5	2a	---		---		---	
$OH_3^{2+\cdot}$	-89.6	2a	20.7	2a	23.2	2a	24.5	2a	23.5	2a
OH_4^{2+}	-59.0	2a	39.6	2a	---		---		---	

$SiH^{2+\cdot}$	-33.0	20c	24.7	20c	18.2[g]/16.0[h]	20b	18.3[g]/16.1[h]	20b	18.8[g]/ 16.5[h]	20b
SiH_2^{2+}	-20.0	20b	---		16.5	20b	17.5	20b	16.7	20b
$SiH_3^{2+\cdot}$	-48.0	20b	---		19.6/15.9	20b	20.9/17.1	20b	15.4	20b
SiH_4^{2+}	-26.1	20b	---		---		---		---	
$SiH_5^{2+\cdot}$	-13.4	20b	---		---		---		---	
PH^{2+}	-16.9	15a	64.5	15a	18.7	15a	---		---	
$PH_2^{2+\cdot}$	-51.6	15a	---	15a	19.2	15a	---		---	
PH_3^{2+}	-25.9	15a	60.9	15a	18.0	15a	---		---	
$PH_4^{2+\cdot}$	-69.0	15a	17.0	15a	20.3	15a	---		---	
$SH^{2+\cdot}$	-69.7	15a	36.3	15a	21.9	15a	---		21.2	13c
SH_2^{2+}	-61.8	15a	39.3	15a	20.8	15a	---		21.0	13c
$SH_3^{2+\cdot}$	-53.5	15c	40.5	15a	19.6	15a	---		21.4	13c

a: in kcal/mol; b: in eV; c: from NH_2^+ ($^1A_1$); d: from NH_2^+ ($^3B_1$); e: from $OH_2^{+\cdot}$ ($^2B_2$); f: from $OH_2^{+\cdot}$ ($^2A_1$); g: from SiH^+ ($^1\Sigma^+$); h: from SiH^+ ($^3\Pi$).

4. DICATIONS CONTAINING SEVERAL "HEAVY" ATOMS

4.1. C_n-Dications

The removal of two electrons from methane leads to a drastic change of the molecular structure: CH_4^{2+} (1) is predicted to prefer a planar anti-van't Hoff D_{4h} symmetry, as already mentioned in the preceding section /14/. The acetylene dication, $HCCH^{2+}$, has been found as a stable dication in CS experiments by Beynon et al. /3d,13a/ with a Q_{min} value of 20.8 eV. In an ab initio MO investigation Pople et al. /23/ showed that the ground state of $HCCH^{2+}$ is a linear triplet ($^3\Sigma_g^-$) which corresponds to a double ionization of acetylene with electrons removed from each of the two π-bonding orbitals. The corresponding $^1\Delta_g$ state with two π -electrons lies 26.2 kcal/mol higher in energy. The barrier for proton loss is calculated to be 65 kcal/mol. However, the deprotonation leads to the $^3\Sigma^-$ state of HC_2^+ and not the $^3\Pi$ ground state. Thus the $^3\Pi$ and $^3\Sigma^-$ surfaces must cross at some distance separating the H^+ and C_2H^+ fragments. At a rather low level of theory, Pople et al. /23/ demonstrated that this crossing will take place at a $H^+ \cdots C_2H^+$ distance longer than the 2.697 Å found for the $^3\Sigma^-$ barrier. Hence, these authors conclude, that the $^3\Sigma^-$ saddlepoint represents a true transition structure (Figure 3).

The situation for the diacetylene dication, $HCCCCH^{2+}$, is quite similar. According to calculations of Lammertsma et al. /24/ the linear triplet is the global minimum, followed by a cyclic singlet structure, some 13 kcal/mol above the linear isomer. The theoretically predicted stability of the $C_4H_2^{2+}$ ion is in agreement with its experimental observation in CS experiments by Rabrenovich and Beynon /25/.

The ethylene dication has been thoroughly studied, both theoretically /26/ and experimentally /13a,27/. Again, the removal of two electrons from neutral C_2H_4 leads to fundamental geometry changes. High level ab initio calculations of Lammertsma et al. predict a perpendicular, anti-van't Hoff D_{2d} equilibrium structure for the dication 10, which is 28 kcal/mol lower in energy than the planar, D_{2h} $C_2H_4^{2+}$ dication 11 /26/.

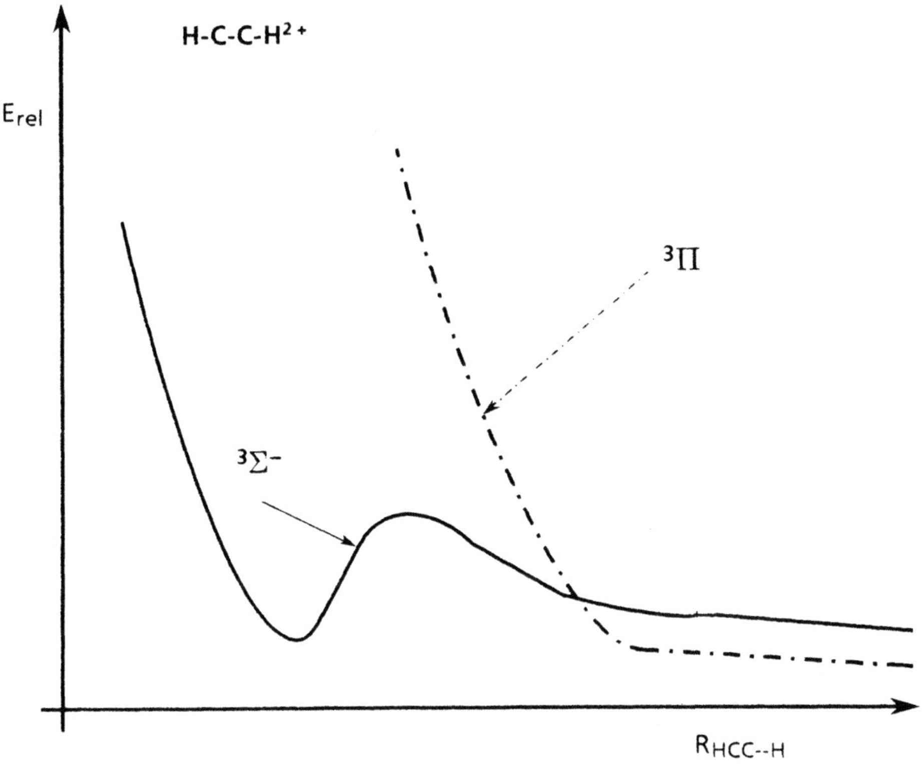

Figure 3: Potential energy diagramme for deprotonation of $HCCH^{2+}$

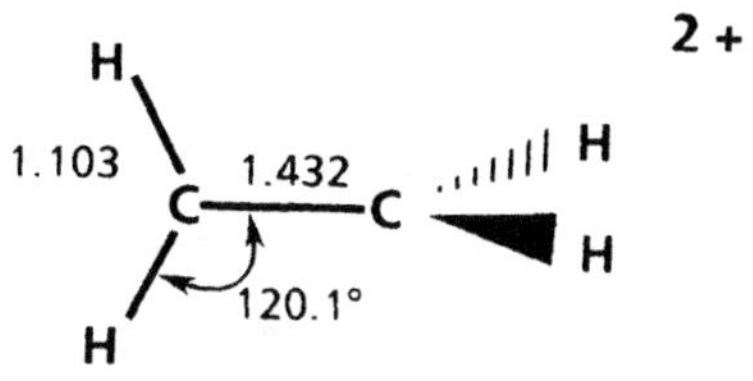

10

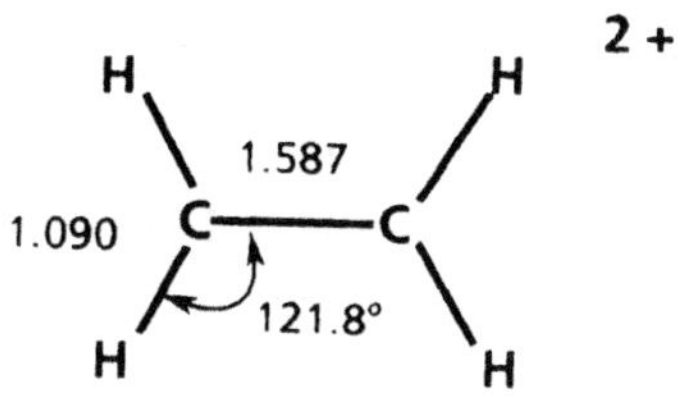

11

11 represents the transition structure for the C-C rotation in 10. The explanation put forward for the preference of the anti-van't Hoff structure is straightforward : In the D_{2d} conformer 10 the two formally vacant p-orbitals of each carbon atom are orthogonal and each interact hyperconjugatively with the corresponding vicinal CH_2 groups. The best charge distribution results. Furthermore, the positive charge resides

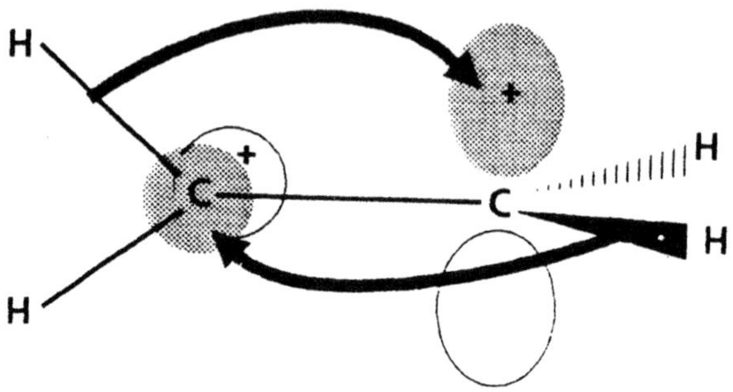

10

Figure 4: Hyperconjugation in perpendicular $C_2H_4^{2+}$ (10)

primarily on the more electropositive hydrogens, which in the perpendicular arrangement are as far apart as possible. Although the deprotonation reaction as well as the C-C cleavage of $C_2H_4^{2+}$ are exothermic by -23 and -9.4 kcal/mol, respectively, the barriers are again high, being 68.8 and 88.4 kcal/mol, respectively /26/. Consequently, it is no surprise that the ethylene dication has been observed in CS experiments /13a/.

Theoretical calculations on the $C_2H_6^{2+}$ potential energy surface led to the tempting suggestion, that the $C_2H_6^{2+}$ species might have the structure of a carbenium-carbonium ion 12 /28/. Charge stripping experiments, performed by Stahl and Maquin, starting from ethane lead to a stable dication and the energy for removing the electron from $C_2H_6^{+\cdot}$ was measured to 18.4 eV /29/. Beynon et al. reported the existence of two distinguishable $C_2H_6^{+\cdot}$ mono-cations /30/, one being formed by electron impact ionization of C_2H_6 and the other generated via ion/molecule

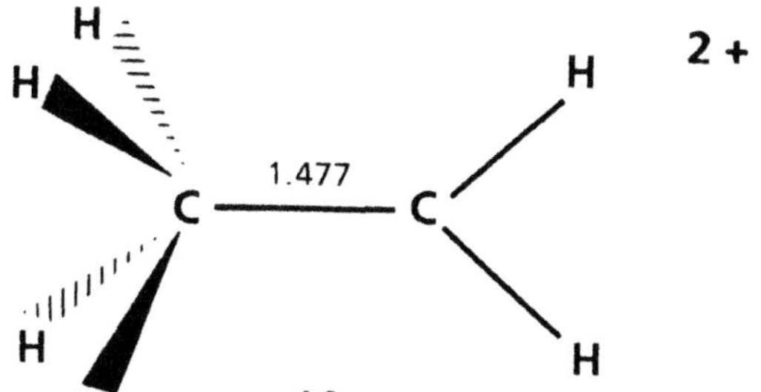

12

reactions of $CH_4^{+\cdot}$ with CH_4. The latter was believed to yield a radical cation having a carbenium-carbonium ion structure, comparable to 12. In fact, the measured ionization energy of the m/z=30 ion formed in the course of the $CH_4^{+\cdot}$ + CH_4 reaction was different (Q_{min}=19.4 eV) from that found for the oxidation of ionized ethane (18.4 eV), thus suggesting that two structurally different m/z 30 ions were monitored. However, in a very recent combined theoretical/experimental re-investigation by Radom and co-workers /31/, it was convincingly demonstrated that the only potential minimum on the $C_2H_6^{+\cdot}$ potential energy surface, corresponding to a $H_4C\text{-}CH_2^{+\cdot}$ radical cation, is an extremely loosely bound complex 13, which is bound by only 0.24 kcal/mol with respect to dissociation to $C_2H_4^{+\cdot}$ + H_2. Such a species is very unlikely to be observable under the conditions of the Rabrenovich and Beynon experiments. In fact, the careful re-investigation of their experiments revealed

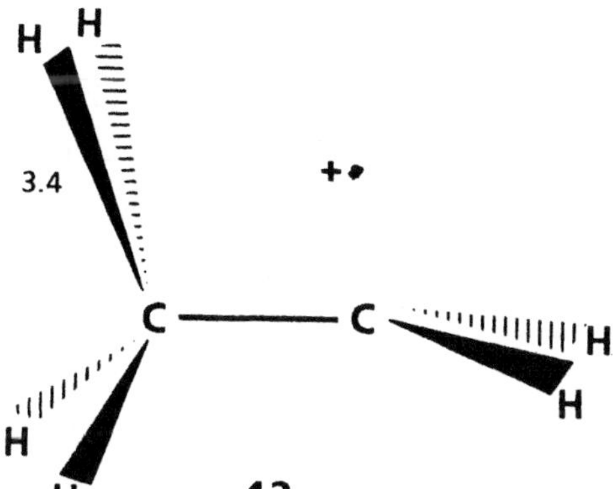

13

that the observed m/z=30 ion did not correspond to $C_2H_6^{+\cdot}$ but rather to the isobaric $^{12}C^{13}CH_5^+$ ion /31/. Moreover, the measured Q_{min} value of 19.4 eV for the formation of m/z 15 dications, assigned to be typical for the reaction 13 ⟶ 12 by Rabrenovich and Beynon is, within experimetal error, identical with the Q_{min} value found for the formation of $C_2H_5^{2+}$ (19.1 ± 0.5 eV) /31/. Thus, there is for the time being no experimental evidence for the existence of doubly charged $C_2H_6^{2+}$ ions having the theoretically predicted structure 12.

The $C_4H_4^{2+}$ dication has been for quite some time of theoretical interest /32/, since a planar, cyclic structure would constitute a formal 2e-Hückel aromatic system. However, calculations by Schleyer and collaborators indicate, that the ground state of $C_4H_4^{2+}$ prefers a puckered, D_{2d} symmetry 14, which is some 9 kcal/mol more stable than

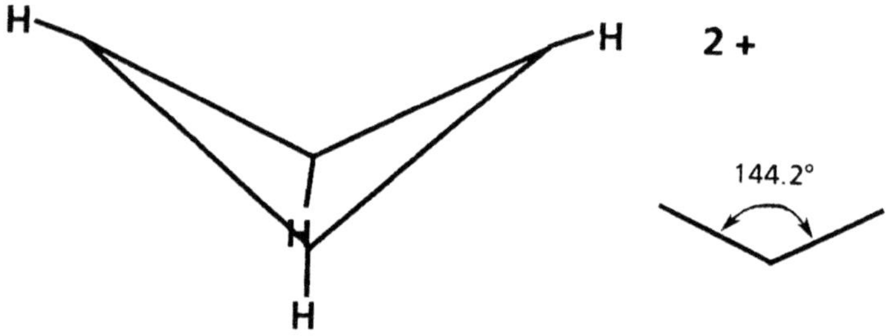

14

the isomer having square, planar geometry /32b,c/. The reason for this puckering is twofold: (i) repulsive, anti-bonding cross-ring interactions are diminished, (ii) stabilization due to orbital mixing is not possible in the higher D_{4h} symmetry. These effects are much more important than the loss in π-energy on puckering. There is, however, no experimental evidence supporting these tempting suggestions.

The dication generated from benzene is a commonly observed gas phase ion /33/. However, the minimum structure does not correspond to a six-membered ring structure but to the fulvene-like dication (15), with the allylcyclopropenium ions (16) and (17) as second and third most stable isomers /34/. Among the benzene-like structures the calculations do not allow a clear cut decision whether a D_{6h} symmetric

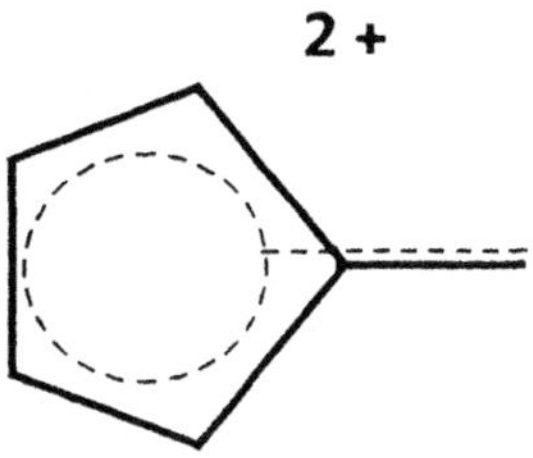

15

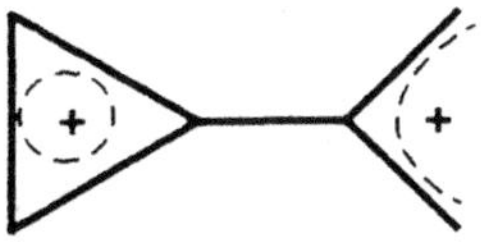

16

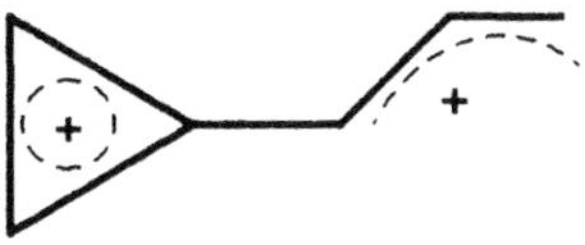

17

triplet or a distorted, lower symmetry singlet structure is the lowest energy isomer; both are, however, roughly 20 kcal/mol less stable than 15 /34/.

4.2 Carbon/Nitrogen Containing Dications

The mono-cations of the elemental composition $CH_3N^{+\cdot}$ are believed to play an important role in interstellar chemistry /35/. It is now well established that the lowest energy isomer is the aminomethylene cation, $HCNH_2^{+\cdot}$, which is about 9 kcal/mol lower in energy than the second stable isomer, the methylene imine cation, $H_2CNH^{+\cdot}$ /36/. There exists also a third isomer, $H_3N\text{-}C^{+\cdot}$, which is, however, 60.5 kcal/mol above the global minimum /37/.

The removal of a further electron leads to the CNH_3^{2+} dications, about which no experimental information are yet available. Theory predicts /37/ the aminomethylene dication (18) to be the by far most stable isomer, 46 and 49 kcal/mol more stable than the other two stable structures, 19 and 20, respectively. Thus, the aminomethylene isomers

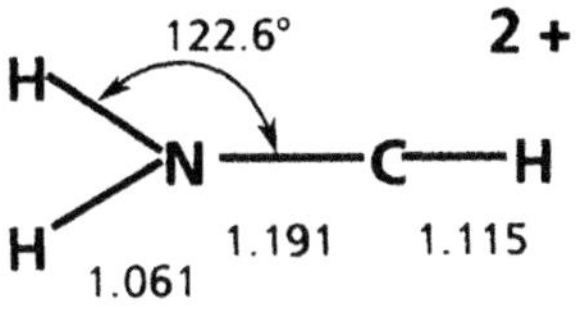

18

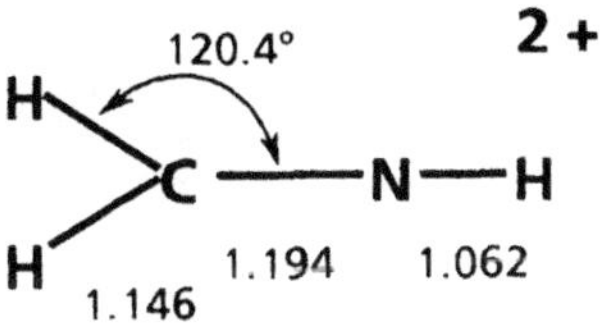

19

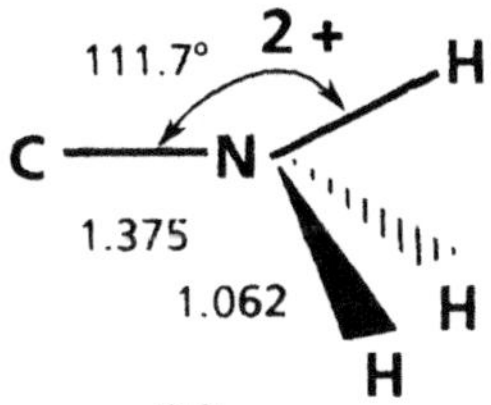

20

gain substantial stability relative to the methylene imine by ionization from the singly to the doubly-charged ion. Both cations are characterized by remarkably short CN bonds (1.191 Å for 18 and 1.194 Å for 19) which are even significantly shorter than for the respective monocharges radical ions (1.251 Å for $H_2NCH^{+\cdot}$ and 1.243 Å for $H_2CNH^{+\cdot}$). In terms of valence bond structures, this shortening may be due to resonance structures like 18a or 19a. Only marginally less stable than 19 is the dication 20, which can be viewed as a donor-acceptor complex between NH_3 and C^{2+} (^1S). This interpretation is further supported by

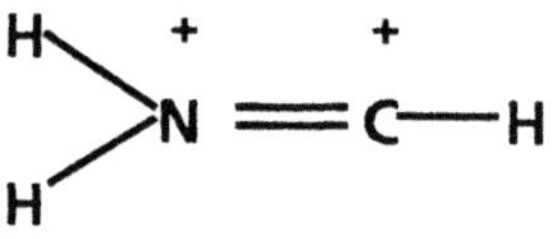

18a

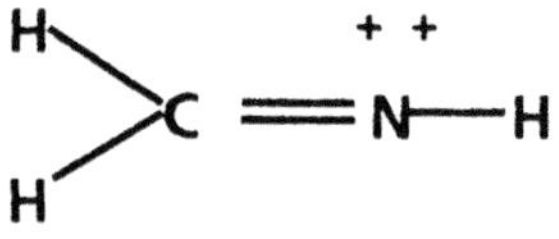

19a

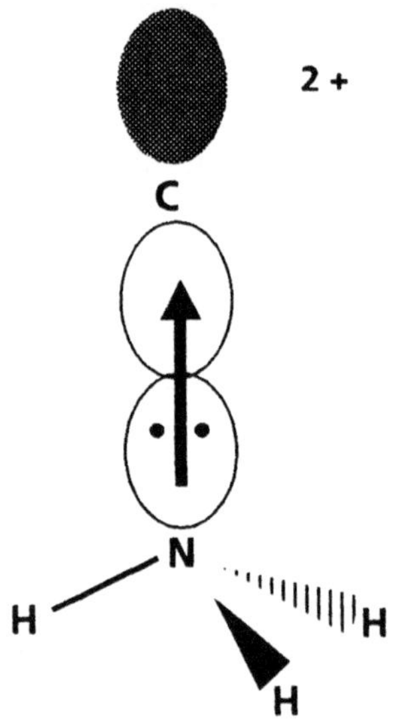

20

the fact, that the geometry of the NH_3 unit in 20 is almost the same as in isolated, neutral ammonia (NH_3: $\measuredangle$XNH=111.6°, r_{NH}=1.002 Å; 20: $\measuredangle$CNH=117.7°, r_{NH}=1.062 Å). The calculation of the possible isomerization and dissociation pathways of the CNH_3^{2+} dications revealed, that the methylene imine dication (19) is not very likely to be a stable species. While the transition state for the isomerization 18 $\longrightarrow$ 20 is associated with a barrier of 74.3 kcal/mol (relative to 18), and thus high enough to prevent facile isomerization, the barrier connecting 18 and 19 is merely 0.1 kcal/mol above the latter. Thus, this ion will rearrange spontaneously to the aminomethylene dication (18). The charge separation processes for the remaining CH_3N^{2+} isomers, 18 and 20, are exothermic (with the exception of the reaction $H_2NCH^{2+} \longrightarrow NH_2^+$ ($^1A_1$) + CH^+, which is endothermic by 64 kcal/mol), but characterized by substantial barriers, thus preventing spontaneous dissociation to monocharged species /37/ (Figure 5).

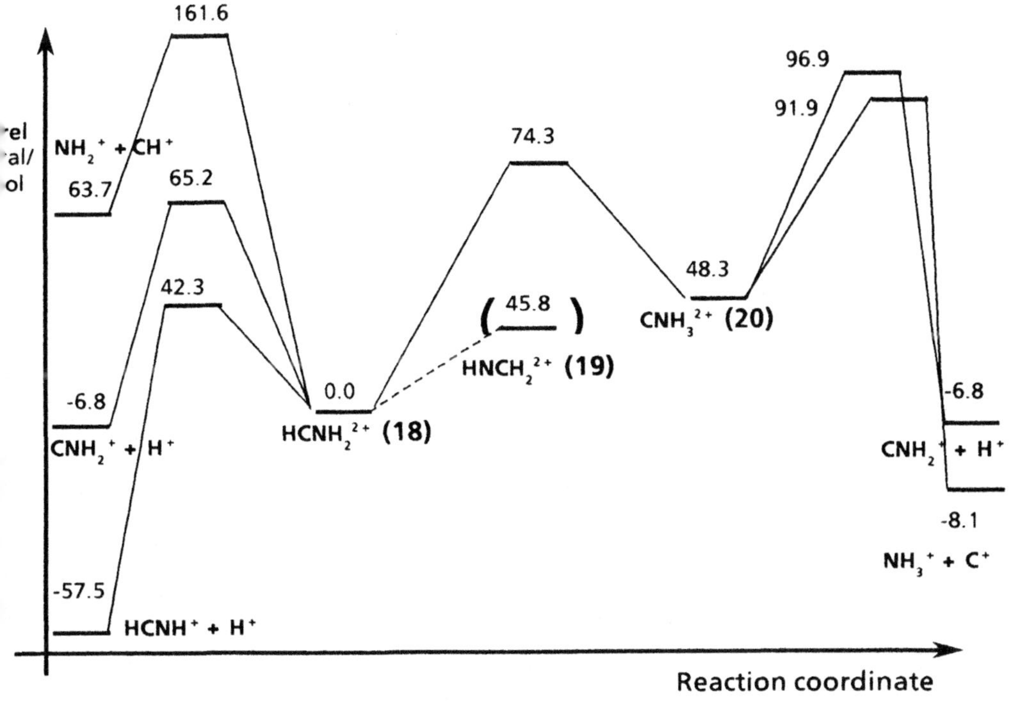

Figure 5: Potential energy diagramme for reactions of CNH_3^{2+} dications.

Charge stripping from the ylide cation radical $H_2C\text{-}NH_3^{+\cdot}$ leads to a stable dication; the Q_{min} value has been determined to be 18.9 eV, while ionization from the methylamine radical cation, $H_3C\text{-}NH_2^{+\cdot}$ needs 17.7 eV to remove a further electron /38/. Ab initio calculations /39/ of the CH_5N^{2+} potential energy surface allow a further characterization of the chemistry of these ions and provide remarkable results: The only minimum found by Radom and co-workers corresponds to the distonic dication 21 which lies in a deep potential well. The most favourable charge separation reactions leading to $CH_2^{+\cdot} + NH_3^{+\cdot}$ and $CH_2NH_2^{+} + H^{+}$ are exothermic, but substantial barriers of the order of 70 kcal/mol prevent spontaneous dissociations to occur. The classical $H_3CNH_2^{2+}$ isomer could not be located as a potential minimum; it is predicted to rearrange spontaneously to the $H_2CNH_3^{2+}$ ion (21). This result must be

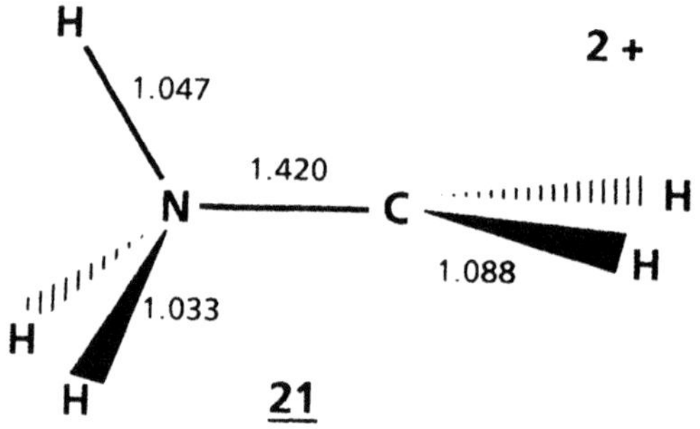

21

compared with those known for the neutral case, where the classical methylamine is the by far most stable species with the distonic isomer lying some 73 kcal/mol higher in energy /40/. For the cation radicals it is now well established, that the distonic isomer gains substantial stabilization upon ionization. The two mono-cationic isomers come quite close in energy, with the $H_2CNH_3^{+\cdot}$ isomer being the more stable one by ca. 2 kcal/mol compared with $H_3CNH_2^{+\cdot}$ /41/. Thus, the stability (instability) of the distonic (classical) isomer is the higher the more electron-deficient the system becomes. This seems to be a general phenomenon and is, indeed, frequently observed. Further examples as well as a rationalization of the unusual stability of ylide dications are given in the subsequent sections.

However, if the $H_3CNH_2^{2+}$ dication does not exist at all, how can the observation of an admittedly weak signal under charge stripping conditions for the oxidation of $H_3CNH_2^{+\cdot}$ be explained ? It is believed, that the removal of an electron from $H_3CNH_2^{+\cdot}$ is accompanied or followed by H-migration to give the stable $H_2CNH_3^{2+}$ ion. One suggestion is, that collision of $CH_3NH_2^{+\cdot}$ with the collision gas leads initially to a Rydberg state (indicated by X in Fig. 6), from which the electron is ejected, followed by rearrangement to form $H_2CNH_3^{2+}$ /38/.

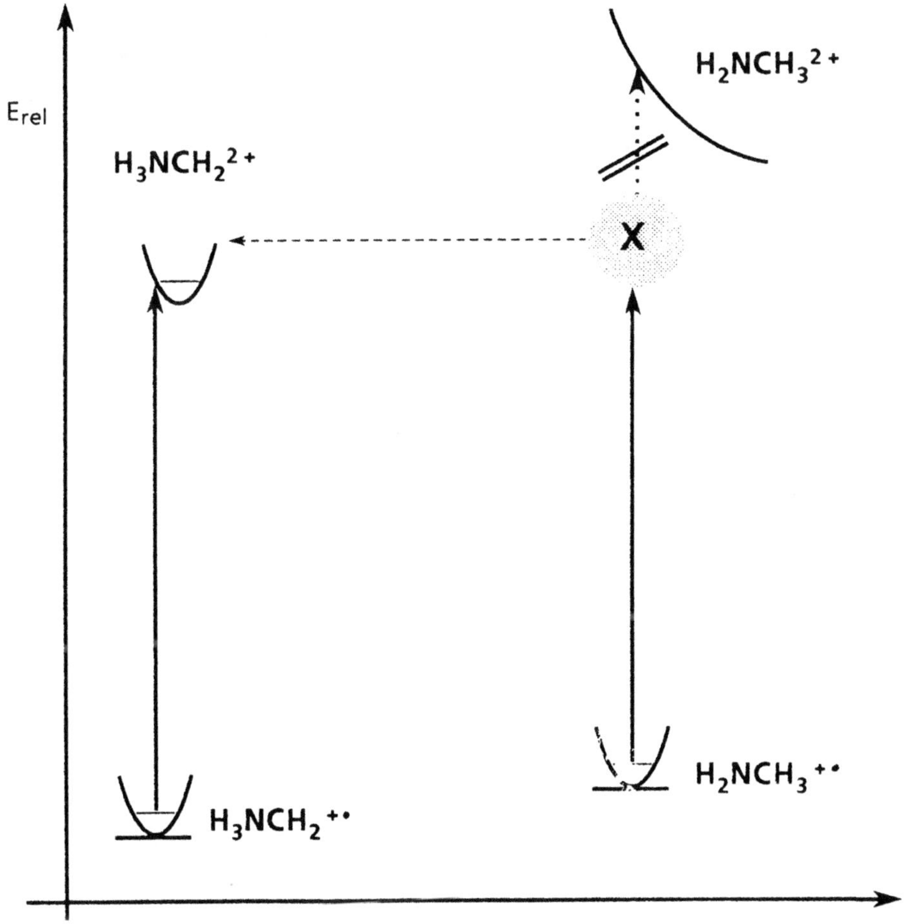

Figure 6: Schematic representation of the ionization of $H_3NCH_2^{+\cdot}$ and $H_2NCH_3^{+\cdot}$

The last nitrogen containing dication, which will be discussed in this section, is the aminoborane ion, BNH_4^{2+}, which is isoelectronic with the ethylene dication. Very recently, experimental evidence for the existence of such a dication has been provided /42/. Ab initio calculations of the BNH_4^{2+} potential energy surface showed, that the most stable isomer is $HBNH_3^{2+}$ (22), which is 60 kcal/mol more stable than the second isomer, i.e. the perpendicular $H_2BNH_2^{2+}$ species (23) /42/.

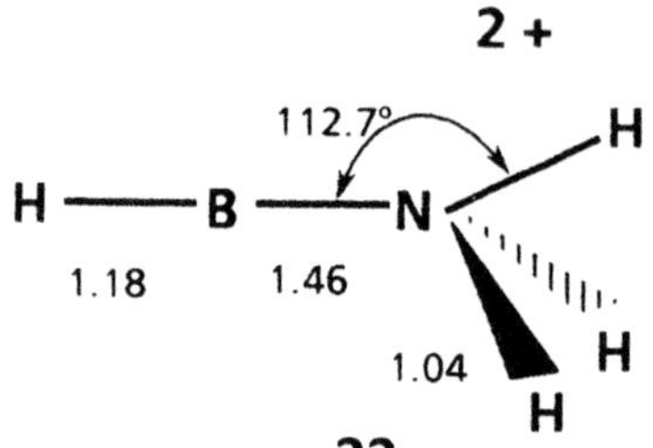

22

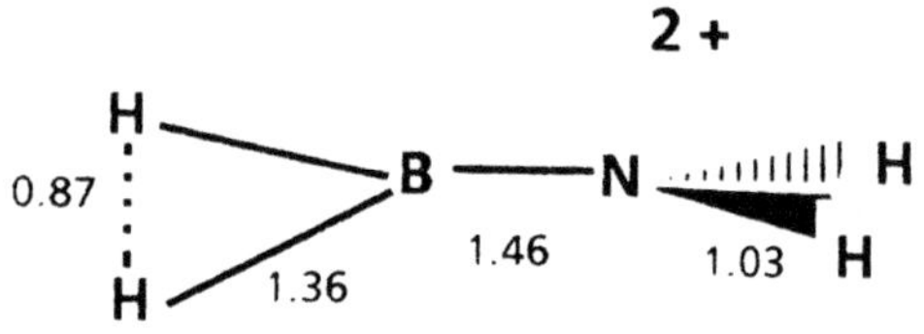

23

This is in remarkable contrast to the results found for $C_2H_4^{2+}$, where the perpendicular $H_2CCH_2^{2+}$ (10) is the only minimum /26/. The ethylidene dication, H_3CCH^{2+}, analogous to 22 is theoretically predicted not to exist at all in a potential well but to rearrange to $H_2CCH_2^{2+}$ without any barrier /26/. This different behaviour is due to different sources of stabilization in the two systems. While for $C_2H_4^{2+}$ hyperconjugation plays the decisive role, this is not the case for 23. The latter structure is best described as a H_2N-B fragment complexed by H_2. On the other hand, the $HBNH_3^{2+}$ dication 22 is stabilized by a powerful donor-acceptor interaction of a formal BH^{2+}, acting as acceptor, and NH_3 which serves as donor, a situation reminiscent to the one already discussed for the isoelectronic CNH_3^{2+} isomer 20.

The barrier for the isomerization 22 $\rightleftarrows$ 23 is computed to be 18 kcal/mol above 23. All studied fragmentation reactions of BNH_4^{2+} are associated with high barriers, in line with the experimentally proven /42/ existence of BNH_4^{2+} dications (Fig. 7)

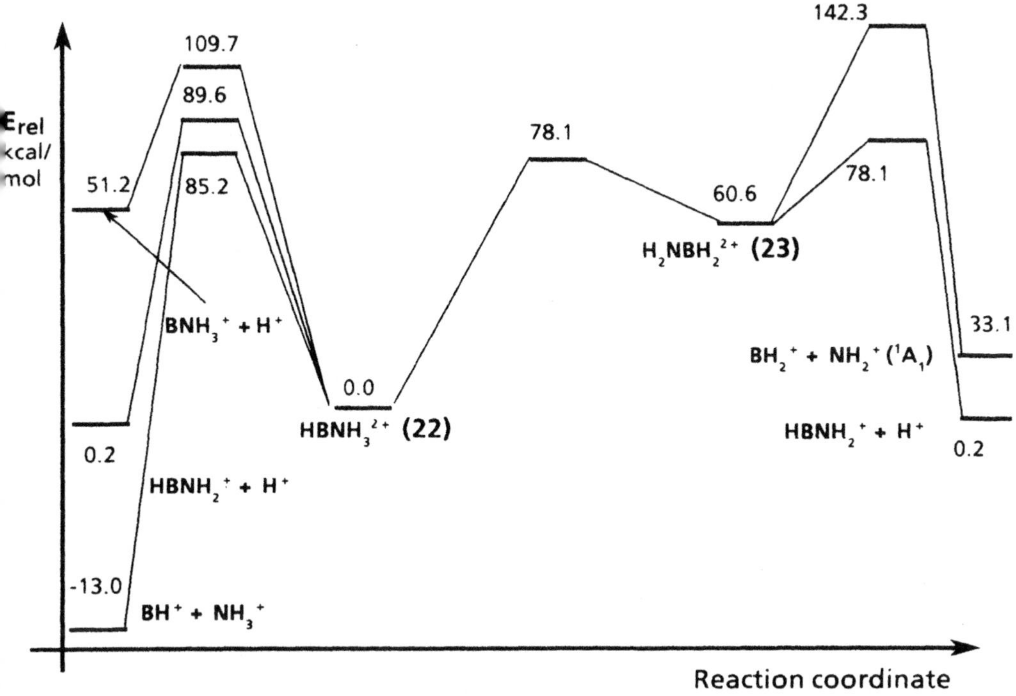

Figure 7: Potential energy diagramme for BNH_4^{2+}

4.3 Carbon/Oxygen Containing Dications

In remarkable agreement with theoretical calculations performed by Bouma and Radom /43/ the hydroxymethylene dication, $HCOH^{2+}$, which can be described as a di-cationic analogue of acetylene, has been found as a stable gas phase species, while the formaldehyde dication, H_2CO^{2+}, is metastable /44/. Labelling experiments indicate that the deprotonation reaction of $HCOH^{2+}$ only takes place from the hydroxyl group; cleavage of the C-H bond is not observed, as predicted by theory. The calculated barriers for the O-H and C-H bond cleavage processes of $HCOH^{2+}$ are 22 and 46 kcal/mol /43/. An analysis of the peak shape and the kinetic energy released in the deprotonation reaction of $HCOH^{2+}$ allows to estimate the barrier as well as the intercharge distance of the transition structure. Again, pleasing agreement between theory and experiment is obtained (Table II) /44/.

Table II:
Experimental and theoretical results for H_2CO^{2+}

	Exp. /44/	Theory /43/	
H_2CO^{2+}	does not exist	no potential minimum	
$HCOH^{2+}$	stable, ΔH_f=641 kcal/mol	stable, ΔH_f=637 kcal/mol	
$HCO\cdots H^{2+}$	E_a=21 kcal/mol; r=4 Å	E_a=22	r=4.08 Å
$HOC\cdots H^{2+}$	E_a > 21 " "	E_a=46	r=4.53 Å

One of the best studied cations known is that of the singly positively charged CH_3O^+ system. Experiment /45/ and theory /46/ agree that the hydroxymethylene cation, H_2COH^+, represents the global minimum, while the oxoniamethylene cation, $HCOH_2^+$, is theoretically predicted to be 78 kcal/mol higher in energy. The barrier connecting these two isomers is computed to be 30 kcal/mol (relative to $HCOH_2^+$). This stability sequence is reversed if a further electron is ejected. Ab initio MO calculations /47/ leave no doubt that the $HCOH_2^{2+\cdot}$ dication 24 forms the global minimum of the CH_3O^{2+} potential surface. The hydroxymethyl radical dication 25 is predicted to be 22 kcal/mol less stable, separated from 24 by a barrier of 24 kcal/mol. Since all charge separation

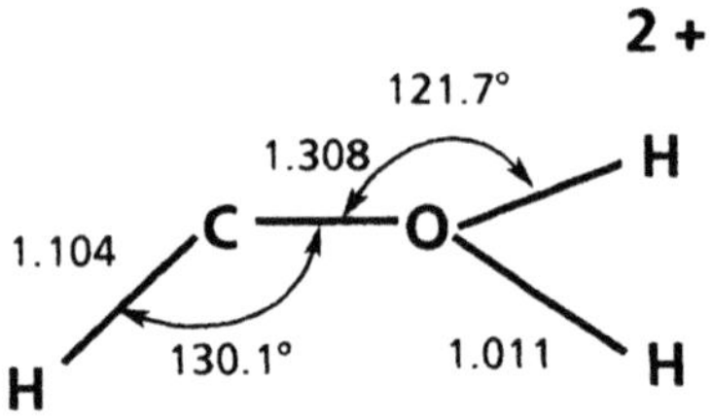

24

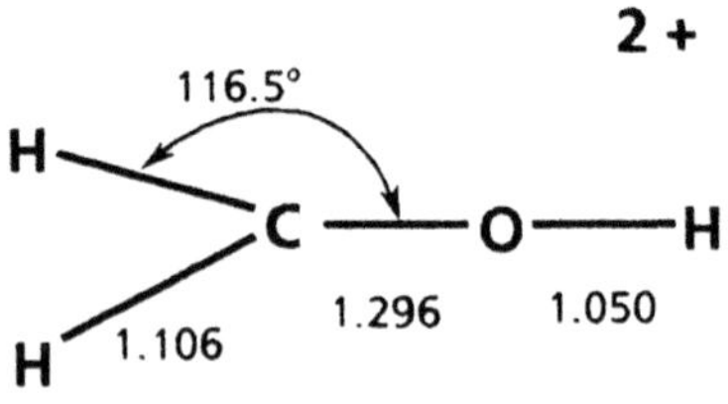

25

reactions considered require activation energies larger than 18 kcal/mol, theory predicts that both these dications should be observable in the gas phase. However, due to the lack of a suitable precursor to generate 24, only the less stable hydroxymethyl dication has been generated in CS experiments /47/. The measured Q_{min} value of 22.4 eV is in excellent agreement with the calculated ionization energy of 22.2 eV.

The methanol dication, CH_3OH^{2+}, is not found to be a stationary point on the CH_4O^{2+} potential energy surface; it dissociates without a barrier to $H_2^{+\cdot}$ + $CHOH^{+\cdot}$ /38,39,43/. On the other hand, the ylide dication $H_2COH_2^{2+}$ (26), a di-cationic pendant to ethylene, resides in a deep potential well.

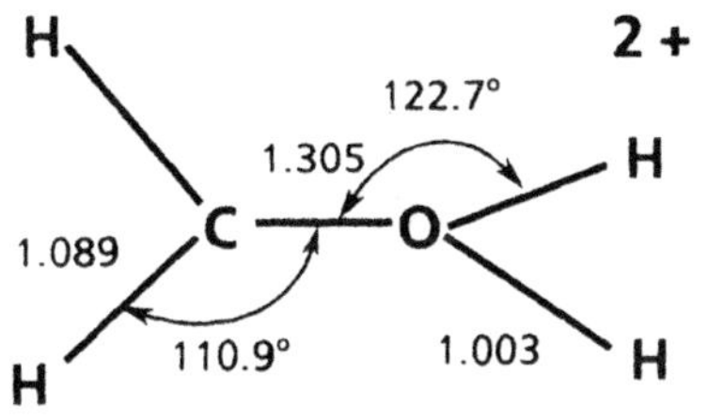

26

All reactions of 26 leading to singly charged fragments are characterized by substantial activation barriers /39,43/. Thus the CH_4O^{2+} system is another example of the extraordinary stability of ylide dications, just like $H_2CNH_3^{2+}$ (21). This similarity also applies for the CS experiments, in which an intense signal for the ionization of $H_2COH_2^{+\cdot}$ has been found, while charge stripping from the methanol cation radical yields a weak peak which again is best assigned in terms of ionization/concomitant H-migration to eventually forms dication 26 /38/.

Among the minima located on the $C_2H_2O^{2+}$ potential energy surface /48/, the ketene dication (27) was found to represent the global minimum followed by the oxirene dication (28) which is predicted to be only 27 kcal/mol less stable than H_2CCO^{2+}. For the neutral system, the energy difference between the two isomers amounts to more than 80 kcal/mol /49/, which reflects the destabilizing cyclic 4π-electron configuration of neutral oxirene. Removal of two π electrons leads to a formal

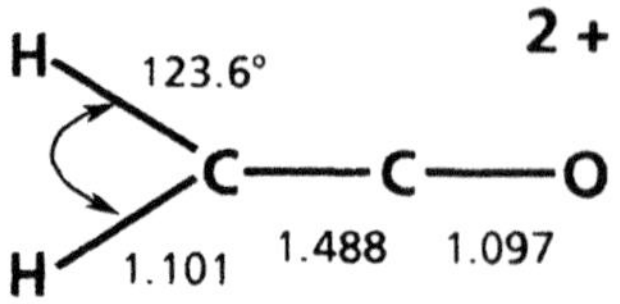

27

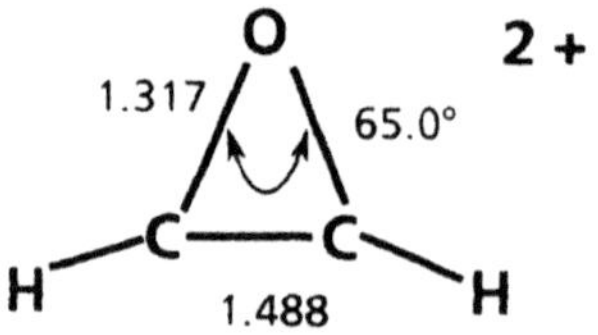

28

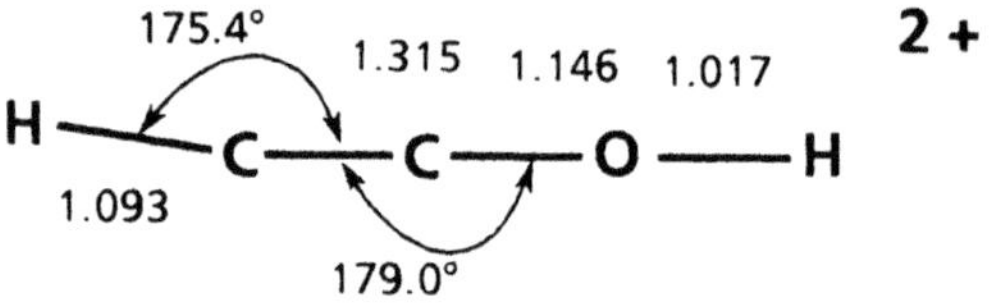

29

Hückel-aromatic system, thereby increasing the stability of the oxirene isomer. Charge stripping from $H_2CCO^{+\cdot}$ leads to ketene dications, with the Q_{min} value of 18.6 eV being in perfect agreement with the theoretical estimation of 18.56 eV /48/. Due to the lack of a suitable precursor molecule 28 is still awaiting experimental verification. Very recently, also a stable hydroxyacetylene dication (29) has been observed in the gas phase /50/ in line with calculations which predict this ion to be a potential minimum in its triplet state, being some 36 kcal/mol less stable than H_2CCO^{2+} (27). (The corresponding singlet hydroxyacetylene is a minimum, too, but 22 kcal/mol less stable than the triplet). The calculated ionization energy (18.5 eV) is again in good agreement with the experimental Q_{min} value, which lies between 18.5 and 19 eV.

As for all other dications mentioned in this article, generation of any of the $C_2H_2O^{2+}$ dications in solution is highly unlikely on the grounds of their tendency to either protonate the solvent shell or to attract an electron or an anion from it.

The repeatedly noted reversal in stability when comparing isomers of dications with the corresponding neutral or mono-cationic species holds also for $C_2H_3O^+$ and $C_2H_3O^{2+\cdot}$ ions. While for the mono-cation the acetylium ion, H_3CCO^+ represents the global minimum, followed by the 1-hydroxyvinyl cation, $H_2C{=}C{-}OH^+$ /51/, for the di-cationic species the latter 30 is predicted to be the most stable isomer with the acetyl dication (31) 29 kcal/mol higher in energy /52/. Again, substantial

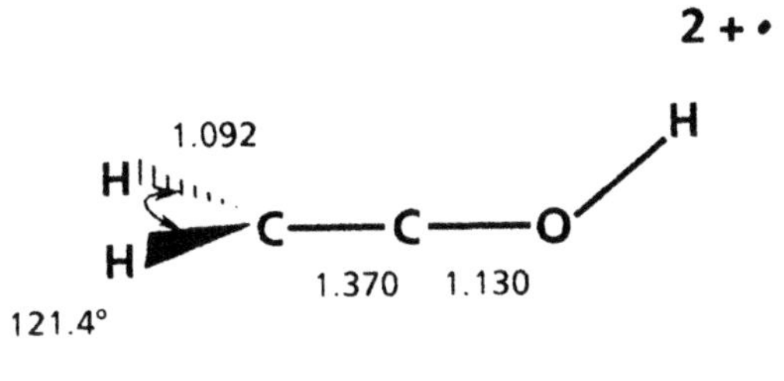

30

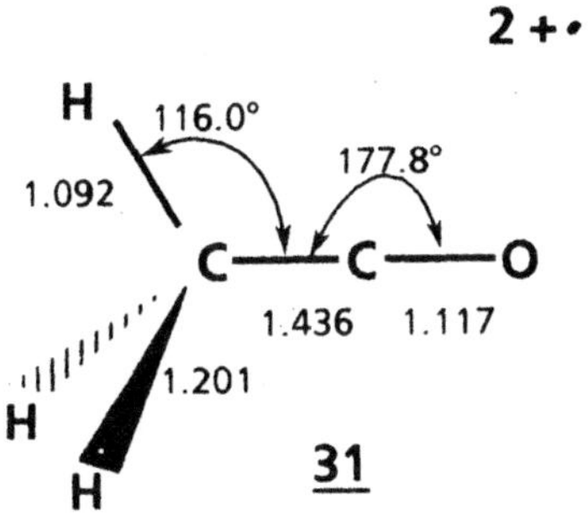

31

activation barriers prevent the spontaneous dissociation of both ions. Charge stripping of $H_2C{=}C{-}OH^+$ leads to 30 with an ionization energy of 18.7 eV, in not too satisfying agreement with the calculated vertical ionization energy of 17.9 eV. Starting from H_3CCO^+ one can also generate a stable $C_2H_3O^{2+\cdot}$ species; there is, however, a huge discrepancy between the measured and the calculated ionization energies (Q_{min} = 18.8 eV; $IE_{calc,vert}$ = 22.6 eV; $IE_{calc,ad}$ = 20.9 eV), which for the time being is not fully resolved. One explanation may be that, like in the charge stripping of $CH_3X^{+\cdot}$ ions, a Rydberg state is involved from which electron ejection is accompanied with or followed by isomerization to $H_2CCOH^{2+\cdot}$ (30).

Significant structural changes associated with successive electron removals were also observed for the C_2H_4O system, which was extensively studied both experimentally and theoretically /53/. While for the neutral species acetaldehyde (32) forms the global minimum, on the $C_2H_4O^{2+}$ surface the acetaldehyde dication does not exist at all. Electron removal from $H_3CCHO^{+\cdot}$ (38) by charge stripping leads to spontaneous C-C bond cleavage, to give H_3C^+ and CHO^+. On the other hand, the yet unknown neutral hydroxymethyl-carbene, $H_3C{-}C{-}OH$ (35) is, upon removal of two electrons transformed to a dication (41) which corresponds to the global minimum on the $C_2H_4O^{2+}$ potential energy surface. Similarly, the unknown ylide 36 is predicted to form a stable dication 45 and the same applies to the very unusual dication, $H_4C\text{-}CO^{2+}$ (42) which has recently been described theoretically /54/. The stability ordering for these and a few more C_2H_4O isomers is pictorially summarized in Chart I for the neutral, mono- and di-cationic species. Most of the $C_2H_4O^{2+}$ dications are prevented by barriers from both facile interconversion and unimolecular dissociation. Thus experimental observation of $C_2H_4O^{2+}$ should be feasible. In fact, CS mass spectrometry provided evidence that H_3CCOH^{2+} (41), $H_2COCH_2^{2+}$ (43) and H_2CCHOH^{2+} (44) can indeed be generated, and a comparison of the experimentally derived and theoretically predicted vertical ionization energies showed good agreement for $H_2COCH_2^{2+}$ (16.3 versus 15.6 eV) and H_2CCHOH^{2+} (17.1 versus 17.2 eV). For H_3CCOH^{2+} (41) the ionization energy could not be measured due to sensitivity problems /53a/. The theoretical investigation /55/ of the $CO_2H_2^{2+}$ potential energy surface indicates that the dihydroxy carbene dication (46) is the most stable isomer, 11 kcal/mol lower in energy than the second isomer $H_2O\text{-}CO^{2+}$ (47) which, in a formal sense, can be described as a di-cationic complex of H_2O^+ and $CO^{+\bullet}$. The dication of formic acid is not found to exist as a minimum structure; the calculations reveal that this dication will most probably isomerize without a barrier to $C(OH)_2^{2+}$ (46). Thus, again a dramatic change of the stabilities of the structural isomer takes place upon ionization. The considered charge separation reactions are, as expected highly exothermic, but are associated with significant barriers of at least 38 kcal/mol. Experimental generation and detection of these dications should therefore be feasible.

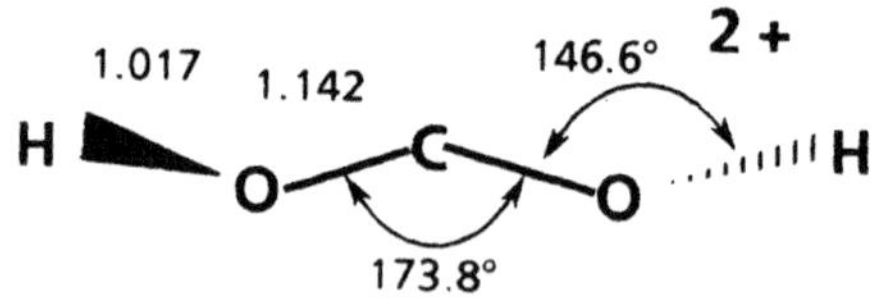

46

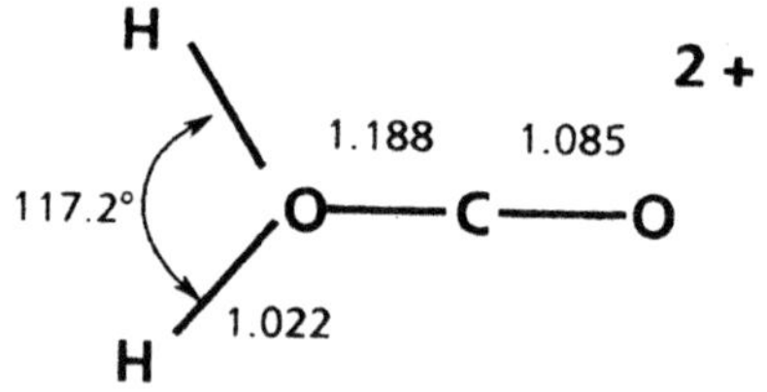

47

Chart I

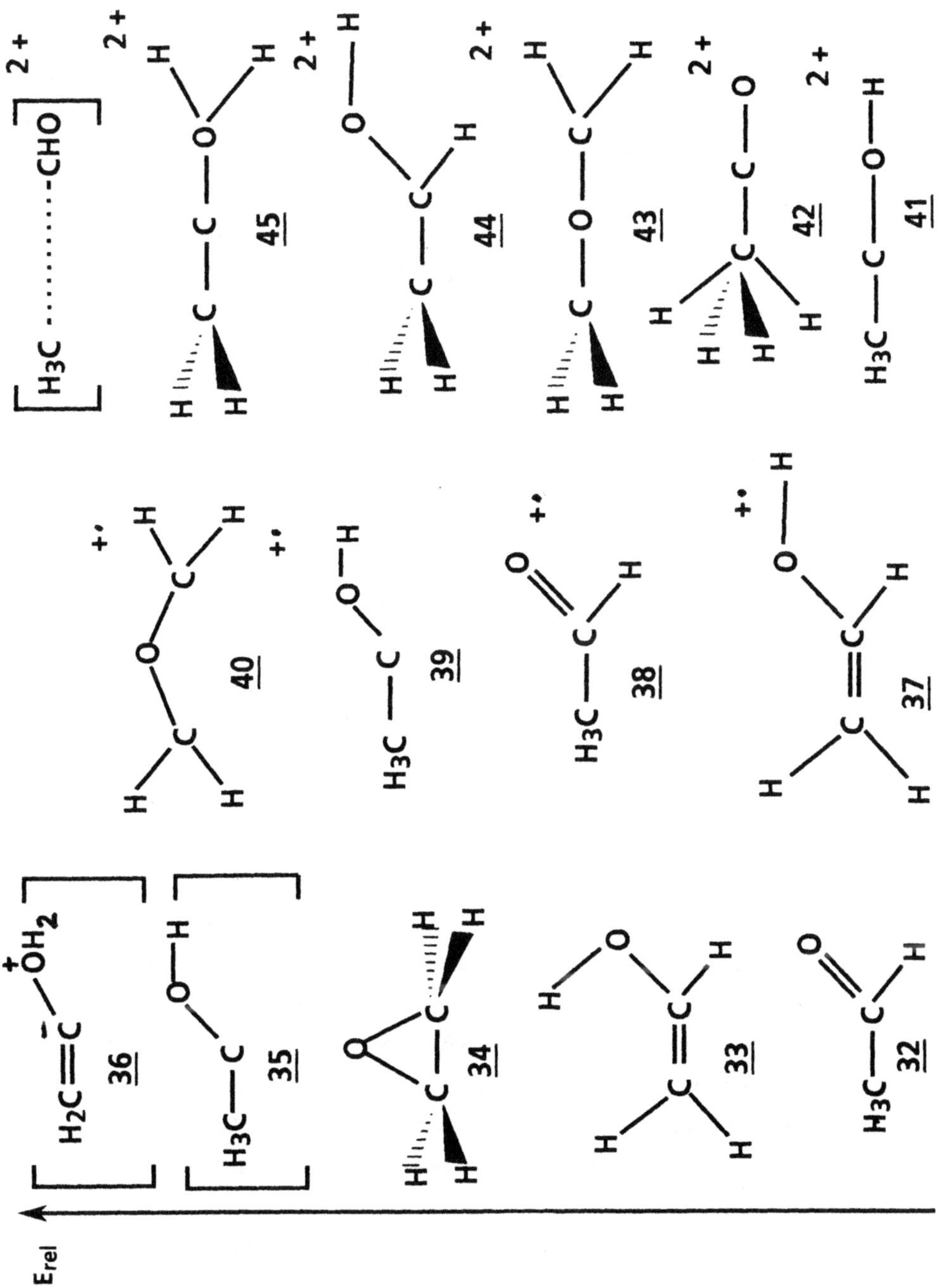

4.4 Carbon/Fluorine Containing Dications

The small dications $CF^{2+\cdot}$ and CF_2^{2+} have formed the subject of extensive experimental /21,56/ and theoretical /57/ investigations. Charge stripping experiments performed by Beynon and colleagues point to the existence of stable CF^+ cations in two distinct electronic states, with the long lived excited state some 5 eV above the ground state. The measured ionization energy to produce $CF^{2+\cdot}$ from ground state CF^+ amounts to 26.0 $\pm$ 0.4 eV. This result is in pleasing agreement with theoretical studies which predicts an ionization energy of 25.5 eV. Beynon et al. also succeeded to generate the CF_2^{2+} dication; however, the reported Q_{min} value of 19.0 eV is connected with large error bars. The corresponding theoretical value is with 20.5 eV substantially higher. The electronic structure and the resulting geometrical aspects of $CF^{2+\cdot}$ and CF_2^{2+} deserve brief mentioning: Upon ionization, the strength of the C-F bond increases substantially which is reflected in shortening of the bond length. While for neutral CF an equilibrium distance of 1.291 Å is predicted, CF^+ is characterized by a bond length of 1.173 Å and $CF^{2+\cdot}$ by a bond of only 1.146 Å, in spite of the increasing electrostatic repulsion. The CF-distance of CF^{2+} is among the shortest CF bonds ever measured or theoretically predicted /57/. This decrease of the CF bond length goes hand in hand with an increase of the harmonic freqency of the CF vibration from 1411 cm^{-1} in neutral CF, 1964 cm^{-1} in CF^+ up to 2209 cm^{-1} for $CF^{2+\cdot}$. Similar, although less pronounced trends are observed for the neutral, mono- and dicationic CF_2 species. The origins of this surprising gain in stability of the dicationic CF bond is the high π-character, due to the interaction of formally occupied p_π orbitals of F with the formally empty carbon orbitals of appropriate symmetry. Thus $CF^{2+\cdot}$ and CF_2^{2+} are best described as $C{=}F^{2+\cdot}$ and $F{=}C{=}F^{2+}$, in close similarity to the neutral, isoelectronic molecules $C{=}N^{\cdot}$ and O=C=O.

Similar to the $CH_3NH_2^{2+}$ and CH_3OH^{2+} ions discussed above, the classical CH_3F^{2+} dication is theoretically predicated to be unstable /39/; proton loss can take place without a barrier. On the other hand the ylide dication, H_2CFH^{2+} (48) represents a kinetically stabilized ion. The calculated /39/ vertical ionization energy of $H_2CFH^{+\cdot}$ is with 18.8 eV in reasonable agreement with the experimental Q_{min} value (17.9 eV) /38/. For the ionization of $CH_3F^{+\cdot}$ (like that for $CH_3NH_2^{+\cdot}$ and $CH_3OH^{+\cdot}$) a huge discrepancy between the theoretical and experimental values is found.

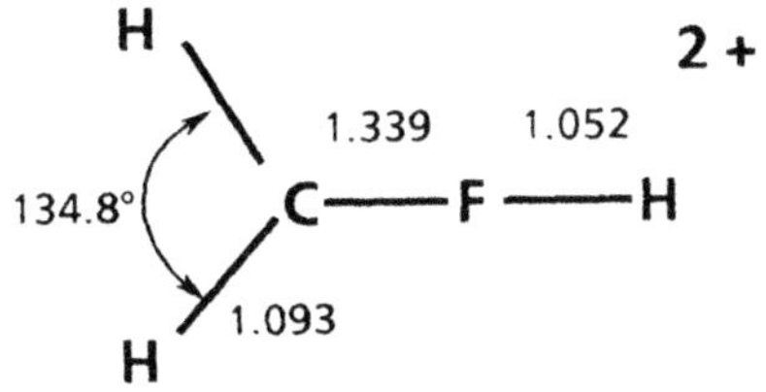

48

As discussed in Chapter 4.1, the ethylene dication (10) prefers an anti-van't Hoff structure due to hyperconjugative stabilization. The replacement of all hydrogens by fluorine atoms leads to drastic changes. It is well known, that CF hyperconjugation has -at best- an only extremely small stabilizing effect. In fact, calculations /58/ indicate, that planar $C_2F_4^{2+}$ (49) is the only minimum while the perpendicular structure 50 represents the transition structure for the C-C rotation; 50 is 2.2 kcal/mol higher in energy than 49. While planar and perpendicular ethylene dications differ significantly in both their geometries and energies (e.g. r_{CC} (plan.) = 1.59 Å, r_{CC}(perp.) = 1.43 Å), the two $C_2F_4^{2+}$ isomers show quite similar geometries: The CC bonds are relatively long, in line with the absence of any conjugative or hyperconjugative interaction; the CF bonds, however, are surprisingly short being much shorter than those calculated for neutral C_2F_4 (r_{CF} = 1.33 Å) or mono-cationic perfluoroethylene (r_{CF} = 1.28 Å).

This structural aspects point to a mesomeric contribution of the lone pair electrons on fluorine interacting with the formally empty p-orbitals on carbon to generate a C-F "double bond" as indicated in terms of valence bond (VB) structures in Fig. 8. The importance of such C-F multiple bonds in dications was already recognized above in the discussion of $CF^{2+\cdot}$ and CF_2^{2+}. However, CF conjugation cannot serve

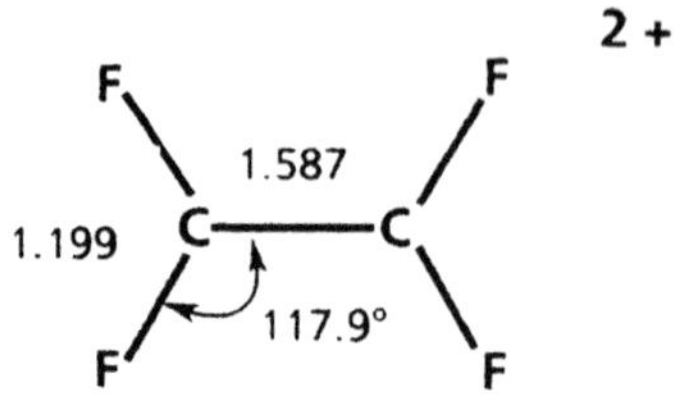

<u>**49**</u>

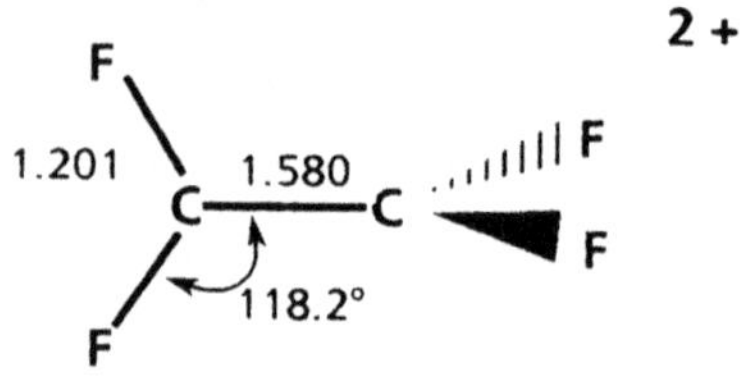

<u>**50**</u>

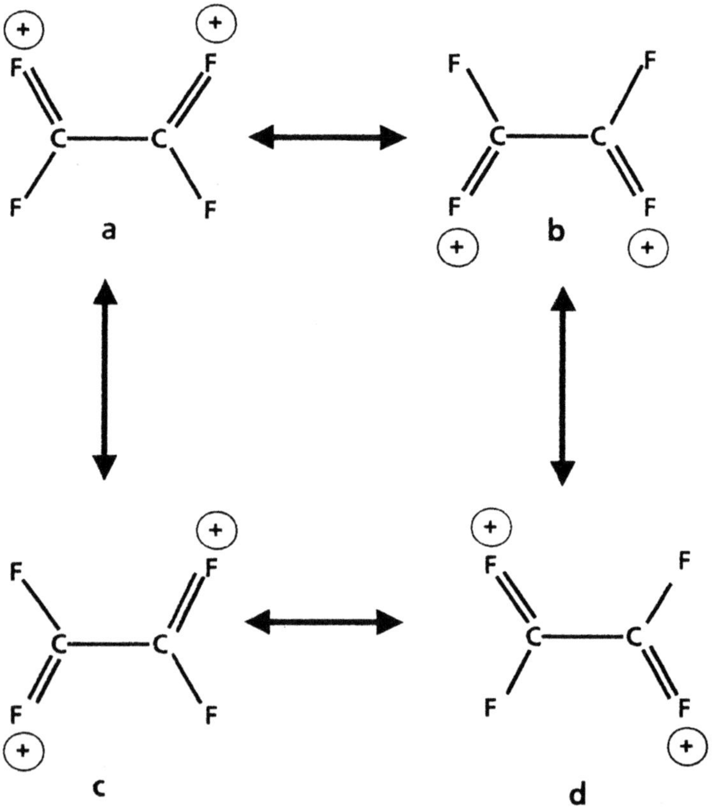

Figure 8: Valence bond structures for $C_2F_4^{2+}$

as an explanation for the preference of the planar $C_2F_4^{2+}$ (49) over the perpendicular isomer (50) since the VB structures a-d are possible for both isomers. The elaborate analysis of this problem /58b/ revealed that π -conjugation across the C-C bond, which is only possible for 49 is most likely the origin of the slightly larger stability of the planar isomer. That this C-C π-conjugation contributes to only a small effect is easily visualized by inspections of the appropriate VB structure e-h (Fig. 9), which are much less important than the VB structures a-d.

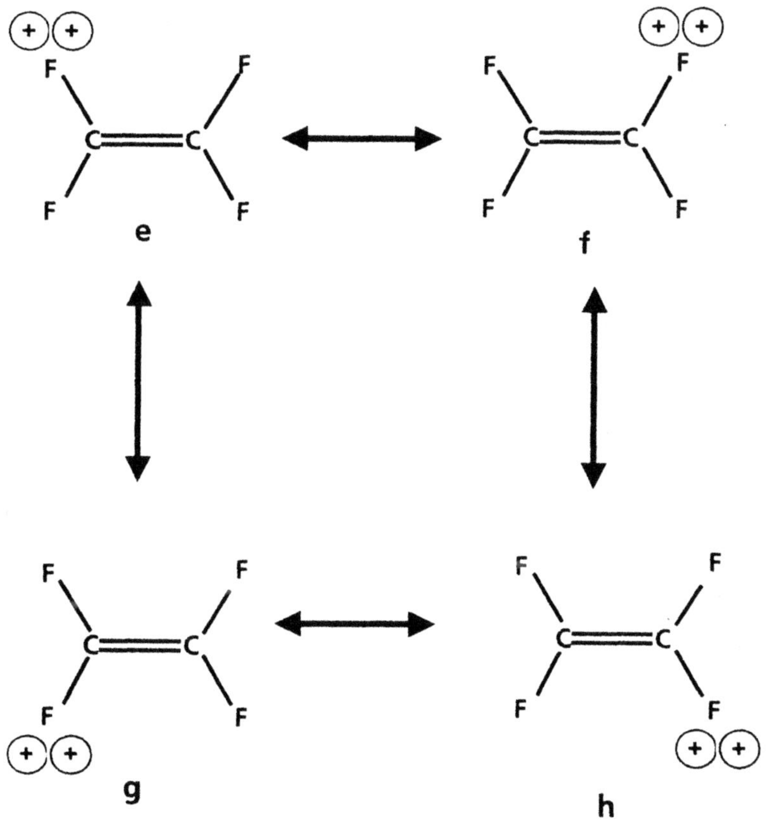

Figure 9: Valence bond structures for $C_2F_4^{2+}$

In the theoretical analysis /58b/ also all other possible $C_2H_nF_{4-n}^{2+}$ isomers were studied. It turned out that the subtle balance between C-H hyperconjugation and fluorine lone pair donation determines which of the two conformers (planar versus perpendicular) represents the minimum for the dications. The calculations predict that $C_2F_4^{2+}$ and $C_2HF_3^{2+}$ prefer the planar arrangement, while for $1.1\text{-}C_2H_2F_2^{2+}$ and $C_2H_3F^{2+}$ the perpendicular form is the more stable one.

The difluoroethylene dications are of particular interest, since five isomers are possible, i.e. 1.1-planar, 1.1-perpendicular, cis- and trans-1,2-planar and 1,2-perpendicular. The global minimum corresponds to a slightly distorted perpendicular 1,2-structure, with the cis and trans-isomers of the planar form merely 0.9 and 0.2 kcal/mol less stable. The perpendicular 1,1-isomer is 3.9 kcal/mol above and the planar 1,1-structure is least stable with 9.1 kcal/mol above the global minimum. This is in remarkable contrast to the neutral and mono-cationic counterparts, where the 1,1-isomer is more stable than cis- and trans difluore ethylene /59/. The observed reversal of the stabilities supports the suggestion that electrostatic interactions are responsible for the energetical preference of the 1,1-isomer compared to the vicinally substituted species in neutral and mono-cationic difluoroethylenes. For both cases, the two carbons bear charges of opposite sign, which leads to an electrostatic attraction, which is impossible, due to symmetry, for the 1,2-species. Contrary to this, both carbons are positively charged and repel each other in the $1,1\text{-}C_2H_2F_2^{2+}$ dication, thus leading to Coulombic destabilization. The various $C_2H_nF_{4-n}^{2+}$ dications have also been generated and observed in CS experiments /58a,60/. Hence, significant barriers must exist for the fragmentation reactions. The agreement of the calculated and measured ionization energies is reasonable, with $C_2HF_3^{2+}$ being the exception. For this system a discrepancy between Q_{min} and IE of 2 eV is found (Table III).

Table III: Calculated vertical ionization energies and Q_{min} values to generate $CH_nF_{4-n}^{2+}$ dications

Precursor ion	$IE_{calc,vert.}$ [a]	Q_{min}
$C_2H_3F^{+\cdot}$	19.1	18.7[c)]
$C_2H_2F_2^{+\cdot}$ [b)]	19.0	18.7[c)]
$C_2HF_3^{+\cdot}$	18.4	20.4[c)]
$C_2F_4^{+\cdot}$	18.2	19.0[d)]

a) Ref. 58b

b) The IE/Q_{min} are almost identical for the isomeric $C_2H_2F_2^{+\cdot}$ species.

c) Ref. 60

d) Ref. 58a

5. DONOR-ACCEPTOR INTERACTIONS AND THE PECULIAR STRUCTURES OF DICATIONS

As repeatedly noted in this article, dications often prefer structures, which are not known or at best extremely unstable for the neutral counterparts. The best studied examples are the ylide dications of which $CH_2NH_3^{2+}$, $CH_2OH_2^{2+}$ and CH_2FH^{2+} have been briefly discussed in the preceding chapters.

In a thorough theoretical investigation Yates et al. /39/ showed that also the second row ylide dications, i.e. $CH_2PH_3^{2+}$, $CH_2SH_2^{2+}$ and CH_2ClH^{2+} are species which reside in deep potential wells, in line with experimental results. The corresponding classical isomers, i.e. $CH_3PH_2^{2+}$, CH_3SH^{2+} and CH_3Cl^{2+} are significantly less stable, both thermochemically and kinetically. In fact, for the CH_3X^{2+} system (X = PH_2,SH,Cl) the calculated barriers towards isomerization and/or fragmentations are either small or do not exist at all.

A second structurally remarkable class of dications is represented by the methonium derivatives, CH_4X^{2+}, which were first studied by

Lammertsma /54/. Systems with X = CH_2 and CO were already briefly mentioned in Chapters 4.1 and 4.3 but X may also comprise some other groups, e.g. FH, OH_2, NH_3, N_2 etc.

The common feature of these two classes of dications, i.e. H_2CX^{2+} and H_4CX^{2+}, is that they may be interpreted as the result of a bonding interaction of a neutral donor molecule X and a di-cationic acceptor, i.e. CH_2^{2+} or CH_4^{2+}, respectively. The expected strength of this interaction as well as the corresponding geometrical consequences can be easily predicted by a simple frontier molecular orbital (FMO) model as recently described by Frenking et al. /61/.

The linear CH_2^{2+} dication possesses two, degenerate vacant p (π) atomic orbitals on carbon. The dominant interaction between CH_2^{2+} and a neutral molecule X (which has a lone pair) is the charge transfer from the lone pair of X into one of the vacant CH_2^{2+} p (π) orbitals. Additionally, many donors have high-lying, occupied π -orbitals, which leads to a further charge transfer into the second p(π) orbital of CH_2^{2+}, thus increasing the total charge donation (Fig. 10).

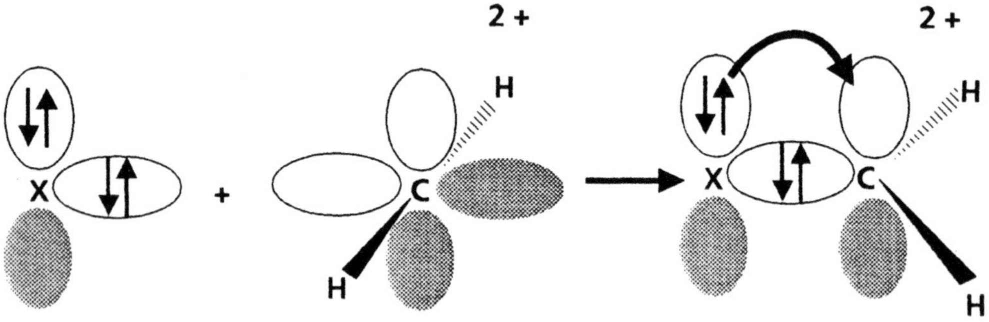

Figure 10: Donor/acceptor interaction between X and CH_2^{2+}

The formation of such a donor-acceptor complex is accompanied with a change of the hybridization of CH_2^{2+} from sp to sp^2. As a consequence, in the course of complexation the linear CH_2^{2+} dication undergoes bending of the CH_2 unit and the amount of bending may serve as an indicator for the strength of the donor-acceptor interaction. The energy, ΔE, of reaction (5) is a direct measure of the stability of the $H_2C\text{-}X^{2+}$ complex relative to its fragments

$$H_2C^{2+} + X \longrightarrow H_2CX^{2+} \qquad (5)$$

As shown in Table IV the increase of bending in the CH_2^{2+} subunit indeed follows the increase of the calculated complexation energies.

Table IV: Bending of the CH_2 fragment and complexation energies, ΔE (Equation 5).

X	HCH (in degrees)	E (kcal/mol)
FH	134.8	116.6
OC	129.6	120.8
N_2	128.2	126.8
CO	125.4	152.9
OH_2	126.2	214.7
NH_3	122.6	249.6
CH_2	119.8	273.2

What determines the donor strength of X in CH_2X^{2+} ? In Figure 11 the eigenvalues of the highest molecular orbital (HOMO) of X are plotted against the stabilization energies, ΔE. The correlation is obvious. Molecules with a higher lying HOMO are better donors and show larger stabilization energies, ΔE, compared to species with a lower lying HOMO.

Very similar observations have been found for CH_4^{2+} acting as acceptor molecule; the correlation of the stabilization energies of reaction (6) is also included in Figure 11.

$$CH_4^{2+} + X \longrightarrow CH_4X^{2+} \qquad (6)$$

This qualitative concept, which stresses the particular importance of donor-acceptor interactions for the structures and stabilities of dications is, however, not limited to ions CH_4X^{2+} or CH_2X^{2+}, but can also be used to explain properties of many other dications. This may be illustrated by the example of the $CH_3O^{2+\cdot}$ dications. As discussed in Chapter 4.3 the most stable isomer is $HCOH_2^{2+\cdot}$ (24) followed by $H_2COH^{2+\cdot}$ (25). A third structure, which is chemically irrelevant and was not mentioned before, is $H_3C\text{-}O^{2+\cdot}$, some 116 kcal/mol less stable than $HC\text{-}OH_2^{2+\cdot}$. These three ions may by considered as donor-acceptor complexes arising from acceptors $CH^{2+\cdot}$, CH_2^{2+} and $CH_3^{2+\cdot}$ and donors OH_2, $OH^{\cdot}$ and O, respectively. The FMO analysis establishes a sequence of donor strength $OH_2 > OH^{\cdot} > O$, while the acceptor strength is found as $CH^{2+\cdot} > CH_2^{2+} > CH_3^{2+\cdot}$. Thus the stability order is in perfect agreement with the strength of the donor-acceptor interaction indicated by the frontier orbitals /61/.

However, it should be kept in mind, that this simple model does, of course, not cover all kinds of possible interactions, and many exceptions may well exist. Nevertheless, the concept is of great value as a guiding rule for predicting structures and stabilities of isomeric dications.

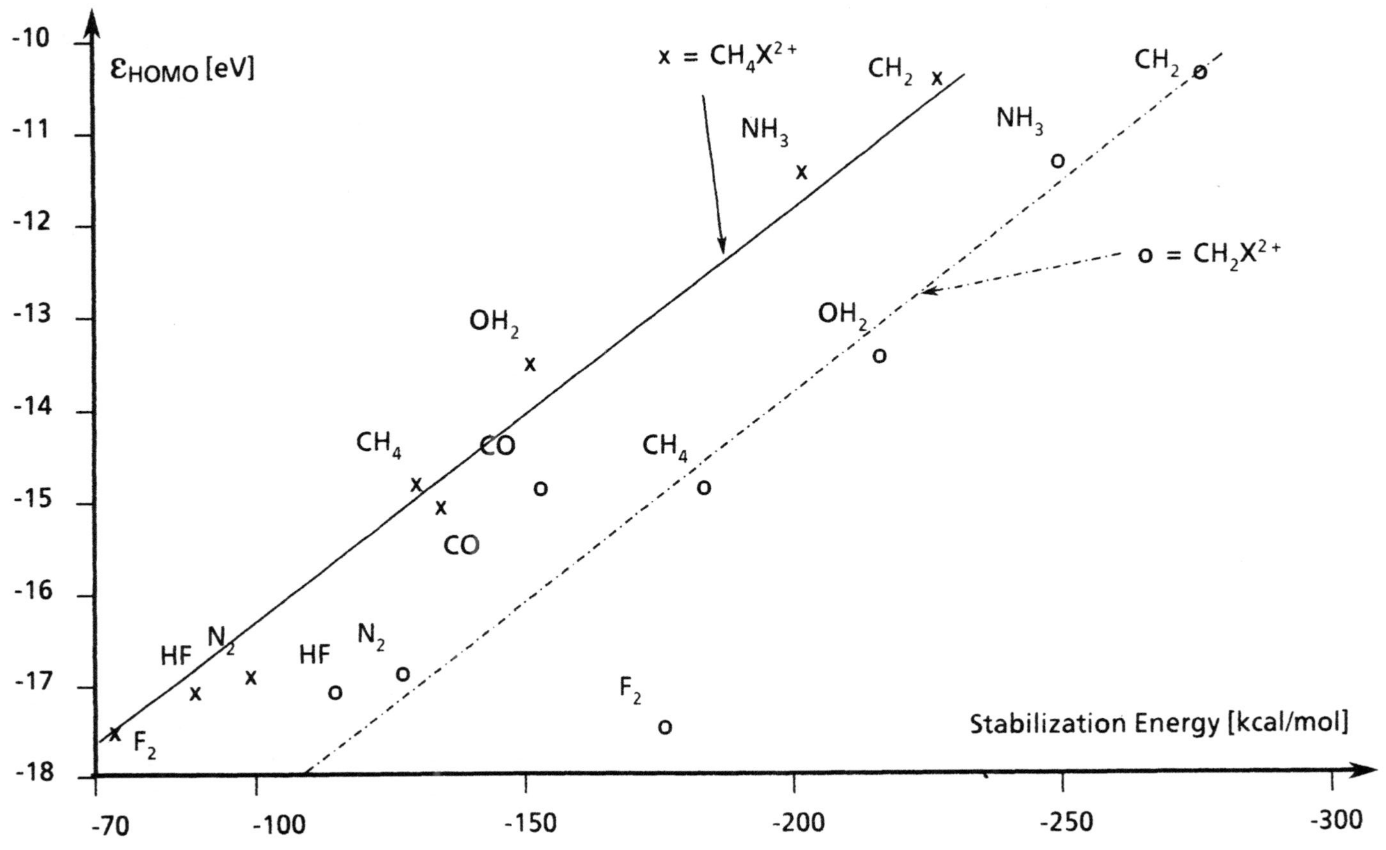

Figure 11: HOMO eigenvalues and stabilization energies for reactions (5) and (6).

6. SUMMARY AND OUTLOOK

The exciting new field of small, doubly-charged gas phase cations serves as a convincing example for the synergistic interplay of theory and experiment. In many combined experimental/theoretical studies the following salient features of dications were observed. (i) The relative stabilities of di-cationic isomers are frequently reversed when compared with their neutral or mono-charged counterparts; (ii) significant structural changes which often favour anti-van't Hoff geometries; (iii) highly exothermic charge separation reactions, which in the gas phase, however, are prevented by substantial barriers from occuring spontaneously, thus making observations of the thermochemically unstable dications feasible; (iv) the prospects of generating in solution any of the dications discussed are quite remote. Proton transfer from the dication to the solvent shell or addition of negatively charge species to the dications will occur avidly.

What are the topics which are likely to be studied in the near future, concerning multiply-charged cations ? We believe, that much effort will be invested in the study of triply and even higher charged cations. There is already compelling evidence for the existence of such, truly remarkable species, like e.g. CS_2^{3+} /62/. Furthermore, a more detailed understanding of the physics and chemistry of the charge stripping process is highly desirable, not the least to remove the discrepancies between calculated and measured ionization energies found for some systems. This should also include rigorous theoretical treatments of these poly-charged ions /10,63/. A thorough spectroscopic characterization, which is now nearly routinely used for many gaseous mono-cations, is also warranted for multiply-charged cations, and for several diatomic dications promising results were recently published /64/.

The study of solvated mono-charged cations and anions in the gas phase has provided a deep inside into both intrinsic properties of charged species and the role differential solvation plays /65/. Related gas phase studies on dications are more or less unknown. As the generation in solution of any of the dications discussed in this article is highly unlikely, studies in the gas phase might be feasible, although it must be admitted that the very same processes which prevent the generation of dications in solution (proton transfer to solvent shell or transfer of negatively charged species from it to the dication) may also operate in the gas phase.

The study of neutral and mono-charged van der Waals complexes and clusters with its many facets belongs undoubtedly to the main lines of basic research in chemistry and physics. The recent reports /66/ on the successful generation and characterization of multiply-charged cluster ions with up to five (!) elementary charges are extremely encouraging and more exciting results are likely to emerge in the future, including also the study of multiply-charged anions /67/.

7. ACKNOWLEDGEMENTS

The continuous support of our work by the Fonds der Chemischen Industrie and the Deutsche Forschungsgemeinschaft is gratefully acknowledged. We are indebted to Dr. Gernot Frenking, SRI International, Menlo Park, and Dr. Daniel Stahl, ETH Lausanne, and his co-workers for a highly successful and enjoyable collaboration.

8. REFERENCES

/1/ P.v.R. Schleyer, Am Chem. Soc. Div. Pet. Chem. Prepr. 28, 413 (1983).

/2/ For a comprehensive literature survey see: a) W. Koch, N. Heinrich, H. Schwarz, F. Maquin, D. Stahl, Int. J. Mass Spectrom. Ion Processes 67, 305 (1985). For recent reviews on stable dications in solution see: b) G.K.S. Prakash, T.N. Rawdah, G.A. Olah, Angew. Chem. 95, 356 (1983); Angew. Chem. Int. Ed. Engl. 22, 390 (1983); c) R.M. Pagni, Tetrahedron 40, 4161 (1984).

/3/ a) R.G. Cooks, T. Ast, J.H. Beynon, Int. J. Mass Spectrom. Ion Phys. 11, 490 (1973); b) D.L. Kemp, R.G. Cooks in R.G. Cooks: "Collision

Spectroscopy", Plenum Press, New York, 257 (1978); c) T. Ast, Adv. Mass Spectrom. 8A, 555 (1980); d) T. Ast, C.J. Porter, C.J. Proctor, J.H. Beynon, Bull. Soc. Chem. Beograd 46, 135 (1981); e) K. Levsen, H. Schwarz, Mass Spectrom. Rev. 2, 77 (1983); f) T. Ast, Adv. Mass Spectrom. 10A, 471 (1986).

/4/ a) G. Dujardin, S. Leach, O. Dutuit, P.-M. Guyon, M. Richard-Viard, Chem. Phys. 88, 339 (1984); b) G. Dujardin, D. Winkoun, S. Leach, Phys. Rev. A 31, 3027 (1985).

/5/ Review: W. Koch, F. Maquin, D. Stahl, H. Schwarz, Chimia 39, 376 (1985).

/6/ M. Guilhaus, A.G. Brenton, J.H. Beynon, M. Rabrenovich, P.v.R. Schleyer, J. Phys. B 17, L605 (1984); J. Chem. Soc. Chem. Commun. 210 (1985).

/7/ H. Yagisawa, H. Sato, T. Watanabe, Phys. Rev. A 16, 1352 (1977).

/8/ A.C. Hurley, J. Molec. Spectrosc. 9, 18 (1962).

/9/ a) F.H. Dorman, J.D. Morrison, J. Chem. Phys. 35, 575 (1961); b) ibid. 39, 1806 (1963).

/10/ a) R.W. Wetmore, R.J. LeRoy, R.K. Boyd, J. Phys. Chem. 88, 6318 (1984); b) R.W. Wetmore, R.K. Boyd, R.J. LeRoy, Chem. Phys. 89, 329 (1984).

/11/ L. Pauling, J. Chem. Phys. 1, 56 (1933).

/12/ H. Halim, N. Heinrich, W. Koch, J. Schmidt, G. Frenking, J. Comput. Chem. 7, 93 (1986).

/13/ a) M. Rabrenovich, C.J. Proctor, T. Ast, C.G. Herbert, A.G. Brenton, J.H. Beynon, J. Phys. Chem. 87, 3305 (1983); b) T. Ast, C.J. Porter, C.J. Proctor, J.H. Beynon, Chem. Phys. Lett. 78, 439 (1981) c) C.J. Proctor, C.J. Porter, T. Ast, P.D. Bolton, J.H. Beynon, Org. Mass Spectrom. 16, 454 (1981); d) M. Rabrenovich, C.G. Herbert, C.J. Proctor, J.H. Beynon, Int. J. Mass Spectrom. Ion Phys. 47, 125 (1983); e) M. Rabrenovich, A.G. Brenton, J.H. Beynon, ibid. 52, 175 (1983); f) D. Stahl, F. Maquin, T. Gäumann, H. Schwarz, P.-A. Carrupt, P. Vogel, J. Am. Chem. Soc. 107, 175 (1985).

/14/ a) A.W. Hanner, T.F. Moran, Org. Mass Spectrom. 16, 512 (1981); b) P.E.M. Siegbahn, Chem. Phys. 66, 443 (1982); c) J.A. Pople, B.

Tidor, P.v.R. Schleyer, Chem. Phys. Lett. 88, 533 (1982); d) K. Lammertsma, M. Barzaghi, G.A. Olah, J.A. Pople, P.v.R. Schleyer, M. Simonetta, J. Am. Chem. Soc. 105, 5258 (1983).

/15/ a) S.A. Pope, I.A. Hillier, M.F. Guest, J. Kendric, Chem. Phys. Lett. 95, 247 (1983); b) S.A. Pope, I.A. Hillier, M.F. Guest, Faraday Symp. Chem. Soc. 19, paper 8 (1984); c) W. Koch, H. Schwarz, Int. J. Mass Spectrom. Ion Processes 68, 49 (1986).

/16/ R.K. Boyd, S. Singh, J.H. Beynon, Chem. Phys. 100, 297 (1985).

/17/ J.S. Wright, V. Barclay, E. Kruus, Chem. Phys. Lett. 122, 214 (1985).

/18/ W. Koch, unpublished calculations.

/19/ G.A. Olah, G.K.S. Prakash, M. Barzaghi, K. Lammertsma, P.v.R. Schleyer, J.A. Pople, J. Am. Chem. Soc. 108, 1032 (1986).

/20/ a) W. Koch, G. Frenking, H. Schwarz, J. Chem. Soc. Chem. Commun. 1119 (1985); b) W. Koch, G. Frenking, H. Schwarz, F. Maquin, D. Stahl, J. Chem. Soc. Perkin II 757 (1986); c) W. Koch, G. Frenking, C.C. Chang, J. Chem. Phys. 84, 2703 (1986).

/21/ C.J. Porter, C.J. Proctor, T. Ast, J.H. Beynon, Croat. Chim. Acta 54, 407 (1981).

/22/ For a more theoretical discussion of this point, see: P.J. Bruna, S.D. Peyerimhoff, Faraday Symp. Chem. Soc. 19, 193 (1984).

/23/ a) J.A. Pople, M.J. Frisch, K. Raghavachari, P.v.R. Schleyer, J. Comput. Chem. 3, 468 (1982); b) For semi-empirical calculations, see: G.W. Burdick, G.C. Shields, J.R. Appling, T.F. Moran, Int. J. Mass Spectrom. Ion Processes 64, 315 (1985).

/24/ a) K. Lammertsma, J.A. Pople, P.v.R. Schleyer, J. Am. Chem. Soc. 108, 7 (1986). b) For semi-empirical calculations on $C_4H_2^{2+}$, see: Ref. 23b.

/25/ M. Rabrenovich, J.H. Beynon, Int. J. Mass Spectrom. Ion Processes 54, 87 (1983).

/26/ K. Lammertsma, M. Barzaghi, G.A. Olah, J.A. Pople, A.J. Kos, P.v. R. Schleyer, J. Am. Chem. Soc. 105, 5252 (1983).

/27/ C. Benoit, J.A. Horsley, Mol. Phys. 30, 557 (1975).

/28/ a) G.A. Olah, M. Simonetta, J. Am. Chem. Soc. 104, 330 (1982);

b) P.v.R. Schleyer, A.J. Kos, J.A. Pople, A.T. Balaban, ibid. 104, 3771 (1982); c) K. Lammertsma, G.A. Olah, M. Barzaghi, M. Simonetta, ibid. 104, 6851 (1982).

/29/ D. Stahl, F. Maquin, Chimia, 37, 87 (1983).

/30/ M. Rabrenovich, J.H. Beynon, a) J. Chem. Soc. Chem. Commun. 1043 (1983); b) Int. J. Mass Spectrom. Ion Processes 56, 85 (1984).

/31/ J. Baker, E.E. Kingston, W.J. Bouma, A.G. Brenton, L. Radom, J. Chem. Soc. Chem. Commun. 1625 (1985).

/32/ a) L. Radom, H.F. Schäfer III, J. Am. Chem. Soc. 99, 7522 (1977); b) K. Krogh-Jespersen, P.v.R. Schleyer, J. A. Pople, D. Cremer, ibid. 100, 4301 (1978); c) J. Chandrasekhar, P.v.R. Schleyer, J. Comput. Chem. 2, 356 (1981).

/33/ a) A. Hustrulid, P. Kusch, J.T. Tate, Phys. Rev. 52, 843 (1937); b) W. Higgins, K.R. Jennings, J. Chem. Soc. Chem. Commun. 99 (1965); c) R. Engel, D. Halpern, B.-A. Funk, Org. Mass Spectrom. 7, 177 (1973); d) B. Brehm, U. Fröbe, H.-P. Neitzke, Int.J. Mass Spectrom. Ion Processes 57, 91 (1984); e) P.J. Richardson, J.H.D. Eland, P. Lablanquie, Org. Mass Spectrom. 21, 289 (1986).

/34/ a) K. Lammertsma, P.v.R. Schleyer, J. Am. Chem. Soc. 105, 1049 (1983); b) M.J.S. Dewar, M.K. Holloway, ibid. 106, 6619 (1984).

/35/ W.D. Watson, Astrophys. J. 188, 35 (1974).

/36/ a) M.J. Frisch, K. Raghavachari, J.A. Pople, W.J. Bouma, L. Radom, Chem. Phys. 75, 323 (1983); b) E. Uggerud, H. Schwarz, J. Am. Chem. Soc. 107, 5046 (1985)c) M.T. Nguyen, A.F. Hegarty, P. Brint, J. Chem. Soc. Dalton Trans. 1915 (1985).

/37/ W. Koch, N. Heinrich, H. Schwarz, J. Am. Chem. Soc. 108, 0000 (1986).

/38/ F. Maquin, D. Stahl, A. Sawaryn, P.v.R. Schleyer, W. Koch, G. Frenking, H. Schwarz, J. Chem. Soc. Chem. Commun. 504 (1984).

/39/ B.F. Yates, W.J. Bouma, L. Radom, J. Am. Chem. Soc. 108, 0000 (1986).

/40/ R.A. Whiteside, M.J. Frisch, J.A. Pople: "The Carnegie-Mellon Quantum Chemistry Archive", 3rd Edition, Pittsburgh, PA (1983).

/41/ B.F. Yates, R.H. Nobes, L. Radom, Chem. Phys. Lett. 116, 474 (1985).

/42/ W. Koch, C.B. Lebrilla, T. Drewello, D. Stahl, H. Schwarz, manuscript in preparation.

/43/ W.J. Bouma, L. Radom, J. Am. Chem. Soc. 105, 5484 (1983).

/44/ D. Stahl, F. Maquin, Chem. Phys. Lett. 106, 531 (1984).

/45/ F.P. Lossing, J. Am. Chem. Soc. 99, 7526 (1977).

/46/ R.H. Nobes, L. Radom, W.R. Rodwell, Chem. Phys. Lett. 74, 269 (1980).

/47/ W. Koch, F. Maquin, D. Stahl, H. Schwarz, J. Am. Chem. Soc. 107, 2256 (1985)

/48/ W. Koch, F. Maquin, D. Stahl, H. Schwarz, J. Chem. Soc. Chem. Commun. 1679 (1984).

/49/ a) W.J. Bouma, R.H. Nobes, L. Radom, C.E. Woodward, J. Org. Chem. 47, 1869 (1982); b) For a review on neutral oxirene, see: E.G. Lewars, Chem. Rev. 83, 519 (1983).

/50/ B.v.Baar, T. Weiske, J.K. Terlouw, H. Schwarz, Angew. Chem. 98, 275 (1986); Angew. Chem. Int. Ed. Engl. 25, 282 (1986).

/51/ a) P.C. Burgers, J.L. Holmes, J.E. Szulejko, A.A. Mommers, J.K. Terlouw, Org. Mass Spectrom. 18, 254 (1983); b) R.H. Nobes, W.J. Bouma, L. Radom, J. Am. Chem. Soc. 105, 309 (1983).

/52/ W. Koch, H. Schwarz, F. Maquin, D. Stahl, Int. J. Mass Spectrom. Ion Processes 67, 171 (1985).

/53/ For many references, see: a) W. Koch, G. Frenking, H. Schwarz, F. Maquin, D. Stahl, Int. J. Mass Spectrom. Ion Processes 63, 59 (1985); b) H. Schwarz, Adv. Mass Spectrom. 10A, 13 (1986).

/54/ K. Lammertsma, J. Am. Chem. Soc. 106, 4619 (1984).

/55/ W. Koch, H. Schwarz, Chem. Phys. Lett. 125, 443 (1986).

/56/ C.J. Proctor, C.J. Porter, T. Ast, J.H. Beynon, Int. J. Mass Spectrom. Ion Phys. 41, 251 (1982).

/57/ W. Koch, G. Frenking, Chem. Phys. Lett. 114, 178 (1985).

/58/ a) W. Koch, G. Frenking, F. Maquin, D. Stahl, H. Schwarz, J. Chem. Soc. Chem. Commun. 1187 (1984); b) G. Frenking, W. Koch, H. Schwarz, J. Comput. Chem. in press.

/59/ For many references, see: a) G. Frenking, W. Koch, M. Schaale, J. Comput. Chem. 6, 189 (1985); b) D.A. Dixon, T. Fukunaga, B.E. Smart, J. Am. Chem. Soc. 108, 1585 (1986).

/60/ D. Stahl, F. Maquin, unpublished results.

/61/ W. Koch, G. Frenking, J. Gauss, D. Cremer, J. Am. Chem. Soc. 108, 0000 (1986).

/62/ For some references, see: a) W. Henkes, Z. Naturforsch A17, 786 (1962); b) S. Meyerson, R.W. van der Haar, J. Chem. Phys. 37, 2458 (1962); c) V.H. Dibeler, F.L. Mahler, R.M. Reese, ibid 21, 180 (1953); d) E.I. Quinn, F.L. Mahler, J. Res. Natl. Bur. Stand. 62, 39 (1959); e) Ref. 9a; f) A.S. Newton, J. Chem. Phys. 40, 607 (1964); g) R.J. van Brunt, M.E. Wacks, ibid. 41, 3195 (1964); h) C.E. Brion, N.L. Paddock, J. Chem. Soc. A 388,392 (1968); i) F. Figuet-Fayard, J. Chiari, F. Müller, J.P. Ziesel, J. Chem. Phys. 48, 478 (1968); j) S. Becker, J.H. Dietze, Int. J. Mass Spectrom. Ion Phys. 51, 325 (1983); k) T.D. Märk, Int. J. Mass Spectrom Ion Processes 55, 325 (1983); l) B. Brehm, U. Fröbe, H.-P. Neitzke, ibid. 57, 91 (1984); m) L. Morvay, I. Cornides, ibid. 62, 263 (1984); n) S. Singh, R.K. Boyd, F.M. Harris, J.H. Beynon, ibid. 66, 167 (1985); o) W. Koch, H. Schwarz, Chem. Phys. Lett. 113, 145 (1985).

/63/ P.R. Taylor, Mol. Phys. 49, 1297 (1983).

/64/ a) P.C. Cosby, R. Möller, H. Helm, Phys. Rev. A28, 766 (1983); b) D.M. Curtis, J.H.D. Eland, Int. J. Mass Spectrom. Ion Processes 63, 241 (1985).

/65/ a) P. Ausloos: "Interactions Between Ions and Molecules (NATO Advances Study Institute Series B: Physics)", Plenum Press, New York (1975); b) M.T. Bowers: "Gas Phase Ion Chemistry", Vol. 1 and 2, Academic Press, New York (1979); c) M.A. Almoster Ferreira: "Ionic Processes in the Gas Phase (NATO Advanced Study Institute)", 118, D. Reidel, Dordrecht (1984); d) W.W. Duley, D.A. Williams: "Interstellar Chemistry", Academic Press, New York (1984).

/66/ a) W. Henkes, G. Isenberg, Int. J. Mass Spectrom. Ion Phys. 5, 249 (1970); b) J. Gspann, K. Körting, J. Chem. Phys. 59, 4726

(1973); c) D. Echt, K. Sattler, E. Recknagel, Phys. Lett. A90, 185 (1982); d) K. Sattler, D. Echt, A. Reyes-Flotte, J. Mühlhack, K. Knapp, P. Pfau, R. Pflaum, P. Höfer, D. Kreisler, E. Recknagel, Proc. 9th Int. Symp. Molecular Beams, Freiburg i. Breisgau, 151 (1983); e) H. Helm, K. Stephan, T.D. Märk, L. Herestis, J. Chem. Phys. 74, 3844 (1981).

/67/ K. Leiter, W. Ritter, A. Stamatovic, T.D. Märk, Int. J. Mass Spectrom. Ion Processes 68, 341 (1986).

STRUCTURE AND REACTIVITY OF GASEOUS IONS STUDIED BY A COMBINATION OF MASS SPECTROMETRIC, NUCLEAR DECAY AND RADIOLYTIC TECHNIQUES

Fulvio Cacace
University of Rome "La Sapienza"
P.le A.Moro, 5
00185 - Rome, Italy

ABSTRACT. A combination of mass spectrometric techniques, e.g. CI, FT-ICR and CID spectrometry, with nuclear decay and/or radiolytic techniques, brought to bear on the same structural, or mechanistic problem, represents a powerful tool in gas-phase ion chemistry. The major advantages of this multipronged approach, i.e. the wide pressure range, which extends downright to the liquid state, and the positive structural characterization of the charged intermediates, are demonstrated by the results of recent applications. The examples discussed, concerning the reaction of vinyl ion with methane, gas-phase alkylation of biphenyls, and gas-phase aromatic nitration, illustrate the relevance of these studies, which provide, inter alia, a more direct link between gas-phase and condensed-phase ionic chemistry.

1. INTRODUCTION

This article aims to illustrate an integrated approach to structural and mechanistic problems in gas-phase ionic chemistry, based on the combination of mass spectrometry with nuclear-decay and/or radiolytic techniques. The general experimental outline involves identification and characterization of the charged reactants and products, and the study of the relevant ionic reactions by suitable mass spectrometric techniques. The distinctive feature of the integrated approach is that the mass spectrometric measurements are complemented by the study of the <u>same</u> ionic processes, in the <u>same</u> gaseous systems, by the nuclear-decay and /or the radiolytic techniques. Both have been developed into powerful tools that allow extension to the gas phase of the classical methods of physical organic chemistry, including competition kinetics, temperature and pressure-dependence studies, and, of special interest, the actual isolation of the neutral reaction products, whose isomeric composition, structure and stereochemical features provide otherwise hardly accessi-

P. Ausloos and S. G. Lias (eds.), Structure/Reactivity and Thermochemistry of Ions, 467–483.

ble information on the structure and the stereochemistry of their gaseous ionic precursors.

Furthermore, much scope exists for extending the study of ion/molecule reactions to virtually unexplored "high" pressure domains, currently beyond the reach of mass spectrometric techniques, yet readily accessible to the radiolytic and the decay methods. In fact, the high frequency of thermalyzing collision typical of dense gases allows stabilization of excited adducts from exothermic ion/molecule reactions, preventing fragmentation and/or isomerization processes occurring at low pressures. Worthy of note is also the intrinsic kinetic interest attached to the high-pressure limit of ion-molecule reactions, in view of the transition from the electrostatic activation mechanism, peculiar of isolated-pair interactions, to ordinary thermal kinetics. In fact, only at the high-pressure limit can the study of gas-phase ionic processes serve one of its main purposes, i.e. to provide generalized and simplified models of the ionic reactions in condensed media, driven by thermal, rather than electrostatic, activation mechanisms.

2. METHODS

The principles, strenghts and weaknesses of the mass spectrometric techniques used, including CI, CID and ICR spectrometry, are well known and require no further illustration.

The decay technique has been the subject of several reviews,[1-4] and only cursory mention of its basic features is in order here. The decay of a covalently bound T atom in a multitritiated precursor, e.g.

$$CT_4 \longrightarrow CT_3^+ + {}^3He + \beta^- + \bar{\nu} \qquad (1)$$

produces a labeled ion of known structure and initial charge location, whose reactions yield radioactive products, conveniently analyzed by radio chromatographic techniques. Owing to the nuclear nature of the ionogenic process, insensitive to environmental factors, the technique can be applied in gases at any pressure, as well as in liquid and solid systems. The decay ions contain excess internal energy of a peculiar kind, arising from the fact that the decay event occurs in a time far too short to allow relaxation of the structure typical of the neutral parent to the most stable geometry of the daughter ion. Thus, the methyl ions from (1) are formed in a pyramidal structure, reminescent of that of methane, rather than in their ground-state planar geometry. Such "deformation energy" has been calculated around 30 Kcal mol^{-1} for the methyl ions from (1) and for the phenylium ions from the decay of tritiated benzene.[5,6]

The radiolytic technique, outlined by Ausloos in 1966,[7] and largely applied since,[8,9] exploits the action of ionizing radiations on carefully selected gaseous systems as the source of the charged reactants. The complication arising from the simultaneous formation of other reactive species, e.g. free radicals, can be overcome by appropriate choice of the system, its preliminary study by CI mass spectrometry, use of radical scavengers and ion interceptors, etc. A typical example of a mechanistically oriented radiolytic study concerns gas-phase aromatic t-butylation. Ionization of neat isobutane gas is known from extensive CI measurements to yield $t\text{-}Bu^+$ as the predominant ion.[10] High-pressure mass spectrometric experiments have shown that $t\text{-}Bu^+$ reacts with gaseous arenes, yielding directly observable $ArH\text{-}R^+$ adducts.[11] Based on these premises, radiolytic alkylation studies involve γ irradiation of isobutane gas (20 - 720 Torr), containing traces (0.1 - 1.0 Torr) of the aromatic substrate(s) and O_2 (5 - 10 Torr) as a radical scavenger. Use of thermostatic irradiation cells allows the radiolysis to be carried out in a wide (0° - 120°C) temperature range, at the lowest doses compatible with analytical requirements, typically below 10^3 Gray. Radiolytically formed t-butylarenes are analyzed by glc and glc/ms, which allow accurate determination of the yields and the isomeric composition of the products. Apart from the presence of a large excess of a radical scavenger, and detection of the relevant charged intermediates in the CI experiments, the ionic nature of the products is independently demonstrated by the decrease, and the eventual suppression of their yields caused by addition to the irradiated gas of increasing concentrations of bases, e.g. NMe_3, which are known to intercept the charged reactant.

3. RESULTS

The following examples, chosen among recent applications, illustrate relevant results of the integrated approach outlined in the previous section.

3.1. Reaction of Vinyl Ion with Methane.

The reactivity of the ground-state vinyl cation, a CH_2^+-substituted singlet methylene, combines features typical of a carbenium ion with those of a carbene, including the peculiar ability of the latter to insert into σ bonds.[12] Such multifaceted reactivity has been the focus of considerable interest, reflected in a number of mass spectrometric studies.[13] Recently, the problem has been addressed by Speranza and coworkers, who resorted to a combination of FT-ICR mass spectrometry with the decay technique, in order to investigate the reactivity of vinyl ion toward simple molecules, e.g. methane.[14,15] Special care was paid in the preliminary ICR experiments to minimize the excess internal energy of the vinyl ions used as the reactant. Thus, a nominal electron beam energy of 12.8 eV, only 0.3 eV above the appearance potential, was used to generate $C_2H_3^+$ from vinyl bromide.

Proton (deuteron) transfer from X_3^+ (X=H,D,T) to acetylene has proved to be an even better route, the excited $C_2X_3^+$ ions formed being allowed to undergo multiple, quasi-resonant X^+ transfer processes to C_2H_2 molecules, a particularly effective sequence to obtain thermal vinyl cations.

The direct study of the insertion reaction into the C-H bonds of methane is not allowed by the FT-ICR technique nor, for that matter, by other mass spectrometric approaches. In fact, unless effectively stabilized by collisional deactivation, the primary $C_3X_7^+$ adduct, excited by the large exothermicity (59 Kcal mol^{-1}) of the reaction, undergoes complete fragmentation into $C_3X_5^+$ ions of unknown structure

$$C_2X_3^+ + CH_4 \longrightarrow [C_3X_7]^+_{exc} \xrightarrow{+M,\ -M^*} C_3X_7^+ \qquad (2a)$$

$$[C_3X_7]^+_{exc} \longrightarrow C_3X_5^+ + X_2 \qquad (2b)$$

In summary, sequence (2b), whose overall exothermicity is 23 Kcal mol^{-1}, is the only process detectable under mass spectrometric conditions. Its specific rate, deduced from FT-ICR measurements,[14] compares well with previous mass spectrometric results, as shown in Table 1.

Application of the decay technique required multitritiated ethylene as the source of labeled vinyl ions. Preparation of the precursor by partial saturation of acetylene with T_2 over a Lindlar catalyst, purification by preparative glc, isotopic analysis and storage represented, per se, a radiochemical feat, in view of the specific activity involved, some $6 \cdot 10^4$ Ci per mol, compounded by the high radiosensitivity of both C_2H_2 and C_2H_4. The sample used in the decay experiments consisted of a 56:44 mixture of $C_2H_2T_2$ and C_2H_3T, diluted with C_2H_4 to a final specific activity of 363 Ci per mol.

The reactivity of labeled vinyl cations from the decay

$$C_2H_2T_2 \longrightarrow C_2H_2T^+ + {}^3He + \beta^- + \bar{\nu} \qquad (1b)$$

was studied in CH_4 (60-720 Torr), containing a radical scavenger (O_2, 4 Torr), a gaseous base (NMe_3, 0.3 Torr) and a suitable nucleophile, e.g. 1,4-$C_4H_8Br_2$, PhH, or MeOH. The latter were used to sample the population of gaseous cations by fast reactions, yielding products whose structure unequivocally reflects that of the intercepted cation, e.g.

$$C_2X_3^+ + 1{,}4\text{-}C_4H_8Br_2 \longrightarrow C_2X_3Br + C_4H_8Br^+ \quad \Delta H° = -40 \text{ Kcal mol}^{-1} \qquad (3a)$$

$$C_3X_5^+ + 1{,}4\text{-}C_4H_8Br_2 \longrightarrow C_3X_5Br + C_4H_8Br^+ \quad \Delta H° = -7 \text{ Kcal mol}^{-1} \qquad (3b)$$

$$i\text{-}C_3X_7^+ + 1,4\text{-}C_4H_8Br_2 \longrightarrow i\text{-}C_3X_7Br + C_4H_8Br^+ \quad \Delta H^\circ = -6 \text{ Kcal mol}^{-1} \qquad (3c)$$

$$i\text{-}C_3X_7^+ + 1,4\text{-}C_4H_8Br_2 \longrightarrow C_3X_6 + C_4H_8Br_2X^+ \quad \Delta H^\circ = \sim 0 \text{ Kcal mol}^{-1} \qquad (3d)$$

where $C_4H_8Br^+$ denotes a tetramethylene bromonium ion. The other nucleophiles used yield as well structurally diagnostic products, e.g. styrene, allylbenzene, cumene and propene from C_6H_6, methyl vinyl ether, allyl methyl ether, isopropyl methyl ether and propene from MeOH, etc. Sampling the same ionic population with different nucleophiles is clearly useful to overcome problems arising from inefficient, or "blind" channels, affecting the reactivity of a specific nucleophile.

The decay experiments have allowed to detect vinyl, allyl, and, for the first time, isopropyl ions in the system investigated. In fact, the high efficiency of collisional stabilization in the range 60-720 Torr makes branch (2a) not only detectable, but largely predominant, as illustrated in Figure 1, which refers to systems containing C_6H_6 as the trapping nucleophile. Analysis of the tritiated products demonstrates that most (>90%) of the $C_3X_7^+$ complexes formed by addition of nucleogenic $C_2X_3^+$ to CH_4 and stabilized by collisional deactivation, have the isopropyl ion structure when trapped by the nucleophile ca.10^{-9}s after

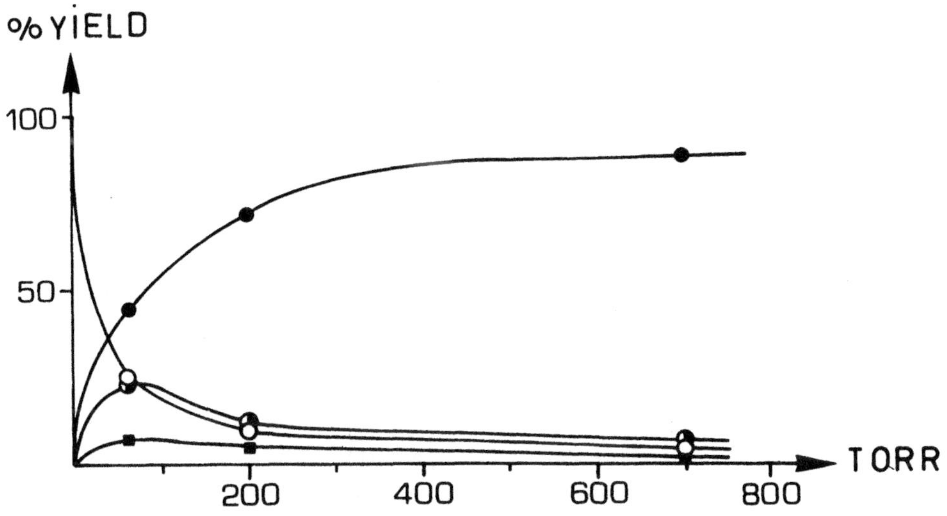

FIGURE 1. YIELDS OF I-PROPYL (●), N-PROPYL (◑), ALLYL (■) AND VINYL (O) PRODUCTS AS A FUNCTION OF CH_4 PRESSURE.

their formation. Since vinyl ion insertion into a C-H bond of methane cannot yield i-$C_3X_7^+$ without some kind of rearrangement, one can speculate as to whether such structural change involves the intermediacy of a protonated cyclopropane moiety, as suggested by the results of an ion-beam scattering study at energies in the 0.5-3.2 eV range.[19] The remarkably low yields of n-propyl derivatives, and the lack of c-C_3X_6 among the products, fail to support this view, at least as long as the formation of a persistent c-$C_3X_7^+$ intermediate (lifetime > 10^{-9}s) is concerned. The products pattern shows that the limited fraction of the $C_3X_7^+$ ions which retain sufficient energy to fragmentate yields allyl, rather than 2-propenyl cations, consistent with the higher activation energy for the dissociation into the 2-propenyl structure, established by metastable transition studies.[20]

Finally, the linear dependance of the $[C_2X_3^+] : ([C_3X_5^+]+[C_3X_7^+])$ ratio on the $[CH_4] : [\text{Nucleophile}]$ ratio, illustrated in Figure 2, allows

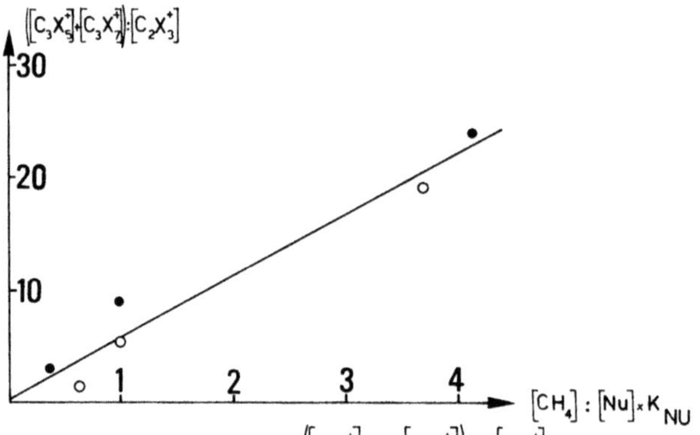

FIGURE 2. LINEAR DEPENDENCE OF THE $([C_3X_5^+] + [C_3X_7^+]) : [C_2X_3^+]$ RATIO ON THE $[CH_4] : [Nu]$ RATIO. Nu=BENZENE (●) AND METHANOL(○).

a direct estimate of the specific rate of the addition of vinyl ion to methane, based on the reasonable assumption that trapping of vinyl, isopropyl and allyl ions by the nucleophiles are relatively fast processes. The k value obtained, $5.5 \pm 1.0 \cdot 10^{-10} cm^3$ molecule^{-1}s^{-1}, is comparable to, if slightly higher than, those deduced from mass spectrometric methods (Table 1), a truly remarkable agreement if one considers the assumptions made in both sets of experiments, and especially the widely different reaction environment. The calculated value corresponds to about one half of the Langevin collision limit, $1.2 \cdot 10^{-9}$ cm^3 molecule^{-1}s^{-1}, pointing to the high efficiency of the reaction.

In conclusion, the decay technique has allowed the first direct study of the insertion of vinyl ion into the bonds of methane, unafa direct estimate of the specific rate of the addition of vinyl ion to methane, based on the reasonable assumption that rapping of vinyl, isopropyl and allyl ions by the nucleophiles are relatively fast processes. The k value obtained, $5.5\pm1.0 \cdot 10^{-10} cm^3$ molecule $^{-1}s^{-1}$, is comparable to, if slightly higher than, those deduced from mass spectrometric methods, a truly remarkable agreement if one considers the assumptions made in both sets of experiments, and especially the widely different reaction environment. The calculated value corresponds to about one half of the Langevin collision limit, $1.2 \cdot 10^{-9}$ cm^3 molecule$^{-1}s^{-1}$, pointing to the high efficiency of the reaction.

In conclusion, the decay technique has allowed the first direct study of the insertion of vinyl ion into the bonds of methane, unaffected by secondary fragmentation, and the characterization of the charged adducts formed, whose structure could only be speculated upon, based on circumstantial mass spectrometric evidence.

3.2 Gas Phase Alkylation of Biphenyls.

A recent study of the methylation of biphenyls[21] provides an example of the synergetic interaction of combined mass spectrometric and radio lytic techniques. The charged reagents, Me_2X^+ ions (X-F,Cl), are via a well-established reaction sequence in MeX. Investigation of biphenyl was suggested by the predominant ortho orientation prevailing in the alkylation of substrates containing substituents with n-electron pairs. A reasonable explanation is that direct attack, ring substitution proceeds via the formation of a preliminary adduct, stabilized by the electrostatic interaction of the cation with the n-electrons of the substituent, which causes an increase of the local concentration of the electrophile at the ortho positions.[22] Biphenyl provides a suitable model to ascertain whether an analogous electrostatic interaction can involve a delocalized π system, rather than a n-electron pair. Furthermore, the dihedral angle of the biphenyl rings can be modified by suitable ortho substituents, thus affecting their conjugation and mutual electronic effects, which may represent an useful mechanistic tool.

Reaction of Me_2X^+ with biphenyls and methylbiphenyls, investigated under typical CI conditions at MeX pressures up to 0.5 Torr, gave abundant methylated adducts, $C_{12}H_{10}Me^+$ and, respectively, $C_{12}H_9Me_2^+$ ions, whose structure was probed by CID spectrometry. The required model ions were obtained from isomeric methylbiphenyls, and respectively dimethylbiphenyls, protonated with $t\text{-}Bu^+$ ions under isobutane CI conditions.

The model ions from the protonation of methylbiphenyls can be discriminated by CID, while the model ions from the protonation of dimethylbiphenyls can be divided by CID into two structural classes depending whether the Me groups are bound to two different rings (Class A), or to the same ring (Class B). The CID spectra of the adducts from the exothermic methylation process characterize them as arenium ions.

However, the structural resolution of the CID experiments does not allow an evaluation of the isomeric composition of the methylated adducts from biphenyl (R=H). The CID spectra of the adducts from methylbiphenyls (R-Me) are linear combinations of Class A and Class B spectra, with a definite predominance of the latter, i.e. alkylation of the Me-substituted ring seems preferred.

Radiolytic methylation involved γ irradiation (37.5°C, 10^4 Gray per hour) of the MeX gas (200-600 Torr), containing traces (< 1 Torr) of the aromatic substrate(s), together with O_2 (5 Torr). Methylated aromatic products were identified by glc and glc/ms, the ionic nature of their formation processes being established by the criteria illustrated in the introductory section. At 600 Torr, the reactivity of biphenyl toward Me_2F^+ is characterized by the following orientation: ortho = 74%, meta = 14%, para = 12%. The radiolytic results on the methylation of methylbiphenyls are consistent with those from CID spectrometry, demonstrating a general preference for the alkylation of the Me-substituted ring, to an extent which increases in the same order deduced from the CID spectra, i.e.

4-methylbiphenyl < 2-methylbiphenyl < 3-methylbiphenyl

The orientation measured in MeF gas at 600 Torr is compared in Figure 3 with that calculated, assuming the additivity of the substituent effects, from the partial rate factors deduced from the

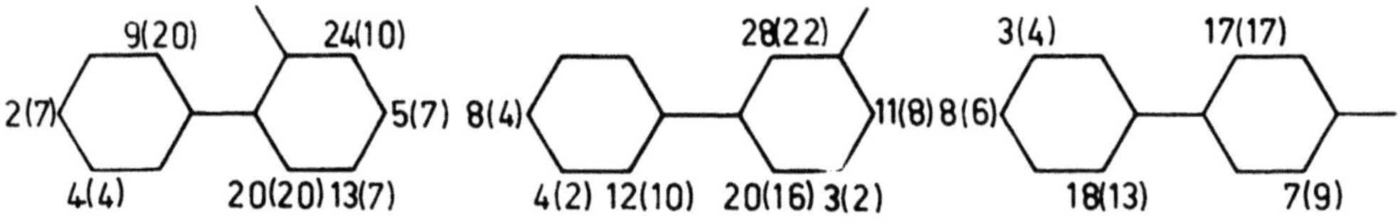

FIGURE 3. EXPERIMENTAL AND CALCULATED (FIGURES IN PARENTHESES) ISOMERIC COMPOSITION OF METHYLATED PRODUCTS FROM METHYLBIPHENYLS.

methylation of biphenyl and of toluene. The agreement is reasonable, except for the 2-methylbiphenyl, where the dihedral angle of the rings is considerably increased by the ortho substituent. The order of the gas-phase basicity, measured by FT-ICR spectrometry,[23] and given in kcal mol^{-1}

biphenyl <	2-methylbiphenyl <	4-methylbiphenyl <	3-methylbiphenyl
(188.3)	(188.9)	(189.4)	(191.8)

reflects the effects of reduced ring coplanarity. The ortho/2 para ratio of the unsubstituted ring of 2-methylbiphenyl is also anomalous, being <u>lower</u> than unity, in contrast with the 3.1 ratio measured in biphenyl. These, and other mechanistic features whose discussion is beyond the scope of this review, are consistent with a model involving the formation of an early, electrostatic complex, where the positively charged Me groups of the symmetrical, bidentate electrophile interact with both rings of biphenyl, enhancing the local concentration of the electrophile at the ortho positions. Moving the rings apart, as in Ph_2CH_2, or increasing their angle, as in 2-methylbiphenyl, can be expected to prevent or reduce the formation of the adduct, lowering the extent of ortho substitution, in agreement with the experimental findings. Furthermore, the more pronounced ortho orientation in the alkylation by Me_2F^+ than by Me_2Cl^+, which stands in contrast with the expected higher selectivity of the latter, finds a reasonable explanation in the higher degree of charge delocalization in Me_2F^+, which enhances the ability of its Me group to interact with the aromatic rings. In solution, the reactivity of biphenyl is also slightly lower than that of toluene, while para orientation predominates in most cases. Significantly, the abnormally high extent of ortho substitution in certain nitrations has long been interpreted as indicative of a preliminary interaction of the electrophile with both rings.[24]

3.3 Gas Phase Aromatic Nitration

Extension of the study of aromatic nitration to the gas phase is of considerable mechanistic interest, as a general, extremely simplified approach to a reaction that has played, and continues to play, a central role in physical organic chemistry. The results of earlier attempts, based on purely mass spectrometric techniques, have been only moderately encouraging. In fact, under the restricted and peculiar set of conditions typical of ICR, or flow-discharge spectrometry, NO_2^+, long recognized as the nitrating agent in solution, fails to add to arenes, undergoing instead charge and oxygen-atom transfer. The reactivity of the other nitrating agents investigated, e.g. $CH_2ONO_2^+$ and $EtO(NO_2)_2^+$, is also hardly consistent with solution-chemistry patterns. To compound matters, the limited structural discrimination of mass spectrometry is particularly detrimental in this area, preventing evaluation of key parameters such as orientation, partial rate factors, etc.

Given the above, the problem appeared eminently amenable to the application of the combined mass spectrometric/radiolytic approach, which indeed has provided the first mechanistic picture of gas-phase aromatic nitration consistent with common experience in solution chemistry.[25] The electrophile used, protonated methyl nitrate, can be characterized, according to MINDO calculations,[26] as a nitronium ion "solvated" by one methanol molecule, $Me\overset{+}{O}(H)NO_2$, with a binding energy of 34 kcal mol^{-1}. The reagent can readily be obtained, e.g. in CH_4 CI of methyl nitrate, according to the process

$$C_nH_5^+ + MeNO_3 \qquad\qquad C_nH_4 + Me\overset{+}{O}(H)NO_2$$

whose exothermicity is estimated around 52 kcal mol^{-1}(n=1) and 22 kcal mol^{-1}(n=2). The CI spectra of $CH_4/MeNO_3/PhH$ mixtures, in the typical molar ratios 1000:10:1, recorded in the pressure range 0.1-0.5 Torr, display an abundant $PhH \cdot NO_2^+$ adduct. The role of protonated methyl nitrate as the precursor of $[Arene \cdot NO_2^+]$ adducts is confirmed by double-resonance ICR experiments. Finally, the CID spectra of the $PhHNO_2^+$ adduct from the nitration of PhH are indistinguishable from those of the $Ph\text{-}NO_2H^+$ ion from the selective protonation of the NO_2 group of nitrobenzene with $t\text{-}Bu^+$ ions in isobutane CI. The CID results provide direct evidence for the formation of a C-N bond following NO_2^+ transfer from protonated methyl nitrate to benzene, while characterization of the nitrated adduct as a O-protonated species can be explained by subsequent, intramolecular proton shifts (vide infra).

The radiolytic study can be regarded as the extension of the CI experiments to a much higher pressure range. It involved γ irradiation (37.5 °C, 5 · 10^3 Gray) of CH_4 (720 Torr), containing $MeNO_3$ (3-10 Torr), a radical scavenger (O_2, 10 Torr), a thermal electron interceptor (SF_6, 5 Torr) and the aromatic substrate(s) (<1 Torr). Analysis of the irradiated systems by glc and glc/ms demonstrated high yields of the expected nitrated products, formed via ionic processes. The latter conclusion was reached on the grounds of the usual discriminating criteria, i.e. insensitivity to the presence of radical scavengers, inhibiting effects of ion interceptors, and direct detection of the relevant ionic intermediates by CI techniques. Further confirmation was provided by the observation that replacing CH_4 with $i\text{-}C_4H_{10}$ nearly suppressed the yields of nitrated products, which is neatly explained, in the framework of an ionic mechanism, by the failure of $t\text{-}Bu^+$, the major ion from the isobutane radiolysis, to protonate $MeNO_3$, less basic than isobutene.

Application of the radiolytic technique allows evaluation of the substrate and positional selectivity of the gas-phase nitration, reported in Table 1, exactly as in conventional solution-chemistry studies. The results characterize the reaction as a moderately selectively electrophilic aromatic substitution.

The mass spectrometric and radiolytic evidence suggest a nitration mechanism based on the nucleophilic displacement of MeOH by the aromatic substrate, represented, in the case of benzene, by the equation:

$$Me\overset{+}{O}(H)NO_2 + PhH \qquad PhHNO_2^+ + MeOH \qquad (6)$$

TABLE 1 - SELECTIVITY OF AROMATIC NITRATION OF SUBSTITUTED BENZENES BY PROTONATED METHYL NITRATE IN CH_4 AT 37.5 °C.

PhX (X)	$k_{PhX}:k_{PkH}$	Orientation % ortho	% meta	% para
Me	5.1	59	7	34
Et	5.6	47	4	49
n-Pr	7.0	50	4	46
i-Pr	6.0	31	5	64
c-Pr	10.6	72	6	22
t-Bu	8.4	17	8	75
Ph	1.5	40	4	56
OMe	7.6	41	-	59
F	0.15	14	13	73
Cl	0.19	36	10	54
CF_3	0.0028	-	100	-
Mesitylene	8.1	-	-	-

As to the structure of the adduct formed, several isomers deserve consideration

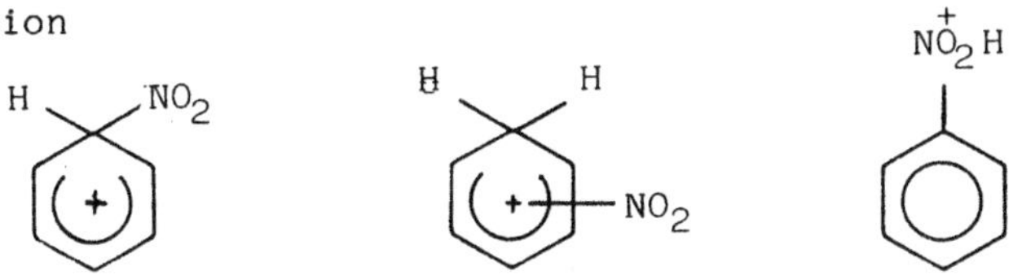

I II III

From theoretically calculated data, i.e. H_f of MeO(H) NO_2 150.5 kcal mol-1,[26] and PA of the "ipso" position of $PhNO_2$ 146 kcal mol^{-1},[27] formation of I would be endothermic by ca.18 kcal mol^{-1}. This is probably an upper limit, if one considers that both the reactants and the products pairs involved in process (6) are stabilized by electrostatic interactions, more effective in the latter, owing to the better solvating properties of MeOH with respect to PhH. Despite such consideration, and the uncertainties affecting theoretically calculated values, the endothermicity of the formation of I is likely to represent a substantial fraction of the energy barrier of the NO_2^+ transfer.

Formation of II would be nearly thermoneutral, depending on the specific isomer considered, while formation of III would be exothermic by ca.30 kcal mol^{-1}, based on the experimentally measured PA of $PhNO_2$, which refers to nitro-group protonation.

While direct formation of III via a concerted mechanism cannot be excluded,[28] the evidence for a modest energy barrier provided by the very selectivity of the nitration makes I the most likely primary intermediate. Is subsequent rearrangement into the more stable isomers II, and especially III, can occur under suitable conditions, as suggested by the results of CID spectrometry, a technique characterized by a relatively long lifetime (ca.10^{-5}s) os isolated ions before structural assay. In this connection, it is tempting to speculate on the "catalytic" role of methanol, suggested by the value of its PA, which happens to be intermediate between those of the "ipso" position and of the nitro group of $PhNO_2$.

Whatever the proton location within the nitrated adduct, its conversion into the neutral nitroarenes formed in the radiolytic experiments is likely to involve deprotonation by any gaseous bases contained in the irradiated system, including the substrate itself, e.g.

$$\text{I, II, III} \xrightarrow[-BH^+]{+B} PhNO_2 \qquad (7)$$

The results of the combined mass spectrometric/radiolytic approach allow useful comparison with solution-chemistry data. With the exception of those of strongly activated substrates (cyclopropylbenzene and anisole), the partial rate factors of monosubstituted benzenes fit the reasonably linear (correlation coefficient 0.961) Hammett's plot illustrated in Figure 4. Its ρ^+ value, 3.9±0.3, characterizes gas-phase nitration as a fairly selective electrophilic substitution. As a comparison, typical ρ^+ values of aromatic nitration range from ca.-6 ($ACONO_2$ in Ac_2O) to ca.-9 (HNO_3 in H_2SO_4).[24] Of particular interest is

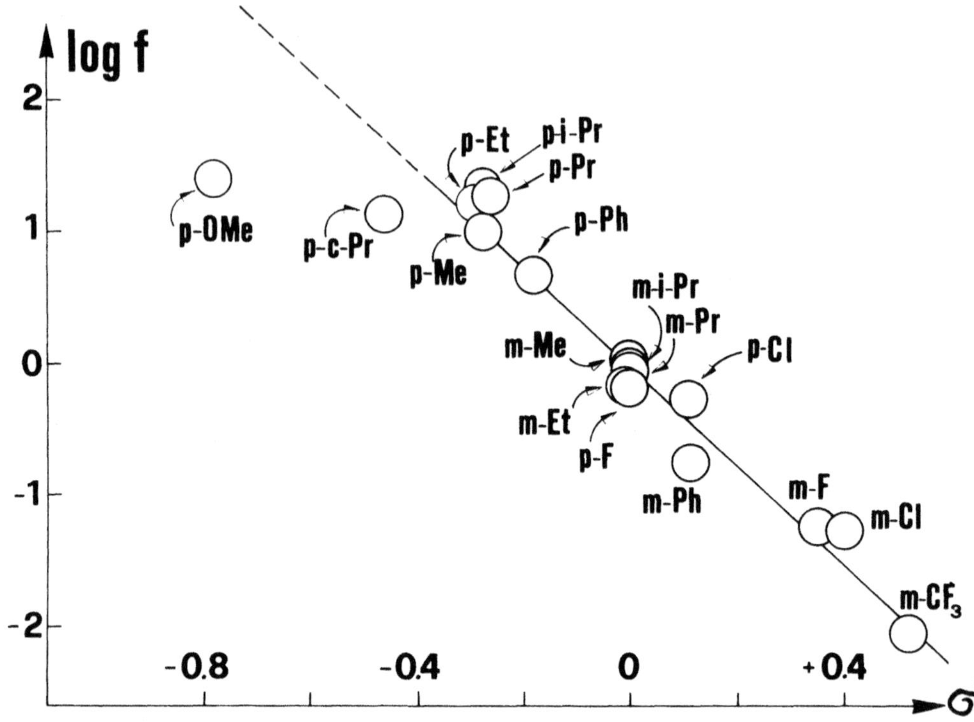

Fig. 4, Log of partial rate factors vs the σ^+ substituent constant in the gas phase aromatic nitration by $Me\overset{+}{O}(H)NO_2$ at 37.5 C.

the anomalously low reactivity display by activated substrates, i.e. the partial rate factors of cyclopropylbenzene and anisole are considerably lower than the values calculated from the correlation obeyed by all other monosubstituted benzenes, and the $k_{Mesitylene}:k_{Benzene}$ ratio is 8.1, also much lower than expected.

Such discrepancies could conceivably reflect the incursion of other reaction channels competing with nitration, e.g. proton transfer from $Me\overset{+}{O}(H)NO_2$ to highly basic substrates, such as anisole and mesitylene, whose PA exceeds 200 kcal mol^{-1}. However, it is tempting to interpret the results as indicative of a general kinetic trend, the gas-phase counterpart of that observed in "encounter-rate" nitrations occurring in solution.[29] It has long been known that in certain condensed media the rate constant of aromatic nitration attains a maximum value that cannot further be increased by enhancing the intrinsic reactivity of the substrate.

Under such conditions the significance of partial rate factors is lost, since substrate and positional selectivity are not determined in the same step. In solution, the preliminary process that becomes rate determining is controlled by microscopic diffusion, clearly inoperative in the gas phase. However, kinetically equivalent effects can derive from entirely different mechanism, i.e. the formation of early, purely electrostatic ion-molecule complexes, a general feature of gas-phase reactions. When activated, highly nucleophilic aromatics are involved, formation of such complexes can become rate-determining, as long as substrate selectivity is involved. In this framework, above a certain activation threshold, the reactivity of the aromatics can be expected to level off, approaching unit collision efficiency without loss of positional selectivity, a picture consistent with the limited evidence presently available. Further investigation of the problem is of interest, since the results could establish a new link between gas-phase and condensed-phase aromatic substitution.

3.4. Temperature Dependence of the Substrate and Positional Selectivity of Gas Phase Aromatic t-Butylation.

The alkylation of benzene and toluene by gaseous t-Bu^+ ions has recently been studied by a radiolytic technique at 720 Torr in the range from 0° to 140 °C.[30] The preparation of the charged reactant, and the criteria to establish the ionic nature of the alkylation have been outlined in the introductory section. The reaction had previously been studied with a mass spectrometric technique by Kebarle and coworkers,[11] whose results, together with those of the radiolytic approach,

(8)

IV, R=H; V,R=Me

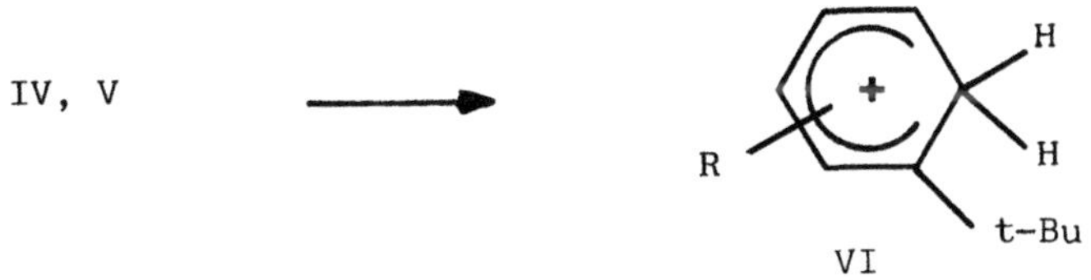

(9)

VI

are consistent with the above scheme. In the radiolytic systems, adducts IV-VI are eventually deprotonated by a gaseous base, yielding the observed t-butylarenes. The temperature dependence of the selectivity, measured by the k_T/k_B ratios, is illustrated in Figure 5. The Arrhenius plot is linear over an extended range (0°-100 °C), where the reactivity of benzene relative to toluene increases regularly with the temperature. Above ca.100 °C the plot displays a noticeable inflection, arising from equilibrium constraints, clearly demonstrated by the data of Kebarle et al.[11] In fact, their mass spectrometric measurements show that the equilibrium constant of the addition (8) to benzene decreases markedly with the temperature, passing from 170 $Torr^{-1}$ at

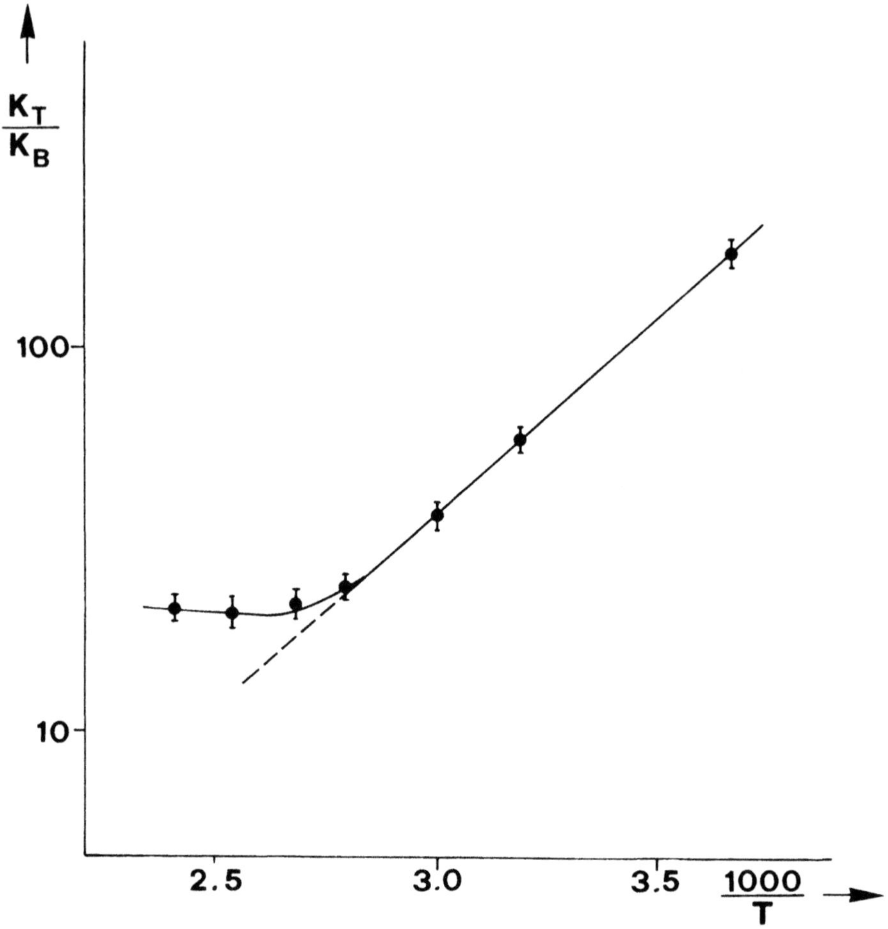

Figure 5. Log of k_T/k_B ratio vs 1/T in the gas-phase alkylation of benzene and toluene by t-Bu^+ at 720 Torr.

32 °C to ca. 0.3 $Torr^{-1}$ at 105 °C. This accounts for the curvature of the Arrhenius plot above 100 °C, since the analogous increase of k_{-1} which affects the t-butylation of toluene as well, becomes significant only at temperatures exceeding the range covered by the radiolytic experiments, owing to the greater stability of V with respect to IV.

Turning to the primary object of the study, its main conclusion is that the temperature dependence of the k_T/k_B ratio conforms to that expected for an ordinary thermal reaction. Regression analysis of the linear portion of the Arrhenius plot leads to a difference of empirical activation energies

$$\Delta E = E_B - E_T = 3.6 \pm 0.4 \text{ kcal mol}^{-1}$$

and to a preexponential factors ratio A_B/A_T crudely estimated around 6.

The positional selectivity changes as well with the temperature, the proportions of m-t-butyltoluene increasing at the expenses of the para isomer at higher temperatures. Assuming that the isomeric composition of products reflects approximately the relative rate of alkylation of the different positions of the toluene ring, one obtains a $E_{meta}-E_{para}$ difference of ca.3 kcal mol^{-1}, not inconsistent with the E_B-E_T difference.

The radiolytic study represents the first attempt to a quantitative evaluation of the Arrhenius parameters of an aromatic substitution by gaseous cations, and the extended temperature range investigated (140 °C) has hardly been approached in previous investigations, irrespective of the reaction environment. Furthermore, the free nature of the ionic reagent and the gaseous medium confer an unusual degree of generality on the results, unaffected by such incidental factors as the solvents, the catalysts, the counterions, etc., that influence Friedel-Crafts reactions in solution.

Finally, the above results provide a clearcut example of the transition from the electrostatic activation mechanism, typical of ion-molecule reactions in the low-pressure range (where for instance aromatic t-butylation displays a negative temperature dependence) to the ordinary thermal kinetics, typical of the high-pressure domain.

4. Summary and Conclusion

An integrated approach to the study of the structure and the reactivity of gaseous ions is outlined, based on the combination of typical mass spectrometric techniques, such as CI, FT-ICR and CID spectrometry, with nuclear-decay and/or radiolytic techniques. The examples reported, concerning the reaction of vinyl ions with methane, gas-phase alkylation of biphenyl, gas-phase aromatic nitration and t-butylation, highlight the advantages of the combined approach, in particular its wide pressure range, the superior structural resolution and the capability of providing mechanistic models directly comparable with those derived from solution-chemistry studies.

ACKNOWLEDGEMENTS. The author is indebted to Italian National Research Council (C.N.R.) and to the Ministry of Pubblica Istruzione for financial support, and to his colleagues and coworkers, whose names are listed in the references, for their contribution to this work.

REFERENCES

(1) Cacace, F., 1970, Adv.Phys.Org.Chem., 8, 79.

(2) Akulov, G.P., 1976, Usp.Khim., 45, 1970.

(3) Cacace, F., 1985,"Treatise on Heavy-Ion Science", Bromley, D.A., Ed., Plenum Press, New York, V.6, p63.

(4) Cacace, F., 1979, "Kinetics of Ion-Molecule Reactions", Ausloos, P., Ed., Plenum Press, New York, p199.

(5) Burdon, J.; Davies, D.W.; del Conde, G., 1976, J.Chem.Soc.Perkin 2 1193.

(6) Schleyer,P.v.R.; Kos, A.J.; Ragavachari, K., 1983, J.Chem.Soc.Chem. Commun., 1296.

(7) Ausloos, P., 1966, Ann.Rev.Phys.Chem., 17, 205.

(8) Speranza, M., 1983, Gazzetta Chim.Ital., 113, 37, and references therein.

(9) Cacace, F., 1982, Radiat.Phys.Chem., 20, 99, and references therein.

(10) For a review, cfr. Harrison, A.G., 1983, "Chemical Ionization Mass Spectrometry", CRC Press, Boca Raton, Fa., p64.

(11) Sen Sharma, D.K.; Kebarle, P., 1982, Can.J.Chem., 60, 2325.

(12) Stang, P.J.; Rappoport, Z.; Hanack, M.; Subramanian, L.R., 1979, "Vinyl Cations", Academic Press, New York.

(13) Batey, J.H.; Tedder, J.M., 1983, J.Am.Chem.Soc.Perkin Trans.2, 1263, and references therein.

(14) Fornarini, S.; Gabrielli, R.; Speranza, M., 1986, Int.J.Mass Spectrom.Ion Proc., in the press.

(15) Fornarini, S.; Speranza, M., 1986, J.Am.Chem.Soc., submitted.

(16) Adams, N.G.; Smith, D., 1977, Chem.Phys.Lett., 47, 383.

(17) Field, F.H.; Franklin, J.L.; Munson, M.S.B., 1963, J.Am.Chem.Soc., 85, 3575.

(18) Fornarini, S.; Speranza, M., 1984, Tetrahedron Lett., 25, 869.

(19) Senzer, S.N.; Lim, K.P.; Lampe, F.W., 1984, J.Phys.Chem., 88, 5314.

(20) Bowen, R.D.; Williams, D.H., 1976, J.Chem.Soc.Perkin 2, 1479.

(21) Cacace, F; de Petris, G.; Fornarini, S.; Giacomello, P., 1986

J.Am.Chem.Soc., submitted.

(22) Attinà, M.; Giacomello, P., 1979, J.Am.Chem.Soc., 101, 6040.

(23) Speranza, M., unpublished results.

(24) Taylor, R., 1966, J.Chem.Soc.B, 727.

(25) Attinà, M.; Cacace, F., 1986, J.Am.Chem.Soc., 108, 318.

(26) Dewar, M.J.; Shanshal, M.; Worley, S.D., 1969, J.Am.Chem.Soc., 91 3590.

(27) Jañez, M., 4-31 G calculations, to be published.

(28) However, the lack of a measurable isotopic effect in the nitration of C_6H_6 and C_6D_6 (k_H/k_D ratio of 1.00 ± 0.05) speaks against a concerted mechanism involving the fission of a C-H bond in the rate-determining step.

(29) Schofield, K., 1980, "Aromatic Nitration", Cambridge University Press, Cambridge, p44.

(30) Cacace, F.; Ciranni, G.; J.Am.Chem.Soc., 1986, 108, 887.

INDEX

GPSR Compliance

The European Union's (EU) General Product Safety Regulation (GPSR) is a set of rules that requires consumer products to be safe and our obligations to ensure this.

If you have any concerns about our products, you can contact us on ProductSafety@springernature.com

In case Publisher is established outside the EU, the EU authorized representative is:

Springer Nature Customer Service Center GmbH
Europaplatz 3
69115 Heidelberg, Germany

Batch number: 10816188

Printed by Printforce, the Netherlands